NEUROENGINEERING

NEUROENGINEERING

Edited by

Daniel J. DiLorenzo, MD, PhD, MBA
Joseph D. Bronzino, PhD

CRC Press
Taylor & Francis Group
Boca Raton   London   New York

CRC Press is an imprint of the
Taylor & Francis Group, an **informa** business

CRC Press
Taylor & Francis Group
6000 Broken Sound Parkway NW, Suite 300
Boca Raton, FL 33487-2742

Library of Congress Cataloging-in-Publication Data

Neuroengineering / [edited by] Daniel J. DiLorenzo and Joseph D. Bronzino.
   p. ; cm.
  Includes bibliographical references and index.
  ISBN 978-0-8493-8174-4 (hardcover alk. paper)
  1. Biomedical engineering. 2. Neural networks (Neurobiology) I. DiLorenzo, Daniel J. II. Bronzino, Joseph D., 1937-
  [DNLM: 1. Biomedical Engineering. 2. Nervous System Diseases--rehabilitation. 3. Biomedical Technology. 4. Prostheses and Implants. 5. Trauma, Nervous System--rehabilitation. WL 140 N49108 2008]

R856.N48 2008
610.28--dc22
                                                        2007015735

Visit the Taylor & Francis Web site at
http://www.taylorandfrancis.com

and the CRC Press Web site at
http://www.crcpress.com

Contents

Background

Neural Modulation

Neural Augmentation

Sensory Prostheses

Motor Prostheses—Electrode Technologies and Command Signal Extraction

Motor Prostheses—Effector Subsystem Technologies

Fundamental Science and Promising Technologies

Foreword

This is an exciting time for neuroengineering. It is at the confluence of fields that are advancing at an ever increasing rate.

The main catalyst has been the spectacular growth of computer science, which makes it possible to analyze and digest the feast of information that marks the complexity of the nervous system. It has been just 60 years since the introduction of the Eniac, a vacuum tube computer that occupied the entire floor of a building at the University of Pennsylvania, capable of fewer calculations than a simple pocket calculator can now accomplish. It has been approximately 30 years since the birth of the Altair 8800, the first very basic personal computer, which led to the formation of Apple Computer the following year and the release five years later of the first IBM personal computer. It has been the development of computer science that has at last made possible the search for the significant but seemingly random response of a group of neurons. The use of myriad calculations to localize activity deep within the brain has become feasible. The identification of a subtle response evoked by electrical or natural stimulation can be deduced despite the host of random activity in which it is buried.

A natural outgrowth of the quest to make computers smaller and faster has been the development of miniaturized electronic components, which are used in myriad devices implanted in or attached to specific neurological sites. It is just 50 years since the first transistors were invented, which led to the introduction of integrated circuits within the decade. Techniques employed in designing and building miniature circuits for computers are used for the development of circuits that can be implanted in or on the brain, muscle or nerves to become part of the biological circuitry.

Advances in imaging during the past 35 years, again dependent of computers, have made it possible to see within the brain without invading or disrupting it. Not only can the anatomy be visualized, but the location of functions can be visualized, making it possible to target specific subcortical sites with immense accuracy on the basis of a rationally devised hypothesis. Abnormal activity can often be visualized without invasion, to help with understanding of the pathophysiology and diagnosis, or to provide a measure of therapeutic result. Once localized, these structures invite study by introducing probes of an immense array of sizes to analyze the metabolism or chemistry, record the electrical activity, or stimulate either acutely or chronically for investigation, diagnosis, or therapy.

Not only are new therapies developing rapidly, but older modalities are being improved and refined with consequent progress in neuromodulation — referring to "older modalities" in the modern field of neuromodulation that is just 35 years old, the oldest discipline presented herein. First peripheral nerve stimulation, then in rapid succession spinal cord and deep brain stimulation were introduced 35 to 40 years ago. At that time, the need for an implanted stimulator was imposed by the use of spinal cord stimulation for pain and soon after for dystonia and spasticity, but it is only in the past five years that we have had widespread use of deep brain stimulation for motor disorders. We are still in the earliest stages of this field, much of which is based on empirical observation, but now an increasingly scientific approach is increasing our knowledge of its mechanisms, and hence the mechanisms of the diseases under management and the rational for such treatment.

The field of neural augmentation has had spectacular recent development. I was present at just about the birth of visual prostheses (it had already been conceived), when Bill Dobelle brought his traveling neurophysiology laboratory to the operating room at the University of Arizona in 1972 to stimulate the occipital cortex of a patient undergoing a craniotomy under local anesthesia, the second patient in his series and the first in the United States. We were thrilled when the patient could localize individual phosphenes in the visual field. Although we could fantasize, we could not visualize the production of useful vision or even the restoration of hearing with a cochlear prosthesis.

The development of motor prostheses has recently taken on a new urgency, with a plethora of war trauma. Replacing lost neural function or duplicating the function of a lost extremity has become a reality, but much is still to be done. The interface between the patient and the prosthesis has become intimate, and will become increasingly more so.

Consequently, this volume comes at an opportune time when the field is about to mushroom because of the geometric rate of progress in those scientific and engineering fields that have united to become neuromodulation. This defines the state of the present art. Despite the excitement generated by advances in the past decade, the next decade promises to be even more exciting.

Philip L. Gildenberg, MD, PhD

Preface

We are at the cusp of a revolution in both technology and clinical practice that holds promise to have more of an impact on human health, disease, and quality of life than any other development in history. Applying and advancing upon a foundation of fundamental neuroscience, clinical observation and experimentation, multidisciplinary engineering technologies, and innovations in surgical technique, we are now in an era where it is conceivable to expect that the blind will see and the paralyzed will walk. The convergence of neuroscience and technology is facilitating the development of therapies that not long ago would have seemed of biblical proportion. This is just the beginning; sooner than many realize, we may see clinical acceptance and adoption of neuromodulation treatments for depression, obesity, high blood pressure, headaches, and many other conditions that compromise health and quality of life.

We have seen how the development of implantable pacemakers and defibrillators over the last half century has redefined life for patients with heart disease. We are about to witness a far more dramatic revolution. The convergence of decades of neuroscience research and technology development has now reached critical mass and is propelling the neural engineering field into the commercialization phase. First-generation pacemakers, surface stimulators and single-channel neurostimulator implants were miniaturized pulse generators, and the engineering challenges involved in their successful commercialization were biocompatible and biosurvivable encapsulation, power consumption and battery life, and component reliability. Now we are conceiving and implanting multichannel implants with closed loop feedback and autonomous control systems, and the engineering challenges push the limits of medical knowledge and human creativity. We are developing a depth knowledge of neural function never before dreamed of… as we must, if we are to be successful in our charge to augment and ultimately to replace neural function. The relentless drive for a deeper understanding of neural function is being propelled more by its application to therapy and not just by scientific curiosity. The payoff to society is enormous: improved quality-of-life and functional restoration following stroke, spinal cord injury, and traumatic brain injury, with extension into deafness, blindness, paralysis and movement disorders, and even some mental illness, seizure disorders, and other systemic conditions.

Neural Engineering has taken longer to reach the cusp of commercialization than many other medical device fields, such as that of pacemakers, largely because of the number and complexity of the multiple fields of knowledge in neuroscience, biomaterials, and engineering required to demonstrate proof of concept and, further, to achieve both acceptable efficacy and safety. But neural engineering has the potential for impact on many disease states that are far from the realm of the neurologist or neurosurgeon, because excitable and hence stimulatable, tissue is implicated in the functioning of virtually every organ system in the body.

This book represents contributions from thought leaders in industry, academia, and clinical medicine and provides an understanding of the history, physiology, and a cross section of the most promising engineering technologies in the industry. This book grew out of the volume I coedited with Cedric Walker several years ago, and I am indebted to him for involving me in that project. A history and overview of the field is presented in which DiLorenzo and Gross relate the extraordinary breadth and duration of investigation and discovery that has led up to the current flood of innovation.

Important clinical applications of neuromodulation are presented. Kiss provides a detailed review of the fundamental science and mechanisms of action underlying deep brain stimulation. Velasco et. al., leaders in the field of neurostimulation for the treatment of epilepsy, describe their experiences with seizure control. Fountas and Smith present an overview of the clinical, surgical, and technological facets of responsive neurostimulation for the treatment of seizures. Burton and Phan provide a thorough review of spinal cord stimulation for pain control. Tcheng and Morrell provide a complementary description of the engineering aspects of the seizure detection and termination technology and clinical studies characterizing the responsive neurostimulation system. Richardson describes his pioneering work eluci- dating central pain pathways and his development of neurosurgical techniques and insights to treat pain disorders. Burton and Phan review the state of the art in technologies and techniques applying neuro- stimulation for pain control. Lad, Chao, and Henderson describe techniques and applications in the use of motor cortex stimulation for the treatment of pain. Assad and Eskandar review the history of surgical therapies for obsessive compulsive disorder (OCD) and describe the current knowledge and future directions for the application of deep brain stimulation for OCD.

Promising technologies and applications for neural augmentation are detailed. Sensory prostheses are described first. Eddington, principal research scientist at MIT and Harvard, describes fundamental mechanisms and technologies behind the cochlear implant, the first FDA approved neural prosthesis. Greenberg, president of Second Sight, provides insight into the development of a breathtaking technology which offers hope to the blind.

Motor prostheses are described in two parts, beginning with the technologies applicable to the extrac- tion of a command signal from the brain of the user. Leuthardt, Ojemann, Schalk, and Moran provide a broad overview of the multifaceted field of brain computer interfaces (BCIs). Bakay, neurosurgeon pioneer and collaborator with Kennedy, reveals insights gained from decades of experience in conducting groundbreaking clinical studies and describes neurosurgical considerations relevant to brain computer interfaces, with a focus on his experience with the Cone electrode. Kennedy, a pioneer in neural interfacing and the first to chronically record neural signals from a human, describes and compares various neural recording technologies, including the Cone electrode which provides long term stable human-machine communication. Donoghue, Serruya, and Kim, describe the promising BrainGate technology, a high channel count brain computer interface, and present their human studies.

The efferent component of motor prostheses are then described, comprising technologies applicable to the delivery of control signals to the neuromotor system. Graupe, an early innovator in the research and clinical application of noninvasive FES for gait restoration discusses the research and commercial- ization efforts behind ParaStep. Davis et. al. have outlined the development, success and difficulties of a multi-channel wired-linked stimulating system for functional restoration in paraplegia. Schulman, a cofounder of Pacesetter systems and now president of The Alfred Mann Foundation, and colleagues describe the BION, a broadly applicable platform technology for minimally invasive neural stimulation and sensing. Guevremont and Mushahwar present their neurophysiological studies and elegant applica- tion of spinal cord neuroscience to the development of a novel technology for delivering motor command signals to the spinal cord.

Fundamental science underlying current neurostimulation techniques as well as new paradigm shifting neuromodulation technologies are presented. Grill, an early leader in functional electrical stimulation, describes the theoretical foundation for stimulation of cells of the nervous system. Wells, Mahadevan- Jansen, Kao, Konrad, and Jansen review the history and fundamental science behind optical stimulation and describe a novel optical stimulation technology. Roth and Zangen describe promising innovations in the noninvasive transcranial magnetic stimulation technology, which is being commercialized for a number of promising indications. Pasley, Jobke, Gudlin, and Sabel present their basic scientific studies in the neural plasticity of the visual system and present a new technology which utilized this plasticity for restoration of vision.

The convergence of disciplines described herein is becoming apparent to researchers, engineers, clinicians, established medical device companies, early stage venture capital investors, and a wave of startup companies. We are witnessing the transition of Neural Engineering from basic research to intense commercialization and widespread clinical application and acceptance. The impact of this industry on medicine and on society promises to be as dramatic as that of the development of antibiotics. We are entering an era in which the nervous system is replaceable and augmentable... the Neurological Age. The future is upon us.

Daniel J. DiLorenzo, MD, PhD, MBA

Editor

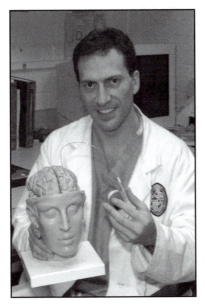

Daniel J. DiLorenzo, MD, PhD, MBA is a Neurosurgeon, Engineer, and Inventor/Entrepreneur. Dan has received several awards for invention, including the 1999 Lemelson-MIT $30,000 Student Innovation Award for his work in the fields of Robotics, Neuroengineering, and Medical Devices. He was named in the 1999 Technology Review TR100 (100 innovators most likely to have an impact on technology) and was recognized by Entrepreneur Magazine as one of 5 "New Geniuses" for his "Innovation" in 2000. He was also named as one of the "40 Under 40" Seattle business leaders by the Puget Sound Business Journal in 2005. He was the founding CEO/CTO of NeuroVista Corporation (formerly BioNeuronics), which he founded while a neurosurgical resident and launched in Seattle to develop innovative technologies to treat epilepsy. Dan is currently on the advisory board for the Smithsonian Institution's National Museum of American History Lemelson Center for the Study of Invention and Innovation. He holds several portfolios of issued and pending patents in the neural engineering and medical device field.

Dan grew up in the Washington DC area, son of Dr. Henry and Marian DiLorenzo, with three siblings, Michael, Nicole, and Christopher. He developed an early fascination with engineering and medicine, and in 8th grade he combined these interests by designing and building an 11 degree of freedom anthropomorphic digital computer controlled robot. This early and formative experience catalyzed his desire to integrate engineering and medicine to create biomedical technologies to treat disease. This has also underscored the importance of early exposure to the sciences, technology, and invention and has been a strong motivator in the initiatives he has become involved with, including the Lemelson Program and its Center at the Smithsonian Institution.

Dan earned his undergraduate and masters degrees in electrical engineering and computer science in 3 and 1 years, respectively, from MIT. He received his MD from the joint Harvard Medical School/MIT Division of Health Sciences and Technology. Concurrently, he completed his PhD at MIT in Mechanical Engineering and focused on implanted devices, neural interfacing, neuroscience, and control theory. He developed technologies for functional electrical stimulation (FES) driven gait restoration in paraplegics, chronic peripheral neural

implants, microfabricated retinal implants, cortical signal processing, and designed several novel wheeled and walking robots. He received an MBA (M.S. in the Management of Technology) at the MIT Sloan School of Management.

Dan began his Neurosurgical residency at Tulane University under the leadership of Dr. Donald Richardson and concurrently served as an Adjunct Assistant Professor of Biomedical Engineering supervising Neuroengineering theses. He passed the US Patent Bar and founded NeuroBionics in 2002 (later renamed BioNeuronics in 2005 and then NeuroVista in 2007). He completed the first several years of his training prior to taking leave in 2004 to focus full-time to raise venture financing, recruit a team, and launch the company. Dan spent 2 years as Chairman and Founding CEO and then as CTO, raising funding, inventing technologies and leading the technology/IP strategy and clinical strategy efforts while working to recruit a world-class team. In 2006, with the venture on solid footing, Dan returned to clinical medicine to continue his Neurosurgery residency at The Methodist Hospital in Houston under the leadership of Dr. Robert Grossman and his world-renowned faculty. In 2005, Dan founded and remains President of DiLorenzo Biomedical, LLC, which is developing and commercializing innovative high-impact technologies to treat and improve the quality of life for people suffering from a spectrum of diseases.

Contributors

Trent Anderson
Department of Clinical
  Neuroscience
University of Calgary
Calgary, Alberta, Canada

Isabel Arcos
Senior Biomedical Engineer
Alfred Mann Foundation for
  Scientific Research
Santa Clarita, California

**Wael F. Asaad, M.D.,
  Ph.D.**
Neurosurgery Resident
Department of Neurosurgery
Massachusetts General Hospital
Boston, Massachusetts

Roy A.E. Bakay, M.D.
Chief of Functional Neurosurgery
Rush University Medical Center
University Neurosurgery
Chicago, Illinois

Andrew Barriskil
Neopraxis Pty. Ltd.
Lane Cove, NSW, Australia

Randal R. Betz, M.D.
Chief of Staff
Shriners Hospital for Children
Philadelphia, Pennsylvania

Bernardo Boleaga, M.D.
Neuroradiologist
CT Scanner de México
Mexico City, Mexico

Allen W. Burton, M.D.
Clinical Medical Director, Pain
  Management Center
Department of Anesthesiology
  and Pain Medicine
University of Texas M.D.
  Anderson Cancer Center
Houston, Texas

Kevin Chao, M.D.
Neurosurgery Resident,
Department of Neurosurgery
Stanford University School of
  Medicine
Stanford, California

Ross Davis, M.D., Ph.D.
Neurosurgeon
Alfred Mann Foundation
Valencia, California

**Daniel DiLorenzo, M.D.,
  Ph.D.**
Neurosurgery Resident,
  The Methodist Hospital
Houston, Texas
Founder, NeuroVista (f.k.a.
  BioNeuronics) Corporation
Seattle, Washington
President, DiLorenzo Biomedical,
  LLC
Houston, Texas

John P. Donoghue, Ph.D.
Professor
Department of Neuroscience
Brown University
Providence, Rhode Island
Co-Founder, Cyberkinetics
  Neurotechnology Systems

**Donald K. Eddington,
  Ph.D.**
Principal Research Scientist,
Cochlear Implant Research
  Laboratory,
Massachusetts Eye and Ear
  Infirmary
Boston, Massachusetts
Research Laboratory of Electronics
Massachusetts Institute of
  Technology
Cambridge, Massachusetts

Emad N. Eskandar, M.D.
Director of Stereotactic and
  Functional Neurosurgery
Massachusetts General Hospital
Harvard Medical School
Boston, Massachusetts

**Kostas N. Fountas, M.D.,
  Ph.D.**
Neurosurgeon
Department of Neurosurgery
Medical College of Georgia
Augusta, Georgia

Philip L. Gildenberg, M.D., Ph.D.
President
Houston Stereotactic Concepts
Houston, Texas

Daniel Graupe, Ph.D.
Professor
Department of Electrical
    Engineering
Department of Bioengineering
University of Illinois
Chicago, Illinois

Robert J. Greenberg, M.D., Ph.D.
President & CEO
Second Sight
Sylmar, California

Warren M. Grill, Ph.D.
Associate Professor
Department of Biomedical
    Engineering
Department of Surgery
Duke University
Durham, North Carolina

Robert E. Gross, M.D., Ph.D.
Assistant Professor and Director
    of Functional and Stereotactic
    Neurosurgery
Department of Neurosurgery
Emory University
Atlanta, Georgia

Julia Gudlin, M.Sc.
Institute of Medical Psychology
Otto-von Guericke-University of
    Magdeburg
Magdeburg, Germany

Lisa Guevremont, Ph.D.
Department of Biomedical
    Engineering
University of Alberta
Calgary, Alberta, Canada

Jaimie M. Henderson, M.D.
Assistant Professor and Director
    of Functional and Stereotactic
    Neurosurgery
Department of Neurosurgery
Stanford University School of
    Medicine
Stanford, California

Thierry Houdayer
Biomedical Engineer
International Center for Spinal
    Cord Injury
Kennedy Krieger Institute
Baltimore, Maryland

E. Duco Jansen, Ph.D.
Associate Professor
Department of Biomedical
    Engineering
Vanderbilt University
Nashville, Tennessee

Fiacro Jiménez, M.D., Ph.D.
Head of the Stereotaxic and
    Functional Neurosurgery and
    Radiosurgery Unit
Mexico General Hospital
Mexico City, Mexico

Sandra Jobke, M.Sc.
Institute of Medical Psychology
Otto-von Guericke-University of
    Magdeburg
Magdeburg, Germany

T. E. Johnston
Shriners Hospital for Children
Philadelphia, Pennsylvania

C. Chris Kao, M.D., Ph.D.
Research Associate Professor
Department of Neurosurgery
Vanderbilt University
Nashville, Tennessee

Philip R. Kennedy, M.D., Ph.D.
President & CEO
Neural Signals Inc.
Duluth, Georgia

Sung-Phil Kim, Ph.D.
Postdoctoral Research Associate
Department of Computer
    Science
Brown University
Providence, Rhode Island

Zelma Kiss, M.D., Ph.D.
Assistant Professor
Department of Neurosurgery
University of Calgary
Calgary, Alberta, Canada
Canadian Institutes for Health
    Research
Alberta Heritage Foundation for
    Medical Research
Edmonton, Alberta, Canada

Peter E. Konrad, M.D., Ph.D.
Associate Professor
Department of Neurosurgery
Vanderbilt University
Nashville, Tennessee

Mauricio Kuri, M.D.
Head of the Anesthesiology
   Department
Hospital Ángeles de las Lomas
Mexico State, Mexico

**Shivanand P. Lad, M.D.,
Ph.D.**
Neurosurgery Resident
Department of Neurosurgery
Stanford University School of
   Medicine
Stanford, California

Eric C. Leuthardt, M.D.
Assistant Professor
Departments of Neurological
   Surgery and Biomedical
   Engineering
Director, Center for Innovation
   in Neuroscience and
   Technology
Washington University
St. Louis, Missouri

**Anita Mahadevan-Jansen,
Ph.D.**
Associate Professor
Department of Biomedical
   Engineering
Vanderbilt University
Nashville, Tennessee

J. Phil Mobley
Senior Vice President for
   Research and Development
Alfred Mann Foundation for
   Scientific Research
Santa Clarita, California

Daniel W. Moran, Ph.D.
Assistant Professor
Department of Biomedical
   Engineering
Washington University
St. Louis, Missouri

Martha Morrell, M.D.
Clinical Professor of Neurology
Stanford University
Chief Medical Officer
NeuroPace, Inc.
Mountain View, California

**Vivian K. Mushahwar,
Ph.D.**
Assistant Professor
Department of Biomedical
   Engineering
University of Alberta
Calgary, Alberta, Canada

José María Núñez, M.D.
Stereotaxic and Functional
   Neurosurgery Unit
Mexico City General Hospital
Mexico City, Mexico

Jeffrey G. Ojemann, M.D.
Associate Professor
Department of Neurological
   Surgery, Regional Epilepsy
   Center
University of Washington School
   of Medicine
Harborview Medical Center
Seattle, Washington

Imelda Pasley, M.Sc.
BioFuture Research Group
Leibniz Institut Für
   Neurobiologie
University of Magdeburg
   Medical Faculty
Magdeburg, Germany

Phillip C. Phan, M.D.
Assistant Professor
Department of Anesthesiology
   and Pain Medicine
University of Texas M.D.
   Anderson Cancer Center
Houston, Texas

**Donald E. Richardson,
M.D., F.A.C.S.**
Department of Neurosurgery
Tulane University Medical School
New Orleans, Louisiana

Yiftach Roth, Ph.D.
New Advanced Technology
   Center
Sheba Medical Center
Tel Hashomer, Israel

Bernhard A. Sabel, Ph.D.
Professor
Institute of Medical Psychology
Otto-von Guericke-University of
   Magdeburg
Magdeburg, Germany

Gerwin Schalk, Ph.D.
Research Scientist
Wadsworth Center
New York State Department of
   Health
State University of New York
Albany, New York

Joseph H. Schulman, Ph.D.
President and Chief Scientist
Alfred Mann Foundation for
  Scientific Research
Santa Clarita, California

Mijail D. Serruya, M.D., Ph.D.
Neurology Resident
University of Pennsylvania
Philadelphia, Pennsylvania
Department of Neuroscience
Brown University
Providence, Rhode Island
Co-Founder, Cyberkinetics
  Neurotechnology Systems

B. Smith
Shriners Hospital for Children
Philadelphia, Pennsylvania

Joseph R. Smith, M.D., F.A.C.S.
Professor and Chief, Functional
  and Epilepsy Neurosurgery
  Service
Department of Neurosurgery
Medical College of Georgia
Augusta, Georgia

Thomas K. Tcheng, Ph.D.
Sr. Manager of Research and
  Sr. Scientist
NeuroPace, Inc.
Mountain View, California

Ana Luisa Velasco, M.D., Ph.D.
Head of the Epilepsy Clinic
Stereotaxic and Functional
  Neurosurgery and
  Radiosurgery Unit
Mexico General Hospital
Mexico City, Mexico

Francisco Velasco, M.D.
Head of the Neurology and
  Neurosurgery Service
Mexico General Hospital
Mexico City, Mexico

Marcos Velasco, M.D., Ph.D.
Neurophysiologist
Stereotaxic and Functional
  Neurosurgery and
  Radiosurgery Unit
Mexico General Hospital
Mexico City, Mexico

Jonathon D. Wells, Ph.D.
Biomedical Engineer/Senior
  Scientist
Aculight Corporation
Bothel, Washington

James H. Wolfe
Alfred Mann Foundation for
  Scientific Research
Santa Clarita, California

Abraham Zangen, Ph.D.
Assistant Professor
Department of Neurobiology
Weizmann Institute of Science
Rehovot, Israel

Background

1

History and Overview of Neural Engineering

Daniel DiLorenzo and
Robert E. Gross

1.1 Background

Recent years have witnessed remarkable advances in the development of technology and its practical applications in the amelioration of neurological dysfunction. This encompasses a wide variety of disorders, and their treatment has been the purview of different, partially overlapping, medical and scientific specialties and societies. As a result of these advances, three somewhat independent and longstanding fields with unique origins — stereotactic and functional neurosurgery, neuromodulation, and functional electrical stimulation — are experiencing increasing convergence and overlap, and the distinctions separating them are beginning to blur.

Stereotactic and functional neurosurgery — which takes its origins at the beginning of neurosurgery in the late 19th century, but which as a society dates to the 1940s — has mainly concerned itself with the surgical treatment of nervous system conditions that manifest as disordered function, including movement disorders (e.g., Parkinson's disease, tremor, dystonia), pain, epilepsy, and psychiatric illnesses. In the first part of the last century, the surgical treatment of these disorders usually involved ablation of nervous tissue, but this has almost completely been supplanted by the advent of electrical neurostimulation. Other modalities within the purview of stereotactic and functional neurosurgery include pharmacological and biological therapies that are delivered via surgical techniques.

Neuromodulation broadly involves alteration of the function of the nervous system. The International Neuromodulation Society was established in the 1990s and dedicated itself to the treatment of nervous system disorders — mostly pain and spasticity — with implantable devices, including pumps and stimulators. However, with the advances in the field, this society has broadened its scope to include implantable devices used to treat disorders of sensory and motor functions of the body. This society has been driven by both anesthesiologists and neurosurgeons, but has mainly been clinical in its purview.

Functional electrical stimulation has pertained mostly to the restoration of upper and lower limb function (after injury and ischemia), bowel, bladder, and sexual function; and respiratory function. This field, represented by the International Functional Electrical Stimulation Society, also is interested in auditory and visual prostheses, and has — in contrast to the two other societies — a distinct engineering bias, limited as it has been to electrical stimulation.

It is quite clear that each of these areas has strong overlap, both clinically as well as technologically. While one or the other may have been more clinical or basic in its scientific approach, with the progress in both technology development and its translation into the clinic, these distinctions are becoming increasingly blurred. Advances in functional neurosurgery have been driven by clinical empiricism and advances in pathophysiological understanding of the target disorders, but also by technological advancement (e.g., the development of the stereotactic frame, and then frameless image guidance stereotactic technology, and the development of implantable neurostimulation systems). Neuromodulation — based as it is on the use of implantable devices — has been driven by technical and pharmacological innovation coupled to increased understanding of neurophysiology and the pathophysiological manifestations in states of disease. Finally, functional electrical stimulation (FES) has mostly been an undertaking driven by engineering technology, as from the outset the goals have been restoration of motor function using controlled electrical stimulation. Clinical application has always been the long-term objective; and with the current synergy of technological innovation and increasing clinical experimentalism (e.g., motor prosthesis trials for spinal cord injury), the field is making great strides.

The technology that is a main driver behind each of these overlapping disciplines can be coined "neural engineering." Although this domain might seem like a relatively recent innovation, a view into the historical underpinnings of neuromodulation in the eighteenth century reveals that, in fact, from the start, technological innovation and neuromodulation developed hand in hand, and some of this pioneering foundation building of the past several centuries might reasonably be called neural engineering.

1.1.1   The Early History of Electrical Stimulation of the Nervous System

The use of electricity for therapeutic purposes can arguably date back even to the use of amber and magnetite in jewelry around 9000 BC [1]. The use of the torpedo fish or electric eel was advocated by Scribonius Largus — one of the first Roman physicians in the first millennium — for the treatment of headache and gout [2]. The patient would apply the fish to the painful region until numb (but not too long) [3]. Its use continued into the sixteenth century, by which time the indications were broadened to include melancholy, migraine, and epilepsy.

By the seventeenth century, electricity was identified as a form of energy, and primitive devices to induce electric current were developed. The first electrostatic generator was constructed in 1672, but electrical sparks from a modified electrostatic generator were first used therapeutically to treat paralysis in 1744 [3] (the first example of neural engineering?). In 1745 in Leyden, the capacity of an electrified glass jar and tin foil — the Leyden jar — to store electrical charge was inadvertently discovered when its discharge caused a physiological effect so severe that the experimenter took two days to recover [4]. Electricity and the nervous system were intricately associated early on because the most sensitive electrical detector at the time was, in fact, the nervous system. The new phenomenon was quickly embraced by the medical field, and "cures" of paralysis and other ailments rapidly proliferated, yielding many fantastic books from "electrotherapists" such as John Wesley the divine[3, 4]. By 1752, even Benjamin Franklin was using it to treat a 24-year-old woman with convulsions [5], and various palsies, but he at least was not convinced of any persistent benefits from the electrical treatments [4].

Albrecht von Haller felt that nerves and muscles were sensitive to the effects of electrical current, but he and others did not believe that electricity played a role in normal functioning [6]. In 1791, Galvani published his famous experiments in which he induced muscular contractions in frog leg muscle using a metallic device constructed of dissimilar metals [7]. He reasoned that the electrical current was being "discharged" from the muscle, a view that was countered by Volta who felt that the electrical current flowed from the dissimilar metals [8] and which led him to develop the voltaic bimetallic pile — the first battery — based on this insight [9]. Electrophysiological experiments using voltaic piles to generate "galvanic current" rapidly proliferated, including important experiments by Aldini — Galvani's nephew — who was interested in, among other things, the therapeutic applications of galvanism [6] (more examples of neural engineering). His particular interest was in "reanimation," especially following near-drowning, and this work included electrically stimulating both animals and men — the latter following the not infrequent hangings and decapitations [10]. He was successful at provoking muscular contractions of the somatic musculature, but was disappointed that he could not get similar results with the heart [6]. In fact, Aldini was likely the first to induce facial muscle contractions with direct brain electrical stimulation in both oxen and human cadavers, predating Fritsch and Hitzig's influential work by many years [11]. A particularly dramatic demonstration on a hanged criminal in England garnered widespread lay coverage, possibly contributing to the inspiration behind the Frankenstein myth [6]. He also used the voltaic piles to treat "mental disorders" including depression — a precursor to electroconvulsive therapy — after first applying the pile safely to his own head [10].

The next major advance arose from Faraday's invention of the first electric generator in 1831, which induced alternating or "Faradic" current" by rotating wires inside a magnetic field. Dubois-Reymond in 1848 demonstrated that the time-varying nature of Faradic current was important in efficient stimulation, and he formulated an early expression of the strength-duration curve relating the threshold for activating the neuromuscular system to the intensity and duration of the current pulse [3]. G.B. Duchenne — regarded as the father of "electrotherapy" — established it as a separate discipline ("*De l'Electrisation localisée*", 1855) [3]. He used Faradic current applied through moistened electrode pads, preferring it to Galvanic current because of its warming properties. In 1852, he stimulated the facial nerve for palsy [12]. With Duchenne's influence and the proliferation of induction coils and batteries, therapeutic electrical stimulation spread widely. By 1900, according to McNeal, most physicians in the United States had an "electrical machine" for the treatment of a plethora of ailments — many neurological, including pain. Many different and interesting devices were constructed for use in every part of the body (literally), including the "hydroelectric bath" (not recommended!) [3]. Called the "Golden Age of Medical Electricity," the close of the nineteenth century would only be eclipsed by the close of the following century in the proliferation of the use of electricity in medical therapeutics.

Fritsch and Hitzig [11] are said to have been the first to stimulate the cortex in living animals, and systematic mapping of cerebro-cortical function was described in animals by Ferrier [13]. The first time the human brain was directly stimulated in an awake patient was in 1878 by Barthlow, who introduced a stimulating needle into a patient's brain through an eroding scalp and skull tumor [14]. Low-amplitude faradic current elicited contralateral muscle contractions and an unpleasant tingling feeling, initiating the field of cerebral localization using electrical stimulation. Cushing observed the effects of intraoperative faradic stimulation in 1909 [15], and Foerster's intraoperative work in the 1930s influenced Wilder Penfield, culminating in Penfield and Jasper's seminal publication of *Functional Anatomy of the Human Brain* in 1954 [16].

1.2 Neural Augmentation (Neural Prostheses)

1.2.1 Origins of the Field of Neural Prostheses

The field of neural prostheses encompasses the set of technologies relevant to the restoration of neurological and neuromuscular function, and this includes both motor prostheses and sensory prostheses. The early motor work focused on technologies for restoration of limb motor function, these having clinical relevance to the treatment of both paralyzed and amputee patients. In the past several decades, as microelectronic and microfabrication technologies have progressed, sensory prostheses have reached the state of practicality, and cochlear implants have become a standard of care in the treatment of sensorineural deafness. Research and commercialization efforts continue in remaining areas, including restoration of the motor and sensory functions of the limbs; motor functions of the bladder, bowel, and diaphragm; and sensory functions including tactile, visual, and vestibular.

In the 1950s, a human study showing sound perception arising from electrode implantation inspired researchers around the world to investigate the possibility of a cochlear prosthesis, which, after several decades of development in academia and industry, became the first FDA-approved neural prosthesis [17].

The Neural Prosthesis Program, launched in 1972 and spearheaded by F. Terry Hambrecht, M.D., brought funding, focus, and coordination to the multidisciplinary effort to develop technologies to restore motor function to paralyzed individuals. The initial efforts were in electrode-tissue interaction, biomaterials and neural interface development, cochlear and visual prosthesis development, and control of motor function using implanted and nonimplanted electrodes.

In his 1980 UC Davis Ph.D. thesis, Edell demonstrated chronic recording from peripheral nerve using a multichannel micromachined silicon regeneration electrode array [18, 19]. This was a major achievement, as it proved that chronic recording from the nervous system was possible. Furthermore, his silicon-based implant was proof of concept that biocompatible recording interfaces could be made from silicon using existing etching and microfabrication technology — and could therefore be made to incorporate electrodes, preamplifiers, processing, memory, and telemetry elements on the same silicon substrate, implanted in the nervous system.

1.2.2 Neural Augmentation: Motor Prostheses

1.2.2.1 Neuromuscular Stimulation for Control of Limb Movement

Liberson and coworkers are credited as being the first to utilize functional electrical stimulation (FES) to restore functional control of movement to a paralyzed limb [20]. They treated more than 100 hemiplegic patients with foot drop, using a transistorized stimulator and conductive rubber electrodes placed on the skin overlying the peroneal nerve. A switch under the sole triggered stimulation during the swing phase of gait, causing contraction of the tibialis anterior and dorsiflexion of the foot [21]. All patients were reported to have received some benefit, but acceptance was limited by skin irritation from the electrode, the need for precise electrode placement, hassle in applying the device compared to functional benefit, and electrode lead breakage; Liberson termed this "functional electrotherapy." However, this term did not gain widespread acceptance; rather, "functional electrical stimulation," coined by Moe and Post, did [22].

Long and Masciarelli, intrigued by Liberson's lower extremity work, devised a system for upper extremity functional restoration in high cervical quadriplegic patients [23]. Using FES-controlled finger extension, via extensor digitorum stimulation, coupled with spring-loaded thumb and finger flexion, the device enabled control of grasp in the paralyzed limb. Although patients tolerated it very well, it did not gain widespread clinical acceptance.

Peckham and colleagues at Case Western Reserve University used chronic percutaneous stimulation of forearm muscles to provide hand grasp, including both palmar and lateral prehension and release, in C5 quadriplegic patients [24].

Restoration of some form of assisted gait in spinal cord-injured patients has been the focus of several research groups. In 1960, Kantrowitz reported the use of surface stimulation of quadriceps and gluteal muscles to effect rising and standing for several minutes from a sitting position in a paraplegic patient [25]. In 1973, Cooper reported bilateral implantation of the femoral and sciatic nerves in a T11-12 paraplegic and claimed ambulation of up to 40 feet using a walker for balance assistance [26].

Using implanted stimulators and electrodes on the femoral nerves bilaterally, Brindley was able to achieve in a paraplegic arising from a sitting position and limited gait assisted with elbow crutches but not requiring a walker, braces, or other support [27]. Several more sophisticated devices followed. Kralj et al. described a four- to six-channel skin surface electrode FES system in which patients ambulated with the assistance of parallel bars or a roller walker [28, 29]. At MIT in 1990, DiLorenzo and Durfee implemented a computer-controlled, four-channel skin surface electrode FES system with bilateral quadricep stimulation and peroneal nerve withdrawal reflex stimulation, and achieved 60 feet of handrail balance-assisted gait in a thoracic spinal cord-injured patient [30] and developed adaptive closed-loop controllers that compensated for time-varying muscle performance, including gain changes due to muscle fatigue [31, 32]. Figure 1.1 shows a sequence of still frames from a gait sequence using this four-channel

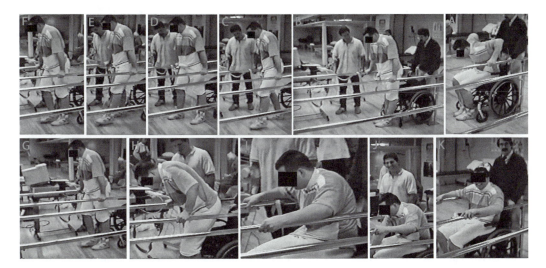

FIGURE 1.1 (See color insert following page 15-4). Series of still images from four-channel FES gait sequence performed by author (DiLorenzo) at MIT in 1990: (A) subject bracing to stand; (B) standing under FES control with computer-controlled stimulator on the far left (left to right: author DiLorenzo and subject Allen Wiegner, Ph.D.); (C) first step; (D) second step; (E) third step; (F) fourth step; (G) preparing to sit; (H) sitting down; (I) seated in wheelchair after walking; (J) repositioning feet; and (K) smiling.

surface FES configuration[30]. Convinced of the greater potential for clinical efficacy and market success with implanted technologies, DiLorenzo pursued related implanted FES work [30, 33–36], and Goldfarb and Durfee continued this line of surface FES work to include a long-legged brace with a computer-controlled friction brake for improved energy efficiency and control [37].

Marsolais reported improved motor function in patients with intramuscular stainless steel electrodes implanted in the quadriceps, hip flexors, extensors, and abductors [38]. Although significant motor torque improvement was achieved in some of these patients, all of whom had some motor function pre-implantation, patients required supervision for ambulation.

Graupe extended the surface FES technology developed by Lieberson and Kralj to include a patient-borne and -controlled system [39, 40]. Graupe then commercialized this technology for clinical use as the ParaStep system. The history of the field of FES for ambulation and the commercialization of ParaStep is described by Graupe in more detail in Chapter 16 of this book.

The first generation of implanted FES systems for upper extremity functional restoration was developed at Case Western Reserve University and Cleveland Veterans Affairs Medical Center for the restoration of hand grasp to a patient with cervical spinal cord injury [41]. This system allowed the paraplegic patient to control hand grasp on an otherwise paralyzed limb by generating movement commands using contralateral shoulder movements. This technology was commercialized by Neuro-Control Corporation, which was founded by Hunter Peckham, Ph.D., Ronald Podraza, and colleagues at Case Western Reserve University in Cleveland, Ohio. NeuroControl received FDA approval in 1997 to market the grasp restoration device, known as the FreeHand System.

1.2.2.2 Chronically Implanted Neuroelectric Interfaces (Recording Arrays)

1.2.2.2.1 Basic Neuroscience

Separate but concurrent research efforts by many investigators in basic neurophysiology were making significant advancements in developing an understanding of how cortical and spinal neural systems control motor function. Extensive research on control of movement by hundreds of neuroscientists has revealed correlations between parameters of limb movement and neural activity in many motor centers, including the primary motor cortex, premotor and supplementary motor cortex, cerebellum, and spinal cord.

The motor cortex was the first area to undergo detailed quantitative study of neuronal activity during whole-arm reaching movements in an awake behaving primate[42]. Georgopoulos et al. first demonstrated a relation between neural cell activity and movement direction [42]. They found that cell firing rates are directionally sensitive in a graded manner; that is, the firing rate is maximal in a so-called preferred direction, and the firing rate tapers off gradually as the direction of movement deviates from this preferred direction. In 1984, Georgopoulos et al. demonstrated the presence of cells in both primary motor cortex and in area 5, the firing rates of which correlate linearly with position of the hand in two-dimensional space [43]. Subsequent work expanded this knowledge to include additional movement parameters and limb movement in three-dimensional space [44–48].

Later research beginning in the 1990s focused on control using neural signals, in both non-human primate and human subjects, lending credence to the notion that sufficient information may be extracted from neural signals to be used as a control signal for a prosthetic system. Schwartz extended this work and characterized neural firing patterns associated with limb movement in three-dimensional space [46] and, with Taylor and Tillary, developed a real-time adaptive prediction algorithm with which a primate was able to control robot arm movement [49]. Donoghue et al. pursued similar lines of research and characterized patterns of behavior from large populations of cells [50]. DiLorenzo showed evidence for the presence of real-time motor feedback error signals in the primate motor cortex, suggesting the presence of feedforward and feedback signals in primary motor cortex [51]. Using the 100-channel intracortical array developed by Normann, Donoghue characterized additional information content in correlations between neuronal activity [52]. Using wire electrode arrays, Nicolelis also characterized real-time prediction of limb trajectory using neural ensembles, and in collaboration with Srinivasan showed remote control of robot movement by a non-human primate [53].

1.2.2.2.2 *Initial Application to Humans*

Kennedy and Bakay demonstrated human two-dimensional control of cursor movement by a paralyzed human patient using the cone neurotrophic electrode [54–56]. This novel electrode facilitates regeneration of neural processes into an insulating glass cone where a stable electrical and mechanical interface forms, providing a long-term, high signal-to-noise ratio. In 1987, Kennedy founded Neural Signals to develop the cone electrode for clinical use; however, the company has refocused its direction on the commercialization of a less-invasive EMG-driven system for locked-in patients.

A spinout from Donoghue and Normann's research efforts, Cyberkinetics Neurotechnology Systems, Inc., has demonstrated in pilot studies human control of cursor movement using a 4-by-4-millimeter, 96-channel intracortical microelectrode array chronically implanted into the primary motor cortex [57]. As of this writing, they have demonstrated the ability to record over 6 months post-implantation.

1.2.3 Neural Augmentation – Sensory Prostheses

1.2.3.1 Sensory Stimulation: Auditory (Cochlear Implant)

Auditory prostheses, which provide patterned stimulation of the eighth cranial nerve, were the first commercially available sensory neural prosthesis. As of this writing, it is the only approved sensory prosthesis. However, visual prostheses are already in clinical trials and are likely to become available in the not too distant future.

An experiment published in 1957 by Djourno and Eyries is credited with inspiring the development of the cochlear implant [17]. Paris otologist Eyries, and neurophysiologist Djourno implanted wires in the inner ear of a deaf patient and were able to elicit sensations of sound with electrical stimulation. This experiment was reproduced in 1964 by Doyle et al. in Los Angeles [58].

Inspired by this 1957 Djourno article, William House, M.D., joined with Jack Urban, president of an electronics company, in Los Angeles to develop a cochlear implant, and they published results on their first patients in 1973 [59]. Working with investigators at the House Ear Institute, founded by his brother Howard House, M.D., also an otolaryngologist, William House pioneered the development of the single-channel cochlear implant and achieved its approval by the FDA [60].

Blair Simmons of Stanford University first evaluated multichannel cochlear stimulation, placing four wires into the auditory nerve [61]. Several years later, Dr. Robin Michelson, an otolaryngologist at the University of California–San Francisco (UCSF), implanted a multichannel cochlear implant in the scala tympani [62]. Michael Merzenich, also of UCSF, led the research and development of this and subsequent multichannel cochlear implants [63]. Al Mann founded Advanced Bionics® Corporation in 1993; licensed the UCSF cochlear implant technology; and recruited Joe Schulman, Ph.D., and Tom Santogrossi to develop this technology into the CLARION cochlear implant [64–67].

In parallel with developments in the United States, in Melbourne, Australia, Graeme Clark initiated a large and extremely productive clinical and research program at the University of Melbourne in Australia [68–71]. The implants developed under Clark are manufactured by Cochlear Ltd., a subsidiary of Nucleus Pty. Ltd. in Sydney, and Cochlear has the largest worldwide population of implanted patients, with over 65,000 patients having been implanted with the Nucleus worldwide [72].

In 1975, Ingeborg and Hochmair began development of a multichannel cochlear implant, which was implanted in Vienna two years later [73, 74]. They continued development of the device for nearly a decade and in 1989 founded MED-EL, which has since become the third major manufacturer of cochlear implants.

There are several factors that have contributed to the success of the cochlear implant:

1. Well developed foundation of basic science, particularly in the area of auditory neurophysiology.
2. Focused and coordinated research programs and funding, including that of the NIH Neural Prosthesis Program.
3. Clinician champions who were active in the research and facilitated its clinical acceptance as early adopters themselves and as thought leaders.
4. Cochlear anatomy, including the tonotopic arrangement of sensory receptors and their protection by a bony encasement, facilitating stable chronic neuroelectric interfacing.

Because of this surgically accessible anatomy, an electrode array can be placed and press chronically against a rigid surface while being in close proximity to neural cells, without the risk of damage to or migration through soft neural tissues. This neural interfacing problem is a major hurdle in the development of visual, somatosensory, and motor prostheses, which generally must interface directly with soft neural tissues, which are subject to perpetually changing forces and displacements.

1.2.3.2 Sensory Stimulation: Visual

The restoration of vision has captivated the imagination of man since biblical times. Now, after 4 decades of modern research, beginning with the stimulation of the visual cortex by Brindley in 1966 [75], visual prostheses are making the transition from basic science research to commercialization. William Dobelle, Ph.D., achieved similar results using cortical stimulation to evoke phosphenes in 1974 [76, 77], and he was a pioneer in the drive to commercialize the visual cortex prosthesis. In 1983, he acquired Avery Labs, a manufacturer of electrodes for brain stimulation. From this and other ventures, he derived funds to advance his vision prosthesis research and development efforts at the Dobelle Institute, located in Portugal. An excellent historical review is provided by Greenberg in Chapter 11 of this book. Figure 1.2 shows an example of an early microfabricated electrode built by DiLorenzo.

1.2.3.3 Sensory Stimulation: Tactile

Electrocutaneous nerve stimulation was discovered in 1745 by von Kleist, who described a shock from an electrostatically charged capacitor [79]. Studies published by von Frey in 1915 [80] and Adrian in 1919 [81] demonstrated the spectrum of sensations elicited by electrocutaneous stimulation, ranging from vibration to prickling pain and stinging pain. In 1966, Beeker described an artificial hand with electrocutaneous feedback of thumb contact pressure. Initial results, although qualitative, were positive [82]. In 1977, Schmidl reported that electrocutaneous sensory feedback of grip strength improved control of an experimental hand. He found electrocutaneous feedback to be less susceptible to adaptation than vibrotactile feedback [83]. In 1979, Shannon showed that electrocutaneous feedback on the skin above

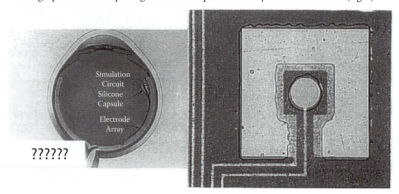

<u>Retinal Implant</u>: Microelectrode array for retinal activation fabricated on flexible substrate for adherence to retinal curvature: Schematic implanted eye (left), photomicrograph of 1 of 10 bipolar gold electrode pairs with 25 μm inner electrode (right)

FIGURE 1.2 Example of an early microfabricated Pt-Ir bipolar electrode for acute preclinical experimentation in retinal stimulation built in 1991 by DiLorenzo [78].

the median nerve is used to encode gripping force in a myoelectrically controlled prosthesis, patients reported an improved level of confidence when using the prosthesis [84].

In 1974 at Duke University, Clippinger implanted a sensory feedback system incorporating an electrode pair to stimulate the median nerve. He provided a frequency-modulated signal that transmitted gripping force at the terminal hook. Patients were able to perceive grip force as well as object consistency [85]. In 1977, Clippinger reported a system using implanted electrodes to provide afferent sensory feedback for upper-extremity amputees [86]. The same year, he reported that postoperative stimulation of the sciatic nerve by a lower-extremity prosthesis offered the additional benefit of postoperative pain reduction; and in 1982, he reported six-year success in a lower-extremity amputee using sciatic nerve stimulation to transmit heel strike force and leg structure bending moment on a single channel [87]. No further work on implanted sensory feedback prostheses is published by Clippinger.

In 1995 at MIT, DiLorenzo and Edell demonstrated the functionality of a chronically implanted, multichannel, intrafascicular peripheral electrode array in an animal model using behavioral and neurophysiological recording, including eye blink reflexes and somatosensory-evoked potentials. These studies demonstrated the functionality of sensory-afferent fibers following transaction, as well as the functionality of a chronically implanted intrafascicular stimulating neuroelectric interface [33–35, 78]. Horch et al. have developed percutaneous flexible microwires for intrafascicular stimulation of neural tissue [88–90].

1.3 Neuromodulation

1.3.1 Chronic Electrical Stimulation of the Nervous System for Functional Disorders

The early history of the use of electrical stimulation of the nervous system was limited to the acute setting because of the lack of availability of implantable stimulators for chronic use. The first progress in this regard was in the area of cardiac pacing — which only recently has come to be viewed, rightfully, as neuromodulation. Although Aldini was motivated to "reanimate" the heart with galvanic stimulation [6], the first successes were by Hyman, who used faradic stimulation to resuscitate the heart in animals and apparently some patients as well, after cardiac arrest [3]. It was not until 1952 that an artificial pacemaker was successfully used by Zoll; and in 1958, a chronic pacemaker was used for 96 days (although it had to be wheeled around on a table!) [91]. An implantable device that had to be charged through the skin was implanted the next year in Sweden [92], followed by a radio-frequency coupled pacemaker [93, 94], and the first fully implantable, self-powered device was implanted in 1960 [95].

1.3.2 Early Development of Deep Brain Stimulation for Psychiatric Disorders and Pain

The availability of implantable pacemakers paved the way for their use in treating nervous system disorders of "function," including movement disorders (tremor, Parkinson's disease, dystonia), pain, epilepsy, and psychiatric disorders. Until this time, a common procedure in the 1950s was subfrontal leucotomy for various psychiatric indications; and in fact, Spiegel and Wycis developed their stereotactic frame mainly for use in this surgery. Pool, at the Neurological Institute at Columbia University, was the first to reason that electrical stimulation might provide a nondestructive, reversible alternative to ablative procedures, in this case for psychiatric indications. He had previously — in 1945 — implanted the first patient with an induction coil for stimulating the femoral nerve for paraparesis (the first example of true functional electrical stimulation) [96] and in 1948 he placed a silver electrode in the caudate nucleus of a patient with Parkinson's disease afflicted with intractable depression, coupled to a permanent mini induction coil placed in the skull. His intent was to activate the caudate with electrical stimulation (although his reasoning is not made clear), and indeed he reported that the patient had some benefits from daily stimulation (with an external primary coil?) for 8 weeks, but a wire broke and therapy was discontinued. Pool also implanted a psychotic patient in 1948 with a cingulate gyrus stimulator.

Neural ablation within the pain-mediating pathways of the brain and spinal cord was the standard treatment for deafferentation (neuropathic) or cancer-related (nociceptive) pain, which was refractory to medical treatment. Thus, the nonablative treatment of pain was advanced when in 1954, Heath at Tulane reported pain relief in schizophrenic patients following electrical stimulation of the septal nuclei via a stereotactic approach (first done in 1950) [97], and similar results were reported by Pool using Heath's technique in a patient treated exclusively for pain (with an externalized electrode wire) [98]. In his 1954 monograph, Pool reported that "focal electrical stimulation of deep midline frontal lobe structures is a new technique that is now being used more and more frequently" [98], which apparently included both himself and Heath in patients with psychiatric disease and/or pain. By this time, Spiegel and Wycis were also implanting chronic electrodes for stimulation, through a stereotactic approach [99]. However, there were as yet no reports of the use of chronic stimulation for the treatment of movement disorders. In the 1960s, levo-dopa for the treatment of Parkinson's disease arrived, and virtually eliminated — for the next several decades — the surgical treatment of Parkinson's disease, except for severe tremor. Moreover, by this time, public outcry eliminated the surgical treatment of psychiatric disorders, which, although mainly directed against leucotomy and lobotomy, also included electrical brain stimulation. Thus, the next advances occurred exclusively in the treatment of pain.

By 1960, Heath and colleagues were stimulating the septal nuclei for pain (without concurrent psychiatric disorder); and in 1961, Mazars and colleagues introduced the sensory thalamus as a target with the idea that stimulation of the deafferented neurons in the thalamus following amputation or stroke, for example, would relieve neuropathic pain [100, 101]. In these and other reports of brain stimulation for pain (septal region [102], caudate [103]), only acute stimulation through externalized wires was used. However, Glenn pioneered the use of his RF-coupled device — first introduced for cardiac pacing in 1959 [94] — to pace the phrenic nerve in a paralyzed patient in 1963 [104]. Shortly thereafter, Sweet and Wepsic [105] implanted an RF stimulator to suppress pain in the peripheral nervous system, and it was used in the first dorsal column stimulator implantation in 1967 by Shealy [106], based on the Melzack and Wall "gate control" theory of pain transmission in the spinal cord [107]. In 1982 at Duke, Nashold, Walker, and Mullen demonstrated a technique for intraoperatively mapping peripheral nerve bundles to accurately place electrodes on individual fasciculi to relieve chronic limb pain [108].

Pursuing an anatomically more central approach, Hosobuchi and colleagues [109] reported successful chronic stimulation of the sensory thalamus for facial pain in 1973, and this was followed by chronic stimulation of sensory thalamus by Mazars [110] and of other regions as well. Notable among these is the periventricular gray region, which — based on the findings of analgesia induced by electrical stimulation in the rat by Reynolds [111] — was chronically stimulated in patients in 1973 by Hosobuchi and colleagues [112] and Richardson and Akil [113, 114]. Ultimately, the fully implantable, battery-powered stimulator was developed for cardiac pacing [95] and then adapted for use in the treatment of pain [14].

1.3.3   The Dawn of Deep Brain Stimulation for Movement Disorders

The earliest use of chronic therapeutic stimulation for movement disorders appears to be that of Bechtereva and colleagues in the erstwhile U.S.S.R. They implanted 24 to 40 electrodes in four to six bundles into the "thalamic – striopallidal nuclei" for patients with hyperkinesias as a result of Parkinson's disease, torsion dystonia, and other causes [115, 116]. Benefits of intermittent stimulation were seen for as long as 3 years, although no implantable stimulator was used. Other early results were from electrical stimulation of sensory thalamus [110, 117] or centromedian–parafascicular complex (CM/Pf) [118] for thalamic deafferentation pain, when improvements in the often associated dyskinesia were noted. Another avenue was that of Cooper and colleagues in the 1970s [119], who stimulated the anterior lobe of the cerebellum to treat various movement disorders, including spasticity and dystonia. Davis and colleagues performed cerebellar stimulation on 316 patients between 1974 and 1981 for spastic motor disorders with good success [120].

By the 1970s, when Parkinson's disease (PD) was mostly being treated pharmacologically, the usual indication for stereotactic surgery in movement disorders was tremor and hyperkinesias (e.g., in dystonia), and the preferred operation was thalamotomy of the ventral intermediate (Vim) nucleus. It had long been appreciated that acute high-frequency, but not low-frequency electrical stimulation of Vim led to immediate tremor arrest, which was used as a final check prior to radiofrequency ablation [121, 122]. The first chronic stimulator implantations directed at movement disorders per se were by Mundinger in 1975 [123]. Brice and McLellan [124] implanted DBS leads in the subthalamic region (a common site for subthalamotomy at the time) in three patients for severe intention tremor resulting from multiple sclerosis. The latter article is of special interest in that it anticipates by decades the idea of a contingent, on-demand system. Andy [125] implanted nine patients with stimulating electrodes in the thalamus but concluded that the likely target was CM/Pf.

At about the same time in 1987, Benabid [126, 127] and Siegfried [128] began implanting deep brain stimulators (DBS) into Vim of the thalamus for tremor from PD and essential tremor. Benabid's report of his large series of Vim stimulators in 1996 [129], followed shortly thereafter by the North American series [130], brought international attention to the field of electrical brain stimulation. The field came full circle as a result of other important trends. The limitations and complications of levo-dopa treatment for PD began to become apparent by the 1980s and, as a result, Laitinen (a student of Leksell, a leader in functional neurosurgery in the 1950s during which time pallidotomies were performed) began to revisit ablative surgery for PD. His influential report in 1992 of the results of pallidotomy refocused attention on the surgical treatment of PD [131]. With a large experience by then in DBS, Siegfried performed the first implantation of a stimulator electrode into the posteroventral internal globus pallidus for PD in 1992 [132]. The next advance resulted from new insights into the pathophysiology of PD: the report of DeLong and colleagues of the amelioration of experimental PD in nonhuman primates by the ablation of the subthalamic nucleus, whose glutamatergic driving actions on the globus pallidus internus was elucidated [133]. Benabid then targeted this novel region for deep brain stimulation in 1993 (the subthalamic region, including white matter projections, had been the target for "subthalamotomies," but never before had the subthalamic nucleus per se been targeted because of concerns over the development of hemiballism) [134].

1.3.4   The Efforts toward Neurostimulation for Epilepsy

In the 1950s, Cooke [135] and Dow [136] described their work in rats and primates, showing that stimulation of the cerebellum had effects on the electroencephalogram and the frequency of seizures. On this basis, in 1973, Cooper and colleagues reported the use of chronic cerebellar stimulation in patients with epilepsy. They also reported benefits for spasticity [119]. Unfortunately, several subsequent clinical series were unable to replicate Cooper's results with epilepsy [137]. Also in the 1970s, Chkhenkeli reported preliminary results of stimulating the caudate nucleus for epilepsy (followed in 1997 by a larger series) (Chkhenkeli and Chkhenkeli, 1997).

Based on the idea that the anterior nucleus of the thalamus (ANT) has widespread connections with and therefore possible modulatory effects on the cortex, Cooper also investigated and reported the results of stimulation of the anterior nucleus of the thalamus for epilepsy [138]. Also around that time, Mirski and Fisher began providing experimental evidence for a role of the anterior nucleus in epilepsy [139, 140], and that electrical stimulation can mitigate seizure activity [141]. This has provided the impetus to revisit this target for epilepsy [142, 143], and current trials are underway. Meanwhile, Velasco reported that DBS of the centromedian nucleus (CM) — an intralaminar nucleus with widespread connections to the striatum and cortex — reduced seizure frequency in medically refractory patients [144]. Although not replicated in a limited clinical trial [145], results continue to be good for generalized epilepsy [12].

Also during the 1980s, the groups of Gale [146] and Moshe [147] independently began to report a role for the basal ganglia in regulating seizures. These and other lines of investigation paved the way for two small clinical series involving electrical stimulation of the subthalamic nucleus [148, 149] that, although providing mixed results, have led to a larger clinical trial currently underway.

As early as the 1960s, the role of electrical stimulation of vagal nerve afferents on the electroencephalogram was appreciated [150]. This observation led Zabara, a physiologist at Temple University, to investigate the effect of vagal afferent stimulation on seizures in animal models [151, 152]. These preclinical findings were the basis for launching Cyberonics, founded by Jacob Zabara and Reese S. Terry, Jr., in 1987, which developed an implantable vagus nerve stimulator (VNS). Cyberonics conducted trials of vagal nerve stimulation for epilepsy in human patients beginning in 1988 [153, 154] and achieved clinical approval in 1997 for its use as an adjuvant therapy.

In the 1990s, Lesser discovered the phenomenon of afterdischarges in the cortex, following electrical stimulation; and he found that subsequent electrical stimulation can terminate these afterdischarges [155].

In 1997, Fishell founded NeuroPace to develop an implant that applies the principle of afterdischarge termination to the treatment of seizures. In 2001, NeuroPace licensed seizure detection technology developed by Brian Litt, Ph.D. Fisher and Pless were recruited as CEO and CTO, respectively and, as of this writing, the company is in clinical trials for its Responsive Neurostimulator (RNS) device.

In the 1990s, DiLorenzo designed a closed-loop neuromodulation technology, and in 2002 he founded NeuroBionics Corporation. In 2003 and 2004 while a neurosurgical resident, he assembled a team and venture financing term sheets to launch the company. In 2004, after a nationwide search, he recruited several executives from Northstar Neuroscience to join the company, which was renamed BioNeuronics and subsequently NeuroVista Corporation; and as of this writing they are in development of a proprietary technology addressing the epilepsy market. After a two-year leave from clinical medicine spent raising financing and recruiting a seasoned team to continue the engineering, clinical, and IP development projects, DiLorenzo returned to complete neurosurgical training and pursue additional opportunities for innovation.

Several trials of deep brain stimulation are currently underway. In addition, direct electrical stimulation of the epileptic focus in the cortex or hippocampus, in some cases coupled to contingent and others to closed-loop stimulation algorithms, are being examined by a number of investigators.

1.3.5   Current State of the Art of Neurostimulation

Electrical stimulation of the central, peripheral, and autonomic nervous systems has reached the point of standard clinical practice for an expanding list of indications (see Table 1.1). New indications are under active investigation, such as Tourette's syndrome [156] and cluster headache [157]. Vagal nerve stimulation is routinely performed for epilepsy but deep brain stimulation and cortical stimulation are actively being studied. As a nondestructive alternative, electrical stimulation is spurring a judicious resurgence of psychosurgery for obsessive compulsive disorder (internal capsule, nucleus accumbens) and depression (same targets as for OCD, plus vagus nerve and area 25 [158]), from which a large number of patients are disabled and treatment-resistant. The treatment of pain continues to lag behind even

TABLE 1.1   Current and Near Term Indications for Neuromodulation and Neural Augmentation

Approval	Neuromodulation	Neural Augmentation
FDA Approved	Pain (SCS)	
	Epilepsy (VNS)	Quadrapaersis (Ambulation)
	Parkinson's disease (DBS)	Quadrapaersis (C5-6 function)
	Essential Tremor (DBS)	
	Dystonia* (DBS)	
	Depression (VNS)	
Investigational	Neuropathic pain (DBS, CS)	
	Epilepsy (DBS)	
	Occipital neuralgia (PNS)	Blindness
	Migraine headache (PNS)	
	Cluster headache (DBS)	Deafness
	Obsessive-compulsive disorder** (DBS)	Foot drop
	Obesity	Bladder dysfunction
	Depression (DBS, CS)	
	Tourette's Disease (DBS)	
	Hypertension	
	Stroke rehabilitation (CS)	
	Minimally-conscious state (DBS)	
	Addiction (DBS)	
	Aphasia (CS)	
	Tinnitus (CS, PNS)	
	Parkinson's disease, tremor (CS)	

DBS, deep brain stimulation; CS, cortical stimulation; VNS, vagus nerve stimulation; PNS, peripheral nerve stimulation; SCS, spinal cord stimulation

* Humanitarian Device Exemption (HDE) status

** HDE status pending

though it was one of the earliest indications for DBS studies. Spinal cord stimulation is frequently performed for various types of peripheral pain, as is peripheral nerve stimulation. Motor cortex stimulation is under active investigation for neuropathic deafferentation pain (as well as for movement disorders [159] and for rehabilitation after stroke [160]).

In addition to advancing the field through expanding indications, technological and scientific advances are promising to increase effectiveness in certain settings. Most promising is the development of closed-loop strategies for the treatment of the epilepsies, and perhaps other disorders (recall the early work of Brice and McLellan [124] with tremor, discussed above). The ability to detect and/or predict [161] seizures is being capitalized on to tailor stimulation or other neuromodulatory interventions. This promises to increase effectiveness, to decrease cellular injury, adaptation, or habituation, and to decrease battery drain. Other advances may include new electrode designs and stimulation strategies aimed at optimizing activation (or inhibition) of selected neural elements, such as axons over cell bodies. This work will advance by increasing understanding of the mechanisms of neurostimulation, which has lagged behind empirical, clinically based progress. Of course, progress in battery technology, including rechargeability and miniaturization, will increase long-term ease-of-use and tolerability of neurostimulation devices. With the more sophisticated neurostimulators on the drawing board and in development at several companies, the battery lifetime becomes an even more important issue than in first-generation devices.

At the turn of the twentieth century, the remarkable results seen with deep brain stimulation in movement disorders have led to the great resurgence in the use of electrical stimulation for the treatment of a wide variety of neurological disorders. Not since the turn of the nineteenth century has interest in this therapy been so widespread. The difference, hopefully, is that we are now also in the age of evidence-based medicine so that the seemingly striking benefits of electrical neuromodulation have been and will continue to be subjected to objective standards of evaluation, and that the enormous potential neural engineering holds will be responsibly harnessed for the patient's best interest.

References

1. Velasco, F., Neuromodulation: an overview. *Arch. Med. Res.*, 31(3):232–236, 2000.
2. Largus, S., Compositiones medicae. *Joannes Rhodius recensuit, notis illustrauit, lexicon scriboniaum adiecit.* P. Frambotti: Patavii, 1655.
3. McNeal, D.R., 2000 years of electrical stimulation, in *Functional Electrical Stimulation*, F.T. Hambrecht and J.B. Reswick, Eds. New York: Marcel Dekker, 1977, p. 3–33.
4. Chaffe, E. and R. Light, A method for remote control of electrical stimulation of the nervous system. *Yale J. Biol. Med.*, 7:83, 1934.
5. Evans, C., Medical Observations and Inquiries by a Society of Physicians in London. 1754. I:83–86.
6. Parent, A., Giovanni Aldini: from animal electricity to human brain stimulation. *Can. J. Neurol. Sci.*, 31:576–584, 2004.
7. Galvani, L., De viribus electricitatis in motu musculari, commentarus. De Bononiensi Scientiarum et Artium Instituto atque Academia, 7(363–418), 1791.
8. Volta, A., Account of some discoveries made by Mr. Galvani from Mr. Alexander Volta to Mr. Tiberius Cavallo. *Phil. Trans. Roy. Soc.*, 83:10–44, 1793.
9. Volta, A., On the electricity excited by the mere contact of conducting substances of different kinds. *Phil. Trans.*, 90:403, 1800.
10. Aldini, J., Essai theorique et experimental sur le galvanisme, avec une serie d'experiences faites devant des commissaires de l'Institut National de France, et en divers amphitheatres anatomiques de Londres. Paris: Fornier Fils, 1804.
11. Fritsch, G. and E. Hitzig, Uber die elektrische Erregbarkeit des Grosshirns. *Archiv. Anat. Physiol. wissenschaftl. Med.*, 37:300–322, 1870.
12. Velasco, M. et al., Acute and chronic electrical stimulation of the centromedian thalamic nucleus: modulation of reticulo-cortical systems and predictor factors for generalized seizure control. *Arch. Med. Res.*, 31(3):304–315, 2000.
13. Ferrier, D., The localization of function in the brain. *Proc. R. Soc. Lond.*, 22:229, 1873.
14. Davis, R., Chronic stimulation of the central nervous system, in *Textbook of Stereotactic and Functional Neurosurgery*, P.L. Gildenberg and R.R. Tasker, Eds. New York: McGraw-Hill, 1998, p. 963.
15. Cushing, H., Faradic stimulation of postcentral gyrus in conscious patients. *Brain*, 32:44–53, 1909.
16. Penfield, W. and H.H. Jasper, *Epilepsy and the Functional Anatomy of the Human Brain.* Boston: Little, Brown and Co., 1954.
17. Djourno, A. and C.H. Eyries, Prothese auditive par excitation electrique a distance du nerf sensoriel a l'aide d'un bobinage inclus a demeure. *La Presse Medicale*, 65:1417, 1957.
18. Edell, D., A peripheral nerve information transducer for amputees: long-term multichannel recordings from rabbit peripheral nerves. *IEEE Trans. Biomed. Eng.*, 33(2):203–214, 1986.
19. Edell, D.J., Development of a Chronic Neuroelectronic Interface, Ph.D. thesis, U.C. Davis, Davis, CA, 1980.
20. Liberson, W.T. et al., Functional electrotherapy: stimulation of the peroneal nerve synchronized with the swing phase of the gait of hemiplegic patients. *Arch. Phys. Med. Rehabil.*, 42:101–105, 1961.
21. Liberson, W.T., Functional neuromuscular stimulation: historical background and personal experience, in *Functional Neuromuscular Stimulation: Report of a Workshop.* April 27–28, 1972, M.A. LeBlanc, Ed. Washington, D.C., 1972, p. 147–156.
22. Moe, J.H. and H.W. Post, Functional electrical stimulation for ambulation in hemiplegia. *Lancet*, 82:285–288, 1962.
23. Long, C. and V. Masciarelli, An electrophysiological splint for the hand. *Arch. Phys. Med. Rehabil.*, 44:449–503, 1963.
24. Peckham, P.H., J.T. Mortimer, and E.B. Marsolais, Controlled prehension and release in the C5 quadriplegic elicited by functional electrical stimulation of the paralyzed forearm musculature. *Ann. Biomed. Eng.*, 8:369–388, 1980.

25. Kantrowitz, A., Electronic Physiologic Aids: A Report of the Maimonides Hospital. Brooklyn, NY, 1960, p. 4–5.

26. Cooper, E.B., W.H. Bunch, and J.H. Campa, Effects of chronic human neuromuscular stimulation. *Surg. Forum*, 24:477–479, 1973.

27. Brindley, G.S., C.E. Polkey, and D.N. Ruston, Electrical splinting of the knee in paraplegia. *Paraplegia*, 16:434–441, 1978.

28. Bajd, T. et al., The use of a four-channel electrical stimulator as an ambulatory aid for paraplegic patients. *Physical Therapy*, 63(7):1116–1120, 1983.

29. Kralj, A. et al., Gait restoration in paraplegic patients: a feasibility demonstration using multi-channel surface electrode FES. *J. Rehabil. R & D*, 20(1):3–20, 1983.

30. DiLorenzo, D.J. and W.K. Durfee, Unpublished data on 4-channel surface FES for gait restoration, Massachusetts Institute of Technology, 1990.

31. Durfee, W.K. and D.J. DiLorenzo, Linear and nonlinear approaches to control of single joint motion by functional electrical stimulation. In *Proceedings of the 1990 American Control Conference*. 1990.

32. Durfee, W.K. and D.J. DiLorenzo, Sliding mode control of fns knee joint motion. *Abstracts of the First World Congress of Biomechanics*, 2:329, 1990.

33. DiLorenzo, D.J. et al., Chronic intraneural electrical stimulation for prosthetic sensory feedback. In *IEEE EMBS 1st International Conference on Neural Engineering*. Capri Island, Italy: IEEE, 2003.

34. DiLorenzo, D.J. et al., Multichannel intraneural electrical stimulation for prosthetic sensory feed-back. In *Society for Neuroscience Annual Meeting*. New Orleans, LA, 1997.

35. DiLorenzo, D.J. et al., Multichannel intraneural electrical stimulation for prosthetic sensory feed-back. In *Congress of Neurological Surgeons Annual Meeting*. New Orleans, LA, 1997.

36. DiLorenzo, D.J., Cortical Technologies: Innovative Solutions for Neurological Disease, Masters thesis, Massachusetts Institute of Technology, MIT Sloan School of Management: Management of Technology (MOT) Program, Cambridge, MA, 1999, 72 pages.

37. Goldfarb, M. and W.K. Durfee, Design of a controlled-brake orthosis for FES-aided gait. *IEEE Trans. Rehab. Eng.*, 4(1):13–24, 1996.

38. Marsolais, E.B. and R. Kobetic, Functional walking in paralyzed patients by means of electrical stimulation. *Clin. Orthopaed. Related Res.*, 1983(175):30–36, 1983.

39. Graupe, D. et al., EMG-controlled electrical stimulation. In *Proc. IEEE Frontiers of Eng. & Comp. in Health Care*. Columbus, OH, 1983.

40. Graupe, D. and K.H. Kohn, *Functional Electrical Stimulation for Ambulation by Paraplegics*. Malabar, FL: Krieger Publishing Co., 1994.

41. Keith, M.W. et al., Implantable functional neuromuscular stimulation in the tetraplegic hand. *J. Hand Surg.*, 14A:524–530, 1989.

42. Georgopoulos, A.P. et al., On the relations between the direction of two-dimensional arm move-ments and cell discharge in primate motor cortex. *J. Neurosci.*, 2(11):1527–1537, 1982.

43. Georgopoulos, A.P., R. Caminiti, and J.F. Kalaska, Static spatial effects in motor cortex and area 5: quantitative relations in a two-dimensional space. *Exp. Brain Res.*, 54(3):446–454, 1984.

44. Caminiti, R. et al., Shift of preferred directions of premotor cortical cells with arm movements performed across the workspace. *Exp. Brain Res.*, 83(1):228–232, 1990.

45. Kalaska, J.F. et al., A comparison of movement direction-related versus load direction-related activity in primate motor cortex, using a two-dimensional reaching task. *J. Neurosci.*, 9(6):2080–2102, 1989.

46. Schwartz, A.B., R.E. Kettner, and A.P. Georgopoulos, Primate motor cortex and free arm move-ments to visual targets in three-dimensional space. I. Relations between single cell discharge and direction of movement. *J. Neurosci.*, 8(8):2913–2927, 1988.

47. Georgopoulos, A.P., R.E. Kettner, and A.B. Schwartz, Primate motor cortex and free arm move-ments to visual targets in three-dimensional space. II. Coding of the direction of movement by a neuronal population. *J. Neurosci.*, 8(8):2928–2937, 1988.

48. Kettner, R.E., A.B. Schwartz, and A.P. Georgopoulos, Primate motor cortex and free arm movements to visual targets in three-dimensional space. III. Positional gradients and population coding of movement direction from various movement origins. *J. Neurosci.*, 8(8):2938–2947, 1988.

49. Taylor, D.M., S.I. Tillary, and A.B. Schwartz, Direct cortical control of 3D neuroprosthetic devices. *Science*, 296:1828–1832, 2002.

50. Hatsopoulos, N.G. et al., Information about movement direction obtained from synchronous activity of motor cortical neurons. *Proc. Natl. Acad. Sci., U.S.A.*, 95(26):15706–15711, 1998.

51. DiLorenzo, D.J., Neural Correlates of Motor Performance in Primary Motor Cortex, 1999, Ph.D. thesis, Massachusetts Institute of Technology, Department of Mechanical Engineering, Cambridge, MA, 1999, 104 pages.

52. Maynard, E.M. et al., Neuronal interactions improve cortical population coding of movement direction. *J. Neurosci.*, 19(18):8083–8093, 1999.

53. Wessberg, J. et al., Real-time prediction of hand trajectory by ensembles of cortical neurons in primates. *Lett. Nature*, 408:361–365, 2000.

54. Kennedy, P.R., The cone electrode: a long-term electrode that records from neurites grown onto its recording surface. *J. Neurosi. Meth.*, 29:181–193, 1989.

55. Kennedy, P.R. and R.A. Bakay, Restoration of neural output from a paralyzed patient by a direct brain connection. *Neuroreport*, 9(8):1707–1711, 1998.

56. Kennedy, P.R. et al., Direct control of a computer from the human central nervous system. *IEEE Trans. Rehab. Eng.*, 8(2):198–202, 2000.

57. Saleh, M. et al., Case study: reliability of multi-electrode array in the knob area of human motor cortex intended for a neuromotor prosthesis application. In *ICORR — 9th International Conference on Rehabilitation Robotics.* Chicago, IL, 2005.

58. Doyle, J.H., J.B. Doyle, and F.M. Turnbull, Electrical stimulation of eighth cranial nerve. *Arch. Otolaryngol. Head Neck Surg.*, 80:388, 1964.

59. House, W.F. and J. Urban, Long term results of electrode implantation and electronic stimulation of the cochlea in man. *Ann. Otol. Rhinol. Laryngol.*, 82:504, 1973.

60. House, W.F. and K.I. Berliner, Safety and efficacy of the House/3M cochlear implant in profoundly deaf adults. *Otolaryngolog. Clin. N. Am.*, 19:275, 1986.

61. Simmons, F.B. et al., Auditory nerve: electrical stimulation in man. *Science*, 148:104, 1965.

62. Michelson, R.P., Electrical stimulation of the human cochlea: a preliminary report. *Arch. Otolaryngol. Head Neck Surg.*, 93:317, 1971.

63. Merzenich, M.M., D.N. Schindler, and M.W. White, Feasibility of multichannel scala tympani stimulation. *Laryngoscope*, 84:1887, 1974.

64. Kessler, D.K., The CLARION Multi-Strategy Cochlear Implant. *Ann. Otol., Rhinol. Laryngol.*, 177(Suppl.):8–16, 1999.

65. Schindler, R.A. and D.K. Kessler, The UCSF/Storz cochlear implant: patient performance. *Am. J. Otol.*, 8(3):247–255, 1987.

66. Schindler, R.A., D.K. Kessler, and H.S. Haggerty, Clarion cochlear implant: phase I investigational results.[see comment] [erratum appears in *Am. J. Otol.*, 1993 Nov;14(6):627, 1993.]. *Am. J. Otol.*, 14(3):263–272, 1993.

67. Schindler, R.A. et al., The UCSF/Storz multichannel cochlear implant: patient results. *Laryngoscope*, 96(6):597–603, 1986.

68. Clark, G.M., A surgical approach for a cochlear implant: an anatomical study. *J. Laryngol. Otol.*, 89(1):9–15, 1975.

69. Clark, G.M. et al., The University of Melbourne — nucleus multi-electrode cochlear implant. *Adv. Oto-Rhino-Laryngol.*, 38:V-IX, 1987.

70. Clark, G.M. and R.J. Hallworth, A multiple-electrode array for a cochlear implant. *J. Laryngol. Otol.*, 90(7):623–627, 1976.

71. Clark, G.M., R.J. Hallworth, and K. Zdanius, A cochlear implant electrode. *J. Laryngol. Otol.*, 89(8):787–792, 1975.

72. Cochlear Limited. Cochlear Limited Corporate Website. [Corporate Website; cited July 18, 2005]. Available from: http://www.cochlear.com/.

73. Burian, K. et al., Designing of and experience with multichannel cochlear implants. *Acta Oto-Laryngologica*, 87(3–4):190–195, 1979.

74. Burian, K. et al., Electrical stimulation with multichannel electrodes in deaf patients. *Audiology*, 19(2):128–136, 1980.

75. Brindley, G.S. and W.S. Lewin, The sensations produced by electrical stimulation of the visual cortex. *J. Physiol.*, 196(2):479–493, 1968.

76. Dobelle, W.H. and M.G. Mladejovsky, Phosphenes produced by electrical stimulation of human occipital cortex, and their application to the development of a prosthesis for the blind. *J. Physiol.*, 243(2):553–576, 1974.

77. Dobelle, W.H. et al., "Braille" reading by a blind volunteer by visual cortex stimulation. *Nature*, 259(5539):111–112, 1976.

78. DiLorenzo, D.J., Unpublished data on microfabricated multichannel electrode array for retinal stimulation, Massachusetts Institute of Technology, 1991.

79. Canby, E.T., *A History of Electricity*. New York: Hawthorne Books, 1962, p. 21.

80. von Frey, M., Physiological experiments on the vibratory sensation [German]. *Z. Biol.*, 65:417–427, 1915.

81. Adrian, E.D., The response of human sensory nerves to currents of short duration. *J. Physiol. (London)*, 53:70–85, 1919.

82. Beeker, T.W., J. During, and A. Den Hertog, Technical note: Artificial touch in a hand prosthesis. *Med. Biolog. Eng.*, 5:47–49, 1967.

83. Schmidl, H., The importance of information feedback in prostheses for the upper limbs. *Prosthetics Orthotics Int.*, 1:21–24, 1977.

84. Shannon, G.F., A myoelectrically-controlled prosthesis with sensory feedback. *Med. Biolog. Eng. Comput.*, 17:73–80, 1979.

85. Clippinger, F.W., A sensory feedback system for an upper-limb amputation prosthesis. *Bull. Prosthet. Res.*, 10–22:247–258, 1974.

86. Clippinger, F.W., in *Textbook of Surgery: The Biological Basis of Modern Surgical Practice*, D.C. Sabiston, Ed. Philadelphia: W.B. Saunders, 1977, p. 1582–1594.

87. Clippinger, F.W., Afferent sensory feedback for lower extremity prosthesis. *Clin. Orthop.*, 169:202–206, 1982.

88. Malmstrom, J.A., T.G. McNaughton, and K.W. Horch, Recording properties and biocompatibility of chronically implanted polymer-based intrafascicular electrodes. *Ann. Biomed. Eng.*, 26(6):1055–1064, 1998.

89. McNaughton, T.G. and K.W. Horch, Metallized polymer fibers as leadwires and intrafascicular microelectrodes. *J. Neurosci. Methods*, 70(1):103–110, 1996.

90. Lawrence, S.M. et al., Long-term biocompatibility of implanted polymer-based intrafascicular electrodes. *J. Biomed. Mater. Res.*, 63(5):501–506, 2002

91. Furman, S. and J.B. Schwedel, An intracardiac pacemaker for Stokes-Adams seizures. *N. Engl. J. Med.*, 261:943–948, 1959.

92. Senning, A., *Mal. Cardiovasc.*, 4:503–512, 1963.

93. Glenn, W.W. et al., Total ventilatory support in a quadriplegic patient with radiofrequency electrophrenic respiration. *N. Engl. J. Med.*, 286(10):513–516, 1972.

94. Glenn, W.W. et al., *N. Engl. J. Med.*, 261:948–951, 1959.

95. Chardack, W.M., A.A. Gage, and W. Greatbatch, A transistorized, self-contained, implantable pacemaker for the long-term correction of complete heart block. *Surgery*, 48:643–654, October 1960.

96. Pool, J.L., Nerve stimulation in paraplegic patients by means of buried induction coil. Preliminary report. *J. Neurosurg.*, 3:192, 1946.

97. Heath, R.G., Studies in *Schizophrenia: A Multidisciplinary Approach to Mid Brain Relationships*. Cambridge, MA: Harvard University Press, 1954.

98. Pool, J.L., Psychosurgery in older people. *J. Am. Geriatr. Soc.*, 2(7):456–466, 1954.

99. Spiegel, E.A. and H.T. Wycis, Chronic implantation of intracerebral electrodes in humans. In *Electrical Stimulation of the Brain*, D.E. Sheer, Ed. Austin, TX: University of Texas Press, 1961, p. 37–44.

100. Mazars, G., L. Merienne, and C. Ciolocca, Intermittent analgesic thalamic stimulation. Preliminary note. *Rev. Neurol. (Paris)*, 128(4):273–279, 1973.

101. Mazars, G.J., Intermittent stimulation of nucleus ventralis posterolateralis for intractable pain. *Surg. Neurol.*, 4(1):93–95, 1975.

102. Gol, H., Relief of pain by electrical stimulation of the septal area. *J. Neurolog. Sci.*, 5:115–120, 1967.

103. Ervin, F.R., C.E. Brown, and V.H. Mark, Striatal influence on facial pain. *Confinia Neurologica*, 27:75–86, 1966.

104. Glenn, W.W. et al., Diaphragm pacing by radiofrequency transmission in the treatment of chronic ventilatory insufficiency. Present status. *J. Thorac. Cardiovasc. Surg.*, 66(4):505–520, 1973.

105. Sweet, W. and J. Wepsic, Control of pain by focal electrical stimulation for suppression. *Arizona Med.*, 26:1042–1045 1969.

106. Shealy, C.N., J.T. Mortimer, and J.B. Reswick, Electrical inhibition of pain by stimulation of the dorsal columns: preliminary clinical report. *Anesth. Analg.*, 46(4):489–491, 1967.

107. Melzack, R. and P.D. Wall, Pain mechanisms: a new theory. *Science*, 150(699):971–979, 1965.

108. Mullen, J.B., C.F. Walker, and B.S. Nashold, Jr., An electrophysiological approach to neural augmentation implantation for the control of pain. *J. Bioeng.*, 2(1–2):65–7, 1978.

109. Hosobuchi, Y., J.E. Adams, and B. Rutkin, Chronic thalamic stimulation for the control of facial anesthesia dolorosa. *Arch. Neurol.*, 29(3):158–161, 1973.

110. Mazars, G., L. Merienne, and C. Cioloca, [Use of thalamic stimulators in the treatment of various types of pain]. *Ann. Med. Interne (Paris)*, 126(12):869–871, 1975.

111. Reynolds, D.V., Surgery in the rat during electrical analgesia induced by focal brain stimulation. *Science*, 164:444–445, 1969.

112. Hosobuchi, Y., J.E. Adams, and R. Linchitz, Pain relief by electrical stimulation of the central gray matter in humans and its reversal by naloxone. *Science*, 197(4299):183–186, 1977.

113. Richardson, D.E. and H. Akil, Pain reduction by electrical brain stimulation in man. 2. Chronic self-administration in the periventricular gray matter. *J. Neurosurg.*, 47(2):184–194, 1977.

114. Richardson, D.E. and H. Akil, Pain reduction by electrical brain stimulation in man. 1. Acute administration in periaqueductal and periventricular sites. *J. Neurosurg.*, 47(2):178–183, 1977.

115. Bechtereva, N.P. et al., Therapeutic electrostimulation of deep brain structures. *Vopr. Neirokhir.*, 1:7–12, 1972.

116. Bechtereva, N.P. et al., Method of electrostimulation of the deep brain structures in treatment of some chronic diseases. *Confin. Neurol.*, 37(1–3):136–140, 1975.

117. Mazars, G., L. Merienne, and C. Cioloca, Control of dyskinesias due to sensory deafferentation by means of thalamic stimulation. *Acta Neurochir. Suppl. (Wien.)*, 30:239–243, 1980.

118. Andy, O.J., Parafascicular-center median nuclei stimulation for intractable pain and dyskinesia (painful-dyskinesia). *Appl. Neurophysiol.*, 43(3–5):133–144, 1980.

119. Cooper, I.S. et al., Chronic cerebellar stimulation in cerebral palsy. *Neurology*, 26(8):744–753, 1976.

120. Davis, R. et al., Update of chronic cerebellar stimulation for spasticity and epilepsy. *Appl. Neurophysiol.*, 45(1–2):44–50, 1982.

121. Hassler, R. et al., Physiological observations in stereotaxic operations in extrapyramidal motor disturbances. *Brain*, 83:337–350, 1960.

122. Jurko, M.F., O.J. Andy, and D.P. Foshee, Diencephalic influence on tremor mechanisms. *Arch. Neurol.*, 9:358–362, 1963.

123. Mundinger, F. and H. Neumuller, Programmed stimulation for control of chronic pain and motor diseases. *Appl. Neurophysiol.*, 45(1–2):102–111, 1982.

124. Brice, J. and L. McLellan, Suppression of intention tremor by contingent deep-brain stimulation. *Lancet*, 1(8180):1221–1222, 1980.

125. Andy, O.J., Thalamic stimulation for control of movement disorders. *Appl. Neurophysiol.*, 46(1–4):107–111, 1983.

126. Benabid, A.L. et al., Combined (thalamotomy and stimulation) stereotactic surgery of the VIM thalamic nucleus for bilateral Parkinson disease. *Appl. Neurophysiol.*, 50(1–6):344–346, 1987.

127. Benabid, A.L. et al., Long-term suppression of tremor by chronic stimulation of the ventral intermediate thalamic nucleus. *Lancet*, 337(8738):403–406, 1991.

128. Siegfried, J., Therapeutical neurostimulation—indications reconsidered. *Acta Neurochir. Suppl. (Wien.)*, 52:112–117, 1991.

129. Benabid, A.L. et al., Chronic electrical stimulation of the ventralis intermedius nucleus of the thalamus as a treatment of movement disorders. *J. Neurosurg.*, 84(2):203–214, 1996.

130. Koller, W. et al., High-frequency unilateral thalamic stimulation in the treatment of essential and parkinsonian tremor. *Ann. Neurol.*, 42(3):292–299, 1997.

131. Laitinen, L.V., A.T. Bergenheim, and M.I. Hariz, Leksell's posteroventral pallidotomy in the treatment of Parkinson's disease. *J. Neurosurg.*, 76(1):53–61, 1992.

132. Siegfried, J. and B. Lippitz, Bilateral chronic electrostimulation of ventroposterolateral pallidum: a new therapeutic approach for alleviating all parkinsonian symptoms. *Neurosurgery*, 35(6):1126–1129; discussion 1129–1130, 1994.

133. Bergman, H., T. Wichmann, and M.R. DeLong, Reversal of experimental Parkinsonism by lesions of the subthalamic nucleus. *Science*, 249(4975):1436–1438, 1990.

134. Benabid, A.L. et al., Acute and long-term effects of subthalamic nucleus stimulation in Parkinson's disease. *Stereotact. Funct. Neurosurg.*, 62(1–4):76–84, 1994.

135. Cooke, P.M. and R.S. Snider, Some cerebellar influences on electrically-induced cerebral seizures. *Epilepsia*, 4:19–28, 1955.

136. Dow, R.S., A. Fernandez-Guardiola, and E. Manni, The influence of the cerebellum on experimental epilepsy. *Electroencephalogr. Clin. Neurophysiol.*, 14:383–398, 1962.

137. Krauss, G.L. and R.S. Fisher, Cerebellar and thalamic stimulation for epilepsy. *Adv. Neurol.*, 63:231–245, 1993.

138. Upton, A.R. et al., Suppression of seizures and psychosis of limbic system origin by chronic stimulation of anterior nucleus of the thalamus. *Int. J. Neurol.*, 19–20:223–230, 1985.

139. Mirski, M.A. and J.A. Ferrendelli, Interruption of the mammillothalamic tract prevents seizures in guinea pigs. *Science*, 226(4670):72–74, 1984.

140. Mirski, M.A. and J.A. Ferrendelli, Anterior thalamic mediation of generalized pentylenetetrazol seizures. *Brain Res.*, 399(2):212–223, 1986.

141. Mirski, M.A. et al., Anticonvulsant effect of anterior thalamic high frequency electrical stimulation in the rat. *Epilepsy Res.*, 28(2):89–100, 1997.

142. Hodaie, M. et al., Chronic anterior thalamus stimulation for intractable epilepsy. *Epilepsia*, 43(6):603–608, 2002.

143. Kerrigan, J.F. et al., Electrical stimulation of the anterior nucleus of the thalamus for the treatment of intractable epilepsy. *Epilepsia*, 45(4):346–354, 2004.

144. Velasco, F. et al., Electrical stimulation of the centromedian thalamic nucleus in the treatment of convulsive seizures: a preliminary report. *Epilepsia*, 28(4):421–430, 1987.

145. Fisher, R.S. et al., Placebo-controlled pilot study of centromedian thalamic stimulation in treatment of intractable seizures. *Epilepsia*, 33(5):841–851, 1992.

146. Gale, K., Mechanisms of seizure control mediated by gamma-aminobutyric acid: role of the substantia nigra. *Fed. Proc.*, 44(8):2414–2424, 1985.

147. Moshe, S.L. and B.J. Albala, Nigral muscimol infusions facilitate the development of seizures in immature rats. *Brain Res.*, 315(2):305–308, 1984.

148. Benabid, A.L. et al., Deep brain stimulation of the corpus luysi (subthalamic nucleus) and other targets in Parkinson's disease. Extension to new indications such as dystonia and epilepsy. *J. Neurol.*, 2001. 248(Suppl. 3):III37–47.

149. Loddenkemper, T. et al., Deep brain stimulation in epilepsy. *J. Clin. Neurophysiol.*, 18(6):514–532, 2001.

150. Chase, M.H., M.B. Sterman, and C.D. Clemente, Cortical and subcortical patterns of response to afferent vagal stimulation. *Exp. Neurol.*, 16(1):36–49, 1966.

151. Zabara, J., Time course of seizure control to brief, repetitive stimuli. *Epilepsia*, 28:604, 1985.

152. Zabara, J., Control of hypersynchronous discharge in peilepsy. *Electroencephalogr. Clin. Neurophysiol.*, 61:162, 1985.

153. Penry, J.K. and J.C. Dean, Prevention of intractable partial seizures by intermittent vagal stimulation in humans: preliminary results. *Epilepsia*, 31(Suppl. 2):S40–S43, 1990.

154. Uthman, B.M. et al., Treatment of epilepsy by stimulation of the vagus nerve. *Neurology*, 43(7):1338–1345, 1993.

155. Lesser, R.P. et al., Brief bursts of pulse stimulation terminate afterdischarges caused by cortical stimulation. *Neurology*, 53(9):2073–2081, 1999.

156. Visser-Vandewalle, V., et al., Chronic bilateral thalamic stimulation: a new therapeutic approach in intractable Tourette syndrome. Report of three cases. *J. Neurosurg.*, 99(6):1094–1100, 2003.

157. Franzini, A. et al., Stimulation of the posterior hypothalamus for treatment of chronic intractable cluster headaches: first reported series. *Neurosurgery*, 52(5):1095–1099; discussion 1099–1101, 2003.

158. Mayberg, H.S. et al., Deep brain stimulation for treatment-resistant depression. *Neuron*, 45(5):651–660, 2005.

159. Pagni, C.A., S. Zeme, and F. Zenga, Further experience with extradural motor cortex stimulation for treatment of advanced Parkinson's disease. Report of 3 new cases. *J. Neurosurg. Sci.*, 47(4):189–193, 2003.

160. Brown, J.A. et al., Motor cortex stimulation for enhancement of recovery after stroke: case report. *Neurol. Res.*, 25(8):815–818, 2003.

161. Litt, B. and J. Echauz, Prediction of epileptic seizures. *Lancet Neurol.*, 1(1):22–30, 2002.

Neural Modulation

2

Cellular Mechanisms of Action of Therapeutic Brain Stimulation

Zelma Kiss and
Trent Anderson

2.1 Introduction

Deep brain stimulation (DBS) has revolutionized the treatment of several movement disorders, notably Parkinson's disease (PD) (Krack et al., 2003); dystonia (Vidailhet et al., 2005); and tremor (Schuurman et al., 2000). In addition, it has been advocated for other conditions as diverse as obsessive–compulsive disorder (Nuttin et al., 2003); depression (Mayberg et al., 2005); epilepsy (Kerrigan et al., 2004; Theodore and Fisher, 2004); pain (Kumar et al., 1997); cluster headache (Schoenen et al., 2005; Leone et al., 2001); and even obesity (Benabid et al., 2005). Although the site of electrode implantation varies depending on the condition, the common feature of all DBS therapy is that the electrical pulses must be high frequency (>100 Hz) to be effective. Exactly how DBS works and its effects on neuronal functioning within the stimulated nucleus and downstream structures are the focus of this chapter.

This chapter briefly reviews the clinical use of DBS in thalamus for the treatment of tremor to illustrate some of the fundamental effects of DBS that must be explained by cellular models. Most of the chapter is devoted to the proposed mechanisms of action of DBS and the author's investigations of the underlying cellular effects. The similarities and differences in cellular effects of DBS in thalamus are compared to similar *in vitro* studies in subthalamic nucleus (STN). Finally, the potential of therapeutic electrical stimulation is discussed based on our present understanding of its mechanisms of action.

2.2 DBS for Essential Tremor

Essential tremor (ET) is a common movement disorder in which patients exhibit a postural or action-induced 3- to 8-Hz tremor. ET is believed to result from dysfunction in the olivo-cerebellar pathway, based on experimentally induced tremor in animals (Llinas and Volkind, 1973; Kralic et al., 2005) and imaging studies in patients (Wills et al., 1994). The site of DBS electrode implantation for ET is in the ventral lateral (VL) thalamus (Benabid et al., 1996), a part of the thalamus that receives projections from the deep cerebellar nuclei and projects to the primary motor cortex (Hassler, 1959). It is equivalent to the ventro-intermedius or Vim nucleus, which is the terminology used for humans [see Hirai and Jones (1989) for review of nomenclature]. Electrical stimulation is applied as charge-balanced square wave monopolar or bipolar pulses (60 to 120 μs) at 125 to 185 Hz, 1 to 4 V for 12 to 24 hours a day. There are several unique features of thalamic DBS for tremor that are worthy of further discussion in the context of its mechanism of action.

The first two features are closely related. The first is the similarity of thalamic DBS to thalamotomy (Schuurman et al., 2000), a lesion in the VL thalamus (Kiss et al., 2003c); and the second is the importance of location on clinical outcome for both (Papavassiliou et al., 2004; Atkinson et al., 2002). The target for both stimulation and thalamotomy are "tremor cells," neurons that fire synchronously to tremor, and may or may not respond to movements (Lenz et al., 1994). Thalamic lesions are designed to eliminate these tremor cells. Therefore, an early thought about how DBS could work was that electrical stimulation also injured these cells. However, there is no evidence that chronic stimulation damages surrounding brain tissue (Haberler et al., 2000), and the effects of stimulation are reversible (Benabid et al., 1996). Incorrect electrode placement will result in poor outcome for DBS (Papavassiliou et al., 2004), as will thalamotomy (Lenz et al., 1995; Tasker and Kiss, 1995), although DBS is more forgiving because the stimulation parameters can always be changed and the electrode spans a larger region of thalamus than a lesion would (Tasker et al., 1997). The similarity between lesions and DBS suggested that inhibitory mechanisms may play an important role. This mechanism was supported by the clinical finding that muscimol, a $GABA_A$ agonist, reversibly eliminated tremor when microliter quantities were applied to tremor cells-containing thalamus (Pahapill et al., 1999).

Another characteristic is the immediate benefit of thalamic DBS on tremor. Remarkably, within seconds of the stimulator being turned on, the tremor disappears completely or is significantly suppressed (Benabid et al., 1996). Similarly, when the stimulator is turned off, the tremor returns immediately. There seems to be no after-effect of stimulation except that in some patients a rebound worsening of tremor may occur (Hariz et al., 1999).

A fourth critical feature is the frequency dependency of stimulation on tremor: high frequencies (>100 Hz) are required for clinical benefit and low frequencies may worsen movements. Originally it was thought that 5-Hz stimulation of thalamus could induce tremor (Hassler et al., 1960), although this has not been most groups' experience (Benabid et al., 1996; Ushe et al., 2004). Frequencies of 15 Hz (Bejjani et al., 2000) have been reported to induce "myoclonus." We have instead observed significant tremor worsening at 20 Hz, although this requires higher amplitudes than usually applied in patients (Kiss et al., 2003b). In fact, most studies on the frequency effects of stimulation fail to incorporate equivalent current densities in their experiments. Current density is a function of the frequency, pulse width, and amplitude of the square wave electrical pulses applied. While pulse duration and amplitude are the most important features that will determine whether a neural element, such as an axon, will be brought to threshold, frequency likely has a bigger role for somatic excitation. Neural cell bodies have much longer membrane time constants than axons, and therefore firing may require a burst of stimuli to accumulate enough charge to achieve threshold. To compare different frequencies of stimulation, equivalent current densities must be applied.

A fifth important attribute of thalamic DBS is that patients do not experience any involuntary movements or motor disruption with prolonged DBS (Flament et al., 2002). This is despite the VL thalamus projecting directly to primary motor cortex (M1). The major side effect is paresthesia, thought to relate to spread of current into the adjacent somatosensory relay, the ventroposterior nucleus of thalamus (Kiss et al., 2003a). In fact, even when stimulation was applied directly to this sensory thalamic relay, sensory perception was not disturbed (Abbassian et al., 2001).

These key clinical features of thalamic DBS for tremor must be incorporated into the proposed mechanisms of action. DBS effects in thalamus are similar to lesions, immediate/reversible, frequency dependent, and specific to tremor, without adverse effects on motor control. If we do not factor these into our models, we may be studying some epiphenomena of stimulation and not the true mechanisms involved in clinical benefit.

2.3 Mechanisms of Action

There has been a plethora of recent publications that address the mechanisms of action of DBS. To decipher the literature, this section discusses the original data about first local and then distant effects of stimulation. The evidence for each in different models is examined, including results from patients, *in vivo* animals, *in vitro* slices, and computer simulation modeling studies. While these provide some evidence for one or more mechanisms, each technique has limitations; and these are also considered.

2.3.1 Local Effects of DBS

Historically, DBS was thought to mediate its clinical effect by inhibiting neural firing in the stimulated nucleus, because lesions have similar effects to high-frequency stimulation (Schuurman et al., 2000). However, recent evidence using a variety of techniques has brought into question this assumption.

2.3.1.1 Evidence from Humans

Measurement of chronaxie times in patients with DBS electrodes suggested that the neural element responsible for tremor suppression was the axon (Ashby et al., 1995; Holsheimer et al., 2000). Chronaxie is a property of all neural elements that is related to threshold for activation (Ranck, 1975; Nowak and Bullier, 1998). Chronaxie time (C) is defined as the pulse width of the threshold current (I_{th}) having an intensity twice that of the rheobase current (I_{rh}) as per the equation $I_{th} = I_{rh} (1 + C/pw)$. For example, a longer chronaxie time (or pulse duration) is required to activate cell bodies (1 to 10 ms) than axons (50 to 300 μs), and large myelinated axons require shorter pulse widths than small unmyelinated ones (Ranck, 1975; McIntyre and Grill, 1999). While modeling studies questioned the value of chronaxie time measurements using the large 1.5-mm DBS electrode (Miocinovic and Grill, 2004), other studies using microstimulation (with an electrode tip of 15 to 40 μm), which also suppress tremor (Tasker and Kiss, 1995), verified the importance of the axon in mediating effects of electrical stimulation (Kiss et al., 2003a).

Microelectrode recordings in patients undergoing DBS implant provided another opportunity to study the effects of microstimulation on nearby neurons. After high-frequency microstimulation, an initial burst followed by a prolonged inhibition of firing in thalamic cells was observed (Dostrovsky and Lozano, 2002). The authors could not record cellular activity during the high-frequency stimulation due to stimulus artifacts; however, they proposed that thalamic neurons were inhibited during this time and that the burst of spikes immediately after stimulation was due to release from hyperpolarization.

Several other studies have investigated the effects of thalamic DBS using imaging techniques such as positron emission tomography (PET) (Haslinger et al., 2003; Perlmutter et al., 2002) or functional magnetic resonance imaging (fMRI) (Rezai et al., 1999). These are discussed in more detail later when we address the remote effects of stimulation. Locally, DBS increased blood flow in thalamus, suggesting increased neuronal activity. Haslinger et al. (2003) quantified this effect and found that increasing stimulation frequency resulted in a nonlinear, U-shaped response in thalamic blood flow, whereas increasing stimulation amplitude resulted in a linear increase in thalamic blood flow.

The limitations of imaging in humans are related to the indirectness of the method. PET generally measures blood flow and/or metabolism and has limited temporal resolution. Orthodromic/antidromic activation, excitatory and/or inhibitory effects cannot be distinguished (Caesar et al., 2003; Logothetis et al., 2001; Mulligan and MacVicar, 2004; Mathiesen et al., 1998). Therefore, the underlying physiological mechanisms are difficult to decipher. Experiments in animal models have provided further information about the local effects and mechanisms of DBS.

2.3.1.2 Evidence from Animal Models

Relatively little research has been conducted on thalamic DBS *in vivo*. Instead, several studies have focused on DBS in subthalamic nucleus (STN), globus pallidus (GP), or substantia nigra. Consequently, it has been assumed that many of the general findings of these studies may also apply to thalamic DBS.

All types of effects have been reported. STN-DBS produced inhibition within STN (Meissner et al., 2005) and reduced firing at its projection sites (Benazzouz et al., 1995; Benazzouz et al., 2000). Similarly, high-frequency stimulation in the GP of primates reduced the firing frequency of neighboring neurons (Boraud et al., 1996). However, other studies, using a variety of techniques, concluded that efferent outflow from the stimulated nucleus increased during DBS (Anderson et al., 2003; Hashimoto et al., 2003; Maurice et al., 2003; Windels et al., 2000; Windels et al., 2003), and modeling studies of both STN and thalamus have supported this view (see below).

Although *in vivo* animal experiments provide valuable information about the effects of DBS, extracellular recordings have limitations. Action potentials are all-or-none phenomena and can occur in response to direct excitation, disinhibition, inhibition, antidromic or orthodromic invasion, depending on which pathway is being recorded and stimulated. They also fail to account for multiple simultaneous pathway effects. Therefore, our group chose to study DBS directly *in vitro*, to determine the intracellular effects of DBS on thalamic neurons.

2.3.1.3 Evidence from Slice Preparations

Our group has been one of the few studying the effect of DBS on VL thalamic neurons using intracellular techniques. In 2002, we reported that high-frequency stimulation produced depolarization of thalamic neurons in slice (Figure 2.1), and the effects were stimulation frequency and amplitude dependent (Kiss et al., 2002) (Figure 2.2). Two types of neuronal responses to high-frequency stimulation were observed (Anderson et al., 2004). Depolarization could be transient, occurring only at the onset of the stimulus train, or sustained throughout the train, and we referred to these responses as type 1 and 2, respectively (see Figure 2.1). In both cases, depolarization depended on axonal activation and glutamatergic synaptic transmission. Blockers of action potentials, glutamate receptors, and calcium all prevented somatic depolarization. While DBS did not seem to alter the steady-state input resistance of the cells, it did reduce the threshold for action potential generation. We coined the terms "functional deafferentation" and "functional inactivation" to describe the effects of DBS on thalamic neurons. Functional deafferentation is the loss of synaptic input to postsynaptic thalamocortical neurons, as occurred in type 1 responses. Functional inactivation referred to type 2 responses, where depolarization was followed by repetitive spikes, while the cell remained depolarized, and which disrupted the rhythmic pattern of the outgoing signal. These mechanisms, driven primarily by synaptic activation, helped to explain the paradox that lesions, GABA receptor activation with muscimol, and DBS all effectively stop tremor.

Detailed studies of type 1 responses, the transient responses induced by high-frequency stimulation where depolarization occurred only at the onset of the stimulus train, revealed several important features (Anderson et al., 2006). During stimulation, neurons were incapable of firing action potentials or even displaying excitatory postsynaptic potentials (EPSPs). The cells were "functionally deafferented" during the stimulus train (Figure 2.3). Using the principle that two pulses applied 20 to 40 ms apart to the same axonal pathway will potentiate (increase the amplitude of the second EPSP), and two pulses applied to different pathways will not potentiate, we identified two inputs to one neuron in VL slices. Low-frequency stimulation (5 Hz) mimicking afferent tremor input was applied through one electrode (electrode A) and high-frequency stimulation (125 Hz) mimicking DBS was then applied through the same electrode (A). During the simulated DBS train, EPSPs elicited by the 5-Hz stimuli disappeared. DBS seemingly blocked the afferent input from the same pathway reaching the neuron under study. When the 5-Hz tremor-like input was applied with a separate electrode (B) to an independent pathway (B), high-frequency DBS applied though electrode A failed to alter EPSPs elicited through pathway B. This indicated that the suppression of tremor-like afferent activity was specific to the stimulated pathway and did not spill over to another pathway.

We concluded that DBS produced synaptic depression at the presynaptic level for the following reasons. When we applied cyclothiazide, which prevents AMPA receptor desensitization, to the slice, there was no change in the depolarizing effect of simulated DBS. The specificity of the neuronal response to single-pathway stimulation and its rapid time course of depression made the possibility of an intracellular postsynaptic cascade very unlikely. DL-threo-beta-benzyloxyaspartate (DL-TBOA), a glutamate re-uptake

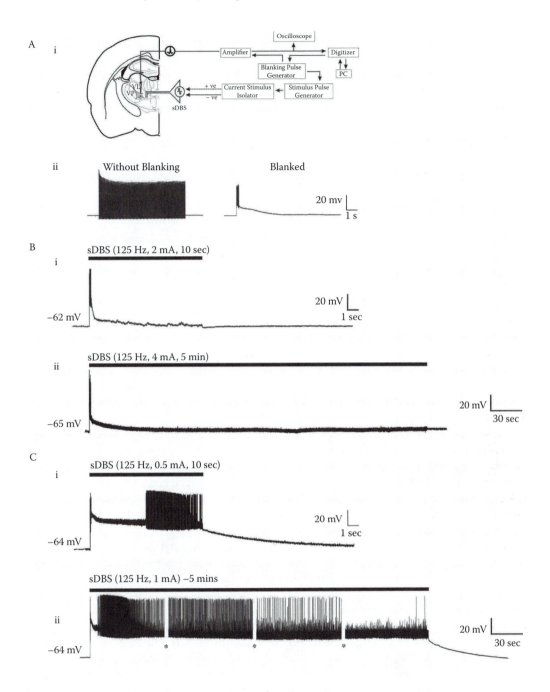

FIGURE 2.1 Experimental setup for intracellular recording from ventral thalamic neurons and two types of membrane responses evoked by DBS. (Ai) Schematic representation of sharp intracellular recording setup in thalamic rat brain slice. (Aii) Intracellular recordings from the same cell with and without the blanking operation activated. (B) Type 1 responses had a large initial depolarization, declining toward a smaller but sustained level of depolarization in response to 10 s (i) or 5 min (ii) of simulated DBS. The black bar indicates stimulus onset and duration. (C) Type 2 responses have a large initial depolarization, which persisted over 10 s (i) or 5 min (ii) and led to varied spike activity. Gaps in the recording indicate times at which current pulse protocols were run during the 5-min stimulus train. (*Source:* Adapted from Anderson et al., 2004. With permission.)

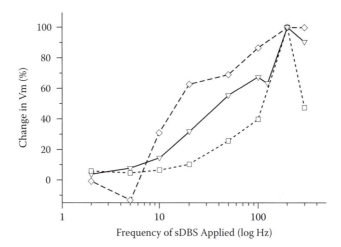

FIGURE 2.2 Frequency dependency of simulated DBS-induced membrane depolarization (Vm). Each line and symbols represent the depolarization recorded in each cell, normalized to its maximum value of nonspiking voltage. Note the effective frequency is about 100–200 Hz. (*Source:* Adapted from Kiss et al., 2002. With permission.)

inhibitor, also failed to alter the effect of DBS in our slice model, making nonspecific extra-synaptic glutamate spillover highly improbable. Synaptic depression was limited to the homosynaptic pathway; α-methyl-4-sulphonophenylglycine or MSPG, a nonselective antagonist of presynaptic metabotropic glutamate receptors, failed to alter responses; and finally, the rapid time course of recovery of homosynaptic EPSPs after DBS was turned off, was in line with known recovery of the readily releasable pool of neurotransmitter (Stevens and Tsujimoto, 1995).

The direct role of extracellular stimulation on neuronal soma was also investigated in the presence of glutamatergic blockade. We used whole cell patch clamp techniques to identify that simulated DBS in thalamic slice activated an I_h current, immediately at the onset of stimulation (Anderson et al., 2006). It failed, however, to inhibit both the persistent Na$^+$ current and the rebound potential related to I_T currents, which in thalamus are made up mainly of the low threshold Ca^{2+} current responsible for the (LTS) rebound bursts (Jahnsen and Llinas, 1984). A activation of I_h did not seem to be responsible for the depolarization induced by simulated DBS.

Slice recordings and rodent models have some limitations. For example, rat thalamus has few GABAergic interneurons (Williams and Faull, 1987) in comparison to primates, and therefore the effect of inhibitory synaptic potentials within the ventral thalamus in response to DBS may be more difficult to elicit. Instead, the rodent nucleus reticularis is overdeveloped and thus provides similar inhibitory input to relay nuclei (1997). Synaptic terminals are preserved in slice; therefore, if GABAergic fibers were the key component of the response to DBS in thalamus, we would expect that they would be seen in our preparation. Experiments performed in ferret geniculate slices in which the nucleus reticularis — thalamocortical network is preserved, yielded similar results to those in rat, finding IPSPs but a predominant effect of depolarization and excitatory neurotransmission (Lee et al., 2005). While brain slices have limitations, the fact that no anesthetic is required and we have complete control over the cellular milieu outweigh its disadvantages. Computer simulation modeling studies have attempted to overcome some of these limitations and further advance our understanding of DBS.

2.3.1.4 Evidence from Modeling Studies

Modeling studies have thus far focused on the immediate responses of local neural elements to extracellular stimulation. Responses to DBS were determined by electrode position, stimulus intensity, waveform duration, and polarity (McIntyre and Grill, 2000), in addition to the proportion of inhibitory and excitatory terminals within the structure (McIntyre et al., 2004). Due to their lower threshold for activation, DBS was predicted to activate axons before cell bodies (Ranck, 1975; Nowak and Bullier, 1998;

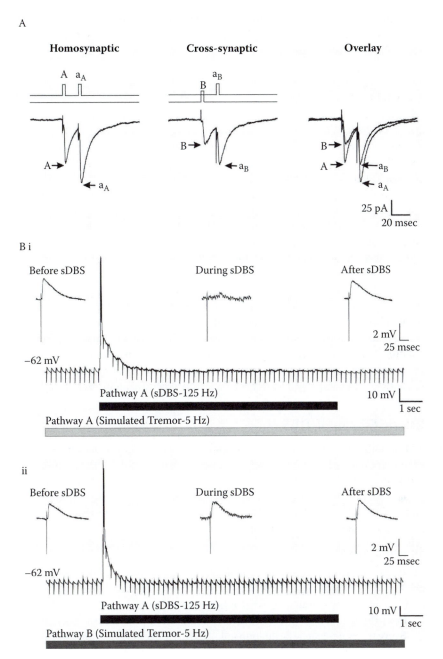

FIGURE 2.3 Pathway specificity of DBS in thalamic slices. (A) Independence of pathways A and B is shown by comparing the paired-pulse ratios of EPSC amplitudes (cells held at −60 mV). The stimulation protocol is illustrated above each voltage-clamp recording. The effect of a homosynaptic prepulse (A) on a second homosynaptic pulse (aA) shows facilitation of the second EPSC, whereas a cross-synaptic prepulse (applied through electrode B) fails to facilitate the second pulse applied to pathway A (aB). (B) Simulated DBS selectively inhibits afferent tremor-like input. Tremor was mimicked in the slice by extracellularly evoking EPSPs at 5 Hz. DBS applied homosynaptically during simulated tremor resulted in complete inhibition of the tremor signal (i) but had no effect on tremor generated along the cross-synaptic pathway (ii). Insets show expanded views of single EPSPs before, during, and after sDBS. Note that the tremor stimuli were delivered at a subharmonic frequency of the sDBS train to ensure noninterference when delivered homosynaptically down the same electrode. To confirm successful delivery, the tremor stimulation artifacts were not blanked. (*Source*: Adapted from Anderson et al., 2006. With permission.)

McIntyre and Grill, 1999), leading to neurotransmitter release from terminals and either inhibition or excitation of postsynaptic neurons, depending on the concentration of excitatory and inhibitory terminals within the stimulated nucleus.

Modeling studies have attempted to reconcile discrepant physiologic findings. Neurons within a stimulated nucleus were reported as being inhibited by DBS (Dostrovsky and Lozano, 2002; Dostrovsky et al., 2000; Boraud et al., 1996; Meissner et al., 2005), whereas, downstream structures that receive input from that nucleus were activated (Anderson et al., 2003; Hashimoto et al., 2003; Maurice et al., 2003; Windels et al., 2000; Windels et al., 2003). McIntyre et al. (2004) proposed decoupling between soma and its axon, where the cell body could be inhibited from firing action potentials, yet the axon that has a lower threshold for activation would fire independently. Two mechanisms can explain such somatic inhibition independent of the axon. First, the cell body may be hyperpolarized while the axon depolarized due to the nature of current spread from the stimulating electrode. Second, inhibitory inputs onto soma, such as inhibitory interneuron terminals, may have a lower threshold for activation compared to the postsynaptic cell body (Rubinstein, 1993; McIntyre et al., 2004).

Finally, modeling studies have proposed that the majority of axons within 2 mm of the cathode will fire action potentials at the stimulation frequency applied (McIntyre et al., 2004). This predicted that axons of passage, with a lower threshold for activation than somata, were the targets of DBS and the downstream effect of DBS in any nucleus was increased afferent input to the nuclei where those axons projected. Only one study even attempted to incorporate multiple nuclei into its model to help explain how DBS might affect the entire network (Rubin and Terman, 2004). These authors also proposed that regular high-frequency stimulation applied to the output of STN would normalize pathologic rhythmic activity in pallidum, thalamus, and cortex. While this study made numerous assumptions, such as faithful conduction of action potentials in axons and unlimited supply of neurotransmitter in synaptic terminals, it leads us to consider and compare the effects of DBS at its downstream projection sites.

2.3.2   Distant Effects of DBS

Modeling studies suggested that the downstream projection sites from the nucleus in which DBS was applied may be more relevant to the mechanism of action than the local effects. Thus, DBS has been investigated at remote sites *in vivo* in humans using imaging techniques and in animals with electrophysiology. Our report on the remote cortical effects of thalamocortical stimulation is the only one performed in slice to specifically address the question of what happens downstream from the thalamus. Each of these types of studies is discussed in turn.

2.3.2.1   Human Studies

Ashby et al. (1995) examined the effects of thalamic DBS on electromyography (EMG). Single thalamic stimuli resulted in sudden short lapses of posture in contracting muscles, similar to a clinical condition called "asterixis" (and perhaps similar to what others have described as "myoclonus" (Bejjani et al., 2000)). EMG from the first dorsal interosseous muscle of the contralateral hand showed a biphasic response at DBS onset: a short-latency facilitation followed by a longer-latency inhibition of EMG activity. Ashby et al. (1995) concluded that the EMG facilitation was due to direct activation of the adjacent corticospinal tract due to its short latency, and because all four DBS electrode contacts produced similar results, indicating that the pathway was parallel to the electrode (Strafella et al., 1997). The effect on EMG was transient, lasting approximately 200 ms, even during prolonged DBS and only occurred with stimulation of the cerebellar-receiving thalamus, not other basal ganglia structures (Strafella et al., 1997). The authors speculated that this effect could be mediated by either (1) disruption of thalamocortical activity, (2) activation of inhibitory processes within the thalamus/reticular nucleus, or (3) orthodromic/antidromic activation of fibers to and from the thalamus.

Several other studies used transcranial magnetic stimulation or functional imaging to investigate how thalamic DBS affects cortical function. Molnar et al. (2004, 2005) recorded motor-evoked potentials in response to transcranial motor cortex magnetic stimulation. The authors found that patients with thalamic DBS had significantly larger motor-evoked potential amplitudes than normal control subjects, and that the cerebello-thalamocortical pathway was facilitated in the ON stimulation state, suggesting that DBS activates rather than inhibits the target area and the motor cortex.

Imaging studies measuring regional cerebral blood flow have also reported that thalamic DBS increases cortical activity. During Parkinsonian tremor, there was increased activity in the sensorimotor cortex, which was reduced by both effective thalamic stimulation (Parker et al., 1992; Wielepp et al., 2001) and thalamotomy (Boecker et al., 1997). Another study found changes in cortical blood flow with 60-Hz ineffective stimulation, and reduction in cerebellar activity with effective 135- to 180-Hz DBS (Deiber et al., 1993). Because Parkinsonian tremor occurs at rest, a better patient group in which to learn the direct effects of thalamic DBS are those with essential tremor. Motor cortical blood flow ipsilateral to effective thalamic DBS increased at frequencies that effectively suppressed tremor in comparison to those that did not suppress tremor (Ceballos-Baumann et al., 2001). Perlmutter et al. (2002) also found an increase in regional cerebral blood flow at thalamocortical terminal fields, although in these cases DBS electrodes were likely located more anterior than usual because the supplementary motor area was mainly affected. Haslinger and colleagues (2003) investigated the effects of varying the amplitude and frequency of stimulation on the cortical blood flow response. Increasing stimulation amplitude resulted in a nonlinear increase in cerebral blood flow in the sensorimotor cortex, whereas increasing the stimulation frequency resulted in a linear increase in blood flow. Therefore, the consensus from human studies is that thalamic DBS results in activation rather than inhibition of thalamic projections. This is in stark contrast to DBS applied to subcortical white matter where adjacent cortical regions exhibited decreased blood flow in patients treated for depression and obsessive–compulsive disorder (Nuttin et al., 2003; Mayberg et al., 2005).

There seems to be no behavioral correlate for motor cortex activation clinically. Aside from 2- to 3-s long, 1- to 2-Hz stimulation producing EMG excitation or inhibition depending on the intensity applied, high-frequency stimulation produces no adverse motor effects (Ashby et al., 1995; Flament et al., 2002). Therefore, one can speculate that perhaps cortical blood flow changes may be an epiphenomenon of thalamic DBS. Another related question is whether the motor cortex is involved in tremor or tremor suppression. While Parkinsonian tremor has been measured from the cortex using electrophysiological techniques (Alberts et al., 1969; Volkmann et al., 1996; Hellwig et al., 2000), essential tremor has been more difficult to consistently demonstrate with coherence measurements between EMG and sensorimotor cortex (Halliday et al., 2000; Hellwig et al., 2001). The role of the cortex in mediating the effects of DBS or thalamotomy remains to be determined.

2.3.2.2 Animal Studies

There have been no *in vivo* experiments specifically examining the effects of VL thalamic DBS on its projection site, the primary motor cortex. However, several studies have applied DBS to other basal ganglia nuclei and recorded activity at the site of their efferent connections.

Two groups have examined the effects of DBS in downstream projection nuclei in nonhuman primates. Anderson et al. (2003) applied DBS to the globus pallidus pars interna and recorded spontaneous activity in pallidal-receiving thalamus. The pallidal input to thalamus is GABA-ergic, and high-frequency stimulation suppressed neuronal firing in thalamus. Hashimoto et al. (2003) instead applied high-frequency stimulation to the STN and recorded spontaneous activity in the globus pallidus. Because the axonal pathway projecting to pallidum from STN is glutamatergic, the findings in this case were excitation and significant increases in mean discharge rate. Both authors agreed that DBS likely interrupted abnormal patterns of thalamic and basal ganglia activity instead of silencing neurons and supported the modeling studies of McIntryre et al. (2004).

While it is agreed that DBS results in modulation of downstream projection sites and therefore, in the case of VL thalamic DBS, cortical activity, exactly what happens in the cortex is not well understood. Therefore, we embarked on a series of experiments in slice to determine the downstream effects of thalamocortical axon stimulation.

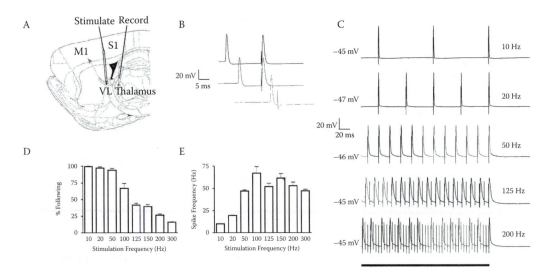

FIGURE 2.4 Complete conduction block did not occur in axons during DBS. (A) Schematic of the slice preparation with location of stimulating electrode in internal capsule and recording electrode in VL thalamus. (B) Thalamic cells were depolarized to evoke somatic spikes, and the stimulating electrode was used to evoke antidromic spikes. Spike collision between somatic and antidromic spikes occurred at short intervals indicating stimulation of the same pathway. (C) At stimulation frequencies <50 Hz, antidromic action potentials followed 1:1 with extracellular pulses throughout the stimulus train, while at higher stimulation frequencies, antidromic spikes could follow 1:1 only at the onset of stimulation. The last 600 ms of a 30-s stimulation train are shown. (D) There was a progressive failure and broadening of antidromic spikes with increasing stimulation frequency. (E) Evoked spike frequency plateaued between 50 and 70 Hz. Antidromic failure and following frequency was measured as the average failure rate or frequency during the last 10 s of a 30-s train. All data were taken from the last spike of a 30-s train and normalized to the first spike of that same train. (*Source*: Adapted from Iremonger et al., 2006. With permission.)

2.3.2.3 Slice Preparations

The first question that must be addressed is the following: does the high-frequency DBS signal applied to thalamocortical axons make it to the cortex? To do so, axonal conduction must be faithful. However, there is increasing evidence to suggest that the axon is not a high-fidelity conductor, but instead an extension of the cell body capable of performing signal filtering and mediating plasticity (Debanne, 2004). Therefore, we first examined the ability of thalamocortical and corticothalamic neurons to follow stimuli applied to the internal capsule in sagittal rat brain slices (Iremonger et al., 2006). After proving that we were antidromically activating the axon of the neuron under study, we applied 10- to 300-Hz stimulation to the capsule. Thalamocortical fibers failed to faithfully conduct action potentials at stimulation frequencies greater than 50 Hz (Figure 2.4). In layer VI cortico thalamic cells, that exhibited antidromic responses to subcortical stimulation, antidromic following decreased with stimulus frequencies greater than 10 Hz and the maximum following frequency peaked at approximately 30 Hz. Despite the inability of both thalamocortical and corticothalamic fibers to follow high-frequency stimulation faithfully, complete conduction block did not occur.

The next stage of processing after the axon is the synapse. Therefore, the next question we asked was whether high-frequency stimulation of thalamocortical axons would affect downstream postsynaptic neurons in the cortex (Iremonger et al., 2006). Motor cortical neurons depolarized in response to subcortical external capsule stimulation in a frequency-dependent manner (Figure 2.5). However, these depolarization responses were not sustained. Intracortical inhibition was not responsible for the return to baseline membrane potential, as $GABA_{A/B}$ antagonists failed to sustain the DBS-induced depolarization. DBS also failed to alter firing rates in motor cortical neurons manually depolarized by intracellular current injection. Instead, synaptic depression was likely responsible for the lack of sustained depolarization in response to high-frequency stimulation (Figure 2.6). There was a marked depression of EPSP/Cs after the

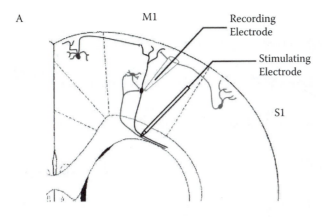

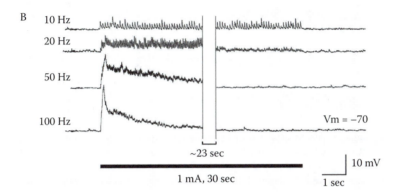

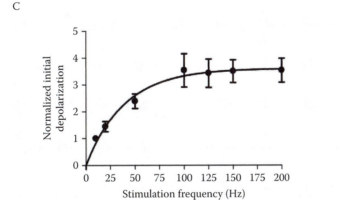

FIGURE 2.5 Frequency-dependent depolarization of primary motor cortex (M1) neurons to subcortical white matter stimulation. (A) Schematic of the slice preparation illustrating stimulating and recording electrode locations. (B) Current clamp traces at resting membrane potential, showing the membrane response to 30 s of stimulation at different frequencies. At 10 and 20 Hz, synaptic transmission could be maintained throughout the 30-s train. At stimulation frequencies >50 Hz, there was a temporal summation of EPSPs at the onset of stimulation, although, after this initial response, stimulation could no longer evoke a postsynaptic response. (C) Amplitude of initial depolarization in response to DBS increases with stimulation frequency up to 100 Hz, after which increasing the stimulation frequency further did not result in a larger depolarization. Depolarization is normalized to the initial depolarization obtained with 10-Hz stimulation. (*Source*: Adapted from Iremonger et al., 2006. With permission.)

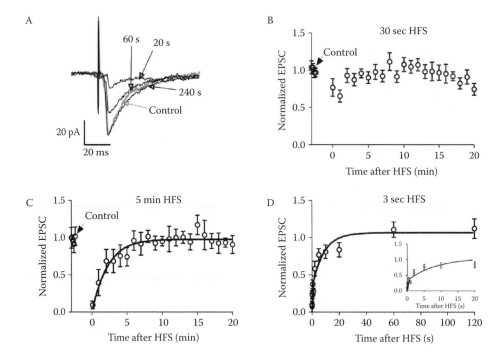

FIGURE 2.6 DBS-induced synaptic depression. (A) Overlying traces showing the control EPSC (thick gray trace) and EPSCs evoked at different times (20, 60, and 240 s) after a 30-s train of DBS (thin black trace). (B and C) Recovery from synaptic depression after 30 s of DBS was faster than that after 5 min of DBS. Slow time course of recovery of the EPSC after 5 min DBS was fitted ($R^2 = 0.91$) with a single-phase exponential with a τ of about 2.1 min. (D) Short-term recovery of the mean EPSC after a 3-s DBS train. Recovery was fitted with an exponential with a τ of 120 ms and 8.6 s, respectively. Inset: expanded time course of recovery from 0 to 20 s. All cells were voltage clamped at −70 mV, and all data were normalized to the mean EPSC amplitude measured before each DBS train. (*Source:* Adapted from Iremonger et al., 2006. With permission.)

DBS train, and its time course depended on the stimulus train length. The time course of recovery fit that described for depletion of the readily releasable pool of transmitter (Rosenmund and Stevens, 1996; Stevens and Wesseling, 1998). Depression of postsynaptic currents with repetitive stimulation was not attributable to desensitization of AMPA receptors because cyclothiazide failed to alter responses to DBS. Synaptic depression was specific to the pathway stimulated. Stimulation of superficial cortical layers projecting to motor cortical neurons continued to produce EPSPs in these neurons during subcortical high-frequency DBS (Iremonger et al., 2006).

Other investigators showed in an intact somatosensory thalamocortical slice that high-frequency thalamic stimulation produced only transient activation of cortex, as measured by voltage-sensitive dye and field potentials (Urbano et al., 2002). They assumed this was related to synaptic depression because such depression is ubiquitous at central nervous system synapses. Its proposed function was to prevent neurons from being saturated with inputs from high-frequency afferents, allowing them to still detect low-frequency afferent input (Abbott et al., 1997).

To summarize, synaptic depression prevented sustained activation of cortical neurons in slice when high-frequency stimulation was applied to subcortical white matter or thalamus. Axons failed to faithfully follow high-frequency stimuli but did not undergo complete conduction block. These results are likely specific to the nuclei and pathway stimulated, and do not necessarily apply to other DBS sites used clinically. In fact, there is accumulating evidence that the properties of the neurons and axons close to where DBS is applied are critically important in determining the cellular response and thereby likely mechanisms involved.

2.3.3   Comparison of Slice Experiments in Thalamus and STN

Reports on the intracellular effects of simulated DBS applied to STN neurons in slice appear to conflict. Some authors found inhibition of ongoing spontaneous activity during high-frequency stimulation (Beurrier et al., 2001; Magarinos-Ascone et al., 2002). Other reports (from some of the same labs) concluded that there was increased activity (Garcia et al., 2003; Lee et al., 2004; Garcia et al., 2005).

The data from thalamus are different from that reported in STN in two important ways: (1) in the role of synaptic transmission in mediating the effects of DBS; and (2) in the direct effects of DBS on specific membrane currents. For example, Do and Bean (2003) observed that in STN neurons deprived of synaptic input (STN neurons in culture), the persistent, transient, and resurgent Na$^+$ currents were inactivated by high-frequency stimulation. Yet most cells fired with each stimulating pulse and even if they did not, they fired at higher rates during stimulation than spontaneously. Beurrier et al. (2001) reported that high-frequency stimulation induced a direct inhibition of membrane currents, specifically the persistent sodium, and the T and L type calcium currents, when experiments were performed at 30 °C. The same team later used a different stimulation method meant to mimic monopolar stimulation using a point-source cathode and ring anode around their slice (Garcia et al., 2003). They described a dual effect of high-frequency stimulation at room temperature, in which spontaneous STN neuronal activity was suppressed and replaced by time-locked spikes. The authors concluded that DBS directly drives STN somatic membranes because neither GABA nor glutamate receptor blockers altered the neuronal responses. Unfortunately, they failed to mention that the somatic driving they observed could have represented antidromic responses for the following reasons. The effect was blocked by tetrodoxin, spikes were consistent with invariable latency and could follow high-frequency stimulation. They suggested that the superimposed high-frequency firing could mask the abnormal basal ganglia firing patterns occurring in Parkinson's disease (Brown, 2003).

Other groups (Magarinos-Ascone et al., 2002; Lee et al., 2004) have investigated the intracellular effects of STN stimulation in slice. Magarinos-Ascone et al. (2002) reported that high-frequency stimulation silenced a specific type of STN neuron, the tonic firing type, by two mechanisms. This team designed experiments to only look at synaptically mediated events. Tonic neurons depolarized in response to DBS initially followed high-frequency stimuli, then burst briefly, and within 20 s of the onset of DBS, were completely silenced ($N = 7$). Eight other cells responded with IPSPs, with a silencing of spontaneous firing. Lee et al. (2004) described only one type of STN neuron and found that bipolar stimulation produced EPSPs, IPSPs, or altered spontaneous firing of action potentials ($N = 16$). These effects were blocked by glutamate and GABA receptor antagonists. Substantia nigra neurons, an STN projection site, also were activated.

While Magarinos-Ascone and Lee reported similar results, which may be somewhat skewed by their small sample size, those of Hammond's group (Beurrier et al., 2001; Garcia et al., 2003) are different. Possible reasons for such discrepant results include the recording chamber used and the type of stimulation applied.

Our studies (Kiss et al., 2002; Anderson et al., 2004, 2006) and those of Garcia et al. (2003) utilized a submerged recording chamber, while Lee et al. (2004) and Beurrier et al. (2001) used an interface chamber. Stimulation within a submerged chamber places the exposed portion of the stimulating electrode in contact with the tissue as well as the bath. This introduces a bath resistance (~50 Ω) in parallel with the tissue resistance (~500 Ω) (Nowak and Bullier, 1996; McIntyre and Grill, 2002). This bath resistance is absent in an interface chamber. Therefore, the increased current shunting into the bath using a submerged chamber vs. an interface chamber results in an applied tissue current of approximately 10% of the total current applied. Figure 2.7 summarizes these differences. While both Garcia et al. (2003) and Lee et al. (2004) applied similar current intensities, given the probable current shunting in Garcia's submerged system, their true applied intensity would only be 10% of that applied in Lee's study. Consequently, this may account for their lack of observed synaptic activation (Garcia et al., 2003). Our current intensities in thalamic tissue (estimated mean of 327 μA when taking into account current shunting) were similar to those utilized by Lee in STN (10 to 500 μA).

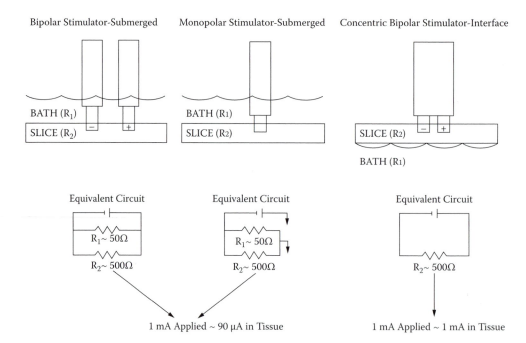

FIGURE 2.7 Illustration of submerged vs. interface slice recording chambers and effects of stimulating configuration. Stimulation within a submerged chamber places the exposed portion of the stimulating electrode in contact with the tissue as well as the bath. This introduces a bath (~50 Ω) resistance in parallel with the tissue resistance (~500 Ω). This bath resistance is absent in an interface chamber. Therefore, the increased current shunting into the bath using a submerged chamber vs. an interface chamber results in an applied tissue current of ~10% of the total current applied.

In addition to the effects of current intensity in different slice chambers, the method of stimulation also may have contributed to the discrepant results. To examine this issue in our thalamic preparation, we compared the response induced by bipolar simulated DBS to that induced by the "monopolar-ring" stimulation of Hammond's team (Garcia et al., 2003). Bipolar stimulation produced a large depolarization in VL-thalamic neurons; however, "monopolar-ring" stimulation, at the same current strength, did not, resulting instead in a minimal membrane depolarization of approximately 2 to 6 mV. Only by increasing the current applied to 3 to 5 times that used with the bipolar configuration could similar depolarizations be elicited.

Finally, the properties of thalamic and STN neurons differ. The resurgent sodium current is prominent in STN but not in thalamic neurons (Do and Bean, 2003). These currents promote rapid recovery from inactivation and thereby allow high-frequency firing and pacemaking. While I_h currents exist in STN neurons, they are less important for pacemaking. Thalamic neurons lack resurgent Na+ currents and instead rely mainly on I_T (T-type Ca^{2+} currents) to generate rhythmic rebound potentials and the characteristic LTS burst (Jahnsen and Llinas, 1984).

2.3.4 Summary and Functional Implications

The cellular mechanisms of DBS are related to frequency-dependent membrane depolarization, which in thalamus is synaptically mediated. Thalamic DBS does not alter intrinsic membrane currents (I_T, I_{Nap}) at physiologic potentials but does have a small effect on I_h. In STN, synaptic transmission likely also plays a role but there are more obvious direct effects of stimulation on membrane properties, such as the resurgent Na+ current. Neurons are either functionally deafferentated, unable to fire action potentials due to synaptic depression, or inactivated, abnormal firing patterns are masked by an artificial firing pattern. The effects are specific to the pathway stimulated so that functional deafferentation can limit propagation of pathophysiological tremor signals without disrupting information from other pathways.

Thalamic stimulation, despite seemingly activating thalamocortical axons projecting to the motor cortex, does not disrupt motor control. Synaptic depression and axonal filtering prevent remote cortical excitation during high-frequency subcortical DBS. Spontaneous cortical firing is not altered, as the synaptic depression is specific to the stimulated pathway, alone.

Our *in vitro* data may explain many of the clinical features important for thalamic DBS in patients. The similarity of thalamic DBS to thalamotomy is likely due to functional deafferentation. Tremor cells are unable to propagate EPSPs or action potentials and thereby the tremor signal does not make it to the motor output level. Functional deafferentation occurs immediately upon application of DBS and is rapidly reversible. DBS is site-specific, both clinically in patients with tremor and in the slice model. The cellular effects of stimulation are frequency dependent, with high frequencies (>60 Hz) required for tremor suppression. Finally, patients do not experience any involuntary movements or motor disruption with prolonged DBS. Axonal filtering at the level of thalamocortical axons and synaptic depression in the motor cortex prevent such disruption. The effects are specific to the pathway stimulated and do not spill over to other pathways in thalamus or cortex.

There remain several features of DBS that have yet to be explained. These include how to reconcile electrophysiological recordings with imaging studies, how modeling can explain the exacerbation of tremor (and other movement disorders) seen with specific low frequencies of DBS, what the role of glial cells are in buffering of released extracellular K^+ (Bikson et al., 2001), and how tremor rebound can be explained by synaptic depression. Of course, much of the slice work should be confirmed in whole animals. Finally, the effect of long-term stimulation (days–weeks) on synaptic properties must also be evaluated. There is currently no information on morphological synaptic changes that result from long-term continuous DBS. For example, do synaptic terminals stay depleted of neurotransmitter, do dendritic spines retract or expand if there is prolonged depletion of transmitter, and what compensatory mechanisms develop?

2.4   Potential of DBS Therapy

While much of the data discussed in this chapter is specific to thalamic and subcortical white-matter DBS, there are some important principles that can be broadly applied to many forms of nervous system electrical stimulation. No matter where the stimulating electrode is placed, the effect of extracellular stimulation will be axonal/dendritic fiber excitation (McIntyre et al., 2004). However, the effect of axonal activation on postsynaptic cellular activity will depend on the ability of axons and synapses to transmit the signal. If the stimulation is low frequency (<50 Hz), then axons and synapses should convey the signal with high fidelity. However, if the stimulation is high frequency (>100 Hz), conduction failure and transmitter depletion may filter out the high-frequency signal. Effects are pathway specific and do not spill over to nonstimulated tracts.

This frequency-dependent effect of electrical stimulation on specific axons and terminals has tremendous potential. The mainstay of treatment for neurologic and psychiatric conditions at this point is medication. Many of these drugs replace or alter neurotransmitter release, re-uptake, or receptors. Drugs work fairly indiscriminately at all similar nervous system synapses, and they cannot be timed to work only when symptoms require. As a result, medications often produce multiple and unacceptable side effects. DBS has the capability to selectively alter neurotransmitter release in a specific pathway and in a controlled manner, as required by symptoms. The costs of introducing a new drug to the market are ever increasing and are limiting new drug development. The costs of electrical stimulation are inexpensive, in comparison. With an understanding of how DBS works at the cellular level, the full potential of electrical neuromodulation may become realized. With knowledge of the anatomy and physiology of axonal projections and neuronal properties, we may be able to predict benefits and side effects of this technology in other pathways and apply it rationally for new indications.

References

Abbassian, A.H., Shahzadi, S., Afraz, S.R., Fazl, A., and Moradi, F. (2001). Tactile discrimination task not disturbed by thalamic stimulation. *Stereotact. Funct. Neurosurg.*, 76:19–28.

Abbott, L.F., Varela, J.A., Sen, K., and Nelson, S.B. (1997). Synaptic depression and cortical gain control. *Science*, 275:220–224.

Alberts, W.W., Wright, E.W., and Feinstein, B. (1969). Cortical potentials and parkinsonian tremor. *Nature*, 221:670–672.

Anderson, M.E., Postupna, N., and Ruffo, M. (2003). Effects of high-frequency stimulation in the internal globus pallidus on the activity of thalamic neurons in the awake monkey. *J. Neurophysiol.*, 89:1150–1160.

Anderson, T.R., Hu, B., Pittman, Q., and Kiss, Z.H.T. (2004). Mechanisms of deep brain stimulation: An intracellular study in rat thalamus. *J. Physiol.*, 559:301–313.

Anderson, T.R., Hu, B., Iremonger, K.J., and Kiss, Z.H.T. (2006). Selective attenuation of afferent synaptic transmission as a mechanism of thalamic deep brain stimulation induced tremor arrest. *J. Neurosci.*, 26:841–850.

Ashby, P., Lang, A.E., Lozano, A.M., and Dostrovsky, J.O. (1995). Motor effects of stimulating the human cerebellar thalamus. *J. Physiol. (London)*, 489:287–298.

Atkinson, J.D., Collins, D.L., Bertrand, G., et al. (2002). Optimal location of thalamotomy lesions for tremor associated with Parkinson disease: a probabilistic analysis based on postoperative magnetic resonance imaging and an integrated digital atlas. *J. Neurosurg.*, 96:854–866.

Bejjani, B.P., Arnulf, I., Vidailhet, M., et al. (2000). Irregular jerky tremor, myoclonus, and thalamus: a study using low-frequency stimulation. *Mov. Disord.*, 15:919–924.

Benabid, A.L., Chabardes, S., and Seigneuret, E. (2005). Deep-brain stimulation in Parkinson's disease: long-term efficacy and safety — What happened this year? *Curr. Opin. Neurol.*, 18:623–630.

Benabid, A.L., Pollak, P., Gao, D.M., et al. (1996). Chronic electrical stimulation of the ventralis intermedius nucleus of the thalamus as a treatment of movement disorder. *J. Neurosurg.*, 84:203–214.

Benazzouz, A., Piallat, B., Pollak, P., and Benabid, A.L. (1995). Responses of substantia nigra pars reticulata and globus pallidus complex to high frequency stimulation of the subthalamic nucleus in rats: electrophysiological data. *Neurosci. Lett.*, 189:77–80.

Benazzouz, A., Gao, D.M., Ni, Z.G., et al. (2000). Effect of high-frequency stimulation of the subthalamic nucleus on the neuronal activities of the substantia nigra pars reticulata and ventrolateral nucleus of the thalamus in the rat. *Neuroscience*, 99:289–295.

Beurrier, C., Bioulac, B., Audin, J., and Hammond, C. (2001). High-frequency stimulation produces a transient blockade of voltage-gated currents in subthalamic neurons. *J. Neurophysiol.*, 85:1351–1356.

Bikson, M., Lian, J., Hahn, P.J., et al. (2001). Suppression of epileptiform activity by high frequency sinusoidal fields in rat hippocampal slices. *J. Physiol.*, 531:181–191.

Boecker, H., Wills, A.J., Ceballos-Baumann, A.O., et al. (1997). Stereotactic thalamotomy in tremor-dominant Parkinson's disease: an H2(15)O PET motor activation study. *Ann. Neurol.*, 41:108–111.

Boraud, T., Bezard, E., Bioulac, B., and Gross, C. (1996). High frequency stimulation of the internal globus pallidus (GPi) simultaneously improves parkinsonian symptoms and reduces the firing frequency of GPi neurons in the MPTP-treated monkey. *Neurosci. Lett.*, 215:17–20.

Brown, P. (2003). Oscillatory nature of human basal ganglia activity: relationship to the pathophysiology of Parkinson's disease. *Mov. Disord.*, 18:357–363.

Caesar, K., Thomsen, K., and Lauritzen, M. (2003). Dissociation of spikes, synaptic activity, and activity-dependent increments in rat cerebellar blood flow by tonic synaptic inhibition. *Proc. Natl. Acad. Sci., U.S.A.*, 100:16000–16005.

Ceballos-Baumann, A.O., Boecker, H., Fogel, W., et al. (2001). Thalamic stimulation for essential tremor activates motor and deactivates vestibular cortex. *Neurology*, 56:1347–1354.

Debanne, D. (2004). Information processing in the axon. *Nat. Rev. Neurosci.*, 5:304–316.

Deiber, M.-P., Pollak, P., Passingham, R., et al. (1993). Thalamic stimulation and suppression of parkinsonian tremor. Evidence of a cerebellar deactivation using positron emission tomography. *Brain*, 116:267–279.

Do, M.T. and Bean, B.P. (2003). Subthreshold sodium currents and pacemaking of subthalamic neurons: modulation by slow inactivation. *Neuron*, 39:109–120.

Dostrovsky, J.O., Levy, R., Wu J.P., et al. (2000). Microstimulation-induced inhibition of neuronal firing in human globus pallidus. *J. Neurophysiol.*, 84:570–574.

Dostrovsky, J.O. and Lozano, A.M. (2002). Mechanisms of deep brain stimulation. *Mov. Disord.*, 17:S63–S68.

Flament, D., Shapiro, M.B., Pfann, K.D., et al. (2002). Reaction time is not impaired by stimulation of the ventral-intermediate nucleus of the thalamus (Vim) in patients with tremor. *Mov. Disord.*, 17:488–492.

Garcia, L., Audin, J., D'Alessandro, G., et al. (2003). Dual effect of high-frequency stimulation on subthalamic neuron activity. *J. Neurosci.*, 23:8743–8751.

Garcia, L., D'Alessandro, G., Fernagut, P.O., et al. (2005). The impact of high frequency stimulation parameters on the pattern of discharge of subthalamic neurons. *J. Neurophysiol.*, 94:3662–3669.

Haberler, C., Alesch, F., Mazal, P.R., et al. (2000). No tissue damage by chronic deep brain stimulation in Parkinson's disease. *Ann. Neurol.*, 48:372–376.

Halliday, D.M., Conway, B.A., Farmer, S.F., et al. (2000). Coherence between low-frequency activation of the motor cortex and tremor in patients with essential tremor. *Lancet*, 355:1149–1153.

Hariz, M.I., Shamsgovara, P., Johansson, F., et al. (1999). Tolerance and tremor rebound following long-term chronic thalamic stimulation for Parkinsonian and essential tremor. *Stereotact. Funct. Neurosurg.*, 72:208–218.

Hashimoto, T., Elder, C.M., Okun, M.S., et al. (2003). Stimulation of the subthalamic nucleus changes the firing pattern of pallidal neurons. *J. Neurosci.*, 23:1916–1923.

Haslinger, B., Boecker, H., Buchel, C., et al. (2003). Differential modulation of subcortical target and cortex during deep brain stimulation. *Neuroimage*, 18:517–524.

Hassler, R. (1959). Anatomy of the Thalamus. In *Introduction to Stereotaxis with an Atlas of the Human Brain*, Schaltenbrand, G. and Bailey, P., Eds. Stuttgart: Thieme, p. 230–290.

Hassler, R., Riechert, T., Mundinger, F., et al. (1960). Physiological observations in stereotaxic operations in extrapyramidal motor disturbances. *Brain*, 83:337–350.

Hellwig, B., Haussler, S., Lauk, M., et al. (2000). Tremor-correlated cortical activity detected by electroencephalography. *Clin. Neurophysiol.*, 111:806–809.

Hellwig, B., Haussler, S., Schelter, B., et al. (2001). Tremor-correlated cortical activity in essential tremor. *Lancet*, 357:519–523.

Hirai, T. and Jones, E.G. (1989). A new parcellation of the human thalamus on the basis of histochemical staining. *Brain Res. Rev.*, 14:1–34.

Holsheimer, J., Demeulemeester, H., Nuttin, B., and de Sutter, P. (2000). Identification of the target neuronal elements in electrical deep brain stimulation. *Eur. J. Neurosci.*, 12:4573–4577.

Iremonger, K.J., Anderson, T.R., Hu, B., and Kiss, Z.H.T. (2006). Cellular mechanisms preventing sustained activation of cortex during subcortical high frequency stimulation. *J. Neurophysiol.*, 96:613–621.

Jahnsen, H. and Llinas, R.R. (1984). Electrophysiological properties of guinea-pig thalamic neurones: an *in vitro* study. *J. Physiol.*, 349:205–226.

Kerrigan, J.F., Litt, B., Fisher, R.S., et al. (2004). Electrical stimulation of the anterior nucleus of the thalamus for the treatment of intractable epilepsy. *Epilepsia*, 45:346–354.

Kiss, Z.H.T., Mooney, D.M., Renaud, L., and Hu, B. (2002). Neuronal response to local electrical stimulation in rat thalamus: physiological implications for the mechanism of action of deep brain stimulation. *Neuroscience*, 113:137–143.

Kiss, Z.H.T., Anderson, T.R., Hansen, T., et al. (2003a). Neural substrates of microstimulation-evoked tingling: a chronaxie study in human somatosensory thalamus. *Eur. J. Neurosci.*, 18:728–732.

Kiss, Z.H.T., Kirstein, D.D., Suchowersky, O., and Hu, B. (2003b). Frequency dependent effects of deep brain stimulation: clinical manifestations and neural network modelling. *International Functional Electrical Stimulation Society. Proc. of the IFESS*.

Kiss, Z.H.T., Wilkinson, M., Krcek, J., et al. (2003c). Is the target for thalamic DBS the same as for thalamotomy? *Mov. Disord.*, 18:1169–1175.

Krack, P.P., Batir, A., Van Blercom, N., et al. (2003). Five-year follow-up of bilateral stimulation of the subthalamic nucleus in advanced Parkinson's disease. *N. Engl. J. Med.*, 349:1925–1934.

Kralic, J.E., Criswell, H.E., Osterman, J.L., et al. (2005). Genetic essential tremor in gamma-aminobutyric acidA receptor alpha1 subunit knockout mice. *J. Clin. Invest.*, 115:774–779.

Kumar, K., Toth, C., and Nath, R.K. (1997). Deep brain stimulation for intractable pain: a 15-year experience. *Neurosurgery*, 40:736–747.

Lee, K.H., Chang, S.Y., Roberts, D.W., and Kim, U. (2004). Neurotransmitter release from high-frequency stimulation of the subthalamic nucleus. *J. Neurosurg.*, 101:511–517.

Lee, K.H., Hitti, F.L., Shalinsky, M.H., et al. (2005). Abolition of spindle oscillations and 3 Hz absence-seizure-like activity in the thalamus by high frequency stimulation: potential mechanism of action. *J. Neurosurg.*, 103:538–545.

Lenz, F.A., Kwan, H.C., Martin, R.L., et al. (1994). Single unit analysis of the human ventral thalamic nuclear group: tremor-related activity in functionally identified cells. *Brain*, 117:531–543.

Lenz, F.A., Normand, S.L. Kwan, H.C., et al. (1995). Statistical prediction of the optimal site for thalamotomy in Parkinsonian tremor. *Mov. Disord.*, 10:318–328.

Leone, M., Franzini, A., and Bussone, G. (2001). Stereotactic stimulation of posterior hypothalamic gray matter in a patient with intractable cluster headache. *N. Engl. J. Med.*, 345:1428–1429.

Llinas, R.R. and Volkind, R.A. (1973). The olivo-cerebellar system: functional properties as revealed by harmaline-induced tremor. *Exp. Brain Res.*,18:69–87.

Logothetis, N.K., Pauls, J., Augath, M., et al. (2001). Neurophysiological investigation of the basis of the fMRI signal. *Nature*, 412:150–157.

Magarinos-Ascone, C.M., Pazo, J.H., Macadar, O., and Buno, W. (2002). High-frequency stimulation of the subthalamic nucleus silences subthalamic neurons: a possible cellular mechanism in Parkinson's disease. *Neuroscience*, 115:1109–1117.

Mathiesen, C., Caesar, K., Akgoren, N., and Lauritzen, M. (1998). Modification of activity-dependent increases of cerebral blood flow by excitatory synaptic activity and spikes in rat cerebellar cortex. *J. Physiol.*, 512:555–566.

Maurice, N., Thierry, A.M., Glowinski, J., and Deniau, J.M. (2003). Spontaneous and evoked activity of substantia nigra pars reticulata neurons during high-frequency stimulation of the subthalamic nucleus. *J. Neurosci.*, 23:9929–9936.

Mayberg, H.S., Lozano, A.M., Voon, V., et al. (2005). Deep brain stimulation for treatment-resistant depression. *Neuron*, 45:651–660.

McIntyre, C.C. and Grill W.M. (1999). Excitation of central nervous system neurons by nonuniform electric fields. *Biophys. J.*, 76:878–888.

McIntyre, C.C. and Grill, W.M. (2000). Selective microstimulation of central nervous system neurons. *Ann. Biomed. Eng.*, 28:219–233.

McIntyre, C.C. and Grill, W.M. (2002). Extracellular stimulation of central neurons: Influence of stimulus waveform and frequency on neuronal output. *J. Neurophysiol.*, 88:1592–1604.

McIntyre, C.C., Grill, W.M., Sherman, D.L., and Thakor, N.V. (2004). Cellular effects of deep brain stimulation: model-based analysis of activation and inhibition. *J. Neurophysiol.*, 91:1457–1469.

Meissner, W., Leblois, A., Hansel, D., et al. (2005). Subthalamic high frequency stimulation resets subthalamic firing and reduces abnormal oscillations. *Brain*, 128:2372–2382.

Miocinovic, S. and Grill, W.M. (2004). Sensitivity of temporal excitation properties to the neuronal element activated by extracellular stimulation. *J. Neurosci. Methods*, 132:91–99.

Molnar, G.F., Sailer, A., Gunraj, C.A., et al. (2004). Thalamic deep brain stimulation activates the cerebellothalamocortical pathway. *Neurology*, 63:907–909.

Molnar, G.F., Sailer, A., Gunraj, C.A., et al. (2005). Changes in cortical excitability with thalamic deep brain stimulation. *Neurology,* 64:1913–1919.

Mulligan, S.J. and MacVicar, A. (2004). Calcium transients in astrocyte endfeet cause cerebrovascular constrictions. *Nature,* 431:195–199.

Nowak, L.G. and Bullier, J. (1996). Spread of stimulating current in the cortical grey matter of rat visual cortex studied on a new *in vitro* slice preparation. *J. Neurosci. Methods,* 67:237–248.

Nowak, L.G. and Bullier, J. (1998). Axons, but not cell bodies, are activated by electrical stimulation in cortical gray matter. I. Evidence from chronaxie measurements. *Exp. Brain Res.,* 118:477–488.

Nuttin, B., Gabriels, L.A., Cosyns, P.R., et al. (2003). Long-term electrical capsular stimulation in patients with obsessive-compulsive disorder. *Neurosurgery,* 52:1263–1274.

Pahapill, P.A., Levy, R., Dostrovsky, J.O., et al. (1999). Tremor arrest with thalamic microinjections of muscimol in patients with essential tremor. *Ann. Neurol.,* 46:249–252.

Papavassiliou, E., Rau, G., Heath, S., et al. (2004). Thalamic deep brain stimulation for essential tremor: relation of lead location to outcome. *Neurosurgery,* 54:1120–1130.

Parker, F., Tzourio, N., Blond, S., et al. (1992). Evidence for a common network of brain structures involved in Parkinsonian tremor and voluntary repetitive movement. *Brain Res.,* 584:11–17.

Perlmutter, J.S., Mink, J.W., Bastian, A.J., et al. (2002). Blood flow responses to deep brain stimulation of thalamus. *Neurology,* 58:1388–1394.

Ranck, J.B. (1975). Which elements are excited in electrical stimulation of mammalian central nervous system: a review. *Brain Res.,* 98:417–440.

Rezai, A.R., Lozano, A.M., Crawley, A.P., et al. (1999). Thalamic stimulation and functional magnetic resonance imaging: localization of cortical and subcortical activation with implanted electrodes. Technical note. *J. Neurosurg.,* 90:583–590.

Rosenmund, C. and Stevens, C.F. (1996). Definition of the readily releasable pool of vesicles at hippocampal synapses. *Neuron,* 16:1197–1207.

Rubin, J.E. and Terman, D. (2004). High frequency stimulation of the subthalamic nucleus eliminates pathological thalamic rhythmicity in a computational model. *J. Comput. Neurosci.,* 16:211–235.

Rubinstein, J.T. (1993). Axon termination conditions for electrical stimulation. *IEEE Trans. Biomed. Eng.,* 40:654–663.

Schoenen, J., Di Clemente, L., Vandenheede, M., et al. (2005). Hypothalamic stimulation in chronic cluster headache: a pilot study of efficacy and mode of action. *Brain,* 128:940–947.

Schuurman, P.R., Bosch, D.A., Bossuyt, P.M.M., et al. (2000). A comparison of continuous thalamic stimulation and thalamotomy for suppression of severe tremor. *N. Engl. J. Med.,* 342:461–468.

Steriade, M., Jones, E.G., and McCormick, D.A. (1997). *Thalamus.* Amsterdam: Elsevier.

Stevens, C.F. and Tsujimoto, T. (1995). Estimates for the pool size of releasable quanta at a single central synapse and for the time required to refill the pool. *Proc. Natl. Acad. Sci., U.S.A.,* 92:846–849.

Stevens, C.F. and Wesseling, J.F. (1998). Activity-dependent modulation of the rate at which synaptic vesicles become available to undergo exocytosis. *Neuron,* 21:415–424.

Strafella, A.P., Ashby, P., Munz, M., et al. (1997). Inhibition of voluntary activity by thalamic stimulation in man: relevance for the control of tremor. *Mov. Disord.,* 12:727–737.

Tasker, R.R. and Kiss, Z.H.T. (1995). The role of the thalamus in functional neurosurgery. *Neurosurg. Clin. N. Am.,* 6:73–104.

Tasker, R.R., Munz, M., Junn, F.S.C.K., et al. (1997). Deep brain stimulation and thalamotomy for tremor compared. *Acta Neurochir. Suppl.,* 68:49–53.

Theodore, W.H. and Fisher, R.S. (2004). Brain stimulation for epilepsy. *Lancet Neurol.,* 3:111–118.

Urbano, F.J., Leznik, E., and Llinas, R.R. (2002). Cortical activation patterns evoked by afferent axons stimuli at different frequencies: an *in vitro* voltage-sensitive dye imaging study. *Thalamus Related Syst.,* 1:371–378.

Ushe, M., Mink, J.W., Revilla, F.J., et al. (2004). Effect of stimulation frequency on tremor suppression in essential tremor. *Mov. Disord.,* 19:1163–1168.

Vidailhet, M., Vercueil, L., Houeto, J.L., et al. (2005). Bilateral deep-brain stimulation of the globus pallidus in primary generalized dystonia. *N. Engl. J. Med.*, 352:459–467.

Volkmann, J., Joliot, M., Mogilner, A., et al. (1996). Central motor loop oscillations in Parkinsonian resting tremor revealed by magnetoencephalography. *Neurology,* 46:1359–1370.

Wielepp, J.P., Burgunder, J.M., Pohle, T., Ritter, E.P., Kinser, J.A., and Krauss, J.K. (2001). Deactivation of thalamocortical activity is responsible for suppression of parkinsonian tremor by thalamic stimulation: a 99mTc-ECD SPECT study. *Clin. Neurol. Neurosurg.*, 103:228–231.

Williams, M.N. and Faull, R.L. (1987). The distribution and morphology of identified thalamocortical projection neurons and glial cells with reference to the question of interneurons in the ventrolateral nucleus of the rat thalamus. *Neuroscience,* 21:767–780.

Wills, A.J., Jenkins, I.H., Thompson, P.D., et al. (1994). Red nuclear and cerebellar but no olivary activation associated with essential tremor: a positron emission tomographic study. *Ann. Neurol.*, 36:636–642.

Windels, F., Bruet, N., Poupard, A., et al. (2003). Influence of the frequency parameter on extracellular glutamate and gamma-aminobutyric acid in substantia nigra and globus pallidus during electrical stimulation of subthalamic nucleus in rats. *J. Neurosci. Res.*, 72:259–267.

Windels, F., Bruet, N., Poupard, A., et al. (2000). Effects of high frequency stimulation of subthalamic nucleus on extracellular glutamate and GABA in substantia nigra and globus pallidus in the normal rat. *Eur. J. Neurosci.*, 12:4141–4146.

3

Neuromodulation: Current Trends in Interfering with Epileptic Seizures

Ana Luisa Velasco,
Francisco Velasco,
Marcos Velasco,
Bernardo Boleaga,
Mauricio Kuri,
Fiacro Jiménez, and
José María Núñez

3.1   Introduction

Epilepsy is a medical condition that is very frequent around the world. It is estimated that 1% of the population suffers epilepsy. From these patients, only 70% are controlled with antiepileptic medication. The remaining 30% of patients may benefit from surgical intervention. The use of chronic stimulation of the brain, so-called neuromodulation, has been shown to be a reliable procedure in the control of epileptic seizures. In 1970, the first totally implantable stimulating systems became available (Rise, 2000). Based on the work of Cooke and Snider (1955), Cooper et al. (1978) used cerebellar stimulation to control different varieties of epileptic seizures.

Our group has worked on three deep brain stimulation procedures according to the type of epileptic seizures: (1) stimulation of the centromedian nucleus of the thalamus in the control of intractable generalized seizures and atypical absences of the Lennox–Gastaut syndrome (Velasco et al., 1987, 1989, 1993a, 1993b, 1995, 2000a, 2000b, 2002); (2) stimulation of the hippocampus for the control of mesial temporal lobe seizures (Velasco et al., 2000d, 2001b); and (3) stimulation of the supplementary motor area for motor seizures initiated in this zone (unpublished data).

3.2   Electrical Stimulation of the Centromedian Thalamic Nuclei (ESCM)

Although ESCM has been used in cases of difficult-to-control seizures with multifocal onset in the frontal and temporal lobes, as well as in cases of seizures with no evidence of focal onset such as Lennox–Gastaut syndrome, it has proven to have its best result in the latter.

The role of midline and intralaminar thalamic nuclei in the genesis and propagation of epileptic attacks was proposed long ago on the basis of clinical observations (Penfield and Jasper, 1954). Although the controversy surrounding the anatomical initiation of the epileptic attacks remains, there appears to be agreement that the thalamocortical interactions are essential in the development of most of them (Pollen et al., 1963; Gloor et al., 1977; Quesney et al., 1977; Avoli et al., 1983; Steriade, 1990; Velasco et al., 1991).

The decision to stimulate the CM (centromedian thalamus) was based on the idea to interfere with the thalamocortical interactions and thus stop either the genesis or propagation of the seizures. In 1984 our group performed the first ESCM (electrical stimulation of the centromedian nucleus of the median) trial on a 12 year old male who had severe atypical absences refractory to high levels of antiepileptic medication. The CM was chosen as a target because of its relatively large size and close relationship to the conventional stereotaxic landmarks and because the CM is an intralaminar nucleus that forms part of the nonspecific reticular-thalamo-cortical system that transmits and integrates the cerebral inputs of the generalized seizures (Jasper and Droogleever-Fortuyn, 1947; Hunter and Jasper, 1949; Gloor et al., 1977). Since then, several types of seizures were included in the protocol of ESCM (Velasco et al., 1993a). The best results were obtained in patients with generalized seizures of the Lennox-Gastaut syndrome (associated with 2-Hz spike wave complexes and mental deterioration). Other forms of seizures, particularly those with focal origins (as in temporal lobe seizures), are not significantly improved by ESCM to a lesser extent (Velasco et al., 2000b). However, secondary tonic clonic generalization is relieved, suggesting that ESCM interferes with the propagation mechanisms of focally initiated epileptic activity.

The ESCM procedure has been refined progressively by increasing the number of treated patients and lengthening the follow-up period. Our group also has better defined a number of predictor factors that must be taken into account to achieve a good outcome:

1. Selection of responding patients
2. Verifying correct DBS implantation based on definition of the stereotaxic coordinates of the CM optimal effective areas and neurophysiological characterization of the targeted area
3. Performing periodic monitoring of the reliability of ESCM on a long-term follow-up (Velasco et al., 2000b; Velasco et al., 2001a).

3.2.1   Patient Selection

In this review we analyze the last thirteen patients with Lennox–Gastaut syndrome (LGS) selected from patients of the Epilepsy Surgery Clinic of the General Hospital of Mexico on the basis of having generalized difficult-to-control seizures of the LGS type. They underwent ESCM with the idea to correlate seizure type, stereotaxic targeting, and neurophysiologic responses with the final outcome of the patient. The Lennox–Gastaut syndrome is one of the severest forms of childhood epilepsy. It is characterized by drug-resistant generalized seizures, the tonic and atonic seizures, atypical absences, myoclonic attacks, and episodes of nonconvulsive and tonic *status epilepticus* being most characteristic. The peak onset is known to be between 1 and 7 years of age. It is usually preceded by other types of seizure disorders, especially infantile spasms. LGS is accompanied by severe mental deterioration as it progresses. From the electroencephalograph (EEG) standpoint, the diagnosis is based on the presence of slow spike-wave complexes (<2.5 cps) and bursts of rapid (10 Hz) rhythms during slow sleep. The overall prognosis is very severe; 90% of the patients are mentally retarded and 80% continue to have seizures through adulthood (Aicardi, 1994). The selected patients had either secondary LGS with stable or nonprogressive diseases (birth trauma, postencephalitic sequelae, cortical dysplasia, and stable tuberous sclerosis) or primary LGS with no demonstrable lesion in the MRI (magnetic resonance image).

3.2.2   Correct Targeting

To have the optimal results with ESCM, it is extremely important to have good patient selection combined with adequate target localization. The latter should be obtained from two points of view: (1) stereotaxic (defined with ventriculography and MRI) and (2) neurophysiologic definition.

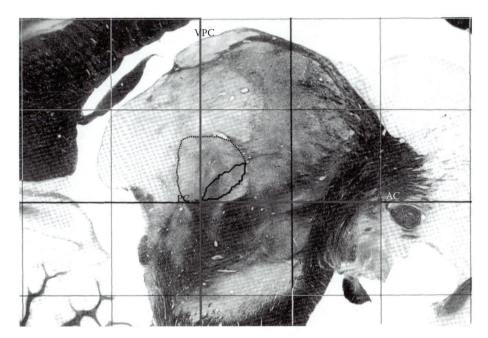

FIGURE 3.1 Optimal targets for seizure arrest located in the basolateral portion of the CM.

3.2.2.1 Ventriculographic Definition of CM Target

Surgical technique: under general anesthesia, electrodes (Medtronic Model 3387 DBS lead Medtronic, Inc., Minneapolis, MN) are stereotactically placed in both left and right CM nuclei through a coronal incision and bifrontal burr holes made at a distance of 10 to 15 mm at each side of the midline at the level of the coronal suture. The CM localization is accomplished by air ventriculography. This method allows us to demonstrate the anterior (AC) and posterior (PC) commissures of the third ventricle with remarkable precision. Two lines are drawn: the AC–PC line and the vertical line perpendicular to the PC (VPC). The target point for the electrode tip was a distance 10 mm from the midline and the intersection of the AC-PC line with the VPC (Velasco et al., 1989, 2000b).

Electrodes are fixed to burr holes using a plastic ring and silastic ring caps (Medtronic). The position of the contacts along the trajectory of the electrode is plotted on the sagittal and frontal sections of the Schaltenbrand and Bailey atlas (1959) according to the standardization technique described elsewhere (Velasco et al., 1975).

The optimal targets for seizure arrest are located in the basolateral portion of the CM which corresponds to the parvocellular portion; this is an area of maximal neuronal population (Mehler, 1996); see Figure 3.1. The best antiepileptic results are located as follows: LAT = 10.0 ± 2.0 mm from the midline, H = from 2.0 to 7.0 mm above the AC–PC line, and AP = from 3.0 to 5.0 mm in front or PC-VPC intersection. Because the CM is a large nucleus with several subdivisions, it is important to point out that the electrode contacts used for stimulation in patients with good outcome are positioned in the ventrolateral or parvocellular part of the CM considered by anatomists as the core of the CM nucleus of Luys, which is surrounded by denser fiber connections ascending from the brainstem to terminate in other nuclei of the thalamus (Mehler, 1996).

3.2.2.2 MRI Confirmation of the CM Target

Electrodes are left externalized to confirm their position by MRI. Scans are performed using 1.5 T Edge equipment software version 9.3 (Marconi Medical Systems, Cleveland, OH), using T2 weighted fast spin echo sequence (echo time 11 ms; repetition time 4070 ms; field of view 16.0 cm; 256 × 256 matrix). Sections are oriented parallel and perpendicular to the AC–PC line for axial and frontal views, and parallel to the midsagittal plane for sagittal sections (Velasco et al., 2000b, 2002).

3.2.2.3 Neurophysiologic Confirmation of the CM Target

Stimuli are delivered by a Grass S8 stimulator and isolation unit attached to the patient by means of a Tektronix CRU and a comparative $10k\Omega$ resistor to monitor the voltage (V), current flow (μA), and impedance ($k\Omega$) of the stimulated contacts within the brain tissue (Velasco et al., 1993a).

The electrical incremental and desynchronizing responses are elicited by unilateral electrical stimulation through adjacent electrode contacts (where the cathode was always the lower contact). Stimuli consist of 5- to 30-s trains of monophasic square pulses of 1.0 ms duration and 6 to 60 Hz frequency. Analysis of the scalp distribution of the incremental spike-wave and desynchronizing electro-cortical responses is made from EEG recordings taken from fronto-polar (FP2, FP1), frontal (F4, F3), central (C4, C3), parietal (P4, P3), occipital (O2, O1), fronto-temporal (F8, F7), and anterior temporal (T4, T3) referred to ipsilateral ears (A2, A1) (Sensitivity = 10 μA/cm; time constant = 0.35 s; paper speed = 15 mm/s).

Low-frequency (6/s), threshold (4–5V = 320–400 μA) unilateral stimulation of CM elicits incremental responses with the typical waxing and waning profile. The incremental responses produced by this stimulation procedure along the CM or other structures are described elsewhere (Velasco et al., 1996). Although there are three types of incremental responses that can be elicited, the Type A one points to the best place to obtain the optimum antiepileptic effect. They are recruiting-like responses elicited by the stimulation of the caudal-basal portions of CM (parvocellular CM close to the nonspecific mesencephalic structures, such as the mesencephalic *tegmentum* tract). They consist of monophasic negative potentials with a latency of 20 ms and a peak latency of 30 to 35 ms (Figure 3.2A). Suprathreshold Type A responses show a bilateral regional scalp distribution with maximal amplitude at the frontal region ipsilateral to the stimulated side (Figure 3.2B) (Velasco et al., 1996).

Unilateral high frequency (60/s) threshold and suprathreshold stimulation of the caudal-basal and central CM elicits a regional EEG desynchronization consisting of an increased frequency of the EEG activity superimposed on a slow negative shift. It also shows a bilateral regional scalp distribution with maximal amplitude at the frontal region ipsilateral to the stimulated side (Figure 3.2C) (Velasco et al., 1996).

3.2.3 Periodic Monitoring of the Reliability of ESCM on a Long-Term Follow-Up

Incremental responses, EEG desynchronization, and slow negative shifts may be useful as biological responses for monitoring the efficiency of ESCM, particularly when the stimulating electrodes are internalized subcutaneously and the physical characteristics of the electrical stimuli cannot be monitored any longer (Velasco et al., 1995). In view of the patient's lack of subjective sensations and the long latency of the antiepileptic effects of ESCM, the reliability of ESCM is questionable. The use of 10 or 60/s and 6 V (800 μA) transcutaneous activation of CM with the Medtronic internalized pulse generator (IPG) by Medtronic, Inc. (Minneapolis) and the recording of scalp incremental responses EEG desynchronization and slow negative shifts is advisable.

3.2.4 Chronic Stimulation Parameters

The parameters recommended for ESCM are as follows: 2 h of daily stimulation sessions. Stimulus consists of 1-min trains of Lilly pulses with an interstimulus interval of 4 min, alternating right and left CM. Such trains consist of 130 Hz frequency, with individual pulses of 450 μs in duration and an amplitude of 400 to 600 μA.

3.2.5 Results

ESCM produces a significant reduction in the number of primary and secondary generalized tonic–clonic seizures and atypical absences of the Lennox–Gastaut syndrome, and also the number of interictal generalized slow spike-wave complexes.

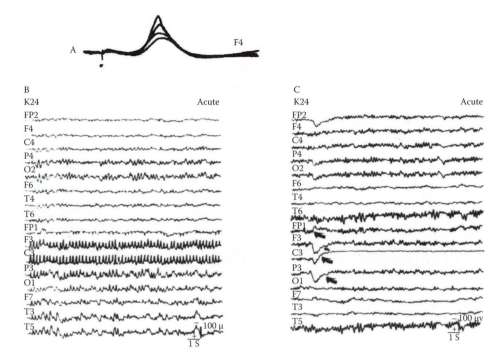

FIGURE 3.2 Neurophysiologic confirmation of the CM target. (A):Type A recruiting-like responses elicited by the stimulation of the caudal-basal portions of CM. They consist of monophasic negative potentials with a latency of 20 ms and peak latency of 30 to 35 ms. (B):Low-frequency (6/s) stimulation elicits suprathreshold Type A responses showing a bilateral regional scalp distribution with maximal amplitude at the frontal region ipsilateral to the stimulated side. (C):Unilateral high-frequency (60/s) threshold and suprathreshold stimulation of the caudal-basal and central CM elicits a regional EEG desynchronization consisting of an increased frequency of the EEG activity superimposed on a slow negative shift with a bilateral regional scalp distribution with maximal amplitude at the frontal region ipsilateral to the stimulated side.

As discussed above, the results depend on both seizure type and correct target selection. Note that the patients with excellent result (i.e., 100% improvement) had generalized seizures and correct stereotaxic placement and electrophysiological responses. All patients who had less than 80% seizure improvement had either another seizure type (e.g., LC, who had residual complex partial seizures) and/or incorrect target selection according to either anatomic (stereotaxic) or physiologic parameters (Table 3.1).

3.3 Electrical Stimulation of the Hippocampus (ESHC)

In the Epilepsy Surgery Clinic of the General Hospital of Mexico, 70% of patients referred for surgery have complex partial seizures arising from the hippocampal formation; other series have reported a similar referral number (Wieser et al., 1993; Williamson, 1993). Resective surgery of the epileptic focus yields very good results (Engel, 1987; Velasco et al., 2000c); nevertheless, there are cases that escape this surgical possibility, that is, patients with bilateral hippocampal foci or patients with epileptic focus located nearby eloquent areas for speech and memory (usually the left side). These latter patients cannot be operated on because it would mean having severe neurological impairment, particularly related to short-term memory. These patients are candidates for neuromodulating procedures. Unfortunately, our experience with ESCM pointed out that CM is not the place to stimulate patients with complex partial seizures originating in the temporal lobe. On the other hand, animal experiments showed that the application of an electrical stimulus to the amygdala or hippocampus following the kindling stimulus produces a significant and long-lasting suppressive effect on seizures (Weiss et al., 1995). For these

TABLE 3.1   The Predictors of Seizure Relief Obtained after ESCM

| | Seizure Type | | | Stereotactic Placement | | Electro Physiological | | Final Improvement |
Initials	GTC	AA	CXP	RCM	LCM	RCM	LCM	%
GA	Y	Y	N	C	C	C	C	100
MAM	Y	Y	N	C	C	C	C	100
AMP	Y	Y	N	C	I	C	C	95
MS	Y	Y	N	C	C	C	C	95
MAPR	Y	Y	N	C	I	C	C	95
JM	Y	Y	N	C	C	C	C	91
JS	Y	Y	N	C	C	C	C	89
IM	Y	Y	N	C	C	C	C	87
LC	Y	Y	Y	C	C	I	I	80
EGV	Y	Y	N	C	C	C	I	70
DC	N	Y	N	C	I	C	C	58
JR	Y	Y	N	C	I	I	I	53
LVAP	N	Y	N	C	C	I	I	30

Note: All patients had generalized tonic clonic seizures (GTC), atypical absences (AA), and only LC also had complex partial seizures (CXP). Correct (C) and Incorrect (I) stereotaxic and electrophysiological parameters are shown. Note that all patients with a seizure improvement 100% had generalized seizures and correct target localization parameters.

Source: From Velasco et al., submitted for publication.

reasons we decided to perform a preliminary study in ten patients with nonlesional temporal lobe epilepsy in whom intracranial electrodes were implanted (either subdural basotemporal grid or hippocampal electrodes) for the detection of the epileptic foci (Velasco et al., 2000c). The study consisted of performing subacute hippocampal stimulation (SAHCS) trial of two to three weeks' duration once the epileptic focus was located and before performing the temporal lobectomy. Two patients had bilateral hippocampal depth electrodes implanted to determine the lateralization of the epileptic focus and eight had unilateral subdural electrode grids on the pial surface of the basotemporal cortex to determine the precise site and extent of the focus (Figure 3.3). In all patients, antiepileptic drugs were discontinued for 72 h prior to SAHCS initiation. Thereafter, SAHCS was applied for a minimum of sixteen days. SAHCS was bipolar and between continuous electrode contacts (cathode attached to the most anterior contact) and consisted of continuous stimulation with biphasic Lilly pulses, 130 Hz frequency, 450 μs in duration, and amplitude of 200 to 400 μA). The reliability of SAHCS was determined by daily measurement of voltage, impedance, and current flow at the intracerebral contacts by means of externalized electrode systems (Velasco et al., 1993a).

To evaluate the antiepileptic effects of SAHCS on temporal lobe epileptogenesis, the number and type of clinical seizures per day and the number of interictal negative EEG spikes at the epileptic focus per 10 s were evaluated. Upon completion of the clinical and EEG studies, SAHCS was discontinued, the electrodes were removed, and an anterior temporal lobectomy was performed ipsilateral to the epileptic focus and SAHCS. Biopsies of mesial and lateral temporal lobe were fixed in 10% formaldehyde buffer solution, embedded in paraffin, and cut in serial coronal sections of 10 μm thickness and perpendicular to the fascia dentata, taken every 1,000 μm and stained with hematoxilin-eosin for perikarion and with Gomori´s technique used for the collagen. Histopathological analysis of the temporal lobe tissue was performed under a light microscope by comparing the contiguous hippocampal tissue at the stimulated vs. nonstimulated tissues.

In seven patients in whom stimulation sites were located within the hippocampal formation and gyrus, there was an evident antiepileptic response. Both complex partial and secondary generalized tonic–clonic seizures were abolished after six days of continuous stimulation and interictal EEG spikes were significantly reduced from days 4 to 11 of SAHCS. The most evident and fastest antiepileptic responses were found in five patients in whom the stimulation contacts were located at either the anterior pes-hippocampus near

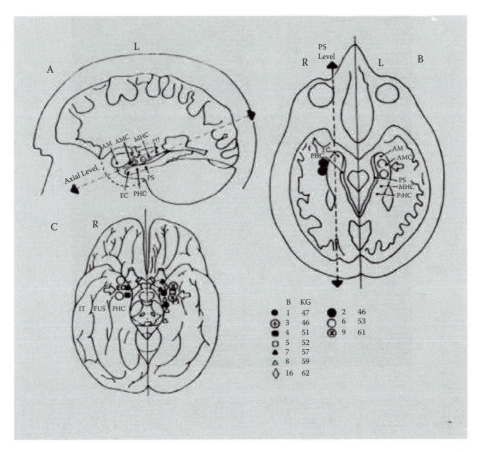

FIGURE 3.3 Diagrams showing the position of the depth and subdural electrode contacts for SAHCS. A and B show parasagittal and axial sections, and C shows basotemporal cortex, all of them showing the position of the stimulation contacts in different patients (indicated by different symbols at the right bottom corner). Arrows indicate sites where SAHCS produced evident and fast antiepileptic responses. Abbreviations: AHC, MHC, and PoHC: anterior, middle, and posterior hippocampus, respectively. AM: amygdala, PS: presubiculum, PHC, EC, FUS, and IT: parahippocampal, entorhinal, fusiform, and inferior temporal gyri, respectively. (*Source:* From Velasco, 2000d.)

the amygdala or at the anterior parahippocampal gyrus near the entorhinal cortex. Three patients did not respond, two of them when SAHCS was accidentally interrupted and the other patient the stimulated contacts were located at the white matter lateral to the hippocampus.

Histopathological analysis of the hippocampal tissues revealed abnormalities due to depth of electrode penetration or lesion due the foreign body electrodes. However, no histopathological differences were found between stimulated and nonstimulated hippocampal tissues. Therefore, the SAHCS effect does not appear to depend on a lesional process but rather to a functional blockage of the hippocampus.

During this preliminary study, we took advantage of this ethically permissible situation and studied some basic mechanisms underlying the beneficial therapeutic effect on seizures due to hippocampal stimulation (Velasco et al., 2000d, 2000e, 2001a, 2001b). Such studies suggest that the antiepileptic effect of hippocampal stimulation is due to an inhibition mechanism, that is, increased threshold and decreased duration of the afterdischarges induced by acute hippocampal stimulation; depression of the paired pulse hippocampal recovery cycles, SPECT hypoperfusion, and autoradiographic increase of the benzo-diazepine receptor binding in the stimulated hippocampal tissue (Cuéllar et al., 1994).

Based on these results, we decided to proceed with a long-term hippocampal stimulation protocol to demonstrate that chronic hippocampal stimulation (CHCS) may produce a sustained antiepileptic effect without undesirable effects on language and memory. Up to now we have stimulated eight patients on

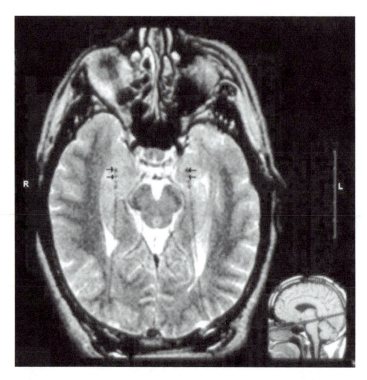

FIGURE 3.4 MRI axial view of patient KG67, showing the placement of bilateral DBS electrodes and the position of the four contacts along the hippocampus. The arrows indicate the contacts used for ESHC. (*Source:* From Velasco et al., 2003.)

a long-term basis, but we will only take into consideration those six patients who have a follow-up period of at least one year (1 to 4 years).

3.3.1 Study Design

Candidates were selected from patients of the Epilepsy Surgery Clinic of the General Hospital of Mexico on the basis of having difficult-to-control temporal lobe seizures. All of them underwent a careful clinical history with special emphasis on the seizure type (complex partial seizures), adequate antiepileptic medication with adequate blood levels, four serial EEGs, magnetic resonance, SPECT, neuropsychological exam, and psychiatric evaluation. All of them had bilateral hippocampal transitory 8 contact depth electrodes (SD 8P, Ad Tech Medical Instrument Co., Racine, WI) implanted to be able to assess the epileptic foci. Two of them had bilateral independent hippocampal foci and the other four patients were selected on the basis of having their hippocampal focus on the dominant hemisphere, with neuropsychological evidence of verbal memory situated here.

   Once the precise epileptic focus was defined, the transitory electrodes were replaced by four contact depth brain stimulation electrodes (3789 DBS and IPG by Medtronic, Inc., Minneapolis, MN) (Figure 3.4) and connected to an independent internalized pulse generator system that was placed in a subcutaneous subclavicular pocket on each side. The target of the electrode contacts was the site of maximal interictal and ictal activities. All antiepileptic drugs were withdrawn to avoid any possible interference with the neuromodulation procedure (Velasco et al., 2000e) and were replaced with phenytoin.

3.3.2 Chronic Stimulation Parameters

The parameters recommended for ESHC are as follows: daily stimulation sessions with 1-min trains of Lilly pulses with an interstimulus interval of 4 min. Such trains consist of a 130 Hz frequency,

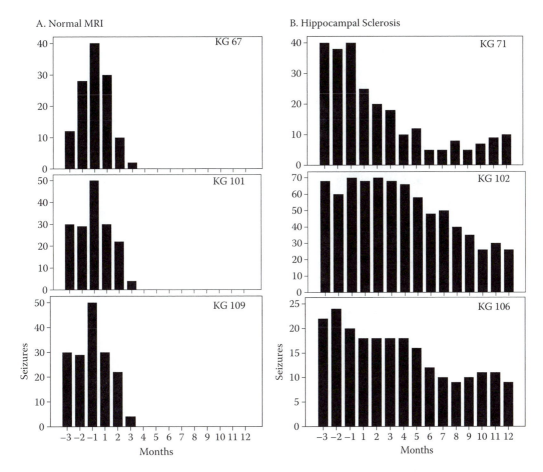

FIGURE 3.5 Seizure reduction per month in six patients with ESHC. Graphs show 3-month baseline period, stimulation started at the fourth month (arrow) after that, 12 month stimulation follow-up. (A): Three patients with normal MRI. (B): Three patients with left hippocampal sclerosis.

with individual pulses of 450 μs in duration and amplitude of 400 to 600 μA. In those patients with bilateral electrodes, the stimulation has the same characteristics but the stimulation alternates right and left hippocampus.

3.3.3 Results

In analyzing the antiepileptic effect of ESHC, the six patients were divided into two distinct groups: (1) three patients had normal MRIs and (2) the other three patients had ipsilateral hippocampal sclerosis. The best results were obtained in the first group with normal MRIs. The antiepileptic effect was evident since the stimulation started, and the patients were seizure-free after the first three to six months of stimulation. Our longest follow-up period is on patient KG67, who had bilateral hippocampal stimulation and has now been seizure-free for four years. Patient KG101 is under left hippocampal stimulation and has been seizure-free for fifteen months; and the most recent patient KG109 also has left hippocampal stimulation and has been seizure-free for a year. The neuropsychological tests in all of them became normal after six months of stimulation (Figure 3.5A).

The patients that constitute the second group have unilateral left mesial temporal sclerosis. Patient KG71 had bilateral epileptic foci. Although he had a 75% seizure reduction, he persisted with auras during the ten-month follow-up. Unfortunately, he had to undergo explantation because of skin erosion. Patient KG102 had 60% seizure improvement, which could be observed very slowly and took nine months to reach its actual seizure number (follow-up twenty months). KG106 has only had 50% seizure improvement after one year stimulation (Figure 3.5B). In all patients in this group, there has only been a slight improvement in the neuropsychological tests. The three patients in this group improved only slightly in their memory tests after more than six months of stimulation.

Although these are preliminary results, we can say that stimulating the hippocampal epileptic focus is effective in the control of mesial temporal lobe seizures. The best cases are those which have normal MRIs, in whom seizure control can be achieved up to 100%. This result is of extreme importance because these patients with intractable seizures and normal MRIs are the ones who are excluded as candidates for temporal lobectomy and are left with no other alternative. Patients with mesial temporal lobe sclerosis do not do so well and their follow-up shows only 50 to 75% seizure reduction. If the patient has temporal sclerosis and the seizures are starting within the sclerotic hippocampus, lobectomy could be risked. Patients with bilateral hippocampal foci and unilateral sclerosis, however, should be considered for neuromodulation because they could have better results, and the risk of having residual seizures after unilateral lobectomy is high.

3.4   Neuromodulation of the Supplementary Motor Area (SMA)

The supplementary motor area (SMA) plays a crucial role in movement organization, basically in the sequential timing and planning of motor tasks (Morris, 1993). It is located in the mesial surface of the superior frontal gyrus.

In contrast with temporal lobectomy, which yields great benefit for patients with complex partial seizures originating in the hippocampus, ablation of the SMA renders limited results and is associated with a high incidence of neurological deficit, impacting the patient's quality of life (Mihara et al., 1996; Olivier, 1996; Smith and King, 1996). These results have limited surgery in patients who show no evolving lesions or with apparently normal MRI.

For this reason and the good results we obtained stimulating the hippocampal foci, we started a new project for neuromodulation of the SMA. We present here the first case.

A 17-year-old male was studied. He started seizing at 14 years old. The seizures were characteristic supplementary motor ones, that is, brief, with abrupt posturing of left arm and sudden version of the head to the left, with preserved consciousness and occasional secondary tonic–clonic seizures. He had up to ten seizures a day, 80% of them being released by sound. Conduct abnormalities with perseverance and verbal aggressiveness were present. The EEG showed frontal parasagittal epileptic activity, while the MRI was normal. Bilateral 20 contact grids were implanted in right and left SMA (Figure 3.6A). AEDs (anti-epileptic drugs) were tapered and daily depth recording was performed for epileptic focus location. Ictal EEG activity showed a mesial focus located in the right SMA. Patient reinitiated AEDs. Grids were explanted and replaced by a four-contact electrode for chronic stimulation (Resume, Medtronic, Inc.) connected to a DBS system (Figure 3.6B). Stimulation was started with the following parameters: bipolar continuous stimulation of 130 Hz, 3.0 V (350 μA).

Follow-up showed an immediate decrease in seizure occurrence, and this effect has been maintained so for nine months now (Figure 3.7). QOL Scale has improved also (Figure 3.8). No adverse effects have been reported up to now.

3.5   Conclusion

Neuromodulation constitutes an innovative neurosurgical technique in the treatment of difficult-to-control seizures. Stimulation should be targeted according to the seizure type: ESCM for generalized seizures and ESHC for mesial temporal lobe epilepsy. Neuromodulation is a method that has several advantages: it is nonlesional, with reversible effect when turned off, does not interfere with the functioning eloquent areas, and even improves neuropsychological performance. Unfortunately, it has some disadvantages: neurostimulating systems are expensive; it needs periodic follow-up visits to verify that

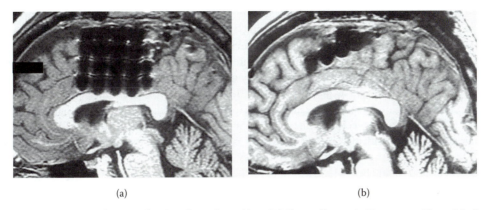

(a)   (b)

FIGURE 3.6 MRI sagittal images showing electrode position. (A) Shows diagnostic 20 contact grid on right hemisphere; (B) shows definite four-contact electrode position. Contacts 2 and 3 where epileptic focus was located are currently being stimulated.

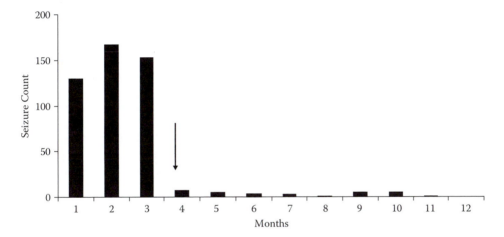

FIGURE 3.7 Graph shows seizure occurrence; months 1, 2, and 3 show baseline period. Arrow indicates SMA chronic stimulation onset. Note the important seizure reduction that occurs during neuromodulation.

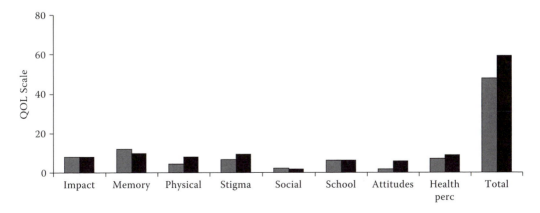

FIGURE 3.8 QOL scale shows in gray bars the patient's baseline and black bars show the results after 9 months of stimulation. Note the increase after chronic neuromodulation.

it is functioning adequately; the stimulators must be exchanged when the batteries wear down, which implies a surgical procedure. Probably the most important disadvantage is the skin erosion due to the stimulating system that often leads to explantation mainly in young children.

Challenges remain. For the epileptologists, there are a number of variables that influence the antiepileptic effect of neurostimulation that we are probably missing, and there are also other seizure types that are subject to be studied to know the antiepileptic effects of neuromodulation (i.e., occipital or frontal epilepsy). For the engineers, there is the need to improve neuromodulation systems so that they are less invasive (so they are not rejected by the patients and can be implanted in small children), include rechargeable batteries, and maybe even be remote controlled so that the use of extension cables is discontinued. The systems must be low cost so they can be accessed by a larger number of epileptic patients. They also should be more reliable and include user-friendly software so the patient can check it himself to avoid so many follow-up visits to guarantee reliable functioning.

References

Aicardi, J., Lennox–Gastaut Syndrome. In *Epilepsy in Children*, 2nd ed. New York: Raven Press, 1994, p. 44–66

Avoli, M., Gloor, P., Kostopoulus, G., and Gutman, J., An analysis of penicillin-induced generalized spike and wave discharges using simultaneous recordings of cortical and thalamic single neurons. *J. Neurophysiol.*, 1983, 50:819–837.

Cooper, I.S., Rickland, M., Amin, I., and Cullinan, T., A long term follow up study of cerebellar stimulation for the control of epilepsy. In Cooper, I., Ed., *Cerebellar Stimulation in Man.* New York: Raven Press, 1978, p. 19–27.

Cooke R.M. and Snider, R.S., Some cerebellar influences on electrically induced cerebral seizures. *Epilepsia*, 1955, p. 4–19.

Cuéllar-Herrera, M., Velasco, M., Velasco, F., Velasco, A.L., Jiménez, F., Orozco, S., Briones, M., and Rocha, L., Evaluation of GABA system and cell damage in parahippocampus of patients with temporal lobe epilepsy showing antiepileptic effects alter subacute electrical stimulation. *Epilepsia* 2004, 45:459–466.

Engel, J., Jr., Outcome with respect to epilepsy seizures. In Engel, J., Jr., Ed., *Surgical Treatment of the Epilepsies.* New York: Raven Press, 1987, p. 553–569.

Gloor, P., Quesney, L.F., and Zumstein, H., Pathophysiology of generalized penicillin seizures in the cat. The role of cortical and subcortical structures. II. Topical application of penicillin to the cerebral cortex and to sucortical structures. *Electroencephalogr. Clin. Neurophysiol.*, 1977, 48:79–94.

Hunter, J. and Jasper, H.H., Effects of thalamic stimulation in unanesthetized animals. The arrest reaction and petit mal-like seizures activation patterns and generalized convulsions. *Electroenceph. Clin. Neurophysiol.*, 1949, 11:305–324.

Jasper, H.H. and Droogleever-Fortuyn, J., Experimental studies of the functional anatomy of petit mal epilepsy. *Res. Public Assoc. Nerv. Ment. Dis.*, 1947, 26:272–298.

Mehler, R.W., Further notes of the centromedian nucleus of Luys. In Purpura, D.P. and Yahr, M.D., Eds., *The Thalamus.* New York: Columbia University Press, 1996, p. 102–111.

Mihara, T., Inoue, Y., and Seino, M., Surgical strategies for patients with supplementary sensorimotor area epilepsy. The Japanese experience. *Adv. Neurol.*, 1996, 70:405–414.

Morris, H., Supplementary motor seizures. In Morris, H., Ed. *The Treatment of Epilepsy, Principles and Practice*, 2nd ed. New York: Lea & Febirger, 1993, p. 541–546.

Olivier, A.: Sugical strategies for patients with supplementary sensorimotor area epilepsy. The Montreal experience. *Adv. Neurol.*, 1996, 70:429–443.

Penfield, W. and Jasper, H., *Epilepsy and the Functional Anatomy of the Human Brain.* Boston, MA: Little Brown, 1954, p. 566–596.

Pollen, D.A., Perot, P., and Reid, K.H., Experimental bilateral spike and wave from thalamic stimulation in relation to the level of arousal. *Electroencephalogr. Clin. Neurophysiol.*, 1963, 15:459–473.

Quesney, L.F., Gloor, P., Kratzemberg, E., and Zumstein, H., Pathophysiology of generalized penicillin seizures in the cat. The role of cortical and subcortical structures. I. Systemic application of penicillin. *Electroencephalogr. Clin. Neurophysiol.*, 1977, 42:640–655.

Rise, M.T., Instrumentation for neuromodulation. *Arch. Med. Res.*, 2000, 34:237–247.

Schaltenbrand, G., Bailey, P., y. Stuttgart, Georg Thieme, 1959, Vol. IV.

Smith, J.R. and King, D.W., Surgical strategies for patients with supplementary sensorimotor area epilepsy. The Medical College of Georgia experience. *Adv. Neurol.*, 1996, 70:415–427.

Steriade, M., Spindling, incremental thalamo-cortical responses and spike-wave like epilepsy. In Avoli, M., Gloor, P., Kustopoulus, G., and Naquet, R., Eds., *Generalized Epilepsy: Neurobiological Approaches*. Boston: Birkhouser, 1990, p. 161–180.

Velasco, F., Velasco, M., and Machado, J.P., A statistical outline of the subthalamic target for the arrest of tremor. *Appl. Neurophysiol.*, 1975, 38:38–46.

Velasco, F., Velasco, M., Ogarrio, C., and Fanghänel, F., Electrical stimulation of the centromedian thalamic nucleus in the treatment of convulsive seizures. A preliminary report. *Epilepsia*, 1987, 28:421–430.

Velasco F., Velasco, M., and Alcalá, H., The electrical stimulation of the thalamus. In Kutt, H. and Resor, S.R., Eds., *Advances in Neurology. Medical Treatment of Epilepsy*. New York: Marcel Decker, 1989, p. 677–780.

Velasco M., Velasco, F., Alcalá, H., Dávila, G., and Díaz de León, A.E., Epileptiform EEG activities in the centromedian thalamic nuclei in children with intractable generalized seizures of the Lennox-Gastaut syndrome. *Epilepsia*, 1991, 32:310–321.

Velasco, F., Velasco, M., Velasco, A.L., and Jimenez, F., Effect of chronic electrical stimulation of the centromedian thalamic nuclei on various intractable seizure patterns. I. Clinical seizures and paroxysmal EEG activity. *Epilepsia*, 1993a, 34:1052–1064.

Velasco, M., Velasco, F., Velasco, A.L., Velasco, G., and Jimenez, F., Effect of chronic electrical stimulation of the centromedian thalamic nuclei on various intractable seizure patterns. II. Psychological performance and background activity. *Epilepsia*, 1993b, 34:1065–1074.

Velasco, F., Velasco, M., Velasco, A.L., Jimenez, F., and Rise, M., Electrical stimulation of the centromedian thalamic nucleus in control of seizures: long-term studies. *Epilepsia*, 1995, 36:63–71.

Velasco, M., Velasco, F., Velasco, A.L., Jimenez, F., Márquez, I., and Rojas, B., Electrocortical and behavioral responses produced by acute electrical stimulation of the human centromedian thalamic nucleus. *Electroenceph. Clin. Neurophysiol.*, 1996, 102:461–471.

Velasco, M., Velasco, F., Velasco, A.L., Jiménez, F., Brito, F., and Márquez, I., Acute and chronic electrical stimulation of the centromedian thalamic nucleus: modulation of reticulo-cortical systems and predictor factors for generalized seizure control. *Arch. Med. Res.*, 2000a, 31:304–315.

Velasco, F., Velasco, M., Jiménez, F., Velasco, A.L., Brito, F., Rise, M., and Carrillo-Ruiz, J.D., Predictors in the treatment of difficult to control seizures by electrical stimulation of the centromedian thalamic nucleus. *Neurosurgery*, 2000b, 47:295–305.

Velasco, A.L., Boleaga, B., Brito, F., Jiménez, F., Gordillo, J.L., Velasco, F., and Velasco, M., Absolute and relative predictor values of some non-invasive and invasive studies for the outcome of anterior temporal lobectomy. *Arch. Med. Res.*, 2000c, 31:62–74.

Velasco, A.L., Velasco, M., Velasco, F., Ménes, D., Gordon, F., Rocha, L, Briones, M., and Márquez, I., Subacute and chronic electrical stimulation of the hippocampus on intractable temporal lobe seizures. *Arch. Med. Res.*, 2000d, 31:316–328.

Velasco, M., Velasco F., Velasco, A.L., Boleaga, B., Jiménez, F., Brito, F., and Márquez, I., Subacute electrical stimulation of the hippocampus blocks intractable temporal lobe seizures and paroxysmal EEG activities. *Epilepsia*, 2000e, 41:158–169.

Velasco, M., Velasco, F., and Velasco, A.L., Centromedian thalamic and hippocampal electrical stimulation for the control of intractable epileptic seizures. *Clin. Neurophysiol.*, 2001a, 18:1–15.

Velasco, F., Velasco, M., Velasco, A.L., Ménez, D., and Rocha, L., Electrical stimulation for epilepsy. 1. Stimulation of hippocampal foci. *Stereotact. Funct. Neurosurg.*, 2001b, 77:223–227.

Velasco, F., Velasco, M., Jiménez, F., Velasco, A.L., Rojas, B., and Pérez, M.L., Centromedian nucleus stimulation for epilepsy: clinical, electroencephalographic and behavioral observations. *Thalamus and Related Systems,* 2002, 34:1–12.

Velasco, F., Velasco, A.L., Velasco, M., Rocha, L., and Ménes, D.: Electrical stimulation of the epileptic focus in cases of temporal lobe seizures. In Lüders, H.O., Ed., *Electrical Stimulation for Epilepsy.* London: Taylor & Francis Health Sciences, 2003, p. 287–300.

Weiss, S.B.R., Li, X.L., Rosen, J.B., Li, H., Heynen, T., and Post, R.M., Quenching: inhibition of development and expression of amygdale kindled seizures with low frequency stimulation. *Neuroreport,* 1995, 4:2171.

Wieser, H.G., Engel, J., Jr., Williamson, P.D., Babb, T.L., and Gloor, P., Surgically remediable temporal lobe syndromes. In Engel, J., Jr., Ed., *Surgical Treatment of the Epilepsies.* New York: Raven, 1993, 49–63.

Williamson, P.D., French, J.A., and Thadani, V.M., Characteristics of medial temporal lobe epilepsy. II. Interictal and ictal scalp electroencephalography, neuropsychological testing, neuroimaging, surgical results and pathology. *Ann. Neurol.,* 1993, 34:781–787.

4

Responsive Neurostimulation for Epilepsy — Neurosurgical Experience: Patient Selection and Implantation Technique

Kostas N. Fountas and
Joseph R. Smith

4.1   Chapter Summary

The concept of seizure abortion after prompt detection employing electrical stimulation is a very appealing one. Several investigators in previous experimental and clinical studies have used stimulation of various anatomical targets with promising results. In this chapter the authors present a brief overview of previous stimulation studies as well as a description of an investigational implantable, local closed-loop RNS™ Neurostimulator System (NeuroPace, Inc., Mountain View, CA). This system consists of a cranially implanted pulse generator, one or two quadripolar subdural strip or depth leads, and an external programmer. The system components and technical characteristics are also presented as well as the criteria for selecting candidates for enrollment in the pivotal investigation and the implantation surgical technique. Although still under investigation, the closed-loop stimulation system appears to be an emerging treatment option that may demonstrate promising results for the management of patients with well-localized, focal, medically refractory epilepsy who are not candidates for surgical resection. However, further clinical validation of the preliminary data from ongoing experimental clinical studies is mandatory before the wide clinical application of this treatment modality.

4.2   Introduction

It is widely accepted that epilepsy is one of the most prevalent severe neurologic disorders across all age groups [45]. It has been previously reported that approximately 1% of the U.S. population develops epilepsy [45], and this percentage is as high as 5% among children and adolescents in North America or Western Europe [22]. The incidence of medically intractable epilepsy has been reported to be approximately 6 out of 100,000 people per year, which translates to 17,000 new cases annually in the United States alone [23]. Although surgical treatment constitutes a valuable alternative to medical treatment in very carefully selected cases, unfortunately a large portion of patients with intractable epilepsy will not have access to surgical therapy. The reasons for this are numerous, including the highly technical complexities and increased costs of preoperative evaluation, significantly limited availability of human and technical resources, and the complicated involvement of the eloquent cortex in the epileptogenic zones [45]. It is apparent that the development of a new treatment modality is of paramount importance for the management of patients with medically refractory epilepsy. The recently increasing employment of deep brain stimulation in the management of movement disorders has rekindled interest in employing electrical stimulation either in an open- or a closed-loop fashion so as to abort or block seizure activity.

4.2.1   Background

The concept of aborting or preventing seizure activity by applying direct electrical, magnetic, or mechanical brain stimulation is not a new one. Approximately twenty centuries ago, Pelops from Alexandria was able to abort something that could be considered a simple partial seizure, by tying a ligature around the affected limb [20, 46]. Later, Brown-Sequard, Jackson, and Gowers, working independently, suggested that *counter-irritation* could be a mechanism for abating seizure activity [5, 21, 25, 26]. A large series of *in vitro* and *in vivo* studies have demonstrated that electrical stimulation effectively reduces and can control synchronized bursting in cortical neurons [4, 19, 31, 44, 51, 63]. Although the effects of electrical stimulation can be accurately predicted at the level of an individual neuron or nerve axon, this is not possible when dealing with highly complex real neuronal networks. The effect of electrical stimulation in a neuronal network is uncertain and its effects on distant neuronal networks are always unpredictable. Although the exact mechanism for action of electrical stimulation remains poorly understood, several mechanisms have been proposed to explain the observed seizure abortive effect [2, 8, 12–14, 26, 32, 42, 44, 46, 48, 50, 54, 55, 64]. Enhancement of inhibition [32], induction of plasticity in the form of short- and long-term depression [8, 12, 50, 55, 64] or depotentiation [26] of synaptic responses, modification of nonsynaptic activity [13], receptor desensitization or down-regulation [2], increases in neuronal synchronization [42], and desynchronization of neuronal networks [14] are only several among the numerous theories proposed.

Electrical stimulation can be delivered in the proximity of an epileptogenic focus (local stimulation model) or in a preselected anatomical target (i.e., anterior thalamic nucleus, centromedian thalamic nucleus, or hippocampus) (remote stimulation model). Local stimulation delivery offers the theoretical advantage of increased spatial selectivity with a minimalization of side effects and electrical current requirements [45]. Furthermore, stimulation can be delivered either in regular predetermined time intervals (open-loop stimulation) or exclusively when specific epileptiform activity is detected (closed-loop or contingent stimulation). The theoretical advantage of contingent stimulation is increased temporal selectivity [45]. However, a closed-loop stimulation system requires the design and development of a seizure detection algorithm that must meet certain criteria: operate in real-time, be highly sensitive and specific, yield rapid detections for timely delivered therapeutic stimulation, and be adaptable so to account for the high inter-individual and occasionally intra-individual variability of seizures [45]. Numerous algorithms seeking to address these criteria have been developed based on different methodologies such as neural network logistic regression analysis [57], electroencephalographic Gaussian mixture models [33], joint sign periodogram transformation [43], neuro-fuzzy inference system [56], nonlinear excitable media [6], autoregressive spectral analysis [29], wavelet decomposition along with feature extraction and data segmentation [49], and a multistage system based on relevance and redundancy analysis [1].

Employment of uncontrolled open-loop, noncontingent stimulation was attempted in several experimental studies and human trials for the control of epilepsy. Stimulation of the cerebellar cortex [9, 10], cerebellar dentate nucleus [7, 53], cerebral cortex [28, 31] anterior thalamic nucleus [11, 24, 27], centromedian thalamic nucleus [7, 53, 59–61], head of caudate nucleus [7, 52, 53], hippocampus [52, 53, 62], and subthalamic nucleus [3, 41] have all been employed with a variety of clinical results. The only controlled clinical studies involved the cerebellar cortex [58] and thalamic centromedian nucleus stimulation [5], and neither showed a significant effect on seizures. However, vagus nerve stimulation, which represents a cyclical type of open-loop stimulation, was shown to reduce seizures in a statistically significant fashion [30]. Open-loop studies on the effect of electrical stimulation on induced afterdischarges (AD) have shown that AD can be aborted [21] and there may be optimal parameters for accomplishing this [28, 35].

The concept of closed-loop stimulation was previously described, and studies utilizing external responsive neurostimulators showed that closed-loop stimulation can affect the duration of spontaneously occurring electrographic seizure activity [18, 34, 38–41, 45, 47]. The results of these initial pilot studies are supported by the findings of a multicenter prospective clinical feasibility study, in which an external RNS (eRNS) was used to demonstrate safety and provide preliminary evidence for the effectiveness of responsive stimulation in suppressing seizures in an acute setting [39]. This study was conducted under a Food and Drug Administration (FDA) investigational device exemption and was also approved by the Institutional Review Board (IRB) of each participating center [38, 39]. The study included twenty-four patients who underwent grid, strip, and/or depth leads implantation for temporary invasive monitoring as part of an evaluation for epilepsy surgery. Preliminary reports are reported for them [38, 39]. A laptop computer was used through wired telemetry for interrogation and programming of the eRNS, and responsive stimulation trials were conducted [38, 39]. In this group of patients, the use of the eRNS system had a positive electroencephalograph (EEG) effect on electrographic seizure activity [38, 39]. No serious adverse effects related to the eRNS occurred in this study [38, 39]. The safety of the strip and depth leads used with the eRNS was additionally confirmed by animal experimental studies [36, 37]

This chapter describes the technical characteristics, the selection criteria, and the surgical technique for implanting an investigational local, closed-loop responsive neurostimulation system (RNS™ Neuro-Pace Inc., Mountain View, CA).

4.3   RNS System Description

The implantable closed-loop RNS™ neurostimulator system (NeuroPace Inc., Mountain View, CA) consists of the following components (Figure 4.1):

- *Pulse generator.* The pulse generator is a hermetically sealed neurostimulator containing the electronics, battery, telemetry coil, and connector hardware to accommodate one or two leads. The dimensions of the pulse generator are 41.5 mm wide, 60.0 mm long, and 8.4 mm thick; its weight is 19.5 g and its volume is 10.5 cc. The pulse generator monitors the patient's electrocorticograms (ECoGs) and triggers electrical stimulation, when specific ECoG characteristics, programmed by the clinician as indicative of electrographic seizures or precursor epileptiform activities, are detected. The pulse generator then stores diagnostic information detailing detections and stimulations, including multichannel stored ECoGs. The pulse generator has a curved shape to facilitate cranial implantation and is positioned extradurally in a tailored cranial defect and held in place with a ferrule or holder (Figure 4.2).
- *Depth lead.* The depth leads are quadripolar leads designed for stereotactic implantation. Depth leads are available with 3.5-mm and 10-mm interelectrode spacings, and in lengths of 30 and 44 cm. Electrodes are composed of platinum and iridium (Figure 4.1).

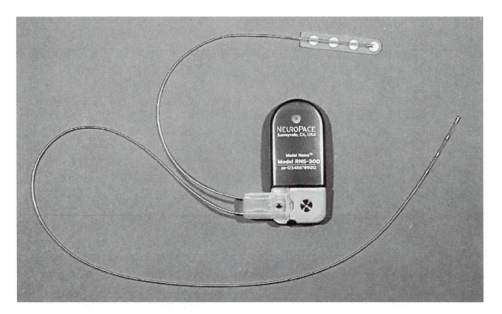

FIGURE 4.1 The implantable Responsive Neuro-Stimulation System (Neuropace, Inc., Mountain View, CA); it is shown in this picture with subdural strip and depth leads.

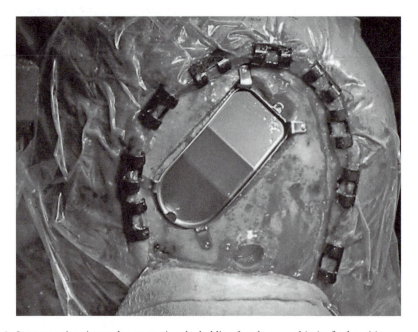

FIGURE 4.2 Intraoperative picture demonstrating the holding ferrule secured in its final position.

- *Strip lead.* The strip leads are quadripolar leads with interelectrode spacings of 10 mm. Leads are available in lengths of 15 and 25 cm. Electrodes are composed of platinum and iridium (Figure 4.1).
- *Programmer.* The programmer is a notebook computer with specialized software and a telemetry wand, which communicates with the pulse generator. The programmer can upload diagnostic and ECoG data from the pulse generator, and which can be used to analyze ECoGs and program pulse generator detection and stimulation parameters. The programmer also has an electrophysiology study mode that allows for real-time stimulation with simultaneous ECoG viewing to test stimulation paradigms.

4.3.1   RNS Technical Characteristics

ECoG storage. The RNS has an ECoG memory buffer. The number of ECoGs stored depends on the number of recording channels and the recording length selected. ECoG storage can be triggered by any of several electrographic events, including seizure onset.

Therapeutic stimulation. The RNS delivers charge-balanced biphasic pulses programmable from 0.5 to 12 mA amplitude, pulse widths programmable from 40 to 1000 μseconds, and frequency programmable from 1 to 333 Hz. Any of the electrode contacts or the pulse generator housing can be programmed as anode or cathode. After delivering a pulse-train therapy, a redetection algorithm determines if epileptiform activity is still present. If so, up to four additional therapies can be delivered per episode. Also, each therapy may consist of one or two bursts. The parameters of each therapy and each burst may be the same or different. The programming software has a built-in charge density limit that will allow programming of no more than 25 μCoulombs/cm²/phase charge density.

4.3.1.1   Subject Selection Criteria for RNS™ Neurostimulator System Pivotal Clinical Investigation

Candidates for RNS implantation should have a history of medically intractable, well-localized, focal, simple partial motor seizures or complex partial seizures, with or without secondarily generalized seizures. In our institution, ictal and interictal surface EEG, video-EEG monitoring, brain MRI, ictal SPECT, and SISCOM studies, detailed neuropsychological evaluation including a WADA testing, and, when necessary, invasive EEG via depth and/or subdural electrodes are employed for localizing the epileptogenic focus. Patients who have been localized to more than two epileptogenic foci are not candidates for implantation. The patients are not candidates for resective surgery either due to the eloquent nature of the involved cerebral cortex, due to the risk of memory impairment following a hippocampal resection because they have bilateral hippocampal seizure onsets, or because they do not desire to undergo resective surgery but would rather undergo surgical implantation of an RNS due to the reversible character of the stimulator implantation. Furthermore, patients who have undergone multiple subpial transections with no satisfactory results could be considered candidates for implantation of RNS.

Candidates are subsequently required to have at least an average of three disabling seizures per month, over a three-month period, before final consideration for RNS implantation. Candidates should be between the ages of 18 and 70 years, and should be able to complete regular office visits and telephone appointments for follow-up purposes per the protocol requirements. Female patients should be using a reliable method of contraception. Patients who have experienced unprovoked status epilepticus in the preceding year or patients with unstable medical conditions as well as active psychosis, severe depression, or suicidal ideation cannot be candidates for implantation. Additionally, patients who are pregnant or are planning to become pregnant in the next two years, and patients who are on a ketogenic diet cannot be considered for implantation. Finally, patients with an active vagus nerve stimulator or who are implanted with an electronic medical device that delivers electrical energy to the head cannot be candidates for implantation. A detailed, written informed consent is routinely obtained prior to participation in the pivotal investigation.

4.3.2   Surgical Implantation

4.3.2.1   Patient Preparation

The area of craniotomy incision will determine the area of skull to be clipped, prepped, and draped. The procedure is usually performed under general endotracheal anesthesia; but with a cooperative patient, neuroleptanalgesia can be used along with field blocks of the scalp. In these cases, control of blood pressure is of paramount importance to minimize potential intraoperative hemorrhage. The location of the epileptogenic focus will dictate the implantation of a depth or strip lead.

4.3.2.2   Implantation of Depth Lead

The implantation of a depth lead is performed with a frame-based stereotactic localization of the target [27]. The MRI/CT-compatible Leksell base-ring and localizer (Elekta AB, Stockholm, Sweden) is fixed

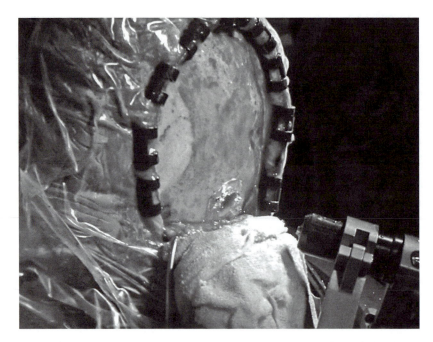

FIGURE 4.3 Intraoperative picture demonstrating the implanted Navigus electrode lead securing device (Image-guided Neurologics, Melbourne, FL) in its final position.

to the patient's skull. Carbon fiber posts and MRI/CT-compatible pins are used. The MRI scan consists of a contrast-enhanced T1-weighted volume acquisition using axial 1.3-mm slices with zero slice-gap. This is followed by a whole-head CT scan using 3-mm slices with zero slice-gap. The two data sets are imported over the local network to a computer workstation (BrainLab, Heimstetten, Germany). After fusing MR and CT data, both target and trajectory are defined. A probe view algorithm is used to maximize the distance between any surface veins and the depth lead at the cortical entry point.

At this point, the patient is appropriately positioned on the operating table by attaching the base-ring to the Mayfield holder. The craniotomy incision is marked with gentian violet, and the surgical area is prepped and draped in a standard sterile fashion. The previously marked incision is now infiltrated with a local anesthetic and a horseshoe-shaped craniotomy flap is turned. The Leksell stereotactic arc system is attached to the base-ring, and a drill guide tube is then advanced through the incision down to the skull and antibiotic irrigation is flushed through the tube. The appropriate burr hole is outlined on the skull and a high-speed, air-driven drill (Midas Rex, Fort Worth, TX) is used. Attention must be given to the diameter of the burr hole to exactly match the diameter of the Navigus (Image-guided Neurologics, Melbourne, FL) securing device, which will be used for securing the implanted depth lead (Figure 4.3). The underlying dura is opened in a linear fashion. At this point, a 2.1 mm inner-diameter guide block is introduced and the dura and pia are cauterized with a mono-polar electrode (AdTech, Racine, WI). Next, a 14-gauge depth electrode cannula (AdTech, Racine, WI) is passed through the same guide block to the target point. Intraoperative fluoroscopy is used to verify proper placement. The cannula and guide tube are then withdrawn. A Navigus cranial base and cap device is then implanted over the burr hole and secured by using the two self-tapping screws provided. Attention must be paid to align the exit groove on the base in the postero-lateral direction, in the same direction that the subcutaneous portion of the lead will later be directed. An insertion tool (AdTech, Racine, WI) is then passed through a large diameter guide block and inserted into the slot and the Navigus device. The insertion tool and guide block are then removed. The depth lead is then carefully inserted to the target point. Fluoroscopy is used again to verify proper position of the

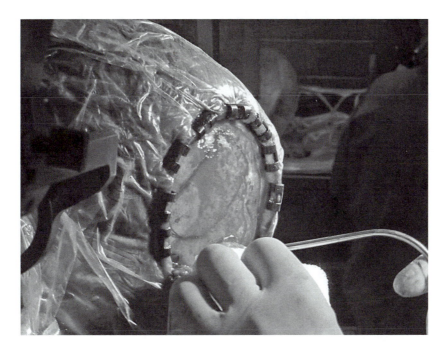

FIGURE 4.4   Intraoperative picture demonstrating the outlining of the pulse generator on the bone.

implanted lead. The stylet of the lead is removed and the distal shaft of the implanted lead is secured into the Navigus device.

4.3.2.3   Subdural Lead Placement

If a subdural strip lead is to be placed, the dura is opened linearly and the 1 × 4 cortical lead is inserted through the dural opening. Fluoroscopy is used to verify proper placement. The distal shaft of the implanted lead is safely secured to the Navigus device.

4.3.2.4   Pulse Generator Implantation

At this point, the provided ferrule is placed on the exposed bone and the desired bony defect is outlined (Figure 4.4). The outlined bone is drilled out by a high-speed, air-driven drill with attention not to traumatize the underlying dura (Figure 4.5). The bony edges are smoothed and meticulous hemostasis is performed if necessary by applying bone wax. Thorough irrigation of the wound is of great importance for removing any residual bone dust. The provided ferrule is then implanted and secured to the adjacent bone at four points with the provided self-tapping mini-screws (Figure 4.2). The pulse generator is connected to the distal end of the already implanted lead or leads, and then is secured in the implanted ferrule (Figure 4.6). At this point, the programmer is used with the sterile covered telemetry wand to interrogate the implanted pulse generator, measure the impedances of all lead contacts, and perform electrocorticography to verify proper function of the implanted RNS™ neurostimulator system. The surgical wound is thoroughly irrigated with bacitracin solution and then is closed in anatomical layers. The wound is covered with a sterile dressing.

The patient is transported to the neurosurgical ward for observation and discharged within two to three days. Before discharge, a postoperative head CT scan and skull plain x-rays (antero-posterior and lateral views) are obtained to provide baseline imaging studies for future reference (Figures 4.7A and 4.7B). The first interrogation and seizure detection programming session is usually performed on the third postoperative day. Further interrogations and programming sessions are usually required for fine adjustment of the implanted RNS™ neurostimulator detection and stimulation parameters.

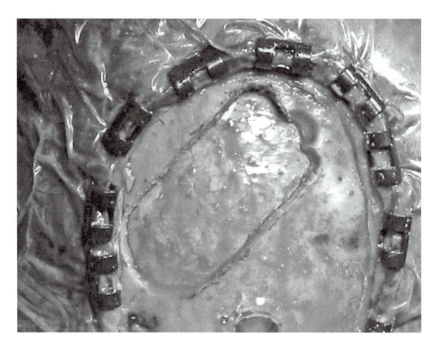

FIGURE 4.5 Intraoperative picture demonstrating the bony defect created for implanting the ferrule and the pulse generator.

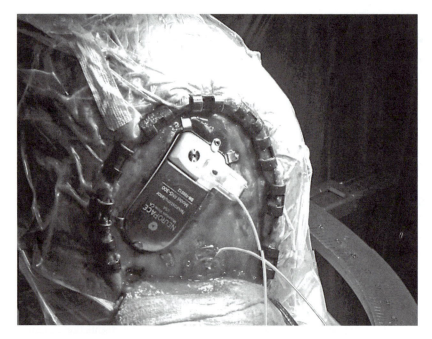

FIGURE 4.6 (See color insert following page 15-4). Intraoperative picture demonstrating the pulse generator connected to an implanted depth lead and secured to the underlying ferrule.

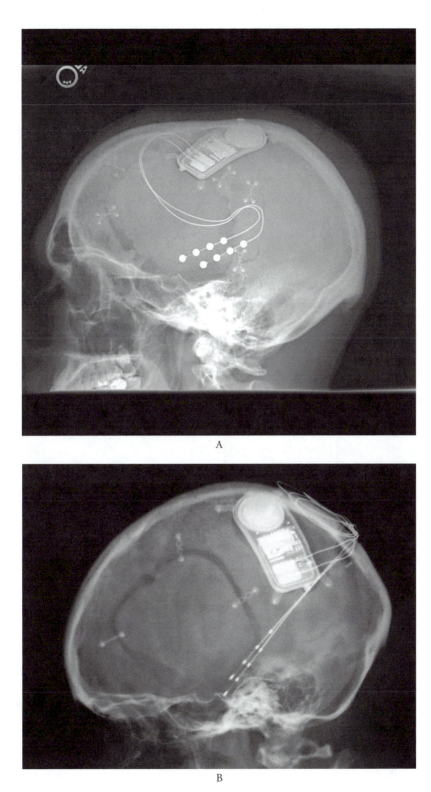

FIGURE 4.7 (A) Postoperative x-rays (lateral view) of one of our patients demonstrating the implanted RNS system with two subdural leads. (B) Postoperative x-rays (lateral view) of one of our patients demonstrating the implanted RNS system with two depth leads.

The operative blood loss is usually minimal (in all of our cases has been maintained at less than 100 cc) while the mean duration of our surgical procedure of implantation is 3.2 hours (range 2 to 4.5 hours).

4.4 Conclusions

The implantable, local, closed-loop RNS™ neurostimulator system is an investigational option in patients with well-localized, focal, medically refractory epilepsy, who are not candidates for surgical resection. A multiinstitutional, prospective, clinical study is underway to evaluate the clinical safety and efficacy of this novel treatment modality. Technical improvement of this system along with accumulation of experience from its clinical use could lead to the development of a system that would accurately detect and efficiently abort any detected epileptiform activity.

Acknowledgments

The authors wish to thank David Greene from NeuroPace, Inc. (Mountain View, CA) for his valuable assistance. The authors also wish to acknowledge their appreciation and thanks to Aaron Barth and Stacy Perry for assistance in the preparation of this chapter.

References

1. Aarabi, A., Wallois, F., and Grebe, R. (2006). Automated neonatal seizure detection: a multistage classification system through feature selection based on relevance and redundancy analysis. *Clin. Neurophysiol.*, 117(2):328–340.
2. Arai, A. and Lynch, G. (1998). AMPA receptor desensitization modulates synaptic responses induced by repetitive afferent stimulation in hippocampal slices. *Brain Res.*, 799(2):235–242.
3. Benabid, A.-L., Koudsie, A., Chabardes, S., et al. (2004). Subthalamic nucleus and substantia nigra pars reticulata stimulation: the Grenoble experience. In Luders, H.O., Ed., *Deep Brain Stimulation and Epilepsy.* Marting Dunitz, London, p. 335–348.
4. Bikson, M., Lian, J., Hahn, P.J., et al. (2001). Suppression of epileptiform activity by high frequency sinusoidal fields in rat hippocampal slices. *J. Physiol.*, 531(Pt 1):181–191.
5. Brown-Sequard, C.E. (1856–1857). Researches on epilepsy: its artificial production in animals, and its etiology, nature and treatment in man. *Boston Med. Surg. J.*: p. 55–57.
6. Chernihovskyi, A., Mormann, F., Muller, M., et al. (2005). EEG analysis with nonlinear excitable media. *J. Clin. Neurophysiol.*, 22(5):314–329.
7. Chkhenkeli, S.A., Sramka, M., Lortkipandze, G.S., et al. (2004). Electrophysiological effects and clinical results of direct brain stimulation for intractable epilepsy. *Clin. Neurol. Neurosurg.*, 106(4):318–329.
8. Contzen, R. and Witte, O.W. (1994). Epileptic activity can induce both long-lasting potentiation and long-lasting depression. *Brain Res.*, 653(1–2):340–344.
9. Cooper, I.S., Amin, I., Riklan, M., et al. (1976). Chronic cerebellar stimulation in epilepsy. Clinical and anatomical studies. *Arch. Neurol.*, 33(8):559–570.
10. Cooper, I.S. and Upton, A.R. (1978). Effects of cerebellar stimulation on epilepsy, the EEG and cerebral palsy in man. *Electroencephalogr. Clin. Neurophysiol. Suppl.*, 34:349–354.
11. Cooper, I.S., Upton, A.R., and Amin, I. (1980). Reversibility of chronic neurologic deficits. Some effects of electrical stimulation of the thalamus and internal capsule in man. *Appl. Neurophysiol.*, 43(3–5):244–258.
12. Dudek, S.M. and Bear, M.F. (1993). Bidirectional long-term modification of synaptic effectiveness in the adult and immature hippocampus. *J. Neurosci.*, 13(7):2910–2918.
13. Dudek, F.E., Yasumura, T., and Rash, J.E. (1998). "Non-synaptic" mechanisms in seizure and epileptogenesis. *Cell. Biol. Int.* 22:793–805.
14. Durand, D.M. and Warman, E.N. (1994). Desynchronization of epileptiform activity by extracellular current pulses in rat hippocampal slices. *J. Physiol.*, 480(Pt. 3):527–537.

15. Estellar, R., Echauz, J., Tcheng, T., et al. (2001). Line length: an efficient feature of seizure onset detection. Proc. 23rd Annu. Conf. IEEE Eng. Med. Bio. Soc., pp. 1707–1709.

16. Fisher, R.S., Uematsu, S., Krauss, G.L., et al. (1992). Placebo-controlled pilot study of centromedian thalamic stimulation in treatment of intractable seizures. *Epilepsia,* 33(5):841–851.

17. Fountas, K.N., Smith, J.R., Murro, A.M., et al. (2005). Implantation of a closed-loop stimulation in the management of medically refractory focal epilepsy. A Technical Note. *Stereotact. Funct. Neurosurg.,* 83:153–158.

18. Fountas, K.N., Smith, J.R., Murro, A.M., et al. (2005). Closed-loop stimulation implantable system for the management of focal, medically refractory epilepsy: implantation technique and preliminary results. *Epilepsia,* 46(S8):240–241.

19. Franaszczuk, P.J., Kudela, P., and Bergey, G.K. (2003). External excitatory stimuli can terminate bursting in neural network models. *Epilepsy Res.,* 53(1–2):65–80.

20. Siegel, R.E. (1976). Galen on the affected parts (de locis affectis). Book III. S. Karger, Basel, Switzerland, pp. 15–90.

21. Gowers, W.R. (1885). Epilepsy and Other Chronic Convulsive Diseases: Their Causes, Symptoms and Treatment. William Wood, New York, p. 235–236.

22. Hauser, W.A. (1995). Epidemiology of Epilepsy in Children. In Adelson, P.D. and Black, P.M., Eds., *Neurosurgery Clinics of North America.* W.B. Saunders Co., Philadelphia, 6(3):419–429.

23. Hauser, W.A. and Hesdorffer, D.C. (2001). Epidemiology of intractable epilepsy. In Luders, H.O. and Comair, Y.G., Eds., *Epilepsy Surgery, 2nd edition.* Lippincott Williams & Wilkins, Philadelphia, p. 55–61.

24. Hodaie, M., Wennberg, R.A., Dostrovsky, J.O., and Lozano, A.M. (2002). Chronic anterior thalamus stimulation for intractable epilepsy. *Epilepsia,* 43(6):603–608.

25. Jackson, J.H. (1868). Case of convulsive attacks arrested by stopping the aura. *Lancet,* 1:618–619.

26. Kang-Park, M.H., Sarda, M.A., Jones, K.H., et al. (2001). Protein phosphatases mediate depotentiation induced by high-intensity theraburst stimulation. *J. Neurophysiol.,* 89:684–690.

27. Kerrigan, J.F., Litt, B., Fisher, R.S., Craunston, S., et al. (2004). Electrical stimulation of the anterior nucleus of the thalamus for the treatment of intractable epilepsy. *Epilepsia,* 45(4):346–354.

28. Kinoshita, M., Ikeda, A., Matsumoto, R., et al. (2004). Electrical stimulation on human cortex suppresses fast cortical activity and epileptic spikes. *Epilepsia,* 45(7):787–791.

29. Kiymik, M.K., Subasi, A., and Ozcalik, H.R. (2004). Neural networks with periodogram and autoregressive spectral analysis methods in detection of epileptic seizure. *J. Med. Syst.,* 28(6):511–522.

30. Labar, D. (2004). Vagal nerve stimulation: effects on seizures. In Luders, H.O., Ed., *Deep Brain Stimulation and Epilepsy.* Martin Dunitz, London.

31. Lesser, R.P., Kim, S.H., Beyderman, L., et al. (1999). Brief bursts of pulse stimulation terminate after-discharges caused by cortical stimulation. *Neurology,* 53(9):2073–2081.

32. Liang, F., Isackson, P.J., and Jones, E.G. (1996). Stimulus-dependent, reciprocal up- and down-regulation of glutamic acid decarboxylase and Ca^{2+}/calmodulin-dependent protein kinase II gene expression in rat cerebral cortex. *Exp. Brain Res.,* 110(2):163–174.

33. Meng, L., Frei, M.G., Osorio, I., Strang, G., and Nguyen, T.Q. (2004). Gaussian mixture models of ECoG signal features for improved detection of epileptic seizures. *Med. Eng. Phys.,* 26(5):379–393.

34. Morrell, M. (2006). Brain stimulation for epilepsy: can scheduled or responsive neurostimulation stop seizures? *Curr. Opin. Neurol.,* 19(2):164–168.

35. Motamedi, G.K., Lesser, R.P., Miglioretti, D.L., et al. (2002). Optimizing parameters for terminating cortical after-discharges with pulse stimulation. *Epilepsia,* 43(8):836–846.

36. Munz, M., Sweasey, R., Barrett, C., et al. (2003). Preclinical testing of an implantable responsive neurostimulator system in a sheep model. In *Society for Neuroscience.* New Orleans, 2003.

37. Munz, M., Sweasey, R., Barret, C., et al. (2003). Implantation and testing of responsive neurostimulator (RNS) system for epilepsy. In *American Society for Stereotactic and Functional Neurosurgery,* New York.

38. Murro, A.M., Park, Y.D., Bergey, G.K., et al. (2003). Multicenter study of acute responsive stimulation in patients with intractable epilepsy. *Epilepsia,* 44(Suppl. 9):326.

39. Murro, A., Park, Y., Greene, D., et al. (2002). Closed-loop neuro-stimulation in patient with intractable epilepsy. In *American Clinical Neurophysiology Society,* New Orleans, LA.

40. Nair, D.R., Matsumoto, R., Luders, H.O., et al. (2004). Direct cortical electrical stimulation in the treatment of epilepsy. In Luders, H.O., Ed., *Deep Brain Stimulation and Epilepsy.* Martin Dunitz, London.

41. Neme, S., Montgomery, E.B., Rezai, A., et al. (2004). Subthalamic nucleus stimulation in patients with intractable epilepsy: the Cleveland experience. In Luders, H.O., Ed., *Deep Brain Stimulation and Epilepsy.* Martin Dunitz, London, p. 349–358.

42. Netoff, T.I. and Schiff, S.J. (2002). Decreased neuronal synchronization during experimental seizures. *J. Neurosci.,* 22(16):7297–7307.

43. Niederhauser, J.J., Esteller, R., Echauz, J., et al. (2003). Detection of seizure precursors from depth-EEG using a sign periodogram transform. *IEEE Trans. Biomed. Eng.,* 50(4):449–458.

44. Oommen, J., Morrell, M., and Fisher, R.S. (2005). Experimental electrical stimulation therapy for epilepsy. *Curr. Opin. Neurol.,* 7(4):261–271.

45. Osorio, I., Frei, M.G., Manly, B.F., et al. (2001). An introduction to contingent (closed-loop) brain electrical stimulation for seizure blockage, to ultra-short term clinical trials, and to multidimensional statistical analysis of therapeutic efficacy. *J. Clin. Neurophysiol.,* 18(6):533–544.

46. Osorio, I., Frei, M.G., Sunderam, S., et al. (2005). Automated seizure abatement in humans using electrical stimulation. *Ann. Neurol.,* 57(2):258–268.

47. Peters, T.E., Bhavaraju, N.C., Frei, M.G., and Osorio, I. (2001). Network system for automated seizure detection and contingent delivery of therapy. *J. Clin. Neurophysiol.,* 18(6):545–549.

48. Poliakov, A.V., Powers, R.K., Sawczuk, A., and Binder, M.D. (1996). Effects of background noise on the response of rat and cat motoneurones to excitatory current transients. *J. Physiol.,* 495(Pt.1):147–157.

49. Saab, M.E. and Gotman, J. (2005). A system to detect the onset of epileptic seizures in scalp EEG. *Clin. Neurophysiol.,* 116(2):427–442.

50. Sheng, M. and Kim, M.J. (2002). Postsynaptic signaling and plasticity mechanisms. *Science,* 298(5594):776–780.

51. Slutzky, M.W., Cvitanovic, P., and Mogul, D.J. (2003). Manipulating epileptiform bursting in the rat hippocampus using chaos control and adaptive techniques. *IEEE Trans. Biomed. Eng.,* 50(5):559–570.

52. Sramka, M., Fritz, G., Gajadosova, D., and Nadvornik, P. (1980). Central stimulation treatment of epilepsy. *Acta Neurochir. Suppl. (Wien.),* 30:183–187.

53. Sramka, M., Fritz, G., Galanda, M., and Nadvornik, P. (1976). Some observations in treatment stimulation of epilepsy. *Acta Neurochir. (Wien.),* (23 Suppl.):257–262.

54. Stacey, W. and Durand, D.M. (2001). Synaptic noise improves detection of subthreshold signals in hippocampal CA1 neurons. *J. Neurophysiol.,* 86(3):1104–1112.

55. Stevens, C.F. and Tsujimoto, T. (1995). Estimates for the pool size of releasable quanta at a single central synapse and for the time required to refill the pool. *Proc. Natl. Acad. Sci., U.S.A.,* 92(3):846–849.

56. Subasi, A. (2006). Application of adaptive neuro-fuzzy inference system for epileptic seizure detection using wavelet feature extraction. *Comput. Biol. Med.,* February 8; [Epub ahead of print].

57. Subasi, A. and Ercelebi, E. (2005). Classification of EEG signals using neural network and logistic regression. *Comput. Meth. Prog. Biomed.,* 78(2):87–99.

58. Velasco, F., Carrillo-Ruiz, J.D., Brito, F., et al. (2005). Double-blind, randomized controlled pilot study of bi-lateral cerebellar stimulation for treatment of intractable motor seizures. *Epilepsia,* 46:1071–1081.

59. Velasco, F., Velasco, M., Jimenez, F., et al.(2001). Stimulation of the central median thalamic nucleus for epilepsy. *Stereotact. Funct. Neurosurg.,* 77(1–4):228–232.

60. Velasco, F., Velasco, M., Ogarrio, C., and Fanghanel, G. (1987). Electrical stimulation of the centromedian thalamic nucleus in the treatment of convulsive seizures: a preliminary report. *Epilepsia*, 28(4):421–430.
61. Velasco, M., Velasco, F., and Velasco, A.L. (2001). Centromedian-thalamic and hippocampal electrical stimulation for the control of intractable epileptic seizures. *J. Clin. Neurophysiol.*, 18(6):495–513.
62. Velasco, F., Velasco, M., Velasco, A.L., et al. (2001). Electrical stimulation for epilepsy: stimulation of hippocampal foci. *Stereotact. Funct. Neurosurg.*, 77(1–4):223–227.
63. Wagenaar, D.A., Madhavan, R., Pine, J., and Potter, S.M. (2005). Controlling bursting in cortical cultures with closed-loop multi-electrode stimulation. *J. Neurosci.*, 25(3):680–688.
64. Zucker, R.S. and Regehr, W.G. (2002). Short-term synaptic plasticity. *Annu. Rev. Physiol.*, 64:355–405.

5

Responsive Neurostimulation for Epilepsy: RNS™ Technology and Clinical Studies

Thomas K. Tcheng and
Martha Morrell

5.1 Introduction and Background

5.1.1 Clinical Market and Relevance of the Therapy

Epilepsy is a neurological disorder that affects 2.3 million people in the United States and as many as fifty million people worldwide [Begley et al., 2000]. Perhaps half have intractable epilepsy; that is, seizures cannot be controlled by antiepileptic drug (AED) therapy, and/or there are side effects from AEDs that adversely impact quality of life. The ketogenic diet, the vagus nerve stimulator, and epilepsy surgery are other treatment options. However, many persons with epilepsy are left without treatment that is efficacious, tolerable, and acceptable. Device-based therapies may provide additional therapeutic options. One approach to treating medically intractable localization-related epilepsy with partial onset seizures is to provide focal stimulation in response to electrographic epileptiform activity. The NeuroPace® RNS™ system includes a cranially implanted responsive neurostimulator that continuously monitors electro-corticographic (ECoG) activity from intracranial electrodes, detects electrographic events of significance according to programmable detection algorithms, and provides responsive stimulation. The intent is to modify abnormal electrographic activity in an effort to prevent or terminate clinically evident seizures.

The RNS™ system includes a fully implantable, patient-specific, field-programmable responsive neuro-stimulator with multiple electrodes surgically placed into selected brain target areas.

Responsive neurostimulation technology is designed to benefit epilepsy patients and their families by potentially improving long-term health, decreasing disability, and reducing healthcare costs. Improved long-term health could result from a reduction in the damaging effects of epileptic seizures, psychiatric disorders, and reproductive dysfunction associated with epilepsy. In addition, patients may also benefit from a reduction or elimination of side effects from concurrent therapies by potentially enabling a reduction or elimination of the dosage of these therapies. Patients and their employers may benefit from increased productivity as they return to the workplace. Epilepsy patients and third-party healthcare payers may benefit from a reduced need for medical care associated with seizures and health complications associated with epilepsy.

NeuroPace's development of responsive neurostimulation technology could also benefit the medical device industry by demonstrating the feasibility and utility of responsive neurostimulation.

5.1.2 Review of Other Technologies

The current practice for treating epilepsy includes AEDs, vagus nerve stimulation, the ketogenic diet, and resective surgery. There are currently no FDA-approved implantable direct brain stimulation devices available to treat epilepsy. In the research setting, direct brain stimulation for epilepsy has had mixed results. The vast majority of this work used noncontingent continuous or on–off cycling stimulation, also referred to as open-loop stimulation.

5.1.2.1 Vagus Nerve Stimulation

Cyberonics currently offers vagus nerve stimulation therapy. The vagus nerve stimulator provides inter-mittent, regularly scheduled electrical stimulation to the left vagus nerve at the level of the external carotid artery. The FDA approved the vagus nerve stimulator therapy system indicated for use as an adjunctive therapy in reducing the frequency of seizures in adults and adolescents over 12 years of age with partial onset seizures, which are refractory to antiepileptic medications.

In two randomized clinical trials in people with intractable partial seizures, therapy with the vagus nerve stimulator was associated with a median percentage reduction in seizures of 23%. In addition, 23 to 30% of patients achieved a 50% or greater reduction in seizures leading to marketing approval, therapy with the vagus nerve stimulator reduced partial seizures [Cyberonics, 2002]. Long-term uncon-trolled follow-up of patients participating in the investigational trials suggests that persons continuing vagus nerve stimulation over one to two years continue to benefit. However, fewer than 5% achieve freedom from seizures. Adverse effects arise in approximately 30% of patients, including hoarseness, headache, muscle pain, throat pain, coughing, and nausea [Schachter and Schmidt, 2001].

5.1.2.2 Thalamic Stimulation

Stimulation of the anterior nucleus of the thalamus for epilepsy is currently being evaluated by an FDA-approved randomized, controlled trial (Medtronic). In this multicenter study, patients with at least six partial seizures per month, with or without secondary generalization, are treated with regular intermittent bilateral high-frequency stimulation of the anterior thalamic nucleus [Oommen et al., 2005]. The results of this investigation were not available at the time this manuscript was prepared.

5.2 Fundamental Neuroscience: Mechanism of Action

5.2.1 Mechanisms and Physiology of Epilepsy

Epilepsy is a complex neurological disease characterized by seizures. Epileptic seizures are sudden, excessive, and temporary neurological events. "Sudden" means a transition from background to abnormal activity. "Excessive" means increased excitation due to a shift in the balance between inhibition and excitation. "Temporary" indicates that the typical seizure is less than 60 seconds. Seizures are usually accompanied by electrographic features that are apparent in electroencephalographic (EEG) recordings. Additionally, abnormal electrographic discharges are often observed interictally (between seizures). In partial onset epilepsy, a seizure onset zone is defined as the brain area where the seizure begins. Seizures can arise in any region of cortex. There are often structural and functional brain abnormalities that can be detected using diagnostic imaging methods. Another common observation is a neurochemical imbalance. These structural, functional, and neurochemical abnormalities are believed to contribute to abnormal electrographic activity that is characteristic of seizures, such as hypersynchronous oscillations.

Direct electrical stimulation can affect neurons and neural systems in different ways, depending on the stimulation parameters and the brain area being stimulated. Continuous high-frequency stimulation is theorized to inhibit neural activity, either by causing preferential release of inhibitory neurotransmitters, by depleting neurotransmitter in the presynaptic terminal, or by causing a depolarization block [Buerrier et al., 2001; Lee et al., 2004; Benabid et al., 2005; Mantovani et al., 2006]. In contrast, short bursts of stimulation delivered in response to detection of a seizure onset are thought to disrupt the oscillatory neural dynamics underlying the seizure process.

5.3 Technological Innovation

5.3.1 Origin of Responsive Neurostimulation Technology

Responsive neurostimulation technology was invented by Robert and David Fischell and neurologist Adrian Upton. In 1997, they founded NeuroPace to design, develop, manufacture, and market the responsive neurostimulator and other implantable devices for the treatment of neurological disorders. Many of the developers and engineers involved in the development of the RNS™ system came from the cardiac defibrillator industry, where pattern detection and responsive stimulation technologies were developed to treat cardiac arrhythmias. These technologies were further developed by NeuroPace to detect and treat the abnormal electrographic brain activity associated with epilepsy. The Half Wave, Line Length, and Area detection tools were based on successful seizure and spike detection tools documented in the literature. These detection tools were simplified and adapted to fit the low power and low computational complexity constraints of an implantable medical device. Stimulation features were developed to be appropriate for direct brain stimulation. These included a current-controlled, charge-balanced, biphasic rectangular waveform, as well as a wide range of stimulation amplitudes, pulse widths, frequencies, and burst durations. Additionally, the developers included the ability to stimulate different brain areas using different stimulation parameters, depending on the pattern or location of the abnormal activity detected. Because responsive stimulation is new technology, a number of safety features were also incorporated into the RNS™ system.

5.3.1.1 Preliminary Stimulation Studies

Responsive stimulation has been effective in terminating seizure activity in a wide range of studies from rat hippocampal preparations to limited human clinical studies. A recurring finding throughout many of these studies is that the earlier the stimulation is applied, relative to the seizure, the greater the likelihood of success [Lesser et al., 1999]. This observation lent support to the hypothesis that responsive stimulation may be an effective therapy for epilepsy.

5.3.1.1.1 Animal Studies of Closed-Loop Stimulation

In animal studies, Psatta [1983] studied interictal spiking activity in epileptic foci in cats. Responsive stimulation was automatically delivered to the caudate nucleus when a 1-sec spike burst was detected, resulting in spike depression. No comparable effects were seen with random stimulation. In a rat hippocampal slice preparation, Nakagawa and Durand [1991] applied subthreshold electrical currents to the stratum pyramidale upon detection of abnormal electrical activity. This resulted in complete suppression of interictal bursts in 90% of the slices. Further confirmation of the potential efficacy of responsive stimulation was demonstrated by Vercueil et al. [1998], who demonstrated the ability of short bursts (5 sec) of high-frequency (130 Hz) bilateral stimulation of the subthalamic nuclei to

FIGURE 5.1 The NeuroPace® RNS™ neurostimulator.

decrease the duration of absence seizures in rats. In their study, no decrease in epileptiform activity was observed when continuous bilateral stimulation was applied. To the extent that animal studies are valid models for human epilepsy, these studies lend support for the use of responsive electrical stimulation to treat epilepsy.

- *Initial closed-loop stimulation studies.* The groundbreaking work of Penfield and Jasper [1954] showed that spontaneous epileptiform activity can be interrupted using electrical stimulation. Since then, progress has been made in the application of responsive stimulation to control seizures in epilepsy patients. Rajna and Lona [1989] observed that epileptic seizures were promptly inhibited in 79 of 139 observations in nineteen patients when acoustical stimuli were delivered at the onset of absence seizures.
- *Human Trials of Closed-loop Stimulation.* Chkhenkeli [1997] reported that manually triggered stimulation of the caudate terminated electrically evoked seizures in humans. Lesser et al. [1999] found that electrical stimulation was effective in terminating epileptiform activity (afterdischarges) caused by brain-mapping stimulation. Osorio et al. [2001] has provided one of the most direct demonstrations of the effectiveness of responsive stimulation in focal or remote locations of the brain using stimulation frequencies higher than 100 Hz in eight patients. As Litt and Baltuch [2001] concluded in their review, past efforts point to intelligent closed-loop brain stimulators as the next logical step in the treatment of epilepsy.

5.3.2 Description and Implementation of Technology

The RNS™ system consists of an implantable responsive neurostimulator (Figure 5.1), electrode leads, a wireless telemetry wand, a patient data transmitter, and a physician programmer. The neurostimulator is designed with a thickness and curvature matching that of the skull so that it replaces an identically shaped section of bone that is surgically removed. During implant, a craniectomy the same size and shape of the neurostimulator is made, and the neurostimulator is placed into a ferrule that is inserted into the craniectomy and anchored to the skull with titanium screws.

Two lead designs can be used with the RNS™ system; either cortical strip leads or depth leads. Each lead contains four electrodes. The leads are surgically positioned so that the electrodes are as close to the seizure focus as possible. Depth leads are usually implanted stereotactically through a burr hole, and cortical strip leads are usually implanted through a craniectomy. The implanted parts of the RNS™ system are shown in Figure 5.2.

The neurostimulator is capable of communicating with an external wand via wireless telemetry. The telemetry wand is connected to either a physician programmer or a patient data transmitter. The programmer is used by clinicians to upload device data (interrogation) and adjust neurostimulator settings (programming), while the patient data transmitter is used by patients for interrogation only.

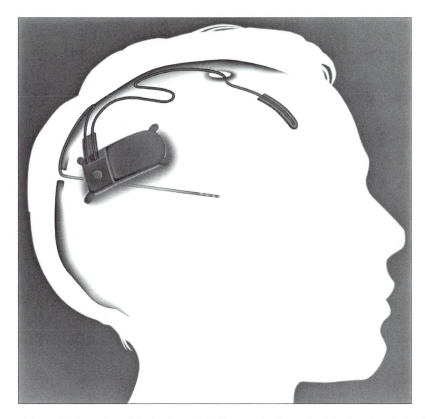

FIGURE 5.2 Schematic illustration of the implanted RNS™ neurostimulator, depth lead, and cortical strip lead.

TABLE 5.1 Summary of RNS™ System Components

RNS™ System Component	Description
Neurostimulator	Fully implantable responsive neurostimulator (42 mm wide, 60 mm long, 8.4 mm thick)
Cortical strip lead	Cortical strip lead with four disk electrodes spaced 10 mm apart
Depth lead	Depth lead with four cylindrical electrodes available with electrode spacings in the range of 3.5 mm to 10 mm apart
Telemetry wand	Hand-held external wireless telemetry transceiver
Physician programmer	Laptop computer with physician programmer software used for interrogating and programming the neurostimulator
Patient data transmitter	Laptop computer with patient data transmitter software used for interrogating the neurostimulator and uploading device data to patient data management system
Patient data management system	Database and clinician website

Both the programmer and patient data transmitter communicate via the Internet with a central patient data management system where device data are stored and can be reviewed by physicians. Patients can interrogate their neurostimulator at home on a daily basis using the wand and patient data transmitter, then upload device data to the patient data management system. This allows physicians to view the latest patient data without an office visit. The components of the RNS™ system are summarized in Table 5.1.

5.3.2.1 RNS™ Neurostimulator Capabilities

The RNS™ neurostimulator is a multifunctional neurostimulation device that includes the following capabilities:

- Electrographic sensing
- Electrographic event detection

- Responsive electrical stimulation
- ECoG storage
- Detection and stimulation event storage
- Electrode impedance measurement
- Battery voltage measurement
- Wireless telemetry

5.3.2.1.1   *Clinical Use Example*

An example of how the RNS™ system's intended clinical use will demonstrate how the different capabilities work together to treat epilepsy. Prior to implant, the patient's seizure focus is localized. This can be done using a range of diagnostic methods, including scalp EEG and intracranial EEG; imaging methods such as magnetic resonance imaging (MRI), positron emission spectroscopy (PET), and single photon emission computed tomography (SPECT); observation of seizure semiology; and neuropsychological testing. If the seizure focus is adequately localized, leads are implanted such that the electrodes are as close as possible to the seizure focus. The proximal ends of the leads are connected to the neurostimulator, which is implanted in the skull. During the implant procedure, the programmer and wand are used to test lead impedance, view real-time ECoG signals, and test neurostimulator functionality. After implant, the programmer and wand are used to program detection settings into the neurostimulator. The neurostimulator then detects electrographic events that meet the programmed detection criteria. When an event is detected, information describing the event is stored, and an ECoG record containing the event may be captured. In between follow-up visits, the patient data transmitter and wand can be used by the patient to upload event information and ECoGs from the neurostimulator to the data transmitter. These device data are subsequently uploaded from the patient data transmitter via the Internet to the patient data management system. Patient device data can then be reviewed by the physician using a Web browser. During follow-up visits, the programmer is used to interrogate and program the neurostimulator. Detection settings can be reviewed and refined based on prior ECoG data uploaded from the neurostimulator using a detection simulator built into the programmer. Once satisfactory detection specificity and sensitivity have been achieved, responsive therapy can be enabled. Prior to enabling responsive therapy, test stimulation can be provided interactively using the programmer to evaluate patient tolerability. Responsive stimulation settings are then configured by the physician and programmed into the neurostimulator to be delivered in response to electrographic event detection. Information describing therapy delivery is also stored in the device and uploaded to the patient data management system along with detection data and ECoGs. During subsequent follow-up visits, the physician can further refine and update detection and therapy settings and optimize therapy to reduce seizure frequency and severity.

5.3.2.2   Electrographic Sensing and ECoG Storage

Electrographic activity is sensed through the cortical strip and depth electrodes using four differential amplifiers. Because two leads can be connected to the neurostimulator, there are eight electrodes that can be used for sensing. Electrodes are commonly assigned to amplifiers in adjacent pairs as shown in Table 5.2. This provides an easily interpretable representation of electrographic neural activity at each of four different locations and informs the selection of stimulation electrodes.

Electrographic data can be continuously collected on all four channels and stored within the neurostimulator as ECoG records. The electrographic data are also used as input to the event detection algorithm. Amplifier channels that are assigned to electrographic event detection also have noise and saturation detectors enabled. An example of one channel from a stored ECoG, along with its spectrogram, is shown in Figure 5.3. ECoGs can be stored in response to a variety of triggers, including detection, therapy, noise, saturation, and time of day.

5.3.2.3   Electrographic Event Detection

Electrographic events are detected using a detection algorithm that continuously monitors electrographic brain activity. The event detection system is capable of specific and sensitive detection. Three detection tools — Half Wave, Line Length, and Area — are used by the detection algorithm. These detection tools

TABLE 5.2 Typical Amplifier Configuration

Electrode	Amp. 1	Amp. 2	Amp. 3	Amp. 4
Lead 1, electrode 1	+			
Lead 1, electrode 2	-			
Lead 1, electrode 3		+		
Lead 1, electrode 4		-		
Lead 2, electrode 1			+	
Lead 2, electrode 2			-	
Lead 2, electrode 3				+
Lead 2, electrode 4				-

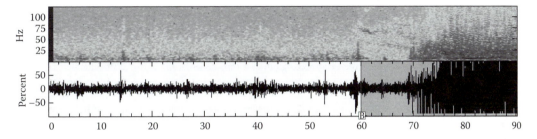

FIGURE 5.3 **(See color insert following page 15-4).** Example of one channel from a stored ECoG, along with its spectrogram. The x-axis is in seconds. The upper panel displays the FFT spectrogram to represent normalized spectral power (red = high, blue = low). The y-axis on the upper panel is frequency (in Hertz). The lower panel displays the time-series trace of the ECoG. The y-axis on the lower panel indicates the relative amplitude of the signal in percent full scale. In the time series trace, the "B" label and the vertical blue line indicate when a detection by a Line Length Tool occurred, the light blue background indicates a detection episode, and the yellow background, barely visible in the lower trace starting at 78 seconds, indicates saturation.

can be applied independently or in combination with a single ECoG channel, and up to two ECoG channels can be configured for detection.

- *Half Wave Tool.* The Half Wave Tool identifies local minima and maxima within the ECoG waveform and uses them to measure half waves. The half wave amplitude is defined as the voltage difference between adjacent minima and maxima, and the half wave duration is defined as the length of time between adjacent minima and maxima. The Half Wave Tool is used to detect electrographic activity with specific frequency and amplitude characteristics.
- *Line Length Tool.* The Line Length Tool estimates a measure of complexity or fractal dimension within the ECoG signal. The line length is calculated as the sum of the unsigned amplitude changes within a time window. Either the absolute difference or the ratio of the average line length between a short window and a long window is used to determine whether signal complexity is increasing or decreasing. The Line Length Tool is sensitive to changes in amplitude or frequency.
- *Area Tool.* The Area Tool estimates a measure of energy within the ECoG signal. The area is calculated as the sum of the unsigned area under the curve within a time window. Either the absolute difference or the ratio of the average area between a short window and a long window is used to determine whether energy is increasing or decreasing. The Area Tool is sensitive to changes in amplitude but minimally sensitive to changes in frequency.

5.3.2.4 Responsive Stimulation

Electrical stimulation is delivered in response to electrographic event detection. A number of stimulation parameters can be adjusted to control stimulation, including current amplitude, pulse width, pulse frequency, and burst duration. Stimulation can be delivered in either a monopolar or bipolar montage. Monopolar stimulation is delivered between any combination of electrodes and the neurostimulator case.

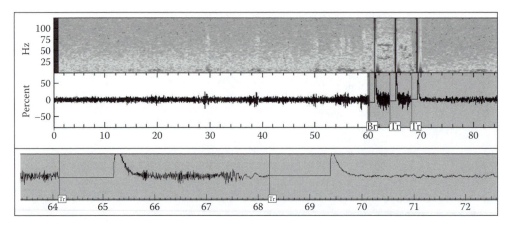

FIGURE 5.4 (See color insert following page 15-4). Example of a possible seizure termination by responsive stimulation. The upper panel shows the FFT spectrogram and time series of the entire ECoG channel. The lower panel shows a close-up of the time series around the time of termination. In the time series plots, detection is indicated by the vertical blue line and the "B" label, a detection episode is indicated by the blue background, and responsive stimulation is indicated by the vertical red lines and the "Tr" labels (the first is partly obscured by the "B" label). These data are from the same subject as in Figure 5.3.

For monopolar stimulation, all the electrodes must be assigned the same polarity, either negative (cathodic) or positive (anodic), while the case is assigned the opposite polarity. For bipolar stimulation, one set of electrodes is defined as cathodic and another set of electrodes is defined as anodic. Stimulation is configured in a patient-specific manner to reduce seizure frequency and severity. In general, stimulation is delivered to electrodes where abnormal electrographic activity is observed. In responsive stimulation, it is important to note that event detection is the primary means by which therapy is allocated over time. One strategy for allocating therapy is to configure detection so that therapy is delivered as early as possible during an electrographic seizure. With redetection during an event, therapy can be delivered repeatedly. An example of where responsive stimulation appears to terminate a seizure is shown in Figure 5.4.

5.4   Clinical Studies

NeuroPace® RNS™ neurostimulation technology has been used in several human clinical trials. An investigation of manually triggered stimulation to terminate afterdischarges produced during brain mapping with intracranial electrodes was completed in 2003. A safety investigation using an external responsive neurostimulator (eRNS) in an epilepsy monitoring unit setting was completed in 2005. A feasibility study using an implanted neurostimulator (model RNS-300) was initiated in 2004. The data from this study provide sufficient evidence of safety and preliminary evidence of efficacy supporting the December 2005 commencement of a pivotal multicenter, double-blind trial of responsive stimulation safety and efficacy using the implanted neurostimulator.

5.4.1   eRNS Study

In the eRNS study, an external responsive neurostimulator provided automatic seizure detection and stimulation. The study subjects were patients being evaluated for epilepsy surgery with intracranial electrodes. Seizure detection algorithms were optimized over several days while electrographic data were acquired for seizure localization. Automatic responsive stimulation was applied to the seizure onset zone at the end of the intracranial study prior to removal of the mapping electrodes. No unanticipated serious device-related adverse events were reported. Four subjects had a perception of stimulation, which was eliminated by adjustment of the stimulation settings. NeuroPace's experience with the eRNS study suggested that it is important for stimulation to occur early in the electrographic event, and that to accomplish this, electrodes should be positioned to capture the earliest electrographic onset. Additionally, multiple electrodes in the vicinity of the seizure onset should be used for delivery of responsive stimulation.

5.4.2 Implantable RNS™ System Feasibility Study

The implantable RNS™ system feasibility study was a multicenter feasibility investigation to assess the safety and preliminary efficacy of a cranially implanted, programmable responsive neurostimulator as adjunctive therapy for medically refractory epilepsy in adults. Automatic detection settings were initially optimized without stimulation, and then responsive stimulation therapy was enabled or subjects were randomized into the stimulation OFF group. In a subsequent open extension period, stimulation was enabled for all subjects. There have been no unanticipated device-related serious adverse events, and no serious surgical complications in this study. NeuroPace has accumulated more than 140 patient years of experience with the implanted responsive neurostimulator system.

5.4.3 Implantable RNS™ System Pivotal Study

The implantable RNS™ system pivotal investigation is currently underway. This study is a randomized, double-blind, multicenter, controlled clinical investigation. The overall objective of this study is to demonstrate that the RNS™ system is safe and effective in reducing the frequency of disabling seizures in adults with medically intractable partial onset seizures who have not responded to two or more antiepileptic drugs. In this study, following completion of the baseline, implant, and subsequent stabilization periods, the patients are randomized to one of two treatment conditions: therapy ON or therapy OFF. The patients are blind to their treatment condition. At the end of the blind evaluation period, the patients enter the Open Label extension period and stimulation can be enabled for all patients in the trial. The intent of this study is to demonstrate safety and effectiveness, and to gather additional information on long-term safety.

5.5   Conclusions, Discussion, and Future Directions

The NeuroPace® RNS™ system is the first fully implantable, closed-loop responsive neurostimulator to be evaluated in clinical trials. Responsive neurostimulation may become an option for adults with medically intractable partial seizures. Because responsive stimulation can be targeted to specific brain regions and therapy is delivered only in response to specific brain activity, responsive stimulation can avoid the systemic side effects of AEDs and can be specifically applied to brain areas where the risks of resective surgery are unacceptably high. Responsive neurostimulation is designed to deliver targeted stimulation less often than continuous or on–off cycling neurostimulation methods. Potential advantages of responsive neurostimulation include:

1. Specifically treating the pathological events
2. Avoiding habituation and desensitization, which may occur with continuous stimulation
3. Reducing side effects of stimulation
4. Preserving battery life reducing follow-up surgical procedures

The technology incorporated into the RNS™ system allows physicians to review and evaluate electrocorticographic activity in their patients, and adjust event detection and responsive stimulation settings in a patient-specific manner. This flexibility allows for evaluation of a range of treatment strategies in the same patient without additional surgery or neurostimulator replacement. The feasibility clinical experience with the RNS™ system provides preliminary data indicating that the neurostimulator and leads can be safely implanted and that responsive stimulation can be safely applied to many regions of the cortex. Epilepsy is the first clinical indication being treated by responsive stimulation technology. The ability to monitor brain activity, detect events of interest, and respond to those events with customizable therapy may be a general strategy that can be applied to a variety of neurological indications.

References

Begley, C.E., Famulari, M., Annegers, J.F., Lairson, D.R., Reynolds, T.F., Coan, S., Dubinsky, S., Newmark, M.E., Leibson, C., So, E.L., and Rocca, W.A. The cost of epilepsy in the United States: an estimate from population-based clinical and survey data. *Epilepsia*, 41:342–351 (2000).

Chkhenkeli, S.A. and Chkhenkeli, I.S. Effects of therapeutic stimulation of nucleus caudatus on epileptic electrical activity of brain in patients with intractable epilepsy. *Stereotact. Funct. Neurosurg.*, 69(1–4 Pt. 2):221–224 (1997).

Cyberonics. *Physician's Manual: NeuroCybernetic Prosthesis SystemNCP® Pulse Generator Models 100 and 101.* Cyberonics (2002).

Lesser, R.P., Kim, S.H., Beyderman, L., Miglioretti, D.L., Webber, W.R., Bare, M., Cysyk, B., Krauss, G., and Gordon, B. Brief bursts of pulse stimulation terminate afterdischarges caused by cortical stimulation. *Neurology*, 53(9):2073–2081 (1999).

Litt, B. and Baltuch, G. Brain stimulation for epilepsy. *Epilepsy and Behavior*, 2(3):S61–S67 (2001).

Nakagawa, M. and Durand, D. Suppression of spontaneous epileptiform activity with applied currents. *Brain Res.*, 567(2):241–247 (1991).

Oommen, J., Morrell, M., and Fischer, R.S. Experimental electrical stimulation therapy for epilepsy. *Curr. Treat. Options Neurol.*, 7(4):261–271 (2005).

Osorio, I., Frei, M.G., Manly, B.F., Sunderam, S., Bhavaraju, N.C., and Wilkinson, S.B. An introduction to contingent (closed-loop) brain electrical stimulation for seizure blockage, to ultra-short-term clinical trials, and to multidimensional statistical analysis of therapeutic efficacy. *J. Clin. Neurophysiol.*, 18(6):533–544 (2001).

Penfield, W. and Jasper, H. *Epilepsy and the Functional Anatomy of the Human Brain*. Boston: Little, Brown (1954).

Psatta, D.M. Control of chronic experimental focal epilepsy by feedback caudatum stimulations. *Epilepsia*. 24(4):444–454 (1983).

Rajna, P. and Lona, C. Sensory stimulation for inhibition of epileptic seizures. *Epilepsia*. 30(2):168–174 (1989).

Schachter, C.S. and Schmidt, D. *Vagus Nerve Stimulation*. London: Taylor & Francis (2001).

Vercueil, L., Benazzouz, A., Deransart, C., Bressand, K., Marescaux, C., Depaulis, A., and Benabid, A.L. High-frequency stimulation of the subthalamic nucleus suppresses absence seizures in the rat: comparison with neurotoxic lesions. *Epilepsy Res.*, 31:39–46 (1998).

6

Deep Brain Stimulation for Pain Management

Donald E. Richardson

6.1   Introduction

Electrical stimulation of the brain has been used as an investigational technique since the days of Horsley and Clarke at the turn of the past century, to activate excitable tissue and determine its function from the resulting physiologic changes (18). The therapeutic use of electrical stimulation in the central nervous system is a relatively recent addition to the armamentarium for the treatment of otherwise intractable pain and suffering, and has been pursued vigorously because of the low potential for side effects and the potential to affect previously untreatable central and deafferentation pain.

There have been reports of sporadic cases of pain relief related to electrical stimulation primarily used for other reasons. Heath, in the early 1950s, found that stimulation of the septal area in a patient with terminal cancer produced pain relief, and also produced dysphoria and euphoria which helped reduce the stress and depression of her terminal illness (17). In his extensive recording of the electrical activity and the effect of electrical activation of multiple sites in the brains of chronic schizophrenics, this was the only patient who had a significant pain problem, but it did demonstrate that pain could be relieved by stimulation. In 1956, Pool et al. also reported that activation of the septal area in the region of the anterior third ventricle produced pain relief (36). Ervin et al. in 1966, reported the relief of pain in a patient by stimulation of the thalamus (11). In 1967, Gol reported a small series of patients treated for pain with electrical stimulation in the septal area, which was only moderately successful (13). In 1969, Reynolds reported that stimulation of the periaqueductal gray produced deep analgesia in the rat sufficient, in fact, to allow surgical procedures to be performed without any evidence of pain or discomfort

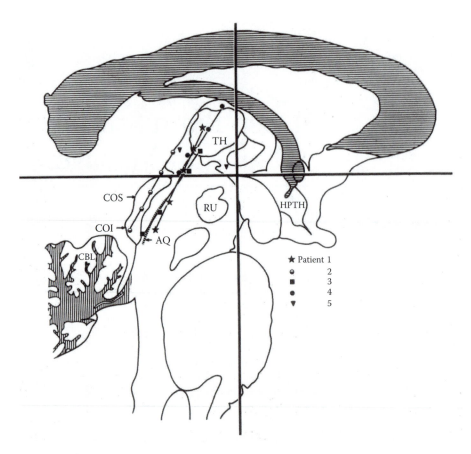

FIGURE 6.1 Stimulation paths through the pariventricular and pariaquiductal gray (PVG, PAG) in five patients; all patients received pain relief in the PVG. AQ = aqueduct of Sylvius, TH = thalamus, COS = superior colliculus, COI = inferior colliculus, RU = nucleus rubber. (*J. Neurosurg.* 47:178–183, 1977.)

(38). In 1971, Mayer et al. in John Liebskind's laboratory at the University of California in Los Angeles reported periaqueductal gray (PAG) stimulation produced analgesia (SPA) in the rat, which spurred a large number of animal studies to investigate the mechanism of stimulation-induced analgesia in animals (27). I had observed that stimulation in one patient during a steriotaxic leisoning procedure for the relief of cancer pain that prelesion stimulation produced pleasant olfactory hallucinations along with pain relief. Since the electrode was misdirected, the exact site of stimulation was never determined however; it did prove to us that SPA was possible in the human (Richardson, unpublished data).

In 1970, based on these studies, Akil and I began a systematic acute stimulation study of sites in the periventricular (PVG) and periaqueductal (PAG) areas in humans to determine sites to be used for pain relief. These acute stimulation studies were carried out in five patients as a prelude to placing stereotaxic lesions in the centrum median for chronic pain. The results led us to believe that the entire area along the wall of the third ventricle, extending from a centimeter above the intercommissural line caudally into the area of the raphae nuclei, could be used for electrical stimulation analgesia in the human (41) (Figure 6.1). We chose a site in the periventricular area for chronic studies because of the minimal side-effects in this area, in contrast to stimulation at sites farther caudal in the brainstem (41, 42). Our chronic studies began in 1971 with the cooperation of Medtronic, Inc., Minneapolis, MN, which provided modified their induction spinal cord stimulation system for our use and later made electrodes for long-term stimulation in human patients (42). Since the early 1970s, many investigators have shown that stimulation in what is now thought to be the endogenous opiate system in the PVG and PAG is effective for relieving certain types of chronic pain.

The second site used for relief of chronic pain has been the primary sensory relay nuclei of the thalamus. Dorsal column stimulation and peripheral nerve stimulation have proven to be effective for relief of pain, and the ventral posterior lateralis (VPL) and ventral posterior medialis (VPM) nuclei of the thalamus are extensions of the main touch and proprioception sensory tracts (lemniscal system) from the dorsal columns into the central nervous system. In 1960, Mazars et al. reported that stimulation of the somatosensory pathways in VPL gave pain relief (28), and our studies in the cat showed that painful stimulation could be reduced by VPL/VPM electrical stimulation (39). Subsequently, Hosobuchi et al. in the early 1970s, began to use chronic VPM and internal capsule (IC) stimulation, initially, for the relief of anesthesia dolorosa following fifth nerve section for trigeminal neuralgia (19). The stimulation of these nuclei produces paresthesia in the area of pain with subsequent loss of the chronic burning paresthetic and aching quality of the pain, characteristic of deafferentation pain.

Young et al. in 1992 reported chronic stimulation of the lateral upper brainstem, more lateral than the periaqueductal gray, in the nuclear mass of Kolliker–Fuse, produced analgesia in three of his patients (56).

6.2 Neurophysiology of Periventricular and Periaqueductal Pain Modulation

Physiologically, stimulation-produced analgesia (SPA) is different for each of the sites selected. The best-understood, but perhaps most complex, is the mechanism for SPA produced by stimulation of the PVG and PAG. Initially, it was thought that stimulation of these sites produced direct effects on pain transmission, but when SPA produced by PVG stimulation was found by Akil and Liebeskind to be inhibited by naloxone, a specific opiate antagonist, understanding the physiology of pain modulation became more complex (4). Watson et al. have shown in the rodent that the cell bodies for the central opiate mechanism (mu system) are in the arcuate, infundibular, and periventricular nucleus of the hypothalamus, with axons extending from these nuclei anteriorly through the septal area and then superiorly and posteriorly through the septal area and medial to the thalamus in the PVG to the raphe nuclei in the ventral periaqueductal gray and inferior to the locus ceruleus, where this tract terminates (52). This promelano-opio-cortin nucleus and tract (also called the beta-endorphin system) has also been identified in the human brain by Pilcher et al (34). We have demonstrated, in patients, that stimulation anywhere along this system produces analgesia but with markedly different side effects. In the periaquiductal gray osilopsia and symptoms identical to the epigastric rising syndrome are so noxious that patients will not allow stimulation longer than a few seconds. Further in the anterior PVG and septal area, mild euphoric is produced, and more inferiorly in the area of the basal forebrain and hypothalamus, elevation of blood pressure is prompt and often dramatic (44) (Figure 6.2). This has given us several potential alternate targets for use in patients who need a different site for electrode placement because of technical reasons, with some obvious restrictions.

The secondary serotonergic fibers from the raphae nuclei and the norepinergic fibers from the locus ceruleus then descend through the dorsolateral funiculus of the spinal cord to impinge on the dorsal horn of the spinal cord (8, 9). Evidence suggests that there is an intermediate opioid interneuron that is activated by these descending tracts that then inhibits pain at the first synapse in the dorsal horn (9).

Thus, the primary effect of stimulation of the PVG/PAG area is to activate the same system that is activated by systemic opioids of the mu type (5, 47), which secondarily activates the descending inhibitory tracts emanating from the locus ceruleus and raphe nuclei and probably the Kolliker–Fuse and parabrachiai nucleus, to inhibit input from somatic origin at the first synapse for pain impulses in the dorsal horn (56).

There is also evidence that this inhibition occurs at the thalamic level and perhaps at the brainstem level in the reticular formation, inhibiting propagation of impulses reaching awareness for pain through both the spinoreticular tracts and the spinothalamic tracts. The evidence for the central mechanism, however, is much less compelling than the evidence for inhibition of input to the spinal cord. Some controversies have been raised about whether activation of these tracts produces analgesia by an opiate mechanism. Our studies with Akil et al. (6, 7) have shown that activation of this system releases endogenous

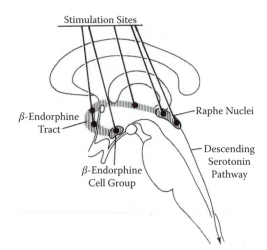

FIGURE 6.2 Stimulation sites tested in the human melano-opio-cortin (beta-endorphin) pathway. (*Appl. Neurophysiol.* 45:116–122, 1982.)

opiates into the ventricular fluid and analgesia is blocked by naloxone in animals (4), and is reduced in humans (21, 42). Questions have been raised by Fessler et al. (12) about the use of contrast material for visualization of the ventricular system during surgery to allow electrode placement, but in our studies the baseline ventricular opiate levels and the post-stimulation opiate levels were both taken long after the administration of contrast material (6, 7). In addition, studies by Yaksh et al. using direct opiate activation of the cells in the periaqueducatal and periventricular gray have shown activation of serotonergic and norepinergic mechanisms in the cord, and the blocking of these monoamine mechanisms chemically reduces the effect of SPA (53, 54, 55) and later with Young et al. in the human patient measured release of beta-endorphin and methionin enkephlelin during PVG stimulation (57, 58). Thus, at the present time, it is thought that activation of target sites in the PVG and PAG activates opiate fibers extending from the periventricular nuclei in the hypothalamus to the raphe nucleus and locus ceruleus, where they terminate and activate serotonergic and norepinergic fibers descending through the dorsolateral funiculus, which terminate in the dorsal horn to inhibit pain at the first synapse, perhaps via an opiate interneuron (Figure 6.3).

The Kolliker–Fuse nucleus, which is in the parabrachial nuclei area, has been described as having norepinephrinergic cell bodies which project inferiorly to the spinal cord. Thus, stimulation of this area may produce direct activation of the descending norepinephrine inhibitor fibers to the dorsal horn with resulting pain relief. Stimulation of areas such as this would, therefore, reduce the chain of events by one synapse, because it directly stimulates the descending norepinephrine fibers and thus is more effective in patients who have a defective opiate mechanism, habituation, or high tolerance to opiates. Young has reported that stimulation produces minimal adverse side effects and was efficacious for pain relief in three of six patients (56).

6.3 Neurophysiology of Ventral Posterior Lateralis, Ventral Posterior Medialis, and Internal Capsule Stimulation

The inhibitory mechanism of stimulation of the main sensory relay nuclei of the thalamus and internal capsule remains obscure, at least in detail. These structures are cephalad extensions of the dorsal columns extending upward through the nucleus Cuniatus and Gracilis from the spinal cord, stimulation of which is known to relieve pain. Our studies in the cat reveal that sectioning of the dorsal columns below the level of activation does not completely obliterate the pain-reducing effects of dorsal column stimulation, indicating an additional more central inhibitory mechanism (39). Complete deactivation of pain

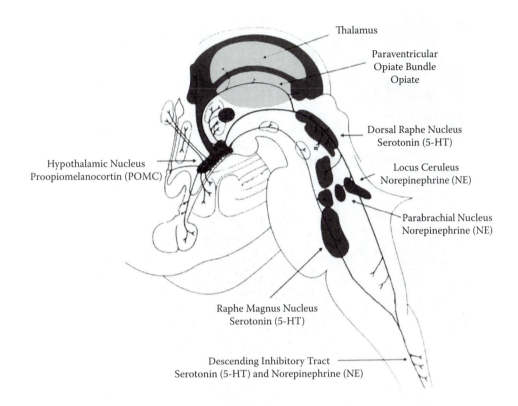

FIGURE 6.3 (See color insert following page 15-4). The beta-endorphin serotonin norepinephrin pain modulation system. (*Neurosurg. Clin. N. Am.* 6(1):135–144, 1995.)

inhibition by dorsal column stimulation can only be obtained by sectioning the dorsal column above and below the level of activation (Richardson, unpublished data). In addition, single-cell recordings in the sensory thalamus reveal an on/off mechanism activated by dorsal column stimulation, which would indicate a complex inhibitory mechanism of interaction between the proprioception, touch, and position sense fibers in the dorsal columns and the pain transmission system in the dorsal horn at the origin of the ventrolateral spino-reticular and spino-thalamic tracts spinal cord, as well as in the synapses in the brainstem, thalamus, and cortex of the brain (40). Mazars et al. (28) in 1960 reported that stimulation of the somatosensory pathways in the cord gave pain relief and Ervin et al. (11) reported pain relief in one patient in 1966 when stimulating the sensory thalamus.

Use of chronic thalamic stimulation was reported by Hosobuchi et al. for anesthesia dolorosa following trigeminal rhizotomy (19). They reported, however, that this stimulation tended to lose its effectiveness over time, and it is necessary to use a programmed type of stimulation with ramping or intermittent stimulation to obtain good pain relief continuously over a long period of time. Placement of electrodes for pain relief in this area is also facilitated by the fact that activation of these centers produces paresthesia in the area of the patient's pain, thus requiring accurate placement of the electrodes.

Hosobuchi et al. (20) suggested using the internal capsule (IC) for neospinothalamic stimulation. I agree that this is more effective and less likely to produce sensory loss following electrode insertion than placement in the main sensory nucleus, the penetration of which may produce small lesions resulting in areas of focal sensory loss, which can be troubling if they involve the face or hand. Insertion of electrodes into the internal capsule does not seem to produce this, although the reason is not clear at this time, but probably indicates that fibers are less vulnerable than cell bodies to the trauma of electrode insertion. This will be addressed in more detail in the discussion of technical development.

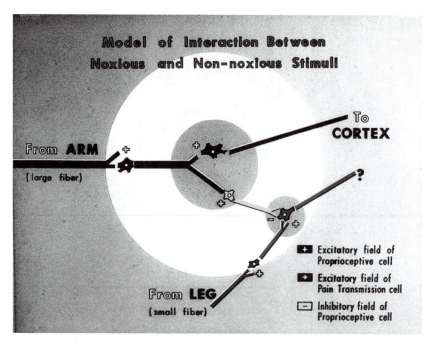

FIGURE 6.4 **(See color insert following page 15-4).** Theoretical interaction of touch/proprioception and pain fibers at each pain synapse. Based on animal studies. (*Surg. Forum* 21:447–449, 1970.)

The inhibitory mechanism of stimulation of the dorsal columns, main sensory relay nuclei of the thalamus and internal capsule which produces paresthesia in the area of the patient's pain is based on the original studies of Melzack and Wall (29) showing that inhibition of pain could be produced by activation of touch and proprioception fibers, originally called the "gate theory of pain transmission." The effect of VPL/VPM/IC as well as dorsal column stimulation are multilevel. The inhibitory mechanism appears to be locally driven at each synapse of the pain pathway through the dorsal horn, brainstem, and thalamus and do not depend on inhibitory mechanisms involving the entry mechanism in the spinal cord alone as in the opiate system (39). Thus, activation of the VPL, VPM, and IC is more effective for relieving deafferentation pain, as well as pain generated in the nervous system due to its ability to inhibit pain neurons at multiple levels, if these structures are intact (Figure 6.4).

6.4   Electrode and Pulse Generator Considerations

There have been several interesting observations over the years during development of the equipment and electrical systems used for stimulation. The original electrode we used on our first patient was solid silver with silver balls (1 mm in diameter) at the tip. We were warned that this electrode would last only a few weeks due to chloridization of the silver ball contacts, which makes them great recording electrodes, but produces degradation of contact between the electrode balls and the brain tissue, as well as the wiring and the silver/silver chloride ball contacts. This makeshift electrode system was connected to a Medtronic external radio-frequency coupled spinal cord stimulation system. Much to our surprise this system lasted between 3 and 4 years before requiring replacement (silver chloride is an excellent electrical conductor) (Figure 6.5).

The first-generation chronic electrode-produced by Medtronic were Schriver-type four-contact pull-down electrodes made from platinum/iridium wire. Soon after their release, we began to notice electrode migration and loss of optimum stimulation; the migration was almost always 5 mm deep to the original placement. We then realized that when the dura is opened, the CSF loss allows the brain to sag; and when CSF volume is restored, the brain floated forward and upward, allowing the semirigid electrode

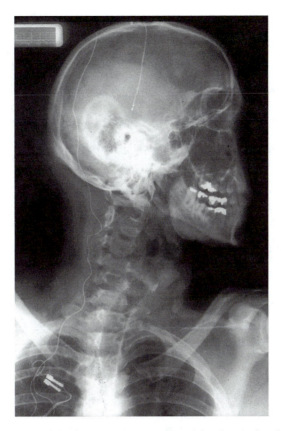

FIGURE 6.5 Post-operative x-ray of the first patient ever implanted for chronic deep brain stimulation, showing silver wiring and electrode adapted to a radio-frequency coupled spinal cord receiver. (*J. Neurosurg.* 47:184–194, 1977.)

to remain in its original position while the brain moved. To solve this problem, the electrode wiring material was changed to pure platinum which is dead soft, allowing the electrode to flex (Figure 6.6).

When electrode replacement was necessary, it became obvious that removal of the four-strand non-encased electrode had became infiltrated with glial tissue between the four wires, and removal was accompanied by a core of scar that if attached to a cerebral vessel could produce vascular injury with resulting intracerebral hemorrhage and hemiplegia or death (personal communications). This required the old electrode to be clipped at the cortex and left in place (Figure 6.7). This problem has been corrected by the most recent revision of the DBS electrode, which is smooth and does not allow glia to infiltrate or adhere to its surface.

Progression from an RF-coupled external pulse generator to a battery-operated internal pulse generator (IPG) has done away with the external transmitter and was a move toward patient acceptance, but for investigational purposes the RF unit is very convenient. In addition, the totally implanted IPG requires replacement for battery depletion; rechargeable battery devices will solve this in the near future.

6.5 Indications and Patient Selection

Patients selected for deep brain implantation first should have chronic intractable pain, defined as unrelieved by the usual procedures directed toward the underlying etiology, that has been present for six months or longer, and which is located in an area not easily controlled by other surgical procedures, such as spinal cord stimulation. Patients with pain in the lower extremities and back may be treated by either spinal cord stimulation or deep brain stimulation, but spinal cord stimulation is a less demanding procedure, and therefore a better choice.

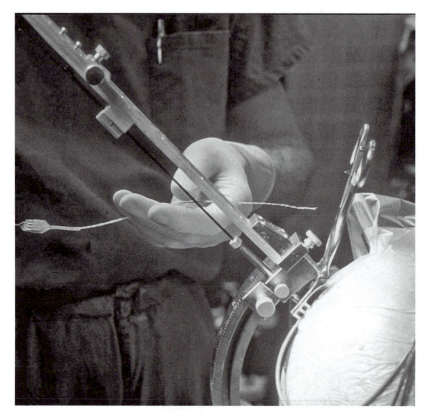

FIGURE 6.6 The original Medtronic Schriver-type electrode at implantation with modified Todd–Wells stereotoxic device.

The selection of deep brain stimulation for the treatment of pain that has not responded to other pain-relieving modalities, such as spinal cord stimulation or other techniques, is an invalid selection criteria. If a patient does not respond to one technique, it does not follow that deep brain stimulation will be effective; the "well, nothing else has been effective so we might as well try it" approach is not appropriate to decision making; but it can be added as another modality if etiology or location predicts that it would be more effective. Deep brain stimulation, while more effective for pain relief than other modalities in some patients, is not significantly more effective than spinal cord stimulation for the relief of pain in the arms, trunk, and lower extremities.

In addition, patients with central pain located in the upper half of the body or patients who have deafferentation pain that cannot be treated by techniques such as spinal cord stimulation are candidates for deep brain implantation. The ideal candidate would be a patient who has deafferentation pain involving the superior quadrant of the body including the face, poststroke pain with the thalamus intact, or spinal cord injury with unilateral pain. These patients can be implanted on the contralateral side of the brain, which would give good coverage of contralateral head, neck, face, arm, and leg.

Patients with bilateral pain secondary to spinal cord injury (i.e., paraplegic pain) are difficult candidates, because VPL/VPM and IC stimulation only gives good pain relief on the side contralateral to stimulation. We have treated patients with paraplegic pain with bilateral VPL or IC electrode implantation, but it is difficult to obtain balanced bilateral stimulation, and patients rarely get more than 50% pain relief in our experience.

One of the most common types of central pain requiring considering DBS is centrally generated pain from brain injury whether from stroke, trauma, or other causes of brain injury. In true thalamic syndrome, however, the target sites may have been damaged; and it can be exquisitely difficult to find an effective site

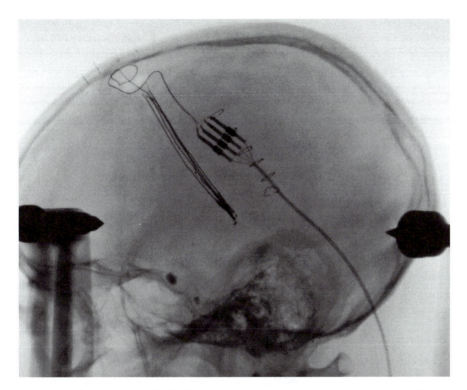

FIGURE 6.7 X-Ray of patient with multiple electrode replacements of Schriver-type electrodes. Old electrodes are cut at cortex. Note retained pantopaque in sellae from previous myelography.

for stimulation except in the most experienced hands. In essence, the usual rules do not apply to patients with damage to the central nervous system. We must remember, in trying to use deep brain stimulation for poststroke pain or other types of centrally generated pain, that we are not dealing with a normal nervous system and the techniques used may have to be extensively modified at times to yield good results.

The factors influencing location of the DBS electrode are more related to the type of pain; neurogenic and deafferentation pain are usually best approached by VPL/VPM/IC stimulation. We have recently reviewed our cases of VPL/VPM/IC stimulation and our results suggest that IC stimulation is more effective for long-term pain relief (45). Somatically generated pain would indicate PVG stimulation as the initial site to try. Somatic pain in the head, face, and neck is a rare condition but an indication for DBS. Neurogenic pain or mixed somatic/neurogenic pain in the face and head is the most common indication for DBS, but it should be remembered PVG activation will often relieve neurogenic pain (47, 57).

Hosobuchi et al., addressing the issue of selection of stimulation sites, devised an intravenous morphine test to determine if the patient's pain was relieved by opiates in an attempt to predict whether the patient would respond to PVG stimulation (opiate sensitive) (19). We have tried this technique in a series of patients and it is quite helpful in some, but in many the results were equivocal.

Technical considerations related to issues such as infection in other areas of potential surgery, arteriovenous malformation of the spinal cord, marked deformity of the spine, generation of pain in the high cervical area or cranium, or spinal cord refractory to stimulation are some reasons for considering DBS. These factories make DBS a viable alternative in some patients.

6.6 Technique

All the patients that we have implanted with deep brain electrodes have had certain basic preoperative studies prior to consideration for surgery. We have a series of psychologic tests performed, which are designed to measure depression, anxiety, psychosis, tendency to exaggerate symptoms, anger, intelligence, and so forth. Our psychologist administers the Minnesota Multi-phasic Personality Inventory, Milton Clinical Multi-axial Inventory, Affect Adjective Stress Test, Beck's Depression Score, and the Beck's or Taylor Anxiety Scale. In addition, our psychiatrist interviews all patients, with all emotional and stress issues being addressed and treated and opiate detoxification carried out. Evidence of psychosis and borderline personality disorder are considered contraindications for surgery. Patients with obsessive–compulsive disorder are in general very difficult to treat with any pain treatment modality because they tend to obsess about their pain, controlled or not.

All patients should be detoxified from all narcotic medication, because exogenous opiates suppress the activity of the endogenous opiate system and thus reduce the effect of activation of this system. From clinical observation, it requires approximately 10 to 14 days for the central nervous system to recover following the last dose of exogenous opiates; therefore, all our patients that we operate on are detoxified and opiate-free for at least 10 to 14 days minimum prior to surgery.

We do not use opiates as premedication, preferring lorazapam 1 mg sublingual as the only premedication. We have used intravenous propofol to produce enough anesthesia and amnesia in patients with high anxiety and apprehension for injection of the local anesthetic and application of the stereotaxic device. After the patient's head has been shaved, local anesthetic infiltrated, and stereotaxic device attached, the propofol anesthesia is allowed to wear off. We prefer to completely shave the head in an attempt to reduce the potential for infection. Propofol anesthesia can also be used for MRI and/or CT scanning but is rarely necessary except in the most severe cases of claustrophobia.

6.6.1 Pariventricular Gray Electrode Placement

We prefer an incision 2.5 cm from the midline at the coronal suture line on the side opposite the more severe pain. Placement closer to the midline was tried early on to allow laying the contacts in the PVG but in some patients subcortical produced leg weakness from traversing the subcortical white matter in the frontal lobe. Our target point ordinarily is 8 to 12 mm posterior to the intercommissural (AC–PC) midpoint and approximately 5 mm lateral to the midline, with the tip of the electrode on or 2 to 5 mm below the AC–PC line, with placement of the tip 1 to 2 mm from the edge of the third ventricle (Figure 6.8). We ordinarily insert a small-diameter exploring electrode and stimulate the area of interest to obtain the results of stimulation. Stimulation of this area usually begins at 250 microsecond pulses at 30 Hz and gradually increases to 8 volts or until side effects are encountered, or pain relief obtained. Side effects from stimulation can be used as an indication of a good target placement. The most reliable indicator is the patient's report that he/she has a sensation of either heat or cold in the contralateral face, which indicates an excellent electrode placement. One of the problems during implantation is that the patient may not be having spontaneous pain while lying on the operating table, therefore, we usually manipulate the part affected, as by straight leg raising, pressure over the painful area, or other manipulation to increase discomfort prior to stimulation. Pinprick sensory testing can be used to determine pain reduction but only one pinprick should be used as a stimulus, because two or three pinpricks always produce localized acute pain. The best response to a single pinprick is, "I couldn't tell, please do it again," which would indicate that the patient has mildly decreased acute pain sensation and indicates a good target placement. A lack of relief of chronic pain, even with decrease in single pinprick testing, denotes good electrode placement but foretells poor results.

Should implantation in the periventricular area not be adequate or be questionable, an electrode can be left in this area and then testing of the internal capsule carried out, or this site can be abandoned and thalamic or IC sites tested.

6.6.2 Thalamic and Internal Capsule Placement

Target sites for thalamic testing were originally taken from the Emmers and Tasker atlas (8) or placed just lateral to the thalamus in the internal capsule, subserving the appropriate portion of the thalamus.

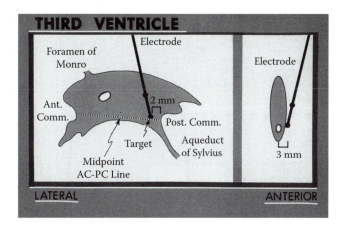

FIGURE 6.8 Target site in the periventricular gray for activation of the beta-endorphin system. (*Neurosurg.* 1:199–202, 1977.)

This atlas is very difficult to translate to the more commonly used Shaltenbrand–Bailey atlas. We have reviewed fifty cases of thalamic and internal capsule placements that were effective for pain relief. From this study we have devised a template of suggested sites for initial electrode testing (Figure 6.9). We usually start testing 5 to 10 mm above the AC–PC line and advance in 2.5 mm increments testing each site, drawing a response chart, and returning to the most effective site. It sometimes takes two or three passes of the electrode at 5 mm offset to obtain good paresthesias in the painful area necessary for good pain relief. Stimulation parameters are different from those used in the periventricular gray; we usually use 60 Hz, 250 μsec pulse width. Care should be exercised while increasing the voltage, because it usually requires less than 2 volts for paresthesia adequate for response.

Once the site of the chronic electrode implantation is determined, the exploring electrode is removed, the chronic electrode insertion tool attached to the stereotaxic device, and the chronic electrode inserted. The electrode is then attached to the skull using a burr hole cap and brought out through a percutaneous puncture site. The percutaneous wire exit site should be inferior and posterior to the burr hole site, so when the patient is lying or standing the drainage is toward the scalp puncture site. A major mistake can be in putting the percutaneous exit site above or anterior to the electrode insertion site, because this may produce stagnation or a lack of drainage with contamination and infection.

Postoperative wound care is critical in preventing infection at the wire exit site, which should be cleaned with peroxide and topical antibiotic ointment applied once or twice daily. Crusting and poor drainage around the wire markedly increase the chance of infection. Using this technique, percutaneous electrodes can be maintained free of infection for long periods of time.

After the electrode has been implanted, it is ill advised to try to test the electrode within 5 to 7 days following surgery, because the edema around the electrode does not allow adequate stimulation, produces poor results, and requires very high stimulation parameters, so our routine is to wait until the seventh postoperative day before stimulation is attempted. If results are not obtained with reasonable stimulation parameters, the testing is put off for several more days.

During the testing period, periventricular area stimulation is usually carried out for 15 minutes every 4 hours, and the degree of pain relief assessed. Stimulation parameters are usually 20 Hz, 250 μsec, and voltage just below side effects level. Over time, the stimulation required to produce good pain relief can usually be decreased, and the duration between stimulation periods can be prolonged. With internal capsule and main sensory nucleus testing, however, stimulation is required on a constant basis or a ramped or frequent intermittent basis. We usually stop stimulation and measure the time for pain to return and then set automatic cycling programmed into the stimulation device.

If the patient obtains good pain relief in testing and wishes to have the system implanted for chronic use, he or she is taken back to surgery for a second procedure under general anesthesia, during which

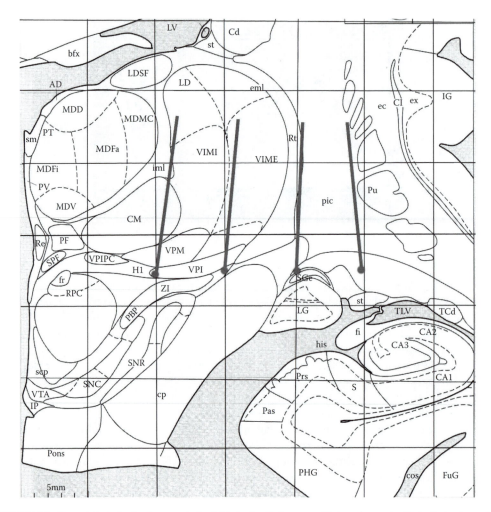

FIGURE 6.9 Suggested electrode placements for trials in the thalamus and internal capsule. Each electrode mark is electrode contact length (1 cm). Electrodes superimposed on atlas in coronal projection 10 mm posterior to commisural midpoint. CM = centrum median, VIMI = ventral posterior medialis, VIME = ventral poster lateralis, pic = posterior limb of internal capsule. (Presented at *European Soc. for Sterotaxic and Funct. Neurosurg.*, Montreux, Switzerland, Oct. 2006.)

the electrode is internalized with a battery-operated internal pulse generator. After the patient recovers from internalization and is off analgesic medications, programming for chronic use can be carried out. We do not allow our patients to have self-programmers, because they tend to deprogram their stimulator from optimal settings and also tend to overstimulate, which reduces battery life. We give our patients only the magnet for turning the device on and off and we program their stimulator parameters to minimize battery drain. If a patient gets 4 hours of pain relief with 10 minutes of stimulation, there is no reason to stimulate every 20 minutes.

6.6.3   Chronic Stimulation

PVG stimulation is usually set at 20 to 30 Hz since this system tends to fatigue if overstimulated, and, since it is a modulation system, it only requires small elevation of its activity. Most patients cannot feel any side effects from stimulation, and care is needed to not overactivate this area. Since patients cannot, tell if the stimulator is on or off, we usually leave it set to self cycle and not to be turned off at night, etc. PVG stimulation rarely requires adjustment.

We keep all patients on a low level of antidepressants, usually a tricyclic; since one of the most common causes of pain control failure is depression. We presume this is related to reduction of brain serotonin and norepinephrine levels. One patient treated with amitrytaline and L-tryptophane to enhance pain relief from PVG stimulation required hospitalization for classic mania. This was resolved after discontinuing medication without sequelae.

Patients with VPL, VPM and IC stimulation often require adjustment of stimulation parameters at varying periods of time; apparently, tolerance to stimulation parameters, not stimulation itself develops. If patients overstimulate, a stimulation holiday for 10 to 14 days can usually reestablish pain control.

6.7   Troubleshooting

Patients will often return saying they are not obtaining good pain relief from deep brain stimulation. The problems fall into two major categories. First, there are changes in the patient that obviate good stimulation pain relief, and second there is a failure of the device.

When a patient reports good side effects, such as paresthesia from internal capsule stimulation, or obtains the usual side effects from periventricular stimulation when the voltage is turned up, this would indicate that the patient's electrode system is functioning properly, but that they have had something has changed in their own internal milieu that would prevent good analgesia. The main reasons for these patients not obtaining good pain relief from stimulation are as follows: First, the use of drugs that inhibit stimulation pain relief, such as a narcotic, would especially prevent good stimulation relief with periventricular stimulation and would require an increase in stimulation intensity for internal capsule and sensory nuclei stimulation. Second, depression may be a marker for depletion of serotonin and norepinephrine, which are necessary for the cascade effect of perivcentricular stimulation for pain relief. Thus, we usually continue our patients over the long term on a small amount of antidepressants, such as trazodone, 100 mg HS, or amitriptyline 50 mg HS, to help prevent tolerance to periventricular stimulation. Third, periods of high stress (anxiety, grief) tend to produce loss of pain relief for unknown reasons. The internal capsule stimulation patients are also maintained on antidepressants, because it has been our observation that patients who become depressed have a decreased response to all types of stimulation analgesia. These patients should first turn their stimulator off for approximately 2 weeks to reduce tolerance and stop medications that will interfere with stimulation of pain relief. A urine drug screen may help at times, and orally administered precursors of norepinephrin and serotonin such as L-dopa to enhance norepinephrine production or L-tryptophan for serotonin production, can be used to enhance analgesia.

Mechanical failure of the stimulation device can occur at several sites, which will be dealt with in order. Constant movement of the system, which is unavoidable, can produce failure of the insulation or electrical wiring, which produces a burning sensation over the wiring or connectors and indicates leakage of current through the insulation. This is usually followed, fairly quickly, by failure of the effective stimulation.

Skin erosion over the electrode burr hole locking device or extension connector to the electrode is a serious complication and requires removal of the hardware because of impending infection. We recommend waiting at least three months before replacing the system.

Fracture of the wiring and failure of electrical contact to the electrode is most commonly adjacent to a fixation point, usually at the locking plug under the scalp, requiring replacement of the electrode. In this situation, the wiring will sometimes show a break on radiography, but often does not, because the electrode and extension wire are multistranded and overlap on the radiographs.

Fracture of the electrode or extension wiring can cause sudden loss of stimulation, but at times can have a stuttering onset and is usually associated with trauma, such as a motor vehicle accident that produces snapping of the head and neck, which can also move the electrode. In this case, checking the impedance of the electrode, through the telemetry system, will usually show that the electrode impedance has increased, which indicates loss of continuity of the wiring.

Patients who have had stimulators for a long time may have battery depletion. This can usually be determined by telemetry, which either returns a response of battery depletion or no response at all, and is usually heralded by intermittent function of the stimulation system.

It is not unusual for patients who have failure of their stimulation system in the periventricular gray to report that their stimulator does not seem to be working but who do not have return of their pain and, therefore, feel that they do not need to have it repaired. This is usually followed in 10 days to 2 weeks by return of their pain to its original level, and they need their system replaced immediately. It is our observation that periventricular stimulation is cumulative and that this pain modulation system probably requires at least 10 days to return to its normal baseline levels, with return of the patient's pain to its prestimulation level.

Patients with VPL/VPM/IC electrodes are acutely aware when their batteries are depleted because of immediate loss of stimulation-induced paresthesias.

6.8   Complications and Side Effects

An extensive list of complications were reported by Levy et al. and I would recommend anyone doing DBS implants to read his reviews (25, 26). Many of these complications are not specifically related to DBS implantation but to surgery in general. But the reported complications and side effects that have been specifically related to DBS are rare and have been reduced by the newer equipment, and reduction of passes necessary for placement of permanent electrodes.

Infection has been reported from 2 to 13% and is less common as more experience has been obtained and with the development of equipment with lower profiles reducing skin erosion over the implant under the scalp. In addition, we have reduced our infection rate following surgery by using intraoperative antibiotics, preventing the implants from touching the skin, irrigation of all incision sites with iodine irrigation solution (1 cc tincture of iodine per liter of solution), and closure with skin clips that do not perforate the skin.

Cerebral hemorrhage with devastating hemiparesis, monparesis, or even death have been reduced by the new electrode which avoids the leukotome effect of the earlier Schreiver-type electrode, and the smooth surface which prevents adhesion to tissue including cerebral vessels and which allows removal of the electrode for replacement without the fear of tearing a cerebral vessel.

Neurological side effects including ptosis, hypesthesia, paresis, dysphoria, dysphasia, confusion, and others are less common with fewer test passes needed for electrode placement, use of MRI/CT instead of ventriculography, and shorter operating time.

Headache following implantation is still an enigma but may be related to proximity to cerebral vessels or pia. We have seen headache in only a few patients in the operating room and have moved the electrode implant site with relief of these symptoms; however, we have one patient with persistent headache that is stimulus parameter sensitive. They are usually unrelated to whether stimulation is on or off, can be seen in any implant site, and may be transient or permanent in our experience (25, 51).

Aggravation of discomfort by stimulation may occur. It may be due to patients describing parethesias as pain; many patients have difficulty describing their symptoms, and care and time are necessary to get an accurate description. We do not have a language for pain and thus rely on "it feels like," which can be very misleading.

6.9   Results

The results of deep brain stimulation for pain relief are difficult to ascertain, because of the wide variety of pain etiology treated by this technique. If we look at the efficacy in general, the results are very good, especially considering that the patient population is made up of otherwise untreatable pain problems. The short-term results of good or excellent pain relief (reduction in pain of more than 50%), in our hands, is 85% of all patients, regardless of etiology, with a gradual reduction to 65% good results after the first year (43). Review of the published results is difficult to analyze because of variation in selection criteria, surgical technique, and preoperative preparation. Results in deafferentation pain, summarized by Gybels and Kupers (14, 15, 16), 14 is 61% initially, deteriorating to 30% good results in 4 years. Gybels (15) found a mean success rate of 57% in his review of the world's literature. Kumar et al (22, 23) reported long-term good results of 63% in 48 patients and suggested poor results in cancer, progressive neurologic disorders, thalamic pain, cauda equina injury, and phantom limb pain, but postlaminectomy syndrome patients did well. Schvarcz (48) found 60% good results in 10 patients with septal area stimulation for pain relief. Plotkin (35) in South Africa obtained 60 to 65% long-term good results in 60 patients with a variety of chronic pain syndromes. Ray and Burton (37) reported 76% good results in 28 patients and suggest intraoperative testing is a good predictor of the long-term results, which has been our observation as well. Mundinger and Salomao (31) report 53% good results in 32 cases, with a variety of stimulation sites and etiologies. Siegfried (49, 50) reports 80% good results in patients with deafferentation pain, but only 50% good results in patients with brachial plexus avulsion or thalamic syndrome. Levy reported in two reviews of DBS for pain that the PVG/PAG stimulation was the only effective site for good results with no effective sites in VPM/VPL/IC sites. We would disagree with that view, having found good relief in those areas in patients with neurogenic and deafferentation pain (25, 26). These patients require frequent changes is stimulation parameters to avoid tolerance to the stimulation parameters but not to DBS itself (45). Kumar reported 63% long-term good results in a mixed group of patients with PVG or sensory thalamus stimulation, which agrees with our results.

The results of DBS for pain control seem to be related to etiology and stimulation site. Good results in deafferentation pain, with sensory nucleus or internal capsule stimulation, and in patients with somatically generated pain, such as postlaminectomy pain, with periventricular gray stimulation. The markers for poor results follow: First, the use of drugs which inhibit stimulation pain, such as narcotics, would especially prevent good stimulation relief with periventricular stimulation and would require an increase in stimulation intensity for internal capsule and sensory nuclei stimulation. Second, depression may be a marker for depletion of serotonin and norepinephrine, which are necessary for the cascade effect of periventricular stimulation for pain relief. Thus, we usually continue our patients over the long term on a small amount of antidepressants, such as trazodone, 100 mg HS, or amitriptyline, 50 mg HS, to help prevent tolerance to periventricular stimulation. Third, periods of high stress (anxiety, grief) tend to produce loss of pain relief for unknown reasons. The internal capsule stimulation patients are also maintained on antidepressants because it has been our observation that patients who become depressed have a decreased response to all types of stimulation analgesia. These patients should first turn their stimulator off for approximately 2 weeks, and we also make sure that they are off all medications that will interfere with stimulation of pain relief. A urine drug screen may need to be used. Orally administered L-dopa can be used to enhance norepinephrine production or L-tryptophan for serotonin production.

Mechanical failure of the stimulation device can occur at several sites, which will be dealt with in order. Constant movement of the system, which is unavoidable, can produce failure of the insulation of electrical wiring, which can produce secondary erosion of the electrode or extension wiring and failure of electrical contact. A burning sensation over the wiring or connectors may indicate leakage of current through the insulation, and is usually followed, fairly quickly, by failure of the device owing to erosion of the wiring secondary to electrolysis. In this situation, the wiring will sometimes show a break on radiography, but often does not, because the electrode and extension wire are multistranded and overlap on the radiographs.

Fracture of the electrode or extension wiring can cause sudden loss of stimulation, but at times can have a stuttering onset and is often associated with trauma, such as a motor vehicle accident that produces snapping of the head and neck, which can also move the electrode. In this case, checking the impedance of the electrode, through the telemetry system, will usually show that the electrode impedance has increased, which indicates loss of continuity of the wiring.

6.10 Discussion

Long-term deep brain stimulation for pain relief was the impetus for the development of a technique for permanent implantation of chronic stimulating electrodes and the necessary equipment to make this possible. This has led to a whole new method of treating many other diseases such as Parkinson's disease, dystonia, essential tremor, Tourette's syndrome, mental diseases, and a whole new field of functional neurosurgery. The need for DBS to relieve pain has diminished in the past few years with development of intrathecal pumps for opiate delivery and with spinal cord stimulation. But there is still a small subgroup of patients where the location and character of their pain leaves no other choice for reduction of severe pain.

Indications for DBS are primarily devoted to relief of pain of neurogenic origin located in the superior quadrant of the body or central pain states from damage to the brain itself.

References

1. Adams, I.E., Hosobuchi, Y., and Fields, H.L. Stimulation of internal capsule for relief of chronic pain. *J. Neurosurg.* 41:740–744, 1974.
2. Akil, H. and Liebeskind, J.C. Monoaminergic mechanisms of stimulation-produced analgesia. *Brain Res.* 94:279–296, 1975.
3. Akil, H. and Richardson, D.E. Contrast medium causes the apparent increase in beta-endorphin levels in human CSF following brain stimulation [letter]. *Pain* 23:301–304, 1985.
4. Akil, H., Mayer, O.J., and Liebeskind, J.C. Antagonism of stimulation-produced analgesia by naloxone, a narcotic antagonist. *Science* 191:961–962, 1976.
5. Akil, H. and Liebeskind, J.C. Monoaminergic mechanisms of stimulation-produced analgesia. *Brain Res.* 94:279–96, 1975.
6. Akil, H., Richardson, D.E., Barchas, J.D., and Li, C.H. Appearance of beta-endorphin-like immunoreactivity in human ventricular cerebrospinal fluid upon analgesic electrical stimulation. *Proc. Nat. Acad. Sci.* 75:5170–5172, 1978.
7. Akil, H., Watson, S.J., Sullivan, S., and Barchas, J.O. Enkephalin like material in normal human CSF: Measurement and levels. *Life Sci.* 23:121–125, 1978.
8. Basbaum, A.I. Opioid regulation of nociceptive and neuropathic pain. *Clin. Neuropharmacol.* 15:372A, 1992.
9. Basbaum, A.I., Clanton, C.H., and Fields, H.L. Opiate and stimulus-produced analgesia: Functional anatomy of a medullospinal pathway. *Proc. Nat. Acad. Sci.* 73:4685–4688, 1976.
10. Emmers, R. and Tasker, R.R. *The Human Somathetic Thalamus.* New York, Raven Press, 1975.
11. Ervin, F.R., Brown, C.E., and Mark, V.H. Striatal influence on facial pain. *Confinia. Neurol.* 27:75–86, 1966.
12. Fessler, R.G., Brown, F.D., Rachlin, J.R., et al. Elevated beta-endorphin in cerebral spinal fluid after electrical brain stimulation: Artifact of contrast infusion? *Science* 224:4652, 1984.
13. Gol, A. Relief by electrical stimulation of the septal area. *J. Neurosci.* 5:115–120, 1967.
14. Gybels, J.M. Electrical stimulation of the central gray for pain relief in humans: A critical review. *Adv. Pain Res. Ther.* 3:0–89004, 1978.
15. Gybels, J.M. Analysis of clinical outcome of a surgical procedure [editorial]. *Pain* 44:103–104, 1991.
16. Gybels, J. and Kupers, R. Deep brain stimulation in the treatment of chronic pain in man: Where and why? *Neurophys. Clin.* 20:389–398, 1990.
17. Heath, R.G. *Studies in Schizophrenia.* Cambridge, Harvard University Press, 1954, p. 1054.
18. Tan, T.C. and Black, P.M. Sir Victor Horsley (1857–1916): pioneer of neurological surgery. *Neurosurgery* 50(3):607–612, 2002.
19. Hosobuchi, Y., Adams, J.E., and Rutkin, B. Chronic thalamic stimulation for the control of facial anesthesia dolorosa. *Arch. Neurol.* 29:158–161, 1973.
20. Hosobuchi, Y., Adams, J.E., and Rutkin, B. Chronic thalamic and internal capsule stimulation for the control of central pain. *Surg. Neurol.* 4:91–92, 1975.
21. Hosobuchi, Y., Adams, J.E., and Linchitz, R. Pain relief by electrical stimulation of the central gray matter in humans and its reversal by naloxone. *Science* 197:183–186, 1977.

22. Kumar, K., Wyant, G.M., and Nath, R. Deep brain stimulation for control of intractable pain in humans, present and future: A ten-year follow-up. *Neurosurgery* 26:774–781, 1990.

23. Kumar, K., Toth, C., and Nath, R.K. Deep brain stimulation for intractable pain: a 15-year experience. *Neurosurgery* 40:736–747, 1997.

24. Landau, B. and Levy, R. v1: Neuromodulation techniques for medically refractory chronic pain. *Ann. Rev. Med.* 44:279–287, 1993.

25. Levy, R.M., Lamb, R.N., and Adams, J.E. Treatment of chronic pain by deep brain stimulation: long term follow-up and review of the literature. *Neurosurgery* 21:885–893, 1987.

26. Levy, R.M. Deep brain stimulation for treatment of intractable pain. *Neurosurg. Clin. N. Am.* 14:389–399, 2003.

27. Mayer, O.J., Wotfle, T.L., Akil, H., Carter, B., and Liebeskind, J.C. Analgesia from electrical stimulation in the brainstem of the rat. *Science* 174:1351–1354. 1971.

28. Mazars, G., Roge, R., and Mazars, Y. Stimulation of the spinothalamic fasciculus and their bearing on the physiology of pain. *Rev. Neurology*: 136–138, 1960.

29. Melzack, R. and Wall, P.D. Pain mechanisms: a new theory. *Science.* 150(699):971–979, 1965.

30. Meyerson, B.A. Biochemistry of pain relief with intracerebral stimulation. Few facts and many hypotheses. *Acta Neurochir.* 30 (suppl):229–237, 1980.

31. Mundinger, F. and Salomao, J.F. Deep brain stimulation in mesencephalic lemniscus medialis for chronic pain. *Acta Neurochir.* 30 (suppl):245–258, 1980.

32. Nandi, D., Aziz, T., Carter, H., and Stein, J. Thalamic field potentials in chronic central pain treated by periventricular gray stimulation — a series of eight cases. *Pain* 101:97–107, 2003.

33. Oliveras, J.L., Hosobuchi, Y., Redjemi, F., et al. Opiate antagonist, naloxone, strongly reduces analgesia induced by stimulation of a raphe nucleus (centralis inferior). *Brain Res.* 120:221–229, 1977.

34. Pilcher, W.H., Joseph, S.A., and McDonal, J.V. Immunocytochemical localization of pro-opiomelanocortin neurons in human brain areas subserving stimulation analgesia. *J. Neurosurg.* 68(4):621–9, 1988.

35. Plotkin, R. Results in 60 cases of deep brain stimulation for chronic intractable pain. *Appl. Neurophysiol.* 45:173–178, 1982.

36. Pool, J.L., Clark, W.D., Hudson, P., and Lombardo, M. *Hypothalamus Hypophyseal Interrelationships.* Springfield, Charles C. Thomas, 1956.

37. Ray, C.D. and Burton, C.V. Deep brain stimulation for severe, chronic pain. *Acta Neurochir.* 30 (suppl): 289–293, 1980.

38. Reynolds, D.V. Surgery in the rat during electrical analgesia induced by focal brain stimulation. *Science* 164:444–445, 1969.

39. Richardson, D.E. Autoinhibition in the sensory system of the cat. *Surg. Forum* 21:447–449, 1970.

40. Richardson, D.E. Single unit responses in the neospinothalamic system of the cat. *J. Surg. Res.* 14:472–477, 1973.

41. Richardson, D.E. and Akil, H. Pain reduction by electrical brain stimulation in man. Part I: Acute administration in periaqueducal and periventricular sites. *J. Neurosurg.* 47:178–183, 1977.

42. Richardson, O.E. and Akil, H. Pain reduction by electrical brain stimulation in man. Part 2: Chronic self-administration in the periventricular gray matter. *J. Neurosurg.* 47:184–194, 1977.

43. Richardson, D.E. and Akil, H. Long-term results of periventricular gray self-stimulation. *Neurosurgery* 1:199–202, 1977.

44. Richardson, D.E. Analgesia produced by stimulation of various sites in the human beta-endorphin system. *Appl. Neurophysiol.* 45:116–122, 1982.

45. Richardson, D.E. Deep brain stimulation for the relief of chronic pain. *Neurosurg. Clin. N. Am.* 6(1):135–144, 1995.

46. Richardson, D.E. Intracranial stimulation for the relief of chronic pain. *Clin. Neurosurg.* 31:316–322, 1983.

47. Rhodes, D.L. and Liebeskind, J.C. Analgesia from rostral brain stem stimulation in the rat. *Brain Res.* 143(3):521–532, 1978.

48. Schvarcz, J.R. Chronic stimulation of the septal area for the relief of intractable pain. *Appl. Neurophysiol.* 48:191–194, 1985.

49. Siegfried, J. Sensory thalamic neurostimulation for chronic pain. *Paceing and Clinical Electrophysiol.* 10:209–212, 1987.

50. Siegfried, J. Neurologic pacemakers. A three decades' evaluation (editorial). *Neurochirurgie* 37:81–85, 1991.

51. Veloso, F., Kumar, K., and Toth, C. Headache secondary to deep brain implantation. *Headache* 38:507–515, 1998.

52. Watson, S.J., Barchas, J.D., and Li, C.H. Beta-lipotropin: localization of cells and axons in rat brain by immuno-cytochemistry. *Proc. Nat. Acad. Sci.* 74:5155–5158, 1977.

53. Yaksh, T.L. Direct evidence that spinal serotonin and noradrenaline terminals mediate the spinal antinociceptive effects of morphine in the periaqueductal gray. *Brain Res.* 160:180–185, 1979.

54. Yaksh, T.L. Pharmacology of spinal adrenergic systems which modulate spinal nociceptive processing. *Pharm. Biochem. Behav.* 22:845–858, 1985.

55. Yaksh, T.L. Opioid receptor systems and the endorphins: A review of their spinal organization. *J. Neurosurg.* 67,157–176, 1987.

56. Young, R.F., Tronnier, V., and Rinaldi, P.C. Chronic stimulation of the Kolliker-Fuse nucleus region for relief of intractable pain in humans. *J. Neurosurg.* 76:979–985, 1992.

57. Young, R.F., Chambi, V.I. Pain relief by electrical stimulation of the periaquiductal and periventricular gray matter. Evidence for non-opioid mechanism. *J. Neurosurg.* 66:364–371, 1987.

58. Young, R.F., Flemming, W.B., Van Norman, A.S., and Yaksh, T.L. Release of *beta*-endorphins and methionine-enkephalin into cerebrospinal fluid during deep brain stimulation for chronic pain. *J. Neurosurg.* 79:816–825, 1993.

7

Spinal Cord Stimulation for Pain Management

Allen W. Burton and
Phillip C. Phan

7.1 Introduction

Spinal cord stimulation (SCS) describes the use of pulsed electrical energy near the spinal cord to control pain.[1] This technique was first applied in the intrathecal space and finally in the epidural space as described by Shealy in 1967.[2] At the present time, most commonly neurostimulation involves the implantation of leads in the epidural space to transmit this pulsed energy across the spinal cord or near the desired nerve roots. This chapter concentrates on this modality: spinal cord stimulation, sometimes called dorsal column stimulation, for pain control. This technique has notable analgesic properties for neuropathic pain states, anginal pain, and peripheral ischemic pain. The same technology can be applied in deep brain stimulation, cortical brain stimulation, and peripheral nerve stimulation.[3–5] These techniques are mainly in the realm of the neurosurgeon.

7.1.1 Mechanism of Action

Neurostimulation began shortly after Melzack and Wall[6] proposed the gate control theory in 1965. This theory proposed that painful peripheral stimuli carried by C-fibers and lightly myelinated A-delta fibers terminated at the substantia gelatinosa of the dorsal horn (the gate). Large myelinated A-beta fibers responsible for touch and vibratory sensation also terminated at "the gate" in the dorsal horn. It was hypothesized that their input could be manipulated to "close the gate" to the transmission of painful stimuli. As an application of the gate control theory, Shealy[2] implanted the first spinal cord stimulator device for the treatment of chronic pain. This technique was noted to control pain, and it has undergone

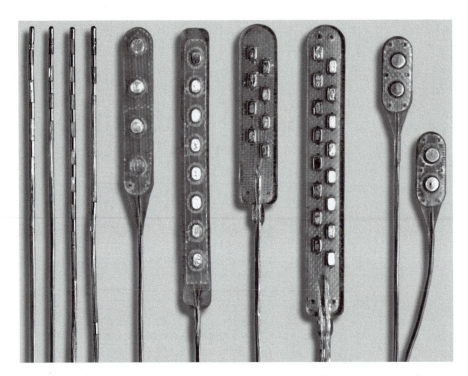

FIGURE 7.1 Neurostimulator leads (left to right): percutaneous type to paddle type. (Courtesy of ANS, Inc.)

numerous technical and clinical refinements in the ensuing years. Although the gate theory was initially proposed as the mechanism of action, the exact underlying neurophysiologic mechanisms remain not clearly understood.

Although the neurophysiologic mechanisms of action of spinal cord stimulation are not completely understood, recent research has given us insight into effects occurring at the local and supraspinal levels, and through dorsal horn interneuron and neurochemical mechanisms.[7, 8] Linderoth and others have noted that the mechanism of analgesia when SCS is applied in neuropathic pain states may be very different from that involved in analgesia due to limb ischemia or angina. Experimental evidence points to SCS as having a beneficial effect at the dorsal horn level by favorably altering the local neurochemistry in that zone, thereby suppressing the hyperexcitability of the wide dynamic range interneurons. Specifically, there is some evidence for increased levels of GABA release, serotonin, and perhaps suppression of levels of some excitatory amino acids, including glutamate and aspartate. In the case of ischemic pain, analgesia seems to derive from restoration of a favorable oxygen supply and demand balance — perhaps through a favorable alteration of sympathetic tone.

7.2   Technical Considerations

SCS is a technically challenging interventional pain management technique. It involves the careful placement of an electrode array (leads) in the epidural space, a trial period, anchoring the lead(s), positioning and implantation of the pulse generator or RF receiver, and the tunneling and connection of the connecting wires. The collaborative effort between a surgeon and anesthesiologist pain specialist is recommended to ensure optimal success with neurostimulation.

Electrodes are of two types, catheter or percutaneous vs. paddle or surgical (Figure 7.1). These electrodes are connected to an implanted pulse generator (IPG) or a radio-frequency (RF) unit (Figures 7.2a, b, c). Currently, three companies (Advanced Bionics, Medtronic, and American Neuro-modulation Systems) make neurostimulation equipment (see Appendix 7A). Interested readers are directed to these companies for further specific information on the equipment.

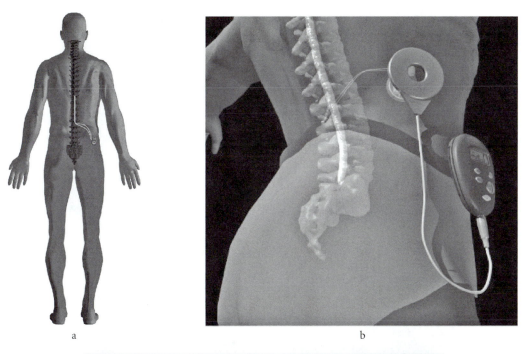

a   b

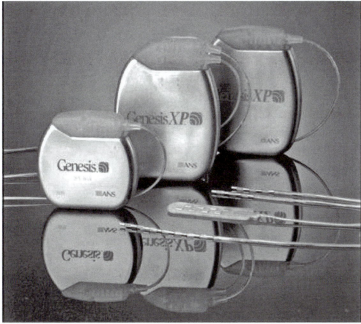

c

FIGURE 7.2   (See color insert following page 15-4). (a) Schematic view of an IPG system implanted; (b) schematic view of an implanted RF SCS system; and (c) representative IPG neurostimulation units with leads. (Courtesy of ANS, Inc.)

A stimulator trial may be accomplished using either one of two techniques: "straight percutaneous" or "implanted lead." In both trial methods, under fluoroscopy and sterile conditions, a lead is introduced into the epidural space with the standard epidural needle placement (Figures 7.3a, b, c). The lead is steered under fluoroscopic imaging into the posterior paramedian epidural space up to the desired anatomic location. Trial stimulation is undertaken to attempt to "cover" the painful area with an

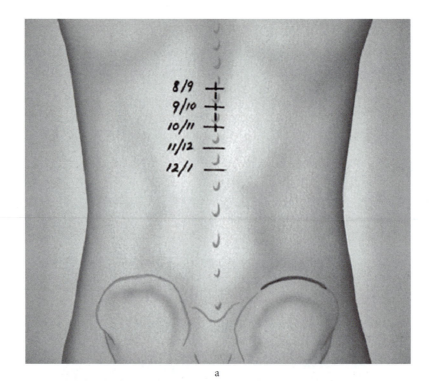

8/9
9/10
10/11
11/12
12/1

a

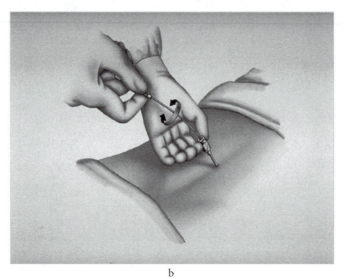

b

FIGURE 7.3 Percutaneous lead placement: (a) marking the interspinous level; (b) percutaneous lead insertion; and (c) dual lead trial. (Courtesy of Medtronic, Inc.)

electrically induced paresthesia. After the painful area is "captured" either with one or two leads, the two techniques diverge.

In the straight percutaneous trial, the needle is withdrawn; an anchoring suture is placed into the skin, and a sterile dressing is applied. When the patient returns after a several-day trial, the dressing is removed, suture clipped, lead removed and discarded *regardless* of the success of the trial. When the patient returns for implant, a new lead is placed in the location of the trial lead and connected to an implanted IPG.

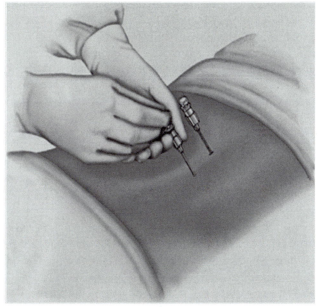

c

FIGURE 7.3 (continued)

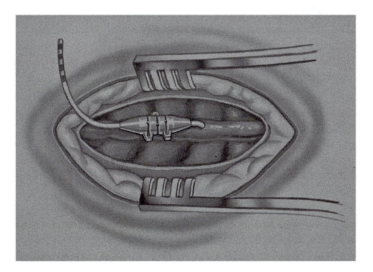

FIGURE 7.4   Anchoring the lead. (Courtesy of Medtronic, Inc.)

In the "implanted lead" trial, after successful positioning of the trial leads, local anesthesia is infiltrated around the needles and an incision is made, cutting down to the supraspinous fascia to anchor the leads securely using nonabsorbable sutures (Figure 7.4). Then a temporary extension piece is tunneled away from the back incision and out through the skin. This exiting percutaneous extension piece is secured to the skin using a suture, with antibiotic ointment and a sterile dressing subsequently applied. If the trial is successful, then at the time of implant, the back incision is opened, the percutaneous extension piece is cut, pulled out through the skin site, and discarded. The permanent leads that were used for the trial are hooked to new extensions and tunneled to an implanted IPG.

The "implanted lead" method has the advantages of saving the cost of new electrodes at implant and ensuring that the implanted lead position matches the trial lead position. Advantages of the "percutaneous lead" approach include avoiding the costs of two trips to the operating room (even for an unsuccessful trial to remove the anchored trial lead); avoiding an incision and postoperative pain during the trial, which may confuse trial interpretation by the patient; and the percutaneous temporary extension piece poses a higher risk for infection. The percutaneous extension must be anchored and meticulously dressed or the risk of infection may be higher than with the straight percutaneous technique.[9] Most consider 50% or more pain relief to be indicative of a successful trial, although the ultimate decision also should include other factors such as activity level and medication intake. To paraphrase, some combination of pain relief, increased activity level, and decreased medication intake is indicative of a favorable trial.

A trial with paddle-type electrodes requires the "implanted lead" approach, with the significant addition of a laminotomy to slip the flat plate electrode into the epidural space. Some physicians trial the patient with the "straight percutaneous" approach and, if successful, will send the patient to a neurosurgeon for a paddle-type implant. The authors' preference is to do a "straight percutaneous" trial with an implant using non-paddle type electrodes.

The IPG/RF unit is usually implanted in the lower abdominal area or in the posterior superior gluteal area (Figure 7.5a, b). It should be in a location that patients can access with their dominant hand for adjustment of their settings with the patient-held remote control unit. The decision to use a fully implantable IPG or an RF unit depends on several considerations. If the patient's pain pattern requires the use of many anode/cathode settings with high power settings during the trial, consideration of an RF unit should be given. The IPG battery life will largely depend on the power settings utilized, but the newer IPG units (Synergy® or Genesis XP®) will generally last several years at average power settings.

7.3 Patient Selection

Appropriate patients for neurostimulation implants must meet the following criterion: the patient has a diagnosis amenable to this therapy (i.e., neuropathic pain syndromes), the patient has failed conservative therapy, significant psychological issues have been ruled out, and a trial has demonstrated pain relief.[10] However, pure neuropathic pain syndromes are relatively less common than the mixed nociceptive/neuropathic disorders, including failed back surgery syndrome (FBSS) (Figure 7.6a, b). Also, many patients with chronic pain will have some depressive symptomatology, but psychological screening can be extremely helpful to avoid implanting patients with major psychological disorders. An interesting study by Olson and colleagues[11] revealed a high correlation between many items on a complex psychological testing battery and favorable response to trial stimulation. This is to say, an overall mood state is an important predictor of outcomes.

A careful trial period is advocated to avoid a failed implant. Trials of different lengths have been advocated; the risks of a longer trial are mainly infection, whereas the risks of too short a trial are misreading success. The author utilizes a five- to seven-day trial with the use of oral antibiotics. We encourage the patient to be as active as possible in their usual environment, with the exception of limiting bending/twisting movements. Despite advances in (1) the understanding of diagnoses that respond to neurostimulation, (2) an increased understanding of and improved psychological screening, and (3) improved multi-lead systems, clinical failures of implanted neurostimulator devices remain too common. Pain practitioners must critically evaluate their own outcomes and adhere to the strict selection criterion outlined above.

7.4 Complications

Complications with spinal cord stimulation range from simple easily correctable problems such as lack of appropriate paresthesia coverage to devastating complications such as paralysis, nerve injury, and death. Prior to the implantation of the trial lead, an educational session should occur with the patient and significant family members. This meeting should include a discussion of possible risks and complications.

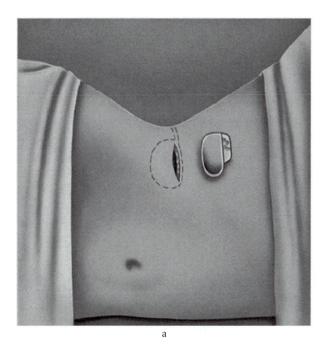

a

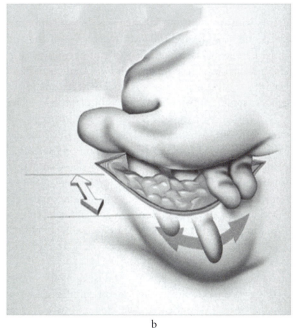

b

FIGURE 7.5 (**See color insert following page 15-4**). Permanent implant: pulse generator internalization (a) in the lower abdominal area or (b) in the posterior superior gluteal area. (Courtesy of Medtronic, Inc.)

In the postoperative period, the caregiver should be involved in identifying problems and alerting the health care team.

North and colleagues[12] reported their experience in 320 consecutive patients treated with SCS between 1972 and 1990. A 5% rate of subcutaneous infection was seen and is consistent with the literature. The predominant complication consisted of lead migration or breakage. This remains the "Achilles' heel" of neurostimulation. In an earlier series, bipolar leads required electrode revision in 23% of patients.

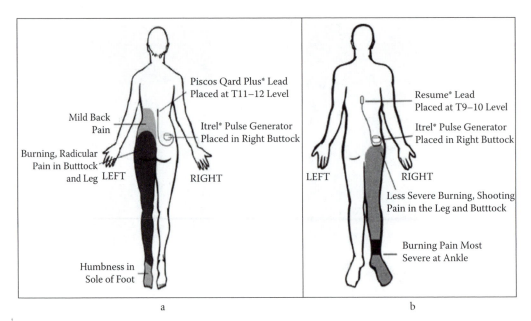

FIGURE 7.6 Ideal candidates: FBSS/CRPS. Note radicular (a) versus axial (b) pain pattern. (Courtesy of Medtronic, Inc.)

The revision rate for patients with multichannel devices was 16%. Failure of the electrode lead was observed in 13% of patients and steadily declined over the course of the study. When analyzed by implant type (single-channel percutaneous, single-channel laminectomy, and multichannel), the lead migration rate for multichannel devices was approximately 7%. Analysis of hardware reliability for 298 permanent implants showed that technical failures (particularly electrode migration and malposition) and clinical failures had become significantly less common as implants had evolved into programmable, "multichannel" devices.

More recent studies by Barolat[13] and May[14] reported lead revision rates due to lead migration of 4.5% and 13.6% and breakage of 0% and 13.6%, respectively. Infections occurred in 7% and 2.5% of cases, respectively. No serious complications were seen in either study. These studies are representative of the complication rate of neurostimulation therapy.

Infections range from simple infections at the surface of the wound to epidural abscess. The patient should be instructed on wound care and recognition of signs and symptoms indicative of infection. Many superficial infections can be treated with oral antibiotics or simple surgical exploration and irrigation. At the authors' center, the standard includes prophylactic intraoperative antibiotics and oral coverage postoperatively for ten days.

If infection reaches the tissues involving the devices, in most cases the implant should be removed. In such cases, one should have a high index of suspicion for an epidural abscess. Abscess of the epidural space can lead to paralysis and death if not identified quickly and treated aggressively. In the case of temporary epidural catheters (somewhat analogous to a percutaneous stimulator trial), Sarubbi[15] discovered only twenty well-described cases. The mean age of these twenty-two patients was 49.9 years, the median duration of epidural catheter use was three days, and the median time to onset of clinical symptoms after catheter placement was five days. The majority of patients (63.6%) had major neurological deficits and 22.7% also had concomitant meningitis. Staphylococcus aureus was the predominant pathogen. Despite antibiotic therapy and drainage procedures, 38% of the patients continued to have neurological deficits. These unusual but serious complications of temporary epidural catheter use require efficient and accurate diagnostic evaluation and treatment, as the consequences of delayed therapy can be substantial. Schuchard reported an infection with Pasturella during an "implanted lead" trial, which required explanting the system.[16]

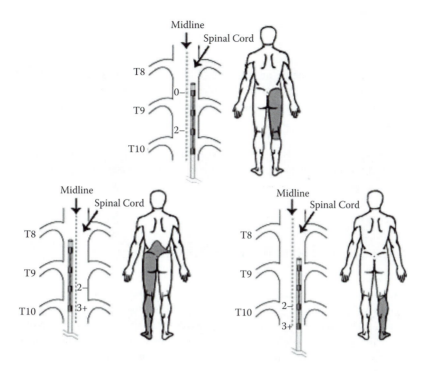

FIGURE 7.7 Typical patterns of coverage using different anodal and cathode combinations. (Courtesy of Medtronic, Inc.)

7.5 Programming

There are four basic parameters in neurostimulation that can be adjusted to create stimulation parasthesias in the painful areas, thereby mitigating the patient's pain (Figure 7.7). They are amplitude, pulse width, rate, and electrode selection.[17]

Amplitude is the intensity or strength of the stimulation measured in volts (V). The voltage can be set from 0 to 10 V, with lower settings typically used over peripheral nerves and with paddle-type electrodes. Pulse width is a measure in microseconds (μs) of the duration of a pulse. Pulse width is usually set between 100 and 400 μs. A larger pulse width will typically give the patient broader coverage. Rate is measured in hertz (Hz) or cycles per second, between 20 and 120 Hz. At lower rates, the patient feels more of a thumping, whereas at higher rates, the feeling is more of a buzzing. Electrode selection is a complex topic that has been the subject of some research by Barolat and colleagues,[18] who provided mapping data of coverage patterns based on lead location in 106 patients. The primary target is the cathode ("–"), with electrons flowing from the cathode(s) "–" to the anode(s) "+." Most patients' stimulators are programmed with electrode selection changed until the patient obtains anatomic coverage, then the pulse width and rate are adjusted for maximal comfort. The patient is left with full control of turning the stimulation off and on, and the voltage up and down to comfort.

An analogy of a stereo system can be used to discuss programming with patients. Amplitude is the "volume control," the pulse width is "how many speakers are on—mono versus surround sound," and the rate is the "bass or treble control." The lowest acceptable settings on all parameters are generally used to conserve battery life. Other programming modes, which save battery life, include a cycling mode during which the stimulator cycles full on/off at patient-determined intervals (minutes, seconds, or hours). The patient's programming may change over time, and reprogramming needs are common. Neurostimulator manufacturing companies can be very helpful to clinicians with patient reprogramming assistance. Many busy pain practices designate a stimulator nurse to handle patient reprogramming needs.

7.6 Outcomes

The most common use for SCS in the United States is failed back surgery syndrome (FBSS); whereas in Europe, peripheral ischemia is the predominant indication. With respect to clinical outcomes, it makes sense to subdivide the outcomes based on diagnosis. In a review of the available SCS literature, most evidence falls within the level IV (limited) or level V (indeterminate) categories due to the invasiveness of the modality and inability to provide blind treatment. Recognition also must be given to the time frame within which a study was performed due to rapidly evolving SCS technology. Basic science knowledge, implantation techniques, lead placement locations, contact array designs, and programming capabilities have changed dramatically since the time of the first implants. These improvements have led to decreased morbidity and much greater probability of obtaining adequate parasthesia coverage with subsequent improved outcomes.[19] Thus, even a level II review study such as the one by Turner[20] with FBSS patients from 1966 to 1994 reported less positive outcomes than Barolat's level IV FBSS study[21] in 2001. This difference in efficacy may represent the effect of improving technology.

7.6.1 SCS Outcomes: Failed Back Surgery Syndrome (FBSS)

There has been one recent prospective, randomized study. North et al.[22] selected fifty patients as candidates for repeat laminectomy. All patients had undergone previous surgery and were excluded from randomization if they presented with severe spinal canal stenosis, extremely large disc fragments, a major neurological deficit such as foot drop, or radiographic evidence of gross instability. In addition, patients were excluded for untreated dependency on narcotic analgesics or benzodiazepines, major psychiatric comorbidity, the presence of any significant or disabling chronic pain problem, or a chief complaint of low back pain exceeding lower extremity pain. Crossover between groups was permitted. The six-month follow-up report included twenty-seven patients. At this point, they became eligible for crossover. Of the fifteen patients who had undergone reoperation, 67% (ten patients) crossed over to SCS. Of the twelve patients who had undergone SCS, 17% (two patients) opted for crossover to reoperation. Additionally, of the nineteen patients who reached their six-month follow-up assessment after reoperation, 42% (eight patients) opted for spinal cord stimulation outside the study. For 90% of the patients, long-term (three-year follow-up) evaluation has shown that spinal cord stimulation continues to be more effective than reoperation, with significantly better outcomes by standard measures and significantly lower rates of crossover to the alternate procedure. Additionally, patients randomized to reoperation used significantly more opioids than those randomized to spinal cord stimulation. Other measures assessing activities of daily living and work status did not differ significantly. The preliminary results have been published in abstract format, but the definitive study has yet to be published.[22]

Two recent, prospective case series have been done. The first, by Barolat[23] examined the outcomes of patients with intractable low-back pain treated with epidural SCS utilizing paddle electrodes and a radio-frequency (RF) stimulator. In four centers, forty-four patients were implanted and followed with the visual analog scale (VAS), the Oswestry Disability Questionnaire, the Sickness Impact Profile (SIP), and a patient satisfaction rating scale. All patients had back and leg pain, and all had at least one previous back surgery, with most (83%) having two or more back surgeries and 51% having had a spinal fusion. Data were collected at baseline, at six months, twelve months, and two years.

All patients showed a reported mean decrease in their ten-point VAS scores compared to baseline. The majority of patients reported fair to excellent pain relief in both the low back and legs. At six months, 91.6% of the patients reported fair to excellent relief in the legs and 82.7% of the patients reported fair to excellent relief in the low back. At one year, 88.2% of the patients reported fair to excellent relief in the legs and 68.8% of the patients reported fair to excellent relief in the low back. Significant improvement in function and quality of life was found at both the six-month and one-year follow-ups using the Oswestry and SIP, respectively. The majority of patients reported that the procedure was worthwhile (92% at six months, 88% at one year). The authors concluded that SCS proved beneficial at one year for the treatment of patients with chronic low back and leg pain.

The second multicenter prospective case series was published by Burchiel in 1996.[24] The study included 182 patients with a permanent system after a percutaneous trial. Patient evaluation of pain and functional levels was performed before and three, six, and twelve months after implantation. Complications, medication usage, and work status also were monitored. A one-year follow-up evaluation was available for seventy patients. All pain and quality-of-life measures showed statistically significant improvement, whereas medication usage and work status did not significantly improve, during the treatment year. Complications requiring surgical interventions were experienced by 17% (twelve of seventy) of the patients.

There have been two systematic review articles on neurostimulation. Turner et al.[25] completed a meta analysis from the articles related to the treatment of FBSS by SCS, from 1966 to 1994. They reviewed thirty-nine studies that met the inclusion criteria. The mean follow-up period was sixteen months with a range of one to forty-five months. Pain relief exceeding 50% was experienced by 59% of patients with a range of 15 to 100%. Complications occurred in 42% of patients, with 30% of patients experiencing one or more stimulator-related complications. However, all the studies were case-control investigations. Based on this review, the authors concluded that there was insufficient evidence from the literature for drawing conclusions about the effectiveness of spinal cord stimulation relative to no treatment or other treatments, or about the effects of spinal cord stimulation on patient work status, functional disability, and medication use.

The second analysis, by North and Wetzel,[26] consisted of a review of case-control studies and two prospective control studies. They concluded that if a patient reports a reduction in pain of at least 50% during a trial, as determined by standard rating methods, and demonstrates improved or stable analgesic requirements and activity levels, significant benefit may be realized from a permanent implant. The authors conclude that the bulk of the literature appears to support a role for spinal cord stimulation (in neuropathic pain syndromes), but also caution that the quality of the existing literature is marginal — largely case series.

7.6.2 SCS Outcomes: Complex Regional Pain Syndrome (CRPS)

Research of high quality regarding SCS and CRPS is limited, but existing data are overwhelmingly positive in terms of pain reduction, quality of life, analgesic usage, and function. Kemler and co-workers[27] published a prospective, randomized, comparative trial to compare SCS versus conservative therapy for CRPS. Patients with a six-month history of CRPS of the upper extremities were randomized to undergo trial SCS (and implant if successful) + physiotherapy vs. physiotherapy alone. In this study, thirty-six patients were assigned to receive a standardized physical therapy program together with spinal cord stimulation, whereas eighteen patients were assigned to receive therapy alone. In twenty-four of the thirty-six patients randomized to spinal cord stimulation along with physical therapy, the trial was successful and permanent implantation was performed. At a six-month follow-up assessment, the patients in the spinal cord stimulation group had a significantly greater reduction in pain; and a significantly higher percentage was graded as much improved for the global perceived effect. However, there were no clinically significant improvements in functional status. The authors concluded that in the short term, spinal cord stimulation reduces pain and improves the quality of life for patients with CRPS involving the upper extremities.

Several important case series have been published on the use of neurostimulation in the treatment of CRPS. Calvillo et al.[28] reported a series of thirty-six patients with advanced stages of complex regional pain syndrome (at least two years' duration) who had undergone successful SCS trial (>50% reduction in pain). They were treated with either spinal cord stimulation or peripheral nerve stimulation, and in some cases with both modalities. Thirty-six months after implantation, the reported pain measured on Visual Analog Scales was an average of 53% better; this change was statistically significant. Analgesic consumption decreased in the majority of patients, and 41% of patients had returned to work on modified duty. The authors concluded that in late stages of complex regional pain syndrome, neurostimulation (with SCS or PNS) is a reasonable option when alternative therapies have failed.

Another case series reported by Oakley and Weiner[29] is remarkable in that it utilized a sophisticated battery of outcomes tools to evaluate treatment response in complex regional pain syndromes (CRPS) using spinal cord stimulation. The study followed nineteen patients and analyzed the results from the McGill Pain Rating Index, the Sickness Impact Profile, Oswestry Disability Profile, Beck Depression Inventory, and Visual Analog Scale. Nineteen patients were reported as a subgroup enrolled at two centers participating in a multicenter study of the efficacy/outcomes of spinal cord stimulation. Specific pre-implant and postimplant tests to measure outcome were administered. Statistically significant improvement in the Sickness Impact Profile physical and psychosocial subscales was documented. The McGill Pain Rating Index words chosen and sensory subscale also improved significantly, as did Visual Analog Scale (VAS) scores. The Beck Depression Inventory trended toward significant improvement. All patients received at least partial relief and benefit from their device, with 30% receiving full relief. Eighty percent (80%) of the patients obtained at least 50% pain relief through the use of their stimulators. The average percent of pain relief was 61%. The authors concluded that patients with CRPS benefit significantly from the use of spinal cord stimulation, based on average follow-up of 7.9 months.

A literature review by Stanton-Hicks of SCS for CRPS consisted of seven case series. These studies ranged in size from six to twenty-four patients. Results were noted as "good to excellent" in greater than 72% of patients over a time period of eight to forty months. The review concluded that SCS has proven a powerful tool in the management of patients with CRPS.[30]

A retrospective, three-year, multicenter study of 101 patients by Bennett et al.[31] evaluated the effectiveness of SCS applied to complex regional pain syndrome I (CRPS I) and compared the effectiveness of octapolar vs. quadrapolar systems, as well as high-frequency and multi-program parameters. VAS was significantly decreased in the group using the dual-octapolar system with reductions in overall VAS approaching 70%. Of the dual-octapolar group, 74.8% used multiple arrays to maximize paresthesia coverage. VAS reduction in the group using quadrapolar systems approached 50%. And 86.3% of quadrapolar systems and 97.2% of dual-octapolar systems continued to be utilized. Overall satisfaction with stimulation was 91% in the dual-octapolar group and 70% in the quadrapolar group ($p < 0.05$). The authors concluded that SCS is effective in the management of chronic pain associated with CRPS I, and that use of dual-octapolar systems with multiple-array programming capabilities appeared to increase the paresthesia coverage and thus further reduce pain. High-frequency stimulation (>250 Hz) was found to be essential in obtaining adequate analgesia in 15% of the patients using dual-octapolar systems (this frequency level was not available to those with quadrapolar systems).

7.6.3 SCS Outcomes: Peripheral Ischemia and Angina

Cook and co-workers[32] reported in 1976 that SCS effectively relieved pain associated with peripheral ischemia. This result has been repeated and noted to have particular efficacy in conditions associated with vasospasm, such as Raynaud's disease.[33] Many studies have shown impressive efficacy of SCS to treat intractable angina.[34] Reported success rates are consistently greater than 80%, and these indications, already widely used outside the United States, are certain to expand within the United States.

7.7 Cost Effectiveness

The cost effectiveness of spinal cord stimulation (in the treatment of chronic back pain) was evaluated by Kumar and colleagues[35] in 2002. They prospectively followed 104 patients with failed back surgery syndrome. Of the 104 patients, 60 were implanted with a spinal cord stimulator using standard selection criteria. Both groups were monitored over a period of five years. The stimulation group annual cost was $29,000 vs. $38,000 in the control group. The authors found 15% returned to work in the stimulation group vs. 0% in the control group. The higher costs in the nonstimulator group were in the categories of medications, emergency center visits, x-rays, and ongoing physician visits.

Bell and North[36] performed an analysis of the medical costs of SCS therapy in the treatment of patients with FBSS. The medical costs of SCS therapy were compared with an alternative regimen of surgeries and other interventions. Externally powered (external) and fully internalized (internal) SCS systems were considered separately. No value was placed on pain relief or improvements in the quality of life that successful SCS therapy can generate. The authors concluded that by reducing the demand for medical care by FBSS patients, SCS therapy could lower medical costs and found that, on average, SCS therapy pays for itself within 5.5 years. For those patients for whom SCS therapy is clinically efficacious, the therapy pays for itself within 2.1 years.

Kemler and Furnee[37] performed a similar study but looked at "chronic reflex sympathetic dystrophy (RSD)" using outcomes and costs of care before and after the start of treatment. This essentially is an economic analysis of the Kemler RSD outcomes article. Fifty-four patients with chronic RSD were randomized to receive either SCS together with physical therapy (SCS + PT; $n = 36$) or physical therapy alone (PT; $n = 18$). Twenty-four SCS + PT patients responded positively to trial stimulation and underwent SCS implantation. During twelve months of follow-up, costs (routine RSD costs, SCS costs, out-of-pocket costs), and effects (pain relief by Visual Analog Scale, health-related quality of life [HRQL] improvement by EQ-5D) were assessed in both groups. Analyses were carried out up to one year and up to the expected time of death. SCS was both more effective and less costly than the standard treatment protocol. As a result of high initial costs of SCS, in the first year, the treatment per patient was $4000 more than control therapy. However, in the lifetime analysis, SCS per patient was $60,000 cheaper than control therapy. In addition, at twelve months, SCS resulted in pain relief (SCS + PT [−2.7] vs. PT [0.4] [$p < 0.001$]) and improved HRQL (SCS + PT [0.22] vs. PT [0.03] [$p = 0.004$]). The authors found SCS both more effective and less expensive as compared with the standard treatment protocol for chronic RSD.

7.8   Peripheral, Cortical, and Deep Brain Stimulation

Although this chapter concentrates on the technique of SCS, it must be noted that neurostimulation can be used successfully at other locations in the peripheral and central nervous systems to provide analgesia.

Peripheral nerve stimulation was introduced by Wall, Sweet, and others[38] in the mid-1960s. This technique has shown efficacy for peripheral nerve injury pain syndromes as well as CRPS, with the use of a carefully implanted paddle lead utilizing a fascial graft to help anchor the lead without traumatizing the nerve.[39]

Motor cortex and deep brain stimulation are techniques that have been explored to treat highly refractory neuropathic pain syndromes, including central pain, deafferentation syndromes, trigeminal neuralgia, and others (Figure 7.8).[40] Deep brain stimulation has become a widely used technique for movement disorders, and much less so for painful indications, although there have been many case reports of utility in treating highly refractory central pain syndromes.[41]

7.9   Conclusions

Spinal cord stimulation is an invasive, interventional surgical procedure. Linderoth and Meyerson[42] have written some principles of neurostimulation that are cornerstones of SCS theory and practice. The difficulty in randomized clinical trials in such situations is well recognized. Based on the present evidence with two randomized trials, one prospective trial, and multiple retrospective trials, the evidence for spinal cord stimulation in properly selected populations with neuropathic pain states is moderate. Clearly, this technique should be reserved for patients who have failed more conservative therapies. With appropriate selection and careful attention to technical issues, the clinical results are overwhelmingly positive.

References

1. Kumar, K., Nath, R., and Wyant, G.M. Treatment of chronic pain by epidural spinal cord stimulation: a 10-year experience. *J. Neurosurg.*, 5:402–407, 1991.
2. Shealy, C.N., Mortimer, J.T., and Resnick, J. Electrical inhibition of pain by stimulation of the dorsal columns: Preliminary reports. *J. Int. Anesth. Res. Soc.*, 46:489–491, 1967.

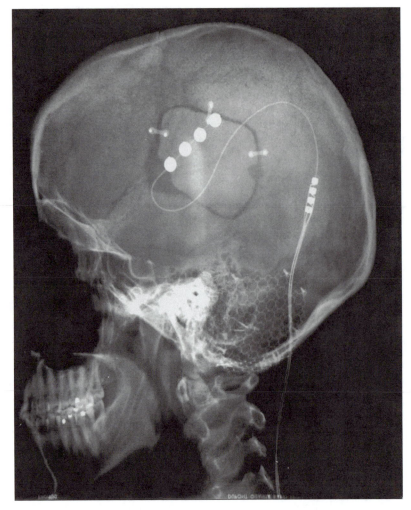

FIGURE 7.8 Radiograph of motor cortex stimulation. (Courtesy of Ali Rezai, M.D., Cleveland Clinic Foundation.)

TABLE 7.1 Principles of Neurostimulation

- SCS mechanism of action is not completely understood but influences multiple components and levels within the CNS with both interneuron and neurochemical mechanisms.
- SCS therapy is effective for many neuropathic pain conditions. Stimulation-evoked paresthesia must be experienced in the entire painful area. No consistent evidence exists for the efficacy of neurostimulation in primary nociceptive pain conditions.
- Stimulation should be applied with low-intensity, just suprathreshold for the activation of the low-threshold, large-diameter fibers, and should be of nonpainful intensity. To be effective, SCS must be applied continuously (or in cycles) for at least 20 min prior to the onset of analgesia. This analgesia develops slowly and typically lasts several hours after cessation of the stimulation.
- SCS has demonstrated clinical and cost effectiveness in FBSS and CRPS. Clinical effectiveness has also been shown in peripheral ischemia and angina.
- Multicontact, multiprogram systems improve outcomes and reduce the incidence of surgical revisions. Insulated paddle-type electrodes *probably* decrease the incidence of lead breakage, prolong battery life, and show early superiority in quality of paresthesia coverage and analgesia in FBSS as compared to permanent percutaneous electrodes.
- Serious complications are exceedingly rare but can be devastating. Meticulous care must be taken during implantation to minimize procedural complications. The most frequent complications are wound infections (approximately 5%) and lead breakage or migration (approximately 13% each for permanent percutaneous leads and 3–6% each for paddle leads).

Source: Modified from Linderoth[42]

3. Kumar, K., Toth, C., and Nath, R.K. Deep brain stimulation for intractable pain: a 15-year experience. *Neurosurgery*, 40(4):736–746, 1997.

4. Nguyen, J.P., Lefaucher, J.P., Le Guerinel, C., et al. Motor cortex stimulation in the treatment of central and neuropathic pain. *Arch. Med. Res.*, 31(3):263–265, 2000.

5. Campbell, J.N. and Long, D.M. Stimulation of the peripheral nervous system for pain control. *J. Neurosurg.*, 45:692–699, 1976.

6. Melzack, R., and Wall, P.D. Pain mechanisms: a new theory. *Science,* 150:971–979, 1965.

7. Oakley, J. and Prager, J. Spinal cord stimulation: mechanism of action. *Spine*, 27:2574–2583, 2002.

8. Linderoth, B. and Foreman, R. Physiology of spinal cord stimulation: review and update. *Neuromodulation*, 3:150–164, 1999.

9. May, M.S., Banks, C., and Thomson, S.J. A retrospective, long-term, third party follow-up of patients considered for spinal cord stimulation. *Neuromodulation*, 5:137–144, 2002.

10. Burchiel, K.J., Anderson, V.C., Wilson, B.J., et al. Prognostic factors of spinal cord stimulation for chronic back and leg pain. *Neurosurgery*, 36:1101–1111, 1995.

11. Olson, K.A., Bedder, M.D., Anderson, V.C., et al. Psychological variables associated with outcome of spinal cord stimulation trials. *Neuromodulation*, 1:6–13, 1998.

12. North, R.B., Kidd, D.H., Zahurak, M., et al: Spinal cord stimulation for chronic, intractable pain: two decades' experience. *Neurosurgery*, 32:384–395, 1993.

13. Barolat, G., Oakley, J., Law, J., et al. Epidural spinal cord stimulation with a multiple electrode paddle lead is effective in treating low back pain. *Neuromodulation*, 4:59–66, 2001.

14. May, M.S., Banks, C., and Thomson, S.J. A retrospective, long-term, third party follow-up of patients considered for spinal cord stimulation. *Neuromodulation*, 5:137–144, 2002.

15. Sarubbi, F. and Vasquez, J. Spinal epidural abscess associated with the use of temporary epidural catheters: Report of two cases and review. *Clin. Infect. Dis.*, 25(5):1155–1158, 1997.

16. Schuchard, M. and Clauson, W. An interesting and heretofore unreported infection of a spinal cord stimulator: smitten by a kitten revisited. *Neuromodulation*, 2:67–71, 2001.

17. Alfano, S., Darwin, J., and Picullel, B. Programming principles. In *Spinal Cord Stimulation: Patient Management Guidelines for Clinicians.* Medtronic, Inc, Minneapolis, MN, 2001, p. 27–33.

18. Barolat, G., Massaro, F., He, J., et al. Mapping of sensory responses to epidural stimulation of the intraspinal neural structures in man. *J. Neurosurg.*, 78(2):233–239, 1993.

19. North, R.B., Kidd, D.H., Zahurak, M., et al. Spinal cord stimulation for chronic, intractable pain: experience over two decades. *Neurosurgery*, 32(3):384–394, 1993.

20. Turner, J.A., Loeser, J.D., and Bell, K.G. Spinal cord stimulation for chronic low back pain: a systematic literature synthesis. *Neurosurgery*, 37(6):1088–1095; discussion 1095–1096, 1995.

21. Barolat, G., Oakley, J., Law, J., et al. Epidural spinal cord stimulation with a multiple electrode paddle lead is effective in treating low back pain. *Neuromodulation*, 2:59–66, 2001.

22. North, R.B., Kidd, D.H., and Piantadosi, S. Spinal cord stimulation versus reoperation for failed back surgery syndrome: a prospective, randomized study design. *Acta Neurochir. Suppl.* (Wien.), 64:106–108, 1995.

23. Barolat, G., Oakley, J., Law, J., et al. Epidural spinal cord stimulation with a multiple electrode paddle lead is effective in treating low back pain. *Neuromodulation*, 2:59–66, 2001.

24. Burchiel, K.J., Anderson, V.C., Brown, F.D., et al. Prospective, multicenter study of spinal cord stimulation for the relief of chronic back and extremity pain. *Spine*, 21(23):2786–2794, 1996.

25. Turner, J.A., Loeser, J.D., and Bell, K.G. Spinal cord stimulation for chronic low back pain: a systematic literature synthesis. *Neurosurgery*, 37(6):1088–1095, 1995.

26. North, R. and Wetzel, T. Spinal cord stimulation for chronic pain of spinal origin. *Spine*, 22:2584–2591, 2002.

27. Kemler, M.A., Barendse, G.A., van Kleef, M., et al. Spinal cord stimulation in patients with chronic reflex sympathetic dystrophy. *N. Engl. J. Med.*, 343(9):618–624, 2000.

28. Calvillo, O., Racz, G., Didie, J., et al. Neuroaugmentation in the treatment of complex regional pain syndrome of the upper extremity. *Acta Orthopeadica. Belgica*, 64:57–63, 1998.

29. Oakley, J. and Weiner, R. Spinal cord stimulation for complex regional pain syndrome: a prospective study of 19 patients at two centers. *Neuromodulation*, 2(1):47–50, 1999.

30. Stanton-Hicks, M. Spinal cord stimulation for the management of complex regional pain syndromes. *Neuromodulation*, 2(3):193–201, 1999.

31. Bennett, D., Alo, K., Oakley, J., et al. Spinal cord stimulation for complex Regional Pain Syndrome I (RSD): a retrospective multicenter experience from 1995–1998 of 101 patients. *Neuromodulation*, 3(2)202–210, 1999.

32. Cook, A.W., Oygar, A., Baggenstos, P., et al. Vascular disease of extremities: electrical stimulation of spinal cord and posterior roots. *NY State J. Med.*, 76:366–368, 1976.

33. Broseta, J., Barbera, J., and De Vera, J.A. Spinal cord stimulation in peripheral arterial disease. *J. Neurosurg.*, 64:71–80, 1986.

34. Eliasson, T., Augustinsson, L.E., and Mannheimer, C. Spinal cord stimulation in severe angina pectoris-presentation of current studies, indications, and clinical experience. *Pain*, 65:169–179, 1996.

35. Kumar, K., Malik, S., and Demeria, D. Treatment of chronic pain with spinal cord stimulation versus alternative therapies: cost-effectiveness analysis. *Neurosurgery*, 51(1):106–115, 2002.

36. Bell, G. and North, R. Cost-effectiveness analysis of spinal cord stimulation in treatment of failed back surgery syndrome. *J. Pain Symptom Manage.*, May 13 (5):285–296, 1997.

37. Kemler, M. and Furnee, C. Economic evaluation of spinal cord stimulation for chronic reflex sympathetic dystrophy. *Neurology*, 59(8):1203–1209, 2002.

38. Wall, P.D. and Sweet, W.H. Temporary abolition of pain in man. *Science*, 155:108–109, 1967.

39. Hassenbusch, S.J., Stanton-Hicks, M., and Shoppa, D. Long-term results of peripheral nerve stimulation for reflex sympathetic dystrophy. *J. Neurosurg.*, 84:415–423, 1996.

40. Tsubokawa, T., Katayama, Y., Yamamoto, T., et al. Chronic motor cortex stimulation in patients with thalamic pain. *J. Neurosurg.*, 78:393–401, 1993.

41. Limousin, P., Krack, P., Pollack, P., et al. Electrical stimulation of the subthalamic nucleus in advanced Parkinson's disease. *New Engl. J. Med.*, 339:1105–1111, 1998.

42. Linderoth, B. and Meyerson, B.A. Spinal cord stimulation: mechanisms of action. In *Surgical Management of Pain*, Burchiel, K., ed. Theime Medical Pub., New York, 2002, p. 505–526.

Appendix 7A

- Advanced Bionics Corporation
  12740 San Fernando Road, Sylmar, California 91342
  800-362-1400
  www.advancedbionics.com
- Medtronic, Inc.
  710 Medtronic Parkway, Minneapolis, Minnesota 55432-5604
  763-514-5604
  www.medtronic.com
- American Neuromodulation Systems, Inc.
  6501 Windcrest Dr., Ste. 100, Plano, Texas 75024
  800-727-7846
  www.ans-medical.com

8

Motor Cortex Stimulation for Pain Management

Shivanand P. Lad,
Kevin Chao, and
Jaimie M. Henderson

8.1   Introduction

Neuropathic pain is one of the most difficult conditions to treat in clinical neurological practice. Central pain is defined as pain initiated or caused by a primary lesion or dysfunction in the central nervous system (CNS). Some of the early work in this area was performed by Penfield and Jasper in the 1950s. During their surgeries on patients with epilepsy, they observed that stimulation of the precentral gyrus elicited sensory responses when the corresponding portion of the adjacent postcentral gyrus was resected. Their extensive mapping and study of the sensory and motor cortex laid the foundation for the understanding that both the pre- and postcentral gyri are involved in pain. In the 1970s, electrical stimulation with bipolar current, targeted in or near the internal capsule, was found to have a positive effect on centrally mediated pain, establishing the concept that thalamic involvement was a key mediator in central pain.

In the 1980s and early 1990s, motor cortex stimulation (MCS) began to be explored by several groups in an effort to find a better way of alleviating neuropathic pain. Interest in MCS has since evolved to provide more effective treatments for central poststroke-related pain, as well as peripheral neuropathic pain conditions, including trigeminal neuropathic pain. Chronic stimulation of the precentral cortex for the treatment of pain was first introduced by the pioneering work of Tsubokawa and colleagues in 1991(Tsubokawa et al., 1991a, 1991b) The articles by Tsubokawa and colleagues in the early 1990s were the first reports on the effectiveness of MCS for neuropathic pain, based on ten years of experience attained in the treatment of twenty-five patients in whom intermittent cortical stimulation was used. Their overall response rate was 75%, with long-term benefits for up to seven months in a patient group that had been resistant to all prior available pain therapies.

Subsequently, there has been growing corroborative evidence derived from case reports and individual case series to suggest that epidural MCS can be an effective treatment for many patients with intractable neuropathic pain. Poststroke pain, phantom limb pain, multiple sclerosis, spinal cord injury pain, postherpetic neuralgia, and neuropathic pain of the limbs or face have all been reported to respond favorably to MCS (Brown and Barbaro, 2003). The majority of clinical studies focus on the use of MCS in poststroke pain, for which there are few other treatments (Tsubokawa et al., 1991a, 1991b, 1993; Katayama et al., 1994, 1998; Nguyen et al., 1997, 2000; Bezard et al., 1999; Carroll et al., 2000; Saitoh et al., 2000, 2001; Smith et al., 2001). Poststroke pain responds variably to MCS, with only approximately 50% of patients achieving adequate relief. However, a few studies have documented excellent results in the treatment of trigeminal neuropathic pain, with 75 to 100% of patients achieving good or excellent pain relief (Meyerson et al., 1993; Ebel et al., 1996; Nguyen et al., 1997; Rainov et al., 1997; Bezard et al., 1999). MCS thus appears to hold great promise for patients with facial neuropathic pain of a peripheral origin, many of whom have failed traditional treatments for trigeminal neuralgia, including microvascular decompression, glycerol or radiofrequency rhizoloysis, stereotactic radiosurgery, or open sectioning of the trigeminal rootlets.

Despite a number of encouraging small studies, MCS has not been rigorously studied in a prospective fashion, and there are still several opinions regarding surgical technique, programming, and patient selection.

8.2 Mechanism of Action

The exact mechanism of action of motor cortex stimulation has yet to be elucidated and is the subject of current study. Positron emission tomography (PET) studies demonstrate that cortical stimulation increases cerebral blood flow (CBF) in the ipsilateral thalamus, cingulate gyrus, orbitofrontal cortex, and brainstem (Peyron et al., 1995; Garcia-Larrea et al., 1999; Saitoh et al. 2004). Some studies have found that regional CBF increases occur in the ipsilateral ventrolateral thalamus, which largely carries cortico-thalamic connections from the motor and premotor areas (Peyron et al., 1995; Garcia-Larrea et al., 1999; Saitoh et al., 2004). Regional CBF increases in this site, however, are less than those seen in the anterior cingulate gyrus, insula, and brainstem. There seems to be a correlation with the extent of pain relief and the increase in cingulate blood flow (Peyron et al., 1995; Garcia-Larrea et al., 1999; Saitoh et al., 2004). Activation of the brainstem periaqueductal gray area has also been suggested to play a role. The results of these CBF studies suggest that the somatosensory cortex is not activated by MCS, nor does there appear to be activation of the downstream motor pathways below the stimulating electrode. However, the exact effects seen may vary, depending on the surgical technique and stimulation parameters chosen.

8.3 Surgical Technique

Following Tsubokawa's initial report (Tsubokawa et al., 1991b), several studies were published using a similar technique of introducing the electrode via a burr hole under local anesthetic (Tsubokawa et al., 1991a,b; Hosobuchi, 1993; Meyerson et al., 1993; Herregodts et al., 1995; Nguyen et al., 1997, 2000; Bezard et al., 1999; Gharabaghi et al., 2005). In several cases, groups that started out using a burr hole technique later switched to performing electrode placement via craniotomy (Meyerson et al., 1993; Nguyen et al., 1997, 2000; Bezard et al., 1999), and the majority of surgeons now perform a craniotomy either under local (Meyerson et al., 1993; Tsubokawa et al., 1993; Katayama et al., 1994, 1998, 2001; Ebel et al., 1996; Nguyen et al., 1997, 2000; Rainov et al., 1997; Bezard et al., 1999; Rainov and Heidecke, 2003; Sharan et al., 2003; Son et al., 2003) or general anesthetic (Carroll et al., 2000; Saitoh et al., 2000, 2001; Roux et al., 2001; Velasco et al., 2002; Brown and Pilitsis, 2005; Nuti et al., 2005; Pirotte et al., 2005). Nearly all investigators place the electrodes epidurally, although subdural placement has been described (Saitoh et al., 2003; Tani et al., 2004). Image-guided neuronavigation is used to precisely identify the motor cortex intraoperatively (Bezard et al., 1999; Mogilner and Rezai, 2001; Sharan et al., 2003; Brown and Pilitsis, 2005; Nuti et al., 2005; Pirotte et al., 2005), and proper placement is confirmed with physiological testing.

Fiducials are placed preoperatively on the scalp and a fine-cut stereotactic MRI is obtained using 1-mm cuts. The central sulcus, Sylvian fissure, and inferior and superior frontal sulci are identified. For facial pain, the optimum target is often identified anterior to the central sulcus adjacent to or below the inferior frontal sulcus. Prophylactic antibiotic medication is given along with anticonvulsant agents. The patient is placed in a supine position with a roll placed beneath the shoulder and the head rotated to the ipsilateral side of the pain. The target is mapped onto the contralateral scalp, and an incision centering over the central sulcus is made.

Once the craniotomy flap is made, a diagnostic electrode array is placed in the epidural space aligned perpendicular to the expected position of the central sulcus, covering both the regions of the pre- and postcentral gyri. Median nerve somatosensory evoked potential may be obtained to identify the N20/P20 phase reversal that occurs across the central sulcus. This provides electrophysiological confirmation of the location of the hand region of the precentral gyrus, located cephalad and medial to the facial region. The target for electrode implantation is marked on the dural surface. Stimulation is performed in a train of three at 200-μsec durations with currents of up to 20 mA. The painful area contralateral to the site of stimulation is observed for muscle contractions during test stimulation, and the threshold for motor response noted. Electromyography electrodes may be placed in appropriate muscle groups to monitor for muscular contractions. Iced saline should be kept available to irrigate the brain in case a seizure is induced.

When the appropriate cortical target has been confirmed, the diagnostic grid is removed, and a four contact paddle electrode is positioned over the motor cortex, parallel to the central sulcus. Some investigators have described placement of the electrode perpendicular to the central sulcus, with the two central contacts of the electrode array over the target point in the motor cortex. A second electrode can be placed posteriorly over the central sulcus to allow for trans-sulcal stimulation and increased programming options. Each electrode is sutured to the outer layer of the dura. If an externalized trial is to be performed, the lead wire is tunneled out through a separate stab incision and the craniotomy bone flap is secured.

The electrode is tested once the patient is fully alert. Testing may continue for several days or until the patient can consistently confirm that stimulation reduces the preoperative pain by at least 50%. If the trial is successful, an implantable pulse generator is placed in the upper chest wall under general anesthesia and attached to the lead(s) with extension cables.

8.4 MCS and Trigeminal Neuropathic Pain

Trigeminal neuropathic facial pain is a syndrome of severe, constant facial pain related to disease of, or injury to, the trigeminal nerve or ganglion. Causes of trigeminal neuropathic pain can include injury from sinus or dental surgery, skull and/or facial trauma, or intentional destruction for therapeutic reasons (deafferentation), as well as intrinsic pathology of any part of the trigeminal system (Burchiel, 2003). Despite extensive studies, no significant advances have occurred in its pharmacological treatment and it continues to be treated with anticonvulsant and antidepressant therapies. Many patients who fail surgical treatment for trigeminal neuralgia will develop trigeminal neuropathic pain (also called trigeminal deafferentation pain (Burchiel, 2003)), for which there are few, if any, effective treatments. Many treatments that are effective for trigeminal neuralgia can, in fact, worsen trigeminal neuropathic pain. Deep brain stimulation of well-defined targets in the sensory thalamus and periaqueductal or periventricular gray matter with stereotactic placement of electrodes has not proven efficacious (Burchiel, 2001; Coffey, 2001). MCS has shown some promise in the treatment of trigeminal neuropathic pain.

Despite the encouraging reports of MCS for neuropathic pain in the literature, there is still some question as to its true efficacy. The effect of MCS can wane over time, requiring reprogramming (Henderson et al., 2004). Even in patients who undergo intensive reprogramming, pain relief cannot always be achieved. There is a risk of seizures during stimulator programming (Meyerson et al., 1993; Ebel et al., 1996; Rainov et al., 1997; Henderson et al., 2004), although the development of epilepsy has not been reported. There thus exists clinical equipoise regarding the true risk/benefit ratio of MCS for the treatment of trigeminal neuropathic pain. In addition, programming parameters vary from investigator to investigator, with some groups achieving pain relief from parameters that other groups find ineffective.

8.5   Stimulation Parameters

There is tremendous variation in reported stimulation parameters for MCS. Pain relief can occur at amplitudes from 0.5 to 10 V, rates from 5 to 130 Hz, and pulse widths from 60 to 450 μs (Bonicalzi and Canavero, 2004). Although most studies have used rates around 40 Hz, others have found higher rates necessary in some cases. There is also no agreement on whether wide or narrow pulse widths provide more effective stimulation. Amplitudes have in many cases been empirically chosen, whereas other investigators base stimulation amplitude on a percentage of motor threshold.

Pain relief is most commonly achieved at amplitudes of 6 V or less, with average amplitudes of 5 V or less in most studies. Amplitudes above 6 V are more likely associated with seizures during programming, with seizures commonly induced at amplitudes approaching 9 V (Henderson et al., 2004).

Many investigators have noted that MCS frequently produces a period of poststimulus pain relief that can range from minutes to hours. Thus, the majority of publications report the use of a cycling mode of stimulation, with ten minutes to three hours on stimulation followed by fifteen minutes to six hours off stimulation. In one study, switching from a continuous to cycling mode (in addition to other programming changes) may have contributed to improvement in pain relief (Henderson et al., 2004).

8.6   Safety

MCS, like any neurosurgical procedure, can be associated with risks and complications, including bleeding, infection, and neurological deficits. Although many studies have reported no adverse events with MCS (Tsubokawa et al., 1991a,b; Hosobuchi, 1993; Herregodts et al., 1995; Sol et al., 2001; Son et al., 2003; Tani et al., 2004; Gharabaghi et al., 2005), there have been some serious complications reported in the literature. Two epidural hematomas have been reported, one small and asymptomatic (Nguyen et al., 1997), the other requiring evacuation and associated with persistent dysphasia (Meyerson et al., 1993). One group reported two patients with devastating cerebral hemorrhages, with one patient dying from this complication and the other remaining in a persistent vegetative state (Saitoh et al., 2001, 2003). It is possible that these complications resulted from this groupís use of subdural rather than epidural placement. Infection of the hardware requiring removal and/or treatment with antibiotics has been reported in a number of studies (Bezard et al., 1999; Carroll et al., 2000; Smith et al., 2001, 2003; Brown and Pilitsis, 2005; Nuti et al., 2005; Pirotte et al., 2005). Some patients have also experienced wound dehiscence that resolved with surgical revision (Bezard et al., 1999). Breakage of the hardware can also occur (Carroll et al., 2000). Two patients in one study experienced transient postoperative neurological deficits (one speech, one motor) (Nuti et al., 2005).

Programming of MCS systems, as well as long-term treatment with MCS, can also be associated with certain risks and side effects. Foremost among these is the risk of seizures, which have been frequently reported in the literature. These have been variously described as brief focal seizures during programming (Nuti et al., 2005; Pirotte et al., 2005), unspecified seizures during programming (Saitoh et al., 2000, 2001; Sharan et al., 2003; Henderson et al., 2004), prolonged focal seizure with postictal speech arrest (Ebel et al., 1996), short-lasting generalized seizures during programming (Katayama et al., 1998) (occurring in the majority of patients in one study (Meyerson et al., 1993)), and generalized seizures with activation of the stimulator (Rainov et al., 1997; Smith et al., 2001). In one recent study of intensive MCS reprogramming, the average threshold for inducing seizures was 8.9 V (Henderson et al., 2004). There is at least one patient who developed severe epilepsy after long-term motor cortex stimulation (Bezard et al., 1999). To further investigate the epileptogenic potential of chronic MCS, Bezard et al. undertook a study in three macaque monkeys who were implanted with MCS electrodes. They found that with stimulation at a rate of 40 Hz and a pulse width of 90 μs, no seizures occurred even at stimulus intensities up to 3 mA above the motor threshold. Higher frequencies and pulse widths induced muscle twitching at lower amplitudes and consequently also induced seizures at lower amplitudes (Bezard et al., 1999).

Other reported side effects from stimulation include painful stimulation of the dura mater (Meyerson et al., 1993; Katayama et al., 1994; Bezard et al., 1999), stimulation-induced dysesthesiae (Nguyen et al., 1997; Katayama et al., 1998; Fukaya et al., 2003), dysarthria (Carroll et al., 2000), and fatigue (Carroll et al., 2000). There are two case reports of unusual side effects: (1) impairment in a motor imagery task (Tomasino et al., 2005), and (2) development of a painful supernumerary phantom arm in poststroke pain patient (Canavero et al., 1999). In addition, there is a suggestion that MCS may adversely affect cognitive function, especially in older patients (Montes et al., 2002).

It is therefore important that this procedure continue to be thoroughly studied and evaluated by experienced groups who have developed techniques to minimize surgical risks and deal with complications in an expeditious manner.

8.7 Efficacy

Published studies have been near uniformly laudatory regarding the efficacy of MCS for the treatment of neuropathic facial pain, beginning with the 1993 publication by Meyerson et al. In a group of ten patients with different types of neuropathic pain, all five patients with trigeminal neuropathic pain obtained between 60 and 90% pain relief at eight to twenty-eight months (Meyerson et al., 1993). A follow-up study by Herregodts et al. (1995) showed 50 to 100% reduction in VAS pain scores in four out of five patients with trigeminal neuropathic pain.

In a 1996 study by Ebel et al. (1996), seven patients with trigeminal neuropathic pain of various etiologies were treated with motor cortex stimulation. Six of the seven patients underwent permanent implantation, with five of six achieving 80% or greater pain relief. Two patients subsequently lost pain relief over the course of several months, leaving three of six patients (50%) having a satisfactory result at last follow-up.

Nguyen et al. have published several descriptions of their surgical technique and programming approach (Nguyen et al., 1997, 2000; Bezard et al., 1999, 2000). In their series, all patients with neuropathic facial pain achieved 40 to 100% pain relief (Nguyen et al., 1997, 2000; Bezard et al., 1999).

More recently, Brown and Pilitsis (2005) reported on ten patients who underwent a trial of MCS for facial pain of various etiologies, including trigeminal neuropathic pain, postherpetic neuralgia, and central post-stroke pain. All eight patients with a peripheral neuropathic mechanism for their pain underwent placement of a permanent system after successful trial; 88% of these patients obtained immediate pain relief of more than 50%, and 75% experienced sustained reduction in pain at three- to twenty-four-month follow-up. A review of the literature corroborated these results, showing twenty-nine (76%) of thirty-eight patients with neuropathic facial pain achieved greater than 50% pain relief (Brown and Pilitsis, 2005). All patients in the implanted cohort experienced a decrease in their medication requirements by more than 50%.

The definition of "success" motor cortex stimulation is not clear from the published literature. Several studies have suggested that a 30% reduction in pain intensity, or a decrease of 2 points on a 1 to 10 scale represents a minimum clinically meaningful reduction in pain (Farrar et al., 2000; Farrar et al., 2001; Cepeda et al., 2003). Most studies of therapeutic interventions in pain regard a 50% reduction in baseline VAS pain ratings as clearly significant.

8.8 Conclusion

Despite the reported success with MCS for the treatment of trigeminal neuropathic pain, there have been no large, controlled, prospective, randomized trials of this modality. We face a situation similar to that experienced with deep brain stimulation (DBS) for pain in the 1970s and 1980s. The procedure was widely utilized with little strong evidence for efficacy until two prospective trials were eventually performed in the 1990s. These trials showed that DBS for pain could be effective, but suggested a very low percentage (13.5 to 17.8%) of patients could be proven to have clinically significant pain relief at long-term follow-up (Coffey, 2001). The lesson learned from this study was that future trials of analgesic devices follow structured protocols for patient selection and utilize uniform implantation and treatment paradigms. It is imperative that MCS be subjected to this type of scrutiny prior to its widespread adoption as a potential standard therapy for chronic pain.

MCS is a promising therapy for use in the treatment of complex central and neuropathic pain syndromes refractory to medical treatment. Ongoing basic and clinical evaluation will provide additional information into the mechanisms, surgical technique, indications, and long-term treatment effectiveness of MCS for patients who suffer from a variety of challenging pain syndromes.

References

Bezard, E. et al. (1999). Cortical stimulation and epileptic seizure: a study of the potential risk in primates. *Neurosurgery*, 45(2):346–350.

Bonicalzi, V. and Canavero, S. (2004). Motor cortex stimulation for central and neuropathic pain (Letter regarding Topical Review by Brown and Barbaro). *Pain*, 108(1–2):199–200; author reply 200.

Brown, J.A. and Barbaro, N.M. (2003). Motor cortex stimulation for central and neuropathic pain: current status. *Pain*, 104(3):431–435.

Brown, J.A. and Pilitsis, J.G. (2005). Motor cortex stimulation for central and neuropathic facial pain: a prospective study of 10 patients and observations of enhanced sensory and motor function during stimulation. *Neurosurgery*, 56(2):290–297; discussion 290–297.

Burchiel, K.J. (2001). Deep brain stimulation for chronic pain: the results of two multi-center trials and a structured review. *Pain Med.*, 2(3):177.

Burchiel, K.J. (2003). A new classification for facial pain. *Neurosurgery*, 53(5):1164–1166; discussion 1166–1167.

Canavero, S. et al. (1999). Painful supernumerary phantom arm following motor cortex stimulation for central poststroke pain. Case report. *J. Neurosurg.*, 91(1):121–123.

Carroll, D. et al. (2000). Motor cortex stimulation for chronic neuropathic pain: a preliminary study of 10 cases. *Pain*, 84(2–3):431–437.

Cepeda, M.S. et al. (2003). What decline in the pain intensity is meaningful to patients with acute pain? *Pain*, 105:151–157.

Coffey, R.J. (2001). Deep brain stimulation for chronic pain: results of two multicenter trials and a structured review. *Pain Med.*, 2(3):183–192.

Ebel, H. et al. (1996). Chronic precentral stimulation in trigeminal neuropathic pain. *Acta Neurochir. (Wien.)*, 138(11):1300–1306.

Farrar, J.T. et al. (2000). Defining the clinically important difference in pain outcome measures. *Pain*, 88(3):287–294.

Farrar, J.T. et al. (2001). Clinical importance of changes in chronic pain intensity measured on an 11-point numerical pain rating scale. *Pain*, 94(2):149–158.

Fukaya, C. et al. (2003). Motor cortex stimulation in patients with post-stroke pain: conscious somatosensory response and pain control. *Neurol. Res.*, 25(2):153–156.

Garcia-Larrea, L. et al. (1999). Electrical stimulation of motor cortex for pain control: a combined PET-scan and electrophysiological study. *Pain*, 83(2):259–273.

Gharabaghi, A. et al. (2005). Volumetric image guidance for motor cortex stimulation: integration of three-dimensional cortical anatomy and functional imaging. *Neurosurgery*, 57(1 Suppl.):114–120; discussion 114–120.

Henderson, J.M. et al. (2004). Recovery of pain control by intensive reprogramming after loss of benefit from motor cortex stimulation for neuropathic pain. *Stereotact. Funct. Neurosurg.*, 82(5–6):207–213.

Herregodts, P. et al. (1995). Cortical stimulation for central neuropathic pain: 3-D surface MRI for easy determination of the motor cortex. *Acta Neurochir. Suppl.*, 64:132–135.

Hosobuchi, Y. (1993). Motor cortical stimulation for control of central deafferentation pain. *Adv. Neurol.*, 63:215–217.

Katayama, Y. et al. (1994). Chronic motor cortex stimulation for central deafferentation pain: experience with bulbar pain secondary to Wallenberg syndrome. *Stereotact. Funct. Neurosurg.*, 62(1–4):295–299.

Katayama, Y. et al. (1998). Poststroke pain control by chronic motor cortex stimulation: neurological characteristics predicting a favorable response. *J. Neurosurg.*, 89(4):585–591.

Katayama, Y. et al. (2001). Motor cortex stimulation for post-stroke pain: comparison of spinal cord and thalamic stimulation. *Stereotact. Funct. Neurosurg.*, 77(1–4):183–186.

Meyerson, B.A. et al. (1993). Motor cortex stimulation as treatment of trigeminal neuropathic pain. *Acta Neurochir. Suppl. (Wien.)*, 58:150–153.

Mogilner, A.Y. and Rezai, A.R. (2001). Epidural motor cortex stimulation with functional imaging guidance. *Neurosurg. Focus*, 11(3):E4.

Montes, C. et al. (2002). Cognitive effects of precentral cortical stimulation for pain control: an ERP study. *Neurophysiol. Clin.*, 32(5):313–325.

Nguyen, J.P. et al. (1997). Treatment of deafferentation pain by chronic stimulation of the motor cortex: report of a series of 20 cases. *Acta Neurochir. Suppl. (Wien.)*, 68:54–60.

Nguyen, J.P. et al. (2000). Motor cortex stimulation in the treatment of central and neuropathic pain. *Arch. Med. Res.*, 31(3):263–265.

Nuti, C. et al. (2005). Motor cortex stimulation for refractory neuropathic pain: four year outcome and predictors of efficacy. *Pain*, 118(1–2):43–52.

Peyron, R. et al. (1995). Electrical stimulation of precentral cortical area in the treatment of central pain: electrophysiological and PET study. *Pain*, 62(3):275–286.

Pirotte, B. et al. (2005). Combination of functional magnetic resonance imaging-guided neuronavigation and intraoperative cortical brain mapping improves targeting of motor cortex stimulation in neuropathic pain. *Neurosurgery*, 56(2 Suppl.):344–359; discussion 344–359.

Rainov, N.G. et al. (1997). Epidural electrical stimulation of the motor cortex in patients with facial neuralgia. *Clin. Neurol. Neurosurg.*, 99(3):205–209.

Rainov, N.G. and Heidecke, V. (2003). Motor cortex stimulation for neuropathic facial pain. *Neurol. Res.*, 25(2):157–161.

Roux, F.E. et al. (2001). Chronic motor cortex stimulation for phantom limb pain: a functional magnetic resonance imaging study: technical case report. *Neurosurgery*, 48(3):681–687; discussion 687–688.

Saitoh, Y. et al. (2000). Motor cortex stimulation for central and peripheral deafferentation pain. Report of eight cases. *J. Neurosurg.*, 92(1):150–155.

Saitoh, Y. et al. (2001). Motor cortex stimulation for deafferentation pain. *Neurosurg. Focus*, 11(3):E1.

Saitoh, Y. et al. (2003). Primary motor cortex stimulation within the central sulcus for treating deafferentation pain. *Acta Neurochir. Suppl.*, 87:149–152.

Saitoh, Y. et al. (2004). Increased regional cerebral blood flow in the contralateral thalamus after successful motor cortex stimulation in a patient with poststroke pain. *J. Neurosurg.*, 100(5):935–939.

Sharan, A.D. et al. (2003). Precentral stimulation for chronic pain. *Neurosurg. Clin. N. Am.*, 14(3):437–444.

Smith, H. et al. (2001). Motor cortex stimulation for neuropathic pain. *Neurosurg. Focus*, 11(3):E2.

Sol, J.C. et al. (2001). Chronic motor cortex stimulation for phantom limb pain: correlations between pain relief and functional imaging studies. *Stereotact. Funct. Neurosurg.*, 77(1–4):172–176.

Son, U.C. et al. (2003). Motor cortex stimulation in a patient with intractable complex regional pain syndrome type II with hemibody involvement. Case report. *J. Neurosurg.*, 98(1):175–179.

Tani, N. et al. (2004). Bilateral cortical stimulation for deafferentation pain after spinal cord injury. Case report. *J. Neurosurg.*, 101(4):687–689.

Tomasino, B. et al. (2005). Mental rotation in a patient with an implanted electrode grid in the motor cortex. *Neuroreport*, 16(16):1795–1800.

Tsubokawa, T. et al. (1991a). Chronic motor cortex stimulation for the treatment of central pain. *Acta Neurochir. Suppl. (Wien.)*, 52:137–139.

Tsubokawa, T. et al. (1991b). Treatment of thalamic pain by chronic motor cortex stimulation. *Pacing Clin. Electrophysiol.*, 14(1):131–134.

Tsubokawa, T. et al. (1993). Chronic motor cortex stimulation in patients with thalamic pain. *J. Neurosurg.*, 78(3):393–401.

Velasco, M. et al. (2002). Motor cortex stimulation in the treatment of deafferentation pain. I. Localization of the motor cortex. *Stereotact. Funct. Neurosurg.*, 79(3–4):146–167.

9

Deep Brain Stimulation for Obsessive Compulsive Disorder

Wael F. Asaad and
Emad N. Eskandar

9.1 Introduction

Obsessive compulsive disorder (OCD) is a relatively common psychiatric diagnosis, affecting about 0.8% of adults and 0.25% of children of ages 5 to 15 years (Heyman et al., 2003; Wittchen and Jacobi, 2005), although some studies report a prevalence as high as 2 to 3% by early adulthood (Zohar, 1999). As the name suggests, it is characterized by obsessions, which are recurrent thoughts usually related to anxiety-provoking themes such as fear of contamination or of disorganization, and by compulsions, which are behaviors that often appear designed to neutralize the obsessions, but are nevertheless persistently unsuccessful. In its most extreme forms, the powerful obsessions of OCD, coupled with the need to repeatedly perform the compulsive behaviors (for example, washing, checking, or ordering), can significantly impair one's ability to perform the useful and necessary tasks of daily living.

The content of the obsessions are generally plausible, not delusional. Patients suffering from OCD — even children — are usually aware nonetheless that while plausible, these obsessions are unreasonable and the associated compulsions are maladaptive. For some, cognitive-behavioral therapy can build on this understanding and reduce the intrusions of obsessions and compulsions into daily life. Others may require medication; serotonin reuptake inhibitors (SRIs) are among the most effective in counteracting the signs and symptoms of OCD (Abramowitz, 1997; Goodman, 1999). For those in whom the OCD is severely disabling and refractory to medical and psychological intervention, direct manipulation of brain structure or function — psychiatric neurosurgery, as it has been called — is of potential benefit. Indeed, over the past several decades, the targets of neurosurgery for OCD have been repeatedly modified in attempts to maximize the specificity of the surgery for the signs and symptoms of OCD while minimizing unwanted cognitive, emotional, and motor-related side effects. The most novel surgical strategies replace lesion-making approaches with electrical stimulation of targeted subcortical structures by implanted electrodes, a methodology usually referred to as deep brain stimulation (DBS).

9.2   Neurobiology of OCD

OCD is thought to arise from a dysfunction of prefrontal cortex and basal ganglia systems. The prefrontal cortex is that region of cortex which is most elaborated in humans compared to other primates, and underlies many "higher" functions, such as the temporal ordering of behavior and the selection of actions based on learned "rules" (Fuster, 2000; Miller, 2000). The prefrontal cortex can be divided into three main subdivisions consisting of the dorsolateral, orbitofrontal, and medial; these are thought to underlie complementary aspects of the prefrontal contribution to action selection. Two of these three — the orbitofrontal and medial frontal cortices — are most likely involved in the pathogenesis of OCD. In particular, the orbitofrontal cortex is implicated in the processing of reward, and perhaps the affective component of decision making (Roberts, 2006; Schultz, 2006). The medial frontal cortex also is implicated in decision making but particularly by signaling rewards, errors, and conflicts among possible behavioral actions (Botvinick et al., 2004; Rushworth et al., 2004; Williams et al., 2004; Brown and Braver, 2005). The basal ganglia, meanwhile, are thought to play a role in action selection and the development of "habits," that is, learned sequences of actions that are appropriate in a given context (Graybiel, 1998). Indeed, the activity of individual neurons within the basal ganglia (and prefrontal cortex) are modified by learning (Pasupathy and Miller, 2005; Buch et al., 2006; Williams and Eskandar, 2006), and electrical stimulation of the caudate portion of the basal ganglia can selectively enhance learning (Williams and Eskandar, 2006).

The prefrontal cortices project to the basal ganglia, which in turn project to the thalamus and then back to the original prefrontal areas (Graybiel et al., 1994). These "loops" are the substrate for the learning of adaptive behavioral responses and for the selection of a particular response appropriate for an immediate situation. In OCD, the overt problem is repeated selection of the same behavioral sequence, despite the lack of an adaptive goal. The association of OCD with a dysfunction in this circuit is therefore not unreasonable. Structural and functional imaging studies lend support to this association (Friedlander and Desrocher, 2006). In particular, frontal cortical regions show hyperactivity during symptom provocation (Rauch et al., 1994; Breiter et al., 1996), and such hyperactivity decreases with appropriate OCD medication (Nakao et al., 2005). More directly, damage to the basal ganglia has been observed in rare instances to cause new-onset OCD (Berthier et al., 1996; Chacko et al., 2000; Berthier et al., 2001; Coetzer, 2004; Thobois et al., 2004). However, whether the various behavioral forms of OCD (McKay et al., 2004; Mataix-Cols et al., 2005) correspond to particular foci of pathology in this circuit, and conversely whether damage to one of several areas can produce the same disease phenomenology, are unknown.

9.3   Earlier Forms of Surgical Intervention for OCD

In the absence of effective psychotropic medicines, psychosurgery was a relatively popular treatment for psychiatric illness in the early 20th century. Cingulotomy was first undertaken to treat the general ailment of "mental disease" in the early 1950s (Whitty et al., 1952), and has persisted in a more refined form specifically for the treatment of OCD. Recent prospective studies demonstrated a significant alleviation of symptoms in about 40% of patients undergoing this procedure (Dougherty et al., 2002; Kim et al., 2003). The operative technique usually entails making one or two lesions in the cingulate gyri bilaterally, each lesion nearly 2 cm in length, using radiofrequency thermocoagulation. While such methodology is relatively crude, it is nevertheless an improvement over older techniques that relied on the injection of phenol or alcohol to create lesions. Furthermore, modern techniques have the advantage of MRI-guided lesion targeting.

The mechanism by which cingulate lesions alleviate the signs and symptoms of OCD are not well understood; certainly, no one believes the effects are somehow specific to OCD-related behavior (in fact, this procedure has been used in patient's suffering from a wide variety of anxiety and mood disorders). Neurobehavioral testing of patients who underwent cingulotomy for the treatment of OCD revealed mild deficits on a classic assay of frontal lobe function, the Wisconsin Card Sorting Task (Kim et al., 2003). In addition, a study examining patients' abilities to use diminished rewards to alter their responses found new deficits immediately postlesioning, consistent with the notion that neurons in this region encode information about the consequences of one's actions and changing rewards. That cingulotomy ameliorates the condition of OCD suggests that the hyperactivity of the cingulate cortex is somehow directly or indirectly contributing to the neurophysiology of the disease. This is consistent, in a general way, with the observation that cingulate neurons signal behavioral errors and decreased rewards (Ito et al., 2003; Williams et al., 2004); specifically, overactivity of these neurons might give the false signal that a particular behavior just executed did not have its desired effect, and therefore promote its repetition. Conversely, even if baseline hyperactivity of the cingulate is not the etiologic factor, lesioning this structure might simply decrease sensitivity to perceived errors, to whatever extent such a perception contributes to OCD and however it might arise.

Subcaudate tractotomy is a procedure that, like cingulotomy, has been used to treat both mood disorders and anxiety disorders, including OCD. This operation was developed after the introduction of frontal lobotomy and cingulotomy in the hopes of minimizing postoperative frontal lobe dysfunction while still achieving a therapeutic benefit. Its goal is ostensibly to sever the orbitofrontal cortex from limbic subcortical structures, in particular, the amygdala. As noted above, the orbitofrontal cortex is hyperactive in OCD (Rauch et al., 1994; Breiter et al., 1996); this hyperactivity possibly reflects augmented input from those limbic structures. There remains debate, however, as to whether the actual target of this operation is the white matter beneath the head of the caudate nucleus, or the islands of gray matter in this region known as the substantia inominata (Feldman et al., 2001). The mechanism by which this procedure treats OCD is even less clear than its target, but might involve lessening the contribution of anxiety- and fear-related signals to typical OCD behavior.

A third lesion-based treatment involves interrupting the fibers of the anterior limb of the internal capsule, a.k.a., anterior capsulotomy. These fibers were initially observed to have degenerated after frontal lobotomy in postmortem studies, and were later understood to relay thalamic afferents to the medial and orbitofrontal cortices. Bilateral anterior capsulotomy was therefore undertaken as another method to preserve some of the benefit of frontal lobotomy while hopefully minimizing the profound negative consequences of that coarse procedure. Successful alleviation of the signs and symptoms of OCD in greater than 60 to 70% of patients has often been reported (Feldman et al., 2001), although a significant number may nevertheless display the hallmark attributes of frontal damage, including apathy and poor decision making (Ruck et al., 2003).

These three operations, while not an exhaustive list of ablative procedures that at one point or another were attempted for OCD, are nevertheless representative of the ways in which debilitating and medically intractable OCD has been subjected to surgical intervention. They appear to share the common goal of weakening the influence of the medial and/or orbitofrontal cortices on behavior. However, there are both theoretical and experimental reasons to suppose a large role for the basal ganglia in the pathophysiology of this disease (Graybiel and Rauch, 2000). These operative strategies, therefore, may amount to little more than double-negative patches on behavior rather than treatments directed at the primary pathology. Thus, incomplete success and the presence of unwanted behavioral side effects are the rule rather than the exception. Furthermore, these procedures are irreversible. If a better treatment is found in the future, one cannot undo the damage done in order to pursue a fresh course of therapy.

9.4 Application and Efficacy of Electrical Brain Stimulation for OCD

Deep brain stimulation (DBS) has, over the past decade, achieved widespread acceptance for the treatment of nonpsychiatric neurological diseases, namely, Parkinson disease (PD), pain, dystonia, and tremor. The application of DBS for psychiatric disease, while in its infancy, shows much promise as an effective and relatively safe therapeutic tool. Several recent studies and case reports show hope, in particular, for the treatment of severe, medically refractory OCD (Anderson and Ahmed, 2003; Nuttin et al., 2003a,b; Aouizerate et al., 2004; Abelson et al., 2005; Greenberg et al., 2006).

While ablative psychosurgery is considered a last-line treatment for psychiatric disease because of both invasiveness and irreversibility, DBS is often considered a relatively minimally invasive and reversible procedure. Nevertheless, significant complications are possible, from hemorrhage and stroke, to seizures, to infections. Furthermore, the electrode tract itself does represent some minor degree of damage, and the electrical stimulation is potentially injurious to surrounding tissues. Importantly, the true benefit of DBS for the alleviation of psychiatric illness is not yet proven. Therefore, in current practice, patients selected to undergo DBS for OCD must qualify under the same stringent criteria set forth for ablative procedures such as capsulotomy.

In general, to qualify for DBS for OCD, patients must have a severe form of the disease, as measured on the Yale–Brown Obsessive–Compulsive Scale (Y–BOCS), usually 25 to 30 out of 40 points. This score is determined by a patient's subjective answers to standard questions during a structured interview by a psychiatrist. In addition, some centers require a low score on the Global Assessment of Function (GAF) scale, reflecting significant disability. A minimum age requirement (usually 18 years) and a minimum length of illness (often 5+ years) are also common criteria. Crucially, a potential candidate's OCD must have been refractory to appropriate attempts at medical and cognitive-behavioral therapy (CBT). These prior attempts at therapy usually involve adequate courses and doses of a serotonin reuptake inhibitor plus an antipsychotic medication, and CBT of at least several months' duration. Often, the presence of another significant psychiatric illness, except depression, is considered an exclusion criterion. Most centers performing any type of behavioral neurosurgery have multidisciplinary committees, consisting of psychiatrists, neurologists, and neurosurgeons, in place to determine which patients meet these criteria for surgery.

The surgical procedure for DBS is relatively minimally invasive, and usually is done with the patient awake and with his or her head fixed in a stereotaxic frame (although frameless techniques are also used, but with perhaps slightly less accuracy). A small lead containing several cylindrical contacts is directed toward the target brain structure through a burr hole in the skull. While the procedure is "blind" (i.e., the lead is not visible as it traverses the brain), hemorrhage, the most worrisome potential complication, is infrequent, occurring with an incidence of about 1 to 3% per electrode implanted (Binder et al., 2005; Voges et al., 2006). The tail end of these leads is then tunneled beneath the skin to a battery pack/pulse generator usually implanted beneath the clavicle. Depending on stimulation parameters (frequency and current), most batteries require a minor operation to be replaced every five months to two to three years; some newer batteries can be inductively charged through the skin every night (but there may nevertheless be a limit to the number of drain/recharge cycles).

Studies examining the efficacy of DBS for OCD have targeted the anterior internal capsule (Anderson and Ahmed, 2003; Nuttin et al., 2003a,b; Abelson et al., 2005; Greenberg et al., 2006) or the nearby ventral caudate (Aouizerate et al., 2004). These studies report a significant alleviation (more than a 35% reduction) of symptoms as measured subjectively on the Y–BOC (Yale–Brown Obsessive–Compulsive) scale. Functional brain imaging has suggested that DBS of the anterior internal capsule may result in a decrease in frontal metabolism in the 3 to 12 weeks following the onset of stimulation (Nuttin et al., 2003a,b; Abelson et al., 2005). Overall, the number of patients responding to the treatment is somewhat less than what has been reported with comparable lesion-based surgery (~30 to 50% with DBS and 60 to 70% with anterior capsulotomy, although substantially fewer patients have undergone DBS for OCD, thus limiting statistical power). However, some studies administered neuropsychiatric testing pre- and postoperatively (Nuttin et al., 2003a,b; Aouizerate et al., 2004); these found no evidence of a significant decline in frontal cognitive functioning (and, among the small mixed effects, some possible signs of improved functioning). Compared with lesions, therefore, DBS for OCD, in its current state, appears to involve a trade-off between the likelihood of success in any particular patient and the presence of cognitive side effects. Unfortunately, there has not yet been any head-to-head comparison of the efficacy or side effects of these treatments.

9.5 General Properties of DBS

Targeting the anterior internal capsule for electrical stimulation may appear to be a strategy opposite to that of lesioning this structure (anterior capsulotomy); however, the effect of electrical stimulation is not necessarily simply increased neural activity. Indeed, there is evidence that DBS in humans can cause neuronal inhibition (Filali et al., 2004; Welter et al., 2004). Specifically, the neural effects of electrical stimulation depend on target tissue properties, the pattern and intensity of stimulation, and electrode structural and biochemical properties.

The type of tissue stimulated — specifically, its composition of neural elements — determines the response to a particular form of stimulation (current, pulse width, and frequency). The most basic distinction is between white matter (containing axons) and gray matter (which contains a mix of cell bodies, axons, and dendrites). Different elements have different sensitivities to electrical stimulation (Ranck, 1975). Furthermore, neural elements such as a cell body or a dendrite will exhibit particular responses to a given pattern of stimulation depending on its subtype, likely reflecting differences in morphology as well as in the classes and distributions of ion channels (Perlmutter and Mink, 2006). In particular, it has been proposed that high-frequency stimulation of the subthalamic nucleus for the treatment of PD silences cell bodies through an effect on calcium channels, but stimulates efferent axons (Garcia et al., 2005). This would produce an effective blockade on the transmission of neural information through that structure, and impose a new output on its downstream targets (and perhaps also alter activity in the upstream structures via antidromic effects).

Electrode design contributes significantly to the target effect of electrical stimulation. The most commonly used DBS lead contains four cylindrical contacts, each 1.27 mm in diameter and 1.5 mm in height. The lead used in DBS for OCD is similar to that used in PD, but has a larger spacing between the contacts. Most current DBS stimulation paradigms use bipolar stimulation across two of these four contacts. The stimulated volume is therefore a function of the stimulation intensity, the distance between the contacts serving as anode and cathode, and the shape of these contacts. For the standard DBS lead, these factors result in an ellipsoid volume of stimulation (Butson and McIntyre, 2006). Furthermore, when this electrode configuration is used to stimulate white matter tracts, the most susceptible axons are those that lie parallel to the lead, experiencing the greatest voltage gradient (Ranck, 1975). Because the lead trajectory is severely limited by anatomical and surgical considerations (e.g., it should avoid critical areas such as motor cortex and should not traverse the ventricles), there is little control over the particular axon bundles most likely to be stimulated. The current electrode design is therefore unlikely to be optimal in terms of either therapeutic benefit or of limiting unwanted side effects.

The materials comprising the electrodes contribute to the chronic stability of the delivered stimulation and the surrounding tissue response. For example, degradation of the metal contacts and gliosis are major factors influencing electrode impedance over time (Moss et al., 2004; Gimsa et al., 2005). It is therefore possible that long-term changes in DBS efficacy (whether increased or lessened) may in part reflect these chemical and biological processes. DBS electrodes currently in use generally have platinum–iridium contacts; this material is known to provide good conductive contact with brain tissue and to be relatively inert biologically (Geddes and Roeder, 2003).

Thus, determining the neural effect of electrical stimulation on a target tissue is a complex function of the type and organization of neuronal elements within that tissue, the stimulation protocol, and electrode properties. Predicting the functional effect of stimulation requires an added level of understanding regarding the neuro-anatomical circuits underlying the relevant behaviors. Without a doubt, DBS, as it currently exists, optimizes neither the target nor the stimulation parameters, nor the electrode hardware.

9.6 Understanding and Improving DBS for OCD

The utility and methodology of DBS, whether for alleviating the burden of movement disorders, pain, or psychiatric disease, has been determined empirically rather than inductively. Significant questions persist regarding basic issues such as the local neural effects of electrical stimulation on different targets and the mechanism by which altering activity in the target structures influences behavior. Although DBS for PD is a well-established therapy and has been studied more intensively, various technical differences make even qualitative comparisons across these applications difficult. For example, in DBS for PD, stimulation frequencies up to 185 Hz are common. Meanwhile, effective stimulation frequencies for DBS of the anterior internal capsule in OCD have been generally lower (up to 130 Hz) (Nuttin et al., 2003a,b; Abelson et al., 2005). PD stimulation is typically in the range of 0.5 to 5.5 volts (V), whereas up to 10.5 V has occasionally been attempted for OCD (Nuttin et al., 2003a,b; Abelson et al., 2005). Finally, these therapies target grossly different structures — gray matter nucleus in PD versus white matter in OCD — adding further uncertainty to any direct comparison.

In the particular case of DBS for OCD, several fundamental questions remain largely unexplored. For example, some groups have noted a possible delay in the beneficial effects of stimulation of weeks to months (Nuttin et al., 2003a,b; Aouizerate et al. 2004), unlike the immediate and dramatic effects of DBS for PD. Likewise, acute capsular stimulation may result in increased frontal metabolism (Rauch et al., 2006), in contrast to chronic stimulation (Nuttin et al., 2003a,b; Abelson et al., 2005). These findings suggest that the effect of anterior capsule stimulation on behavior may be indirect, perhaps resulting through slow neuronal plasticity at a site or sites distant from the stimulated target. Most crucially, the occasional need to use relatively high voltages to achieve therapeutic effects in OCD compared with PD suggests to some that the "real" target might be some distance away, maybe in the ventral caudate (Aouizerate et al., 2004). Others have reported preliminary evidence suggesting the effective target may be the nearby nucleus accumbens (Tass et al., 2003).

DBS optimization will require, foremost, a more detailed understanding of the pathophysiology of OCD. Not all brain regions exhibiting hyper- or hypoactivity in a disease state are necessarily involved in the etiology of that disease; these alterations could be downstream reflections of a primary pathology situated elsewhere, or they could be compensatory. The ideal target will likely be that which is closest to the root pathology (perhaps a metabolic or structural disturbance not yet identified). Once a favorable target is found, electrodes can be designed such that the electric fields produced will best fill the desired volume (Butson and McIntyre, 2006; Gimsa et al., 2006). In addition, it may be possible to employ carefully selected stimulus waveforms and frequencies to bias the stimulation toward particular neuronal elements (McIntyre and Grill, 2002). Such measures, combined, will improve the anatomical and functional specificity of DBS for OCD.

Perhaps one of the biggest limitations of DBS today is its open-loop design. A DBS electrode provides a constant and unremitting pattern of stimulation without regard to behavioral or neurophysiological variables. A closed-loop DBS system could, in principle, provide more specificity in the timing of stimulation and therefore potentially reap benefits such as decreased side effects from unnecessary stimulation and lower charge utilization, which translates into greater safety and longer battery life. Some investigators argue that because an essential feature of neurophysiological disease might be abnormal synchronization and rhythmicity of neuronal population activity (such as is believed to be the case in PD), DBS systems could monitor these oscillations and then actively stimulate to desynchronize them (Hauptmann et al., 2005); alternatively, this could be achieved in open-loop fashion using a simpler two-step stimulation to phase-reset and then desynchronize without the need for monitoring and on-board processing (Tass et al., 2003). However, pathological oscillations are not yet known to be at the root of OCD. The monitoring of other physiological variables, such as the levels of particular neuromodulators or neurotransmitters, could also be used someday to drive appropriate stimulation. Ultimately, the nature of the stimulation itself may evolve to include chemical as well as electrical feedback.

Clearly, DBS as it exists today is a relatively crude tool to manipulate brain function. While it offers significant theoretical benefits over earlier, lesion-based approaches for the treatment of psychiatric diseases such as OCD, many of these benefits are yet to be realized. Nonetheless, the current deficits of knowledge regarding the mechanisms of DBS and the most suitable targets, as well as the unexplored possibilities for optimization of hardware and software, represent potential opportunities for the development of more effective and specific treatments in the near future.

References

Abelson, J.L. et al. (2005). Deep brain stimulation for refractory obsessive-compulsive disorder. *Biol. Psychiatry*, 57(5):510–516.

Abramowitz, J.S. (1997). Effectiveness of psychological and pharmacological treatments for obsessive-compulsive disorder: a quantitative review. *J. Consult. Clin. Psychol.*, 65(1):44–52.

Anderson, D. and Ahmed, A. (2003). Treatment of patients with intractable obsessive-compulsive disorder with anterior capsular stimulation. Case report. *J. Neurosurg.*, 98(5):1104–1108.

Aouizerate, B. et al. (2004). Deep brain stimulation of the ventral caudate nucleus in the treatment of obsessive-compulsive disorder and major depression. Case report. *J. Neurosurg.*, 101(4):682–686.

Berthier, M.L. et al. (1996). Obsessive-compulsive disorder associated with brain lesions: clinical phenomenology, cognitive function, and anatomic correlates. *Neurology*, 47(2):353–361.

Berthier, M.L. et al. (2001). Obsessive-compulsive disorder and traumatic brain injury: behavioral, cognitive, and neuroimaging findings. *Neuropsychiat. Neuropsychol. Behav. Neurol.*, 14(1):23–31.

Binder, D.K. et al. (2005). Risk factors for hemorrhage during microelectrode-guided deep brain stimulator implantation for movement disorders. *Neurosurgery*, 56(4):722–732; discussion 722–732.

Botvinick, M.M. et al. (2004). Conflict monitoring and anterior cingulate cortex: an update. *Trends Cogn. Sci.*, 8(12):539–546.

Breiter, H.C. et al. (1996). Functional magnetic resonance imaging of symptom provocation in obsessive-compulsive disorder. *Arch. Gen. Psychiatry*, 53(7):595–606.

Brown, J.W. and Braver, T.S. (2005). Learned predictions of error likelihood in the anterior cingulate cortex. *Science*, 307(5712):1118–1121.

Buch, E.R. et al. (2006). Comparison of population activity in the dorsal premotor cortex and putamen during the learning of arbitrary visuomotor mappings. *Exp. Brain Res.*, 169(1):69–84.

Butson, C.R. and McIntyre, C.C. (2006). Role of electrode design on the volume of tissue activated during deep brain stimulation. *J. Neural Eng.*, 3(1):1–8.

Chacko, R.C. et al. (2000). Acquired obsessive-compulsive disorder associated with basal ganglia lesions. *J. Neuropsychiatry Clin. Neurosci.*, 12(2):269–272.

Coetzer, B.R. (2004). Obsessive-compulsive disorder following brain injury: a review. *Int. J. Psychiatry Med.*, 34(4):363–377.

Dougherty, D.D. et al. (2002). Prospective long-term follow-up of 44 patients who received cingulotomy for treatment-refractory obsessive-compulsive disorder. *Am. J. Psychiatry*, 159(2):269–275.

Feldman, R.P. et al. (2001). Contemporary psychosurgery and a look to the future. *J. Neurosurg.*, 95(6):944–956.

Filali, M. et al. (2004). Stimulation-induced inhibition of neuronal firing in human subthalamic nucleus. *Exp. Brain Res.*, 156(3):274–281.

Friedlander, L. and Desrocher, M. (2006). Neuroimaging studies of obsessive-compulsive disorder in adults and children. *Clin. Psychol. Rev.*, 26(1):32–49.

Fuster, J.M. (2000). Executive frontal functions. *Exp. Brain Res.*, 133(1):66–70.

Garcia, L. et al. (2005). High-frequency stimulation in Parkinson's disease: more or less? *Trends Neurosci.*, 28(4):209–216.

Geddes, L.A. and Roeder, R. (2003). Criteria for the selection of materials for implanted electrodes. *Ann. Biomed. Eng.*, 31(7):879–890.

Gimsa, J. et al. (2005). Choosing electrodes for deep brain stimulation experiments–electrochemical considerations. *J. Neurosci. Methods*, 142(2):251–265.

Gimsa, U. et al. (2006). Matching geometry and stimulation parameters of electrodes for deep brain stimulation experiments-numerical considerations. *J. Neurosci. Methods*, 150(2):212–227.

Goodman, W.K. (1999). Obsessive-compulsive disorder: diagnosis and treatment. *J. Clin. Psychiatry*, 60(Suppl 18):27–32.

Graybiel, A.M. (1998). The basal ganglia and chunking of action repertoires. *Neurobiol. Learn. Mem.*, 70(1–2):119–136.

Graybiel, A.M. et al. (1994). The basal ganglia and adaptive motor control. *Science*, 265(5180):1826–1831.

Graybiel, A.M. and Rauch, S.L. (2000). Toward a neurobiology of obsessive-compulsive disorder. *Neuron*, 28(2):343–347.

Greenberg, B.D. et al. (2006). Three-year outcomes in deep brain stimulation for highly resistant obsessive-compulsive disorder. *Neuropsychopharmacology,* 31(11):2384–2393.

Hauptmann, C. et al. (2005). Effectively desynchronizing deep brain stimulation based on a coordinated delayed feedback stimulation via several sites: a computational study. *Biol. Cybern.,* 93(6):463–740.

Heyman, I. et al. (2003). Prevalence of obsessive-compulsive disorder in the British nationwide survey of child mental health. *Int. Rev. Psychiatry,* 15(1–2):178–184.

Ito, S. et al. (2003). Performance monitoring by the anterior cingulate cortex during saccade countermanding. *Science,* 302(5642):120–122.

Kim, C.H. et al. (2003). Anterior cingulotomy for refractory obsessive-compulsive disorder. *Acta Psychiatr. Scand.,* 107(4):283–290.

Mataix-Cols, D. et al. (2005). A multidimensional model of obsessive-compulsive disorder. *Am. J. Psychiatry,* 162(2):228–238.

McIntyre, C.C. and Grill, W.M. (2002). Extracellular stimulation of central neurons: influence of stimulus waveform and frequency on neuronal output. *J. Neurophysiol.,* 88(4):1592–1604.

McKay, D. et al. (2004). A critical evaluation of obsessive-compulsive disorder subtypes: symptoms versus mechanisms. *Clin. Psychol. Rev.,* 24(3):283–313.

Miller, E.K. (2000). The prefrontal cortex and cognitive control. *Nat. Rev. Neurosci.,* 1(1):59–65.

Moss, J. et al. (2004). Electron microscopy of tissue adherent to explanted electrodes in dystonia and Parkinson's disease. *Brain,* 127(Pt 12):2755–2763.

Nakao, T. et al. (2005). Brain activation of patients with obsessive-compulsive disorder during neuropsychological and symptom provocation tasks before and after symptom improvement: a functional magnetic resonance imaging study. *Biol. Psychiatry,* 57(8):901–910.

Nuttin, B.J et al. (2003a). Electrical stimulation of the anterior limbs of the internal capsules in patients with severe obsessive-compulsive disorder: anecdotal reports. *Neurosurg. Clin. N. Am.,* 14(2):267–274.

Nuttin, B.J. et al. (2003b). Long-term electrical capsular stimulation in patients with obsessive-compulsive disorder. *Neurosurgery,* 52(6):1263–1272; discussion 1272–1274.

Pasupathy, A. and Miller, E.K. (2005). Different time courses of learning-related activity in the prefrontal cortex and striatum. *Nature,* 433(7028):873–876.

Perlmutter, J.S. and Mink, J.W. (2006). Deep brain stimulation. *Annu. Rev. Neurosci.,* 29:229–257.

Ranck, J.B., Jr. (1975). Which elements are excited in electrical stimulation of mammalian central nervous system: a review. *Brain Res.,* 98(3):417–440.

Rauch, S.L. et al. (1994). Regional cerebral blood flow measured during symptom provocation in obsessive-compulsive disorder using oxygen 15-labeled carbon dioxide and positron emission tomography. *Arch. Gen. Psychiatry,* 51(1):62–70.

Rauch, S.L. et al. (2006). A functional neuroimaging investigation of deep brain stimulation in patients with obsessive-compulsive disorder. *J. Neurosurg.,* 104(4):558–565.

Roberts, A.C. (2006). Primate orbitofrontal cortex and adaptive behaviour. *Trends Cogn. Sci.,* 10(2):83–90.

Ruck, C. et al. (2003). Capsulotomy for refractory anxiety disorders: long-term follow-up of 26 patients. *Am. J. Psychiatry,* 160(3):513–521.

Rushworth, M.F. et al. (2004). Action sets and decisions in the medial frontal cortex. *Trends Cogn. Sci.,* 8(9):410–417.

Schultz, W. (2006). Behavioral theories and the neurophysiology of reward. *Annu. Rev. Psychol.,* 57:87–115.

Tass, P. A. et al. (2003). Obsessive-compulsive disorder: development of demand-controlled deep brain stimulation with methods from stochastic phase resetting. *Neuropsychopharmacology,* 28(Suppl. 1):S27–S34.

Thobois, S. et al. (2004). Obsessive-compulsive disorder after unilateral caudate nucleus bleeding. *Acta Neurochir. (Wien.),* 146(9):1027–1031; discussion 1031.

Voges, J. et al. (2006). Deep-brain stimulation: long-term analysis of complications caused by hardware and surgery-experiences from a single centre. *J. Neurol. Neurosurg. Psychiatry,* 77(7):868–872.

Welter, M. L. et al. (2004). Effects of high-frequency stimulation on subthalamic neuronal activity in parkinsonian patients. *Arch. Neurol.*, 61(1):89–96.

Whitty, C.W. et al. (1952). Anterior cingulectomy in the treatment of mental disease. *Lancet*, 1(10):475–481.

Williams, Z.M. et al. (2004). Human anterior cingulate neurons and the integration of monetary reward with motor responses. *Nat. Neurosci.*, 7(12):1370–1375.

Williams, Z.M. and Eskandar, E.N. (2006). Selective enhancement of associative learning by micro-stimulation of the anterior caudate. *Nat. Neurosci.*, 9(4):562–568.

Wittchen, H.U. and Jacobi, F. (2005). Size and burden of mental disorders in Europe — a critical review and appraisal of 27 studies. *Eur. Neuropsychopharmacol.*, 15(4):357–376.

Zohar, A.H. (1999). The epidemiology of obsessive-compulsive disorder in children and adolescents. *Child Adolesc. Psychiatr. Clin. N. Am.*, 8(3):445–460.

Neural Augmentation

Sensory Prostheses

10

Cochlear Prostheses: An Introduction

Donald K. Eddington

10.1 Introduction

The improvement in speech reception (Osberger et al., 2000; Anderson et al., 2002; Parkinson et al., 2002) enjoyed by the profoundly hearing-impaired using today's cochlear implant systems easily places them second to cardiac pacemakers as the most successful neural prostheses in use today. While this success is built upon a framework of modern advances in fields such as medicine, auditory physiology and psychophysics, materials science, electrical engineering, and electronic technology, the idea of using electricity to elicit sensations of hearing, dates to the 18th century (for reviews oriented around hearing, see Simmons, 1966; Niparko and Wilson, 2000). The purpose of this chapter is to introduce the rationale for cochlear implantation, discuss the benefits associated with current devices, and provide an example of how researchers approach the task of improving these devices[1].

10.2 Rationale

The top panel in Figure 10.1 presents a simplified and schematic diagram of the human peripheral auditory system. Acoustic signals entering the ear canal are conducted via the middle ear to the cochlea where the mechanical energy of sound is translated to electric spikes on auditory nerve fibers. For example, a high-pitch tone causes the basilar membrane to vibrate mainly near the base of the cochlea. This causes hair cells in that region to release neural transmitter that elicits electric spikes on the nerve fibers making synaptic contact with those hair cells. For a low-pitch tone, the maximum membrane vibration is near the cochlear apex, and the number of electric spikes per unit time conducted over those nerve fibers will tend to increase. For more complex acoustic signals such as speech, the spatiotemporal pattern of spike activity on the array of nerve fibers will be more complex. It is these spatiotemporal patterns of electric spike activity that lead to the sound sensations normal-hearing people use to recognize speech, enjoy music, and identify/localize sound sources.

---

[1] A survey of the literature associated with the topics addressed in this chapter is outside its scope and purpose. Readers interested in pursuing these topics in depth are directed to several books as starting points (Tyler, 1993; Niparko et al., 2000; Zeng et al., 2004).

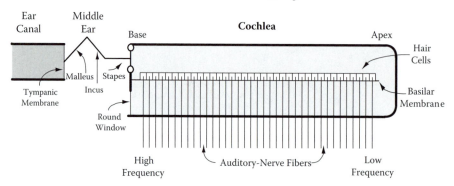

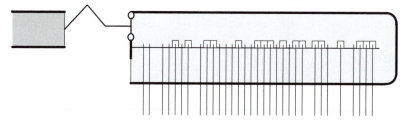

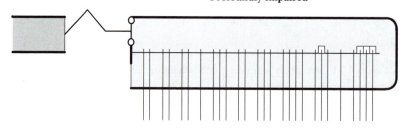

FIGURE 10.1 Schematic diagrams of the human peripheral auditory system. Acoustic signals enter the ear canal at the left and cause the tympanic membrane (ear drum) to vibrate. The three ossicles of the middle ear (malleus, incus, and stapes) conduct the acoustic signal to the base of the cochlea where the footplate of the stapes is secured in the cochlea's oval window by the annular ligament (represented by the two circles on either side of the stapes footplate). The cochlea is a snail-like spiral canal of 2.5 turns (approximately 32 mm in length) in the skull's temporal bone that is filled with fluid. For the purposes of this diagram, the cochlea is unwound and represented by a tube. The round window, a second opening at the base of the cochlea, is covered by a membrane that allows fluid displacement in response to stapes vibration. The basilar membrane runs the length of the cochlea and is mechanically "tuned" with maximum vibrational displacement at the base occurring for high-frequency tones and at the apex for low-frequency tones. The vertical lines represent nerve fibers that form synaptic connections with the auditory system's sensory (hair) cells that sit in the organ of Corti on the basilar membrane. The top panel represents the normal-hearing condition with a full complement of hair cells (~1500) and nerve fibers (~30,000). The number and distribution of hair cells and nerve fibers for the moderate and profoundly impaired conditions are shown in the middle and bottom panels.

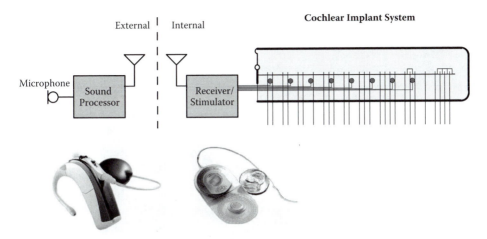

FIGURE 10.2 Schematic diagram of a cochlear implant system. The sound processor often includes a digital signal processing chip used to implement the sound-processing strategy that translates the microphone's output to instructions that control the implanted stimulator. Both the transmitter antenna (the disk connected by a thin cable to the behind-the-ear sound-processor package) and the receiver antenna (just below the nickel) include a small magnet. These magnets hold the transmitter antenna against the skin over the implant. The stimulator delivers electric currents to the array of implanted electrodes as directed by the instructions sent by the sound processor. These currents elicit spikes on the auditory nerve fibers in the region of their respective electrode. (The photograph of a current sound processor and a receiver/stimulator with electrode array are used with permission of Advanced Bionics Incorporated.)

The middle panel of Figure 10.1 represents the loss of hair cells and nerve fibers, which is the cause of most hearing impairment. Because nerve fibers without hair cells will not be excited by acoustic signals, it is not difficult to imagine that the pattern of spike activity seen by the brain in this situation will be a distorted version of that produced by the normal-hearing system. Depending on the severity of hair-cell loss, individuals in this category of impairment can derive great benefit from classic hearing aids that amplify acoustic signals to recruit more of the remaining hair cells.

The condition of deafness or profound hearing impairment is represented in the bottom panel of Figure 10.1. There are very few hair cells and a significant number of auditory-nerve fibers have degenerated. In this case, a hearing aid will not help because there are few (if any) hair cells to translate the sound energy into the patterns of spike activity that carry an acoustic signal's information. This is where a cochlear implant may be beneficial.

The basic components of a cochlear-implant system are shown in Figure 10.2. The sound processor translates the signal from the microphone into instructions that control the receiver/stimulator unit implanted in the bone just under the skin above the patient's ear. The instructions are transmitted to the receiver/stimulator through the skin using a modulated RF (radio frequency) carrier. The energy in the carrier signal is used to power the receiver/stimulator, and the instructions control the electric stimuli delivered to each of the electrodes implanted in the cochlea. Today's implants typically consist of 16 to 24 electrodes.

In the case of a low-pitch tone, the sound processor translates the microphone's output signal into instructions that result in the stimulator driving relatively apical electrodes and exciting remaining nerve fibers in the apical region of the cochlea. For a complex stimulus such as speech, a more complex set of stimuli would be distributed over many of the implanted electrodes, eliciting a more complex pattern of spike activity on the array of remaining nerve fibers. The challenge is to develop sound-processing strategies, stimulators, and electrode arrays that, for any sound at the microphone, work together to elicit patterns of spike activity on the nerve fibers that mimic those that would be generated in a normal-hearing ear. If that can be accomplished, a person who hears normally until age 20, becomes deaf, and is implanted with such a system should understand speech with little if any training. Unfortunately, devices do not provide that level of performance for most implanted patients.

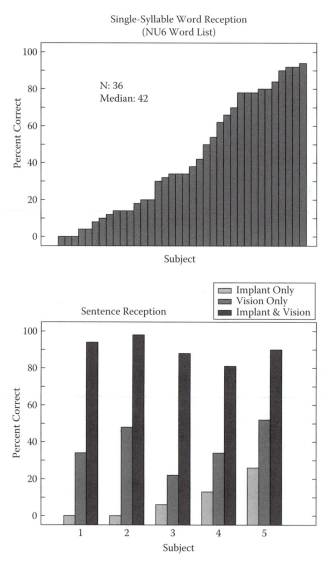

FIGURE 10.3 In the upper panel, each bar represents the single-syllable word score (NU-6 test) of a patient measured without lipreading and after using their cochlear implant for at least three months. Results are for all of 36 patients implanted at the Massachusetts Eye and Ear Infirmary during a recent 8-month period. The lower panel presents sentence-reception scores measured in three different conditions (blue: listening without lipreading; green: lipreading with implant turned off; red: lipreading with implant turned on) for the five poorest-performing patients shown in the upper panel. The CID sentences used in this testing are easier than the single-syllable words because of their high context (e.g., subjects 4 and 5 score substantially higher in the implant-alone case for sentences than for single-syllable words).

10.3 Performance

The upper panel of Figure 10.3 plots performance measured in 36 patients implanted with a recent device using a single-syllable word test. Note the wide range of performance measured in these patients (a normal-hearing person scores 100% on this task). Only those scoring better than about 70% on this task would be able to carry on a relatively fluent conversation without lipreading (e.g., using the telephone). It is clear that these results are far from normal for virtually all implantees and might make one

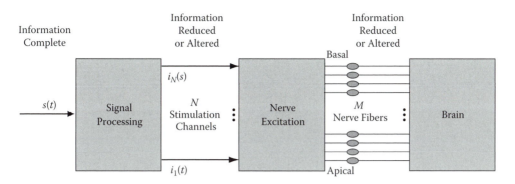

FIGURE 10.4 Diagram illustrating potential sites where information transfer may be limited. Beginning at the left, $s(t)$ represents an acoustic signal (e.g., speech) carrying information that is to be transferred to the brain (rightmost box). The cochlear implant system (represented by the box labeled "Signal Processing") translates $s(t)$ into N channels of electric stimulation, each represented by a horizontal line ($i_1(t), i_2(t), \ldots i_N(t)$). The "Nerve Excitation" box represents the interaction of these stimulus signals with the array of remaining auditory nerve fibers to produce patterns of spike activity that are interpreted by the "Brain."

wonder why even the poorest performers are typically enthusiastic about the benefit they receive and wear their devices all day, every day.

The lower panel of Figure 10.3 is more representative of the benefit gained by these poorer-performing patients. Here, a measure of high-context sentence recognition (the most common daily-life communication task) is used to characterize the performance of the five poorest performing patients from the upper panel for three listening conditions: (1) implant alone, (2) lipreading alone, and (3) listening with the implant while lipreading (the most common listening condition in daily life). Note that when lipreading is combined with the sound sensation from the implant, the benefit to speech reception is generally greater than the sum of the auditory-only and lipreading-alone scores. Even the poorest-performing subjects are able to communicate at a level that makes moving in the hearing world much easier and less stressful.

10.3.1 Improving Performance: An Overall Perspective

While patients using today's cochlear implant systems derive considerable benefit to everyday communication, a number of research groups are working toward identifying factors that limit performance and developing techniques that overcome them. Figure 10.4 diagrams one general perspective that has been successful. The acoustic signal is represented by $s(t)$, which includes all of the information ($I_{s(t)}$) to be transferred to the brain in usable form. The cochlear implant system translates this signal into a number of stimulation signals in the form of current waveforms ($i_k(t), 1 \le k \le N$) that are delivered to the remaining auditory nerve fibers by the implanted electrodes. Depending on the signal processing used, the total information represented in the set of stimulating signals (TI_i) may or may not include all of the information in the original acoustic signal. If $TI_i < I_{s(t)}$, modification of the signal processing to eliminate information distortion would be one place to start in improving performance.

When the electric stimuli interact with the remaining nerve fibers, patterns of spike activity are elicited and delivered to the brain. If the total information carried in the spike patterns of auditory-nerve fibers (TI_n) is less than TI_i, then the process of "nerve excitation" limits information transfer and should be specifically targeted as an opportunity for improving performance.

Even if the total amount of information in the acoustic signal is transferred to the pattern of spikes on the remaining auditory nerve fibers ($I_{s(t)} = TI_i = TI_n$), this does not mean that it will be properly interpreted by the brain. For example, many of today's sound-processing strategies divide the sound spectrum into a number (16 to 24) of analysis bands with each driving a different intracochlear electrode. If the mapping of analysis channel to stimulation channel were reversed (e.g., stimuli

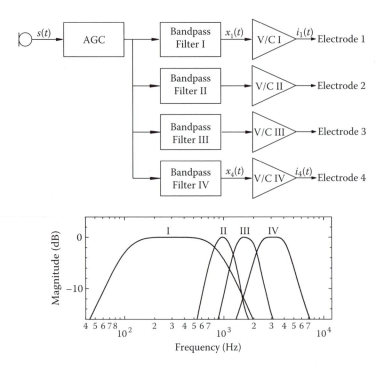

FIGURE 10.5 A block diagram illustrating an analog sound-processing strategy used by the Ineraid cochlear implant system. The automatic gain control (AGC) acts to compress input levels from the microphone above a specified criterion. This is important because sound intensity varies over a range of 120 dB, while the range of stimulus level producing sound sensations from threshold to maximum comfortable loudness for an intracochlear electrode is rarely greater than 24 dB. The magnitude responses of the four analog bandpass filters used to divide the sound spectrum into four segments are shown at the bottom. The output of each filter is converted from a voltage-controlled to a current-controlled waveform by the voltage-to-current (V/C) source converter and then delivered to its respective intracochlear stimulating electrodes. The return path for all intracochlear electrodes is a far-field electrode located in the temporalis muscle.

generated by low-frequency analysis channels being delivered to more basal electrodes and high-frequency channel outputs to more apical electrodes), it is conceivable that the information content would be complete in the stimulus waveforms and in the spatiotemporal pattern of spike activity on the array of nerve fibers, but that the brain would not be able to properly interpret it.

10.3.2 Improving Performance: An Example

As an illustration of how one might identify a factor suspected of distorting information and then implement a change in sound processing designed to correct it, a case is presented in which it was helpful to *decrease* the information in the stimulus waveforms in an effort to eliminate distortions occurring at the point of "nerve excitation" shown in Figure 10.4.

The top panel of Figure 10.5 illustrates the original sound-processing strategy used by the 14 patients we will follow in this example. The information in the acoustic signal that is contained within this processor's bandwidth (100 to 4000 Hz) is substantially preserved when one combines the signals delivered to the electrodes. For example, if the current waveforms are summed and used to drive an audio amplifier, a normal-hearing person will easily score 100% on the single-syllable word test used to generate the data of Figure 10.3.

The analog stimulation waveforms delivered by the Figure 10.5 processing strategy to the implanted electrodes in response to a vowel input are shown in Figure 10.6. The overall magnitude of Channel III is larger than Channels II an IV, and Channel I is the lowest magnitude stimulus. At the time marked

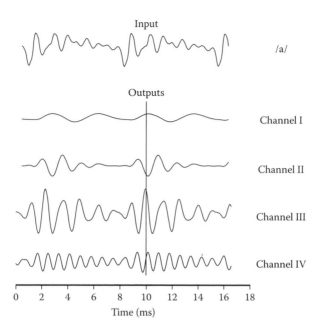

FIGURE 10.6 The top waveform represents an 18-ms segment of the input signal delivered to the AGC of a four-channel processing system (see Figure 10.5) for the vowel /a/ (as in "dah"). The bottom four waveforms represent the current waveforms delivered by each of the system's four analysis channels in response to the input signal. The vertical line marks one point in time that is discussed in the text.

by the vertical line, the stimulus magnitude of Channel II is negative while that of Channel III is positive. Because the electrodes associated with Channels II and III are separated by only 3 mm of highly conductive fluid (70 Ω-cm compared to the surrounding 300 Ω-cm soft tissue and the 5000 Ω-cm bone), it is very likely that the substantial negative potential that would be generated by the Channel-II stimulus alone will be largely cancelled by that of the positive potential that would be generated by the Channel-III stimulus alone. This means that although Channels II and III are both signaling significant within-band energy, the result of potential summation across electrodes is a very small overall stimulus and little neural activity at that point in time. This represents a distortion of the desired magnitude of stimulation and the pattern of neural activity desired. It was therefore hypothesized that this type of electrode interaction could be a factor that limits performance (Eddington et al., 1978; Wilson et al., 1991).

Figure 10.7 is a block diagram of a CIS sound-processing strategy that was designed to minimize potential summation across electrodes (Wilson et al., 1991). In this processing strategy, either rectification followed by low-pass filtering (cut-off frequency between 200 and 400 Hz) or quadrature detection (e.g., using Hilbert transform techniques) are used to extract the envelope ($e_i(t)$) from each bandpass filter's output. The envelope amplitude modulates a carrier (biphasic pulse train; typically 2000 pulses per second [pps]) to produce the stimulus waveform that (after amplitude compression) is delivered to the implanted electrode. Note that the two carrier pulse trains illustrated in Figure 10.7 are offset in time so that only one is pulsing at any time. This is true across all carriers of a CIS processing strategy. Thus a four-channel strategy using 2-kpps carriers limits the duration of the carriers' biphasic pulses to 125 μs to make it possible for the pulses of the four carriers to be interleaved. Because the stimulation pulses do not overlap across channels, the likelihood that the fields generated by individual channels will interact is greatly reduced. This means that the distortions resulting from those interactions should also be reduced.

While the CIS strategy may reduce distortions from field interactions, it accomplishes this at a cost. Significant information is discarded when the envelope is extracted from the output of each bandpass

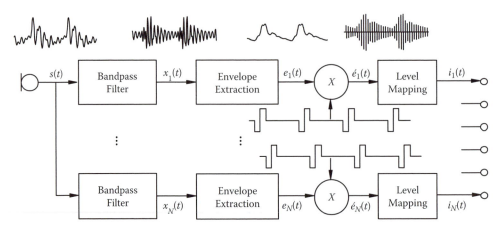

FIGURE 10.7 A block diagram representing a continuous interleaved stimulation (CIS) sound-processing strategy. The microphone output ($s(t)$) is partitioned into N channels by bandpass filters. The envelope signal ($e_i(t)$) extracted from the output signal of each bandpass filter ($x_i(t)$) is used to modulate a train of biphasic pulses. This amplitude-modulated pulse train ($\acute{e}_i(t)$) is then compressed and converted to a current waveform (e.g., $i_i(t) = a_i*\log(\acute{e}_i(t)) + b_i$, where a_i and b_i are selected so that the range of $\acute{e}_i(t)$ is mapped to a range of current levels corresponding to threshold and the current level eliciting a maximum comfortable sensation level). The waveforms at the top illustrate signals at various stages of processing a vowel (/a/) input signal ($s(t)$) for an analysis channel with a bandpass filter center frequency of approximately 1400 Hz.

filter. This is clear when one compares the example waveform for $x_i(t)$ in Figure 10.7 with the example waveform for $e_i(t)$: the fine structure apparent at the output of the bandpass filter is missing from the envelope waveform. The question is whether loss of the information carried in the fine structure will be more than made up for by the elimination of the distortion produced by field interactions.

To test whether the CIS strategy improves performance, we provided a wearable version of the strategy to 14 subjects who had used the analog sound-processing strategy shown in Figure 10.5 for at least 12 months. Before each subject switched to using the CIS system, we measured their ability to understand a list of standard sentences (without lipreading). The result of this test for each subject is shown as a bar in Figure 10.8. The same day each subject switched strategies, sentence recognition was also measured using the CIS processing system. The results for day 1 sentence testing in the CIS condition are represented by open circles in Figure 10.8; note that for most subjects (i.e., 8 of 14), performance decreased substantially with the new sound-processing strategy. Although the initial results were discouraging, each of the subjects agreed to give the new system a chance and wear the device for several months. The results of testing after at least 12 months using the CIS system are marked by the filled circles in Figure 10.8 and show a substantial improvement for 13 out of the 14 subjects.

The foregoing example illustrates one challenge investigators face in improving the performance of cochlear implant systems: that challenge is the promise of a new sound-processing strategy may not be revealed by acute laboratory testing. In at least some cases, it is only after a subject's brain is given the chance to adapt to the new kind of input by using the device for weeks or months that the benefit of the new system is apparent. In the CIS example, the benefit for most subjects was apparent after one or two months of use, but not in the initial laboratory testing. In situations similar to this, a strong theoretical rationale is needed before subjects and investigators invest in the effort needed to conduct a longitudinal evaluation.

10.4 The Future

History suggests that the steady increase in speech-reception performance observed over the past 30 years will continue as investigators continue identifying factors that limit today's performance and focus their

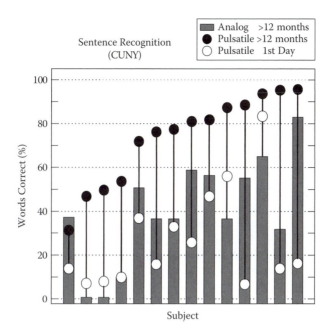

FIGURE 10.8 Plot of the score (percent of words correctly identified) measured for each of 14 subjects in three listening conditions. The word score was measured using CUNY (Boothroyd et al., 1985) (high-context) sentences without lipreading (audio alone). Bar height represents the score measured after at least 12 months' experience with the subject's original analog sound-processing strategy (see Figure 10.5). White circles mark each subject's score measured the day they began using the pulsatile sound-processing strategy, and black circles identify scores measured after at least 12 months of listening to the pulsatile system. Note that when subjects first switched to the new sound-processing strategy, only four scored better than with their original analog system. After 12 months, 13 of the 14 subjects scored substantially better with the pulsatile system. (*Source:* Adapted from Rabinowitz in Eddington et al., 1997.)

efforts on designing new systems that deal with those limitations. One area of current research is to restore the fine-time structure discarded by the CIS processing strategy described previously. The limitation is clear — but designing stimulation systems that restore that information in a form that can be encoded in neural responses to electric stimulation is a challenge that remains.

Considerable effort is also being directed at sound processors for bilateral implants in an effort to restore the benefits normal-hearing people receive from two ears (e.g., localization of sound sources and better speech reception when simultaneous noise and signal sources are spatially separated). Currently, the two implants are treated as two independent and asynchronous systems. New systems are being developed to coordinate stimulation across ears to better represent the 20-µs interaural sensitivity associated with normal-hearing ears.

The coordination of electric stimulation in one ear and acoustic input in the same or unimplanted ear is also being pursued. Some implantees have significant residual hearing that can be used to convey additional information to the brain. Testing in a small number of patients shows that the combination of acoustic and electric stimulation can sometimes improve speech-reception performance in noisy conditions. These results motivate work in electric/acoustic sound-processing strategies and in electrode systems that can be implanted with minimal damage to residual hearing.

In addition to the focus on improved hearing clarity, work also continues to make the physical devices more functional and esthetically pleasing. Devices have moved far from the rather large, clunky boxes worn in belt pouches with cables running underneath clothes to emerge at collar level to connect to microphones and inductive coupling units on the side of the head. Today's units are typically in behind-the-ear hearing aid cases that continue to decrease in size and weight. The next few years will probably see the first totally implantable units with internal batteries that will be recharged while reading or even sleeping.

Taken together, the advances of the past and the energy of today's current research and development teams point to a future that promises even clearer hearing for those suffering from sensorineural hearing loss. The added quality of life associated with these advances will make cochlear implants one of the most successful neural prostheses for many years to come.

References

Anderson, I., Weichbold, V., and D'Haese, P. (2002). Recent results with the MED-EL COMBI 40+ cochlear implant and TEMPO+ behind-the-ear processor. *Ear Nose Throat J.*, 81(4):229–233.

Boothroyd, A., Hanin, L., and Hnath, T. (1985). *CUNY Laser Videodisk of Everyday Sentences.* Speech and Hearing Sciences Research Center, City University of New York.

Eddington, D.K., Dobelle, W.H., Brackmann, D.E., Mladevosky, M.G., and Parkin, J.L. (1978). Auditory prosthesis research with multiple channel intracochlear stimulation in man. *Ann. Otol. Rhinol. Laryngol.*, Suppl. 87(6 Pt. 2):1–39.

Eddington, D.K., Rabinowitz, W.M., Boex-Spano, C., Tierney, J., Delhorne, L.A., Garcia, N., Noel, V., and Whearty, M.E. (1997). Speech processors for auditory prostheses: fith quarterly progress report; NIH contract N01-DC-6-2100. Bethesda, Neural Prosthesis Program, National Institutes of Health: 11.

Niparko, J.K., Kirk, K.I., Mellon, N.K., Robbins, A.M., Tucci, D.L., and Wilson, B.S., Eds. (2000). *Cochlear Implants: Principles and Practices.* Baltimore: Lippincott Williams & Wilkins.

Niparko, J.K. and Wilson, B.S. (2000). History of cochlear implants. In *Cochlear Implants: Principles and Practices.* J.K. Niparko, K.I. Kirk, N.K. Mellon, et al., Eds. Baltimore: Lippincott Williams & Wilkins, pp. 103–108.

Osberger, M.J., Fisher, L., and Kalberer, A. (2000). Speech perception results in children implanted with the CLARION Multi-Strategy cochlear implant. *Adv. Otorhinolaryngol.*, 57:417–420.

Parkinson, A.J., Arcaroli, J., Staller, S.J., Arndt, P.L., Cosgriff, A., and Ebinger, K. (2002). The nucleus 24 contour cochlear implant system: adult clinical trial results. *Ear Hear.*, 23(1 Suppl):41S–48S.

Simmons, F.B. (1966). Electrical stimulation of the auditory nerve in man. *Arch. Otolaryngol.*, 84(1):2–54.

Tyler, R.S., Ed. (1993). *Cochlear Implants: Audiological Foundations.* San Diego: Singular Publishing Group, Inc.

Wilson, B.S., Finley, C.C., Lawson, D.T., Wolford, R.D., Eddington, D.K., and Rabinowitz, W.M. (1991). Better speech recognition with cochlear implants. *Nature*, 352(6332):236–238.

Zeng, F.-G., Popper, A.N., and Fay, R.R., Eds. (2004). Cochlear implants: auditory prostheses and electric hearing. In *Springer Handbook of Auditory Research.* New York: Springer-Verlag.

11

Visual Prostheses

Robert J. Greenberg

11.1 Introduction

An area of recent interest is the development of a visual prosthesis by electrically stimulating the neural tissue of blind patients. Major efforts in this area include electrode arrays implanted on or under the retina, around the optic nerve, and on or in the visual cortex. On February 19, 2000, the Alfred E. Mann Institute for Biomedical Engineering at the University of Southern California, Los Angeles, held a symposium that brought together experts from across the United States to discuss the state of the art in this area of research. This chapter provides historical background in the field and summarizes the presentations from this meeting.

11.2 Historical Background

The possibility of restoring vision to blind patients using electricity began with the discovery that an electric charge delivered to a blind eye produces a sensation of light. This discovery was made by LeRoy in 1755.[1] However, it was not until 1966 that the first human experiments in this field began with Giles Brindley's experiments with electrical stimulation of the visual cortex.[2] He used 180 cortical surface electrodes, which were able to induce perception of spots of light called "phosphenes" but they were ill-defined and could not be combined to make an image. It failed to produce useful vision in these patients. Similar experiments by William Dobelle in 1974 produced essentially the same results.[3, 4]

11.3 Recent Research

Since these early experiments, efforts have been underway to produce penetrating arrays of electrodes that offer the possibility of more closely spaced electrodes and therefore higher resolution cortical devices.[5–8] Richard Normann (University of Utah) has micromachined 100 electrodes from silicon, which was primarily used for recording in the sensory cortex of animals.[5] Another group at the University of Michigan led by Ken Wise also produced micromachined penetrating electrodes for recording.[6] In the 1990s, an effort at the NIH headed by Terry Hambrecht made an array of 38 penetrating microelectrodes, which were implanted in a patient and yielded separable phosphenes at electrode placements closer than had been produced with surface electrodes.[7, 8] Electronics for an implantable cortical prosthesis are being developed (with 1024 channels) at the Illinois Institute of Technology by Philip Troyk (personal communication).

While the cortical stimulation approaches have made progress, these have been hampered by the physiology. The processing that has occurred by the time the neural signals have reached the cortex is greater than at the more distal sites such as the retina. This results in more complex phosphenes being perceived by the patient. The surgery and the implanted prosthesis do provide risks, including intracranial hemorrhage to a blind patient who has an otherwise normal brain. These factors and the lack of availability of implantable electronics have limited the clinical application of these devices.

The limitations of the cortical approach encouraged several groups in United States over the past ten years to explore the possibility of producing vision in patients with an intact optic nerve but with damaged photoreceptors from stimulating the retina.[9–15] Likely candidate diseases are retinitis pigmentosa (RP) or age-related macular degeneration (AMD). It is difficult to determine exactly how many patients are blinded by these diseases because patients often stop seeing their ophthalmologist after being told there is nothing that can be done. However, estimates of legal blindness in the Western (developed) world run as high as 300,000 people with RP and 3 million people with AMD; and 1.2 million people are afflicted (but not yet blind) with RP worldwide and 10 million people are afflicted with AMD in the United States alone.[16]

There have emerged two major approaches to retinal stimulation: (1) epiretinal and (2) subretinal. In the epiretinal approach, electrodes are placed on top of the retina to produce phosphenes. In the subretinal approach, photodiodes are implanted underneath the retina and used to generate currents, which then stimulate the retina. The epiretinal approach has been pursued by a team at Johns Hopkins University led by Eugene de Juan and Mark Humayun,[9, 15] and another at Harvard/MIT Centers led by Joseph Rizzo and John Wyatt,[10] Recently, Rizzo and Wyatt have decided to pursue the subretinal approach. Second Sight, a privately held company in Sylmar, California, is developing a chronically implantable epiretinal prosthesis. Six patients have been implanted with a first-generation device containing 16 electrodes, and a 60-electrode second-generation device should be in patients soon. Patients with the first device have shown the ability to read large letters, locate objects, and detect the direction of motion of objects and light. They have also shown the ability to discriminate multiple levels of gray. The second-generation device is expected to work even better. The subretinal approach has been pursued by the Chow brothers in Chicago — one an ophthalmologist and the other an engineer — who have formed a company called Optobionics (Chicago, Illinois).[14] They implanted ten patients in an initial feasibility study; this study showed some temporary subjective improvements in vision that Optobionics believes was caused by a secondary neurotrophic effect and not direct stimulation by the implant. They have recently implanted twenty more patients at three centers with better vision than the first group. Subretinal and epiretinal implants are also being pursued in Germany by large groups led by Zrenner[11] and Eckmiller,[12] respectively. Two companies have been formed in Germany by these individuals. There is also a group in Japan at Nagoya University led by Tohru Yagi. This group focuses primarily on cultured neuron preparations (personal communication).

Finally, there is a group led by Claude Veraart at the Neural Rehabilitation Engineering Laboratory in Brussels, Belgium, that has implanted a nerve cuff electrode with four electrodes around the optic nerve of a blind patient. That patient is able to identify in which quadrant she sees a phosphene.[17] Recently, an eight-channel device was implanted into a second patient. The new device was implanted inside the ocular orbit and has not performed as well as the first implant.

On February 19, 2000, the inaugural symposium of the Alfred Mann Institute–University of Southern California (AMI–USC) entitled "Can We Make the Blind See? — Prospects for Restoring Vision to the Blind" was held. Lecturers included Dr. Dean Baker, Director of the AMI–USC; Gerald Loeb, an FES researcher at the AMI–USC; Dr. Dean Bok, a retinal physiologist from UCLA; retinal prosthesis researchers Drs. Robert Greenberg, Mark Humayun, Joseph Rizzo, John Wyatt, and Alan Chow; cortical prosthesis researchers Drs. Richard Normann and Philip Troyk; and Dr. Dana Ballard, a visual psychophysicist from the University of Rochester.

Baker provided the welcome and Loeb gave a brief history of neural prosthetics. Bok's talk highlighted biological approaches to inherited retinal degenerations, which result in photoreceptor loss. He chose to talk about two genes (rhodopsin and RDS) whose mutations cause a form of autosomal dominant

FIGURE 11.1 Experimental protocol for intraocular patient testing at the Johns Hopkins Medical Center.

inherited blindness — retinitis pigmentosa.[18] He discussed biological approaches to these diseases. Specifically, he described work by Matthew LaVail, William Hauswirth, and Al Lewin Laboratories, where subretinal injections were performed in transgenic rats carrying one of the rhodopsin mutations (P23H). By injecting viral vectored ribozymes for the selective destruction of mutant mRNA produced by the P23H mutation, there was a dramatic arrest in the photoreceptor degeneration. Bok also spoke about his own work with LaVail and Hauswirth laboratories where a viral vectored secreted form of ciliary neurotrophic factor (CNTF) was injected subretinally. When tested with transgenic rats containing an RDS mutation (peripherin P216L), the photoreceptor loss was again slowed.

After Greenberg gave a brief introduction to retinal prosthetics, Humayun spoke about the epiretinal prosthesis efforts at the Johns Hopkins University.[9, 15] He spoke about recent experiments of intraocular electrical stimulation in RP and AMD patients (Figure 11.1). Under local anesthesia, different stimulating electrodes were inserted through the eye-wall and positioned over the surface of the retina. Data from the ten most recently tested patients were reported. These awake patients reported simple forms in response to pattern electrical stimulation of the retina. A nonflickering perception was created with stimulating frequencies between 40 and 50 Hz. The stimulation threshold also depended on the targeted retinal area (higher in the extramacular region).

Next, Rizzo and Wyatt spoke about their work at The Massachusetts Eye and Ear Infirmary and the Massachusetts Institute of Technology (MEEI–MIT).[10] They reported tests on six humans tested intraocularly — similar to the tests performed at the Johns Hopkins Medical Center. Using microfabricated electrode arrays placed in contact with the retina, five patients blind from retinitis pigmentosa and one volunteer with normal vision were tested. The normal volunteer was having his eye enucleated because of a cancer. Their most significant results included (1) safe contact of the retina with a microfabricated array, (2) determination of strength-duration curves in two volunteers, and (3) creation of visual percepts with crude form. In the best cases, volunteers were able to distinguish two spots of light when two electrodes separated by roughly 2 degrees of visual angle were driven. Thresholds reported exceed the accepted charge density limits for chronic neural stimulation for the electrodes used. It was suggested that the quality of these results would improve with a chronically implantable prosthesis.

Alan Chow from Optobionics spoke about his Artificial Silicon Retina™ (ASR).[14] ASRs are semiconductor-based, silicon chip, microphotodiode arrays (microscopic solar cells) designed for surgical implantation into the subretinal space. The arrays are approximately 2 to 3 mm in diameter, and 50 to 75 μm thick. Chow reported successful electrical stimulation of normal animal retinas.

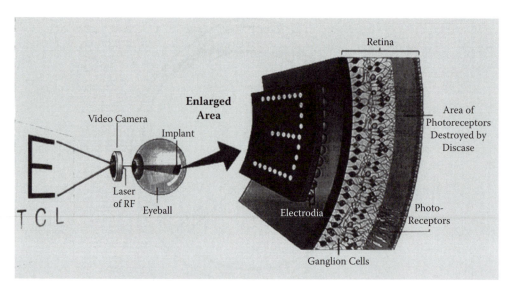

FIGURE 11.2 Concept for an epiretinal prosthesis.

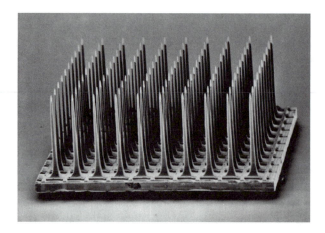

FIGURE 11.3 Penetrating cortical electrode array designed by Normann's lab.

Greenberg spoke about "Second Sight" and its mission of producing a chronically implantable retinal prosthesis. Second Sight has chosen a retinal prosthesis approach over the cortical approach because of concerns of patient safety, although the cortical approach has the potential to treat the largest number of blind patients (because it does not require the patients to have an intact retina or optic nerve). Second Sight has also chosen the epiretinal approach (Figure 11.2) over the subretinal approach because of the belief that the photodiodes used by Chow and Zrenner will not be able to produce enough electrical energy to stimulate abnormal human retinas.

Normann from the University of Utah spoke about his electrode arrays that have been used to record both acute and chronic electrophysiological recordings from various brain structures in monkeys, cats, and rats.[5] The standard array is a 4.2-mm square grid with 100 silicon microelectrodes, 1.0 mm long and a spacing of 0.4 mm (Figure 11.3). Normann also spoke about his new Utah Slant Array (USA) electrodes that have been used to record from the peripheral nerve.

Dr. Philip Troyk from the Illinois Institute of Technology spoke about the issues of implantable hardware. One issue raised was that the next-generation neuroprostheses will be five to ten times denser, electrically and physically, than current neuroprosthetic devices. Troyk discussed the need for heat dissipation by implanted prosthetics, particularly eye-mounted devices. Data were presented where a suspended-carrier, closed-loop Class-E transcutaneous magnetic link was used to generate data transmission rates greater than 1 Mbit/s with a 5-MHz carrier.[19] Troyk also pointed out that the stimulation strategies to produce usable sight are still unknown. When reliable implantable hardware systems become available, testing can begin to devise efficacious image-to-stimulation transformations. It is important that the implantable hardware developed at this phase not restrict the nature of the stimulation sequences from the standpoint of amplitude, pulse-width, frequency, and temporal modulation.

Then, Dana Ballard from the University of Rochester discussed the visual representations that affect sensorimotor task performance.[20] Volunteers were shown videos of a simulated driving environment and their eye movements monitored. By tracking saccadic eye movements, inferences can be drawn describing the underlying cortical processing.[21]

Finally, a panel discussion convened in which the relative merits of the different approaches to visual prosthetics were debated. The session ended with a general consensus that visual prostheses are technically feasible, and that chronically implanted devices and clinical testing are a necessary next step to assess efficacy.

Acknowledgments

This chapter resulted from the Inaugural Symposium of the Alfred E. Mann Institute for Biomedical Engineering at the University of Southern California, Los Angeles, held on February 19, 2000. The Alfred E. Mann Institute for Biomedical Engineering at the University of Southern California (AMI–USC) was established with a $150 million donation from Alfred Mann to the university and has as its goal the transfer of university research to the public sector for the benefit of patients.

References

1. Clausen, J. Visual sensations (phosphenes) produced by AC sine wave stimulation. *Acta Physiol. Neurol. Scand. Suppl.*, 94:1–101, 1955.
2. Brindley, G. and Lewin, W. The sensations produced by electrical stimulation of the visual cortex. *J. Physiol. (London)*, 196:479–493, 1968.
3. Dobelle, W.H. and Mladwovsky, M.G. Phosphenes produced by electrical stimulation of human occipital cortex and their application to the development of a prosthesis for the blind. *J. Physiol.*, 243:553–576, 1974.
4. Dobelle, W.H., Mladejovsky, M.G., Evans, J.K., Roberts, T.S., and Girvin, J.P. 'Braille' reading by a blind volunteer by visual cortex stimulation. *Nature*, 259:111–112, 1976.
5. Rousche, P.J. and Normann, R.A. Chronic recording capability of the Utah Intracortical Electrode Array in cat sensory cortex. *J. Neurosci. Methods*, 82(1):1–15.
6. Hoogerwerf, A.C. and Wise, K.D. A three-dimensional microelectrode array for chronic neural recording. *IEEE Trans. Biomed. Eng.*, 41(12):1136–1146, 1994.
7. Bak, M., Girvin, J.P., Hambrecht, F.T., Kuftar, C.V., Loeb, G.E., and Schmidt, E.W. Visual sensations produced by intracortical microstimulation of the human occipital cortex. *Med. Biol. Eng. Comp.* 28:257–259, 1990.
8. Schmidt, E.M., Bak, M.J., Hambrecht, F.T., Kufta, C.V., O'Rourke, D.K., and Vallabhanath, P. Feasibility of a visual prosthesis for the blind based on intracortical microstimulation of the visual cortex. *Brain*, 119(Pt. 2):507–522, 1996.
9. Humayun, M.S., deJuan, E., Dagnelie, G., Greenberg, R.J., Propst, R., and Phillips, D.H. Visual perception elicited by electrical stimulation of retina in blind humans. *Arch. Ophthalmol.*, 114:40–46, 1996.
10. Wyatt, J. and Rizzo, J. Ocular implants for the blind. *IEEE Spectrum*, 33(5):47–53, 1996.
11. Zrenner, E., Miliczek, K.D., Gabel, V.P., Graf, H.G., Guenther, E., Haemmerle, H., et al. The development of subretinal microphotodiodes for replacement of degenerated photoreceptors. *Ophthalmic Res.*, 29(5):269–280, 1997.

12. Eckmiller, R. Learning retina implants with epiretinal contacts. *Ophthalmic Res.,* 29:281–289, 1997.

13. Greenberg, R.J. Analysis of Electrical Stimulation of the Vertebrate Retina — Work towards a Retinal Prosthesis. Thesis, Johns Hopkins University, Baltimore, MD, 1998.

14. Chow, A.Y. and Peachey, N.S. The subretinal microphotodiode array retinal prosthesis. *Ophthalmic Res.,* 30(3):195–198, 1998.

15. Humayun, M.S., deJuan, E., Weiland, J.D., Dagnelie, G., Katona, S., Greenberg, R., and Suzuki, S. Pattern electrical stimulation of the human Retina. *Vision Res.,* 39:2569–2576, 1999.

16. Davis, R. Future possibilities for neural stimulation. In *Textbook of Stereotactic and Functional Neurosurgery,* McGraw-Hill, New York, 1997, chap. 217, p. 2064–2066.

17. Veraart, C., Raftopoulos, C., Mortimer, J.T., Delbeke, J., Pins, D., Michaux, G., et al. Visual sensations produced by optic nerve stimulation using an implanted self-sizing spiral cuff electrode. *Brain Res.,* 813(1):181–186, 1998.

18. Kedzierski, W., Bok, D., and Travis, G.H. Transgenic analysis of RDS/peripherin N-glycosylation: effect on dimerization, interaction with rom1, and rescue of the rds null phenotype. *J. Neurochem.,* 72(1):430–438, 1999.

19. Troyk, P.R. and Schwan, M.A. Closed-loop class E transcutaneous power and data link for micro-implants. *IEEE Trans. Biomed. Eng.,* 39(6):589–599, 1992.

20. Smeets, J.B., Hayhoe, M.M., and Ballard, D.H. Goal-directed arm movements change eye-head coordination. *Exp. Brain Res.,* 109(3):434–440, 1996.

21. Ballard, D.H., Hayhoe, M.M., Li, F., and Whitehead, S.D. Hand-eye coordination during sequential tasks. *Philos. Trans. R. Soc. Lond. B Biol. Sci.,* 337:331–338, 1992.

Motor Prostheses—Electrode Technologies and Command Signal Extraction

12

General Clinical Issues Relevant to Brain–Computer Interfaces

Eric C. Leuthardt,
Jeffrey G. Ojemann,
Gerwin Schalk, and
Daniel W. Moran

12.1 Introduction

Over the past ten years, the idea of machines that could be controlled by one's thoughts has emerged as a near-term clinical possibility. The most common technical term for such a device is a "brain–computer interface" (BCI). Other synonymous terms include direct brain interface (DBI), brain machine interface (BMI), and motor neuroprosthetics. These are machines that create a new output channel from the brain beyond the natural motor and hormonal commands. BCIs recognize some form of electrophysiological signal in the brain of a subject, and use these signal alterations to either communicate with, or control some element of, the outside world consistent with the intentions of that subject. Examples of such applications would be an electrocorticographic (ECoG) signal controlling a cursor on a computer screen, a prosthetic limb, or one's own limb (e.g., through a bionic implant). These types of devices hold significant potential for improving the quality of life for people with severe motor impairment, including those with spinal cord injury, stroke, neuromuscular disorders, and amputees. These are patients who currently have very few options for any substantive intervention in altering their level of function. Moreover, due to the aging population and improved survival following stroke and trauma, these populations are increasing in size and relevance.

Now, with the improved insight into the electrophysiological principles underlying motor-related cortical function, the development of economical and fast computer processing, and a growing social awareness of the needs of the severely handicapped, the idea of a practical and clinically viable BCI is beginning to deserve serious consideration. It will be critical for the neuroprosthetic community (i.e., scientists, engineers, industry, and medical practitioners) to understand not only what these devices are, but also what their clinical and surgical implications for patient care will be. This will require knowledge of the fundamental framework of how these systems operate, the current BCI platforms and their limitations, the relevant issues when applied clinically, and the anticipated important milestones for their evolution toward entering standard clinical practice.

This chapter provides a reference for evaluating the clinical issues inherent to brain–computer interfaces. We discuss the critical features, functions, and platforms of *output BCIs*; additionally, we define the key surgical elements to consider for an implantable BCI and then critically review the literature of the various platforms relative to these considerations.

12.2   Definition and Essential Features

The *First International Meeting on Brain Computer Interface Technology* (2000) defined a brain–computer interface (BCI) as " a communication system that does not depend on the brain's normal output pathways of peripheral nerves and muscles [70]." Essentially, a BCI is a machine that can decode human intent from brain signal *alone* to create a new communication channel for patients with severe motor impairments [71]. A real-world example of this would entail a subject "locked in" by a brainstem stroke controlling a cursor on a screen with his/her ECoG signal alone. The construct would not require the assistance of overt motor activity. It is important to underline this point. A true BCI allows for a completely novel output pathway from the brain. Wolpaw, in a review of BCI technology, states this principle cogently [71]:

> A BCI changes electrophysiological signals from mere reflections of central nervous system (CNS) activity into the intended products of that activity: messages and commands that act on the world. It changes a signal such as an EEG rhythm or a neuronal firing rate from a reflection of brain function into the end product of that function: an output that, like output in conventional neuro-muscular channels, accomplishes the person's intent. A BCI replaces nerves and muscles and the movements they produce with electrophysiological signals and the hardware and software that translate those signals into actions.

Given this new output channel, the patient using the BCI must have feedback to improve the performance of how they alter their brain signals. Just as a child learns to catch a ball, or an athlete perfecting certain moves, there must be continuous alteration of the subject's neuronal output (whether neuromuscular or electrophysiological) matched against feedback from their overt actions such that the subject's output can be tuned to optimize their performance toward the intended goal. Therefore, the brain must adapt its signals to improve performance. The BCI, additionally, should also be able to evolve to the changing milieu of the user's brain to further optimized functioning. This dual adaptation requires a certain level of training and learning curve, both for the user and the computer. The better the computer and subject are able to adapt, the shorter the training required for control.

There are four essential elements to the practical functioning of a brain–computer interface platform. (Figure 12.1) [71]:

1. Signal acquisition: the BCI system's recorded brain signal or information input.
2. Signal processing: the conversion of raw signal information into a useful device command.
4. Device output: the overt command or control functions that are administered by the BCI system.
4. Operating protocol: the manner in which the system is turned on and off.

All four elements act in concert to actualize the user's intention to his/her surroundings.

Signal acquisition is a measurement of the electrophysiological state of the brain in realtime. This measure of brain activity is usually recorded with electrodes. This is, however, not a theoretical requirement. These electrodes can be either invasive (implanted beneath the skin) or noninvasive (signal acquired externally). The most common types of signals include electroencephalography (EEG), electrical brain activity recorded from the scalp [15, 16, 19, 46, 62, 67], electrocorticography (ECoG) [32, 34], electrical brain activity recorded beneath the skull [32, 34, 54], field potentials, measured by electrodes monitoring brain activity from within the parenchyma [1], and "single units," activity

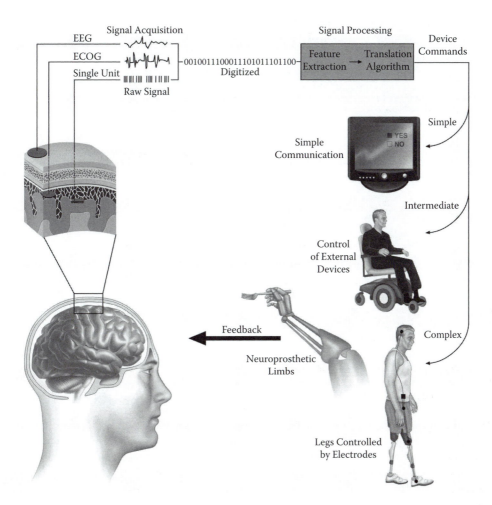

FIGURE 12.1 (See color insert following page 15-4). A schematic of the essential components of a brain–computer interface (BCI) system. A BCI replaces nerves and muscles and the movements they produce with electrophysiological signals (i.e., EEG, ECoG, single-unit action potentials), and the hardware and software that translate those signals into actions. The essential elements to the practical functioning of a BCI platform are illustrated as follows. (1) Signal acquisition, the BCI system's recorded brain signal or information input. This signal is then digitized for analysis. (2) Signal processing, the conversion of raw information into a useful device command. This involves both feature extraction, the determination of a meaningful change in signal, and feature translation, the conversion of that signal alteration to a device command. (3) Device output, the overt command or control functions that are administered by the BCI system. These outputs can range from simple forms of basic word processing and communication to higher levels of control such as driving a wheelchair or controlling a prosthetic limb. As a new output channel, the user must have feedback on their overt device output to improve the performance of how they alter their electrophysiologic signal. All of these elements play in concert to make manifest the user's intention to his or her environment. (Abbreviations: EEG, electroencephalography; ECoG, electrocorticography.) (*Source*: From Leuthardt, E.C., Schalk, G., Moran, D., and Ojemann, G. The emerging world of motor neuroprosthetics: a neurosurgical perspective. *Neurosurgery*, 59:1–14; discussion 11–14, 2006. With permission.)

measured by microelectrodes monitoring individual neuron action potential firing [21, 26, 31, 65]. Figure 12.2 shows the relationship of the various signal platforms in terms of anatomy and population sampled. Other possible signals include MEG, fMRI, PET, and optical imaging. To date, all these types of signals are not currently practical, either due to prohibitive equipment costs or excessively slow time constants. Once acquired, the raw signals are then digitized and sent to the BCI system for further evaluation and decoding.

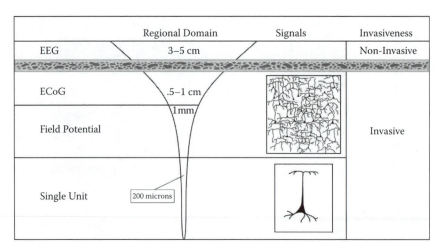

	Regional Domain	Signals	Invasiveness
EEG	3–5 cm		Non-Invasive
ECoG	.5–1 cm		
Field Potential	1mm		Invasive
Single Unit	200 microns		

FIGURE 12.2 The figure demonstrates the relationship of the various signals utilized in BCI operation in regard to the area of cortex distinguishable, the neuronal population, and the level of invasiveness. EEG on one end of the spectrum is the least invasive, while having the lowest signal fidelity due to the large regional signal domain. Single-unit monitoring, on the other hand, has the highest level of signal fidelity by monitoring single neuron action potentials, but is the most invasive of the signal modalities due to the need for cortical penetration. These relationships govern the risk benefit assessment in terms of what level of control is necessary against the level of risk in signal acquisition. The ideal platform has the least amount of risk in device application while maintaining a high level of complex information for device control. (*Source:* From Leuthardt, E.C., Schalk, G., Moran, D., and Ojemann, G. The emerging world of motor neuroprosthetics: a neurosurgical perspective. *Neurosurgery,* 59:1–14; discussion 11–14, 2006. With permission.)

With the signal processing aspect of BCI operation, there are two essential functions: (1) feature extraction and (2) signal translation. The first extracts significant identifiable information from the gross signal, and the second renders that identifiable information into machine commands that control a device. This process of converting raw signal into one that is meaningful requires a complex array of statistical analysis. These techniques can vary from directional cosine tuning of individual neuron action potentials, to the assessment of frequency power spectra, event-related potentials, and cross-correlation coefficients for analysis of signals taken from larger neuronal populations [36, 42, 47]. In short, these statistical methods assess the probability that an electrophysiological event correlates with a given cognitive or motor task. As an example, after recordings are made from an ECoG signal, the BCI system must recognize that a meaningful alteration has occurred in the electrical rhythm that is statistically significant (feature extraction), and then associates that change with a specific cursor movement (translation). As mentioned previously, it is essential that the signal processing be dynamic such that the BCI can adjust to the changing internal signal environment of the user. With regard to the actual device output, this is the overt action that the BCI accomplishes. As in the previous example, this can result in moving a cursor on a screen; other possibilities include controlling external devices such as a wheelchair or robotic arm, or controlling some other intrinsic physiologic process such as moving one's own limb.

It is important to consider the operating protocol when considering BCI for practical clinical application. This refers to the method by which the user controls *how* the system functions. Such illustrative examples include turning on and off the system, controlling what kind of feedback occurs and how fast it is provided, manipulating how quickly the system implements commands, and switching between various device outputs. These examples are features critical for BCI functioning in the real world. Currently, research protocols are very controlled, in that all the parameters are set by the investigator. The researcher turns the system on and off, adjusts the speed of interaction, or defines very limited goals and tasks. In a fully implemented BCI, it is obligatory that users will have control over these steps on their own in an unstructured environment.

12.3 General Clinical Considerations of Brain-Computer Interfaces

With the emergence of these BCI technologies, the neuroprosthetic community should have a matrix upon which to evaluate these new systems as they apply to patients. This matrix should assess the following six questions:

1. Safety
2. Durability: How long will the construct last?
3. Reliability: Will it work without a lot of problems?
4. Complexity of control: Does the BCI system have sufficiently complex control to be useful?
5. Suitability: Is the BCI appropriate for the given patient population?
6. Efficacy: Have there been sufficient experimental and technical studies to support the system's efficacy?

We will review the relevant issues and the implications of each of these questions.

In addition to the processing issues that define the requirements of a BCI system, there is a unique set of factors that a clinician must consider about a given platform when considering application to a patient population. The most fundamental issue is that a BCI system is safe. First, surgical placement must have acceptable clinical risk, and then subsequently over time the implant must be reliable and durable in its ability to acquire signals. Understanding the risks of initial surgical application is reasonably straightforward as they will most likely utilize variants of standard neurosurgical practices. There are equivalent types of technical procedures, which include the placement of deep brain stimulators, placement of cortical stimulators for pain, and placement of grid electrodes. What will require closer examination is the construct's likelihood for ongoing function. This can be affected by how the device is designed (i.e., will the construct break down in a couple years?) and how the patient responds to the implant histopathalogically (i.e., will scar formation prohibit signal acquisition after a period of time?). If the BCI has a short duration of function, this will necessitate removal and reimplantation around areas of eloquent cortex. The risk of injury to those regions will be increased by devices with short operational longevity and which require reoperation and replacement.

In addition to the issues of safety, there are factors related to performance that must be deliberated for a BCI to have practical application. These issues include complexity of control and levels of speed and accuracy. The complexity of the control afforded by a given BCI can be assessed by how many *degrees of freedom* (DOF) of control exist. Degrees of freedom refer to how many processes can be controlled in parallel. This can also be understood in terms of dimensions in space. A minimum of three-dimensional control, or three degrees of freedom, will likely be required for a BCI to be practical clinically, such that it can substantively enable a motor-impaired individual to meaningfully engage in his or her environment. One-dimensional control, or 1 DOF, allows for binary interaction (e.g., yes or no) or proportional control (e.g., move left fast or to the right slowly). Two-dimensional control (2 DOF) allows for moving a cursor on a screen along an x- and y-axis. This level of control, although an improvement from no ability to communicate, remains limited. Three-dimensional control, however, provides a substantive improvement in enabling the BCI user. This level of control equates to a subject controlling an object in three-dimensional space (such as a very basic robotic arm) or controlling an object in two-dimensional space with a parallel switch command function (i.e., controlling a computer mouse with a "click" function). This type of control would allow a given patient to either perform such tasks as operating a Windows-based computer, directing a wheelchair (with a brake function), or performing very basic operations of a prosthetic limb in three-dimensional space. For truly more physiologic approximations of limb function, such as controlling a robotic arm for an amputee or inducing the paralyzed limb to move in a coordinated fashion, would require many more degrees of freedom. As an example, controlling a prosthetic leg in a fashion that approximates normal human use would require a minimum of 7 DOF (i.e., three at the hip, one at the knee, one in the leg, and two at the ankle).

One can measure the overall level of function of a BCI system, or its performance, by its speed and accuracy. These are important considerations for a human neuroprosthesis, which will need to function at a rate acceptable for the patient to interact in real-world scenarios and also be able to operate with a minimum of errors that could potentially lead to dangerous situations (e.g., failure to stop a wheelchair, inability to ask for help, misdirecting a prosthetic limb, etc.). These variables are incorporated into a single value known as the rate of information communicated per unit time, or bit rate (bits per minute) [51]. The bit rate of a BCI system must increase as the complexity of choices increases. Therefore, more information must be communicated when choosing between eight choices and four. As a corollary, the information necessarily increases from one-dimensional control to two-dimensional control, and so on. Also, with regard to the rate of information transfer, it is not simply how many choices are made per unit time, but rather how many choices are made *correctly*. Accuracy has a marked impact on information transfer. As an example, a BCI system that is 90% accurate in a two-choice system conveys the same amount of information as a BCI system that is 65% accurate in a four-choice system [51]. The current bit rate for human BCI systems is approximately 25 bits/minute [70]. This translates to a very simple level of control — that is, being able to answer yes and no, very basic word processing, etc. The information transfer rate for an effective BCI system that reliably and quickly responds to the user's environment must necessarily be higher for more functional levels of control. What bit rate will be required will depend on the task and the patient.

For those patients who may require a BCI, they may be very dissimilar with regard to both their clinical needs and their optimal platform. Patients with spinal cord injury (SCI), amyotrophic lateral sclerosis (ALS), amputations, and stroke may all have some type of motor impairment but they may require very different device outputs relevant to their clinical needs. An ALS and locked-in stroke patient may have needs primarily related to communication. An SCI patient, however, may require a device that allows him to control some type of motorized wheelchair. Further refined function would be needed for an amputee who would require very fine control of a prosthetic limb. Moreover, these patients may vary in what type of signal and implant platform will work best given their pathology. A motor cortical related implant may be optimal for a subject with cord dysfunction or amputation, but may not work well in an ALS or stroke patient where that part of the brain may not be normal. Therefore, it is vital that the patient population and its underlying pathology be taken into consideration for what type of platform may be used and what functions it provides.

As new advances in the motor neuroprosthetic field emerge, one must be able to distinguish between technical demonstrations and practical demonstrations of BCI function. A technical demonstration refers to the first time that something is technically feasible. Examples of these include when Fetz and Finocchio in 1971 first demonstrated that one degree of control could be obtained from the operant training of a monkey to alter the firing rate of a single neuron [17], or when single degree of freedom control in human BCI systems was further demonstrated by Wolpaw et al. in 1991 using EEG signals [74], and then by Leuthardt et al. in 2004 with electrocorticography [34]. These are exciting demonstrations of what is possible. The subsequent step in application to clinical subjects must be a demonstration of feasibility in real-world use. Achieving control with a neuroprosthetic construct is very different in real-world situations with multiple distracters and uncontrolled variables and objectives from that of more constrained experimental conditions. A current example of this is revealed in some of the single-unit-based systems developed by Donoghue, which are now being commercialized by the company Cyberkinetics [57]. In 2002, Serruya et al., using microelectrode arrays in monkeys, were able to achieve two-dimensional control [58]. The highest standard to date is three-dimensional control, accomplished by Taylor et al. in 2002 through the use of microelectrode arrays in primates [65]. Preliminary reports in clinical application seem to indicate that control has been somewhat limited despite optimal results in previous primate paradigms, and there have been limitations with long-term recording (single-unit drop-out) [14, 22]. To date, whether this is due to the subject being in a less controlled environment, a limitation of the signals acquired, or simply the early nature of the human trials, is not clear and will require further investigation.

12.4 Current BCI Platforms

There are essentially three types of platforms that currently have potential for clinical application in the near future. They differ primarily on the signal they utilize for control, namely, EEG, ECoG, and single-unit recording. Each has technical aspects that give them advantages and disadvantages regarding their utility in a clinical setting. The signal platform, its history, and the pros and cons as they relate to patients are reviewed below.

12.4.1 EEG-Based Systems

Human BCI experience until recently has been historically limited to electroencephalographic (EEG) recordings, and studies have mainly investigated the use of *sensorimotor rhythms*, *slow cortical potentials*, and *P300 evoked potentials* derived from the EEG [30, 55, 71].

12.4.1.1 Sensorimotor Cortex Rhythms

In awake individuals, Rolandic (sensorimotor) cortex typically displays 8 to 12-Hz EEG activity when they are not actively involved with motor planning or actions [18, 20, 28, 44]. This "rest" activity, called *mu rhythm* when recorded over sensorimotor cortex, is thought to be produced by thalamocortical circuits [44]. The *beta rhythm* is typically associated with 18 to 26-Hz beta rhythms. There were several features about mu and/or beta rhythms that originally implied they could be useful for BCI-based communication. These rhythms are associated with those regions of cortex that are most closely associated to the brain's normal motor output pathways. Both real and imagined motor movements are typically accompanied by a decrease in mu and beta activity over sensorimotor cortex, in particular contralateral to the movement. In considering BCI operation, this decrement in activity does not require actual motor movements, but notably can be accomplished with imagined movements alone [40, 49]. Thus, these frequency alterations can occur independently of activity in the brain's normal output channels of peripheral nerves and muscles, and could therefore serve as the basis for a BCI. Figure 12.3A shows representative results from a BCI using sensorimotor cortex rhythms. Subjects, including those with amyotrophic lateral sclerosis (ALS) or spinal cord injury [30] have learned to control mu or beta amplitudes in the absence of movement or sensation, and can use this control to move a cursor to select letters or icons on a screen or to operate a simple orthosis [48]. Two-dimensional cursor control and mouse-like sequential reach-and-select control have also been demonstrated [41, 72, 73].

12.4.1.2 Slow Cortical Potentials

Slow cortical potentials (or SCPs) are slow changes in EEG potentials centered at the top of the head, or vertex, and occur over time scales of several seconds. Negative SCPs are usually associated with movement and other cognitive functions involving cortical activation, while positive SCPs are usually associated with a reduction in such activations [2, 52]. Birbaumer and coworkers have shown that subjects can learn to control SCP amplitude [15]. Figure 12.3B shows the typical topography and time course of this phenomenon, which provides the basis for a BCI that Birbaumer, Kubler, and colleagues refer to as a "thought translation device (TTD)" [3, 4, 29]. The Birbaumer TTD has provided basic communication capability [30] and control over simple Internet tasks in numerous patients with end-stage ALS. Given its ability to provide function for actual clinical users who have almost no remaining voluntary movement, this is strong evidence that EEG-based BCIs do not depend on neuromuscular function.

12.4.1.3 P300 Evoked Potentials

When interspersed within routine stimuli, infrequent or particularly significant auditory, visual, or somatosensory stimuli, typically evoke a positive potential in the EEG that peaks at about 300 ms and is centered over parietal cortex [12, 63]. This P300, or "oddball," potential distinguishes the brain's response to intermittent or robust stimuli from its response to routine stimuli. P300 potentials have been used as the basis for a BCI system [13, 16]. The system flashes letters or other symbols in rapid succession.

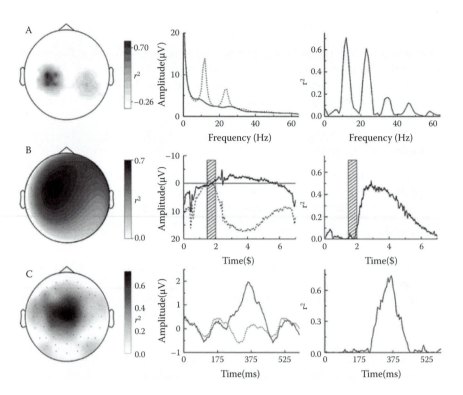

FIGURE 12.3 Three EEG brain signals used for BCIs in humans. (A) Sensorimotor rhythm control of cursor movement. Left: topographical distribution on the scalp (nose on top) of control (measured as r^2, the proportion of the single-trial variance that is due to target position) calculated between top and bottom target positions for a 3-Hz band centered at 12 Hz. Middle: voltage spectra for a location over left sensorimotor cortex (i.e., C3) for cursor movement up (dashed) and down (solid). Right: corresponding r^2 spectrum for top vs. bottom targets. The user's control is sharply focused over sensorimotor cortex and in the mu and beta rhythm frequency bands. (B) Slow cortical potential (SCP) control of cursor movement. Left: topographical distribution of SCP control, calculated between the two tasks of producing cortical negativity (top target) or positivity (bottom target). Center: time courses of the EEG at the vertex for the negativity task (solid line) and for the positivity task (dashed line). Right: corresponding r^2 time course calculated between the two conditions. (C) P300 control of a spelling program. Left: topographical distribution of the P300 potential at 340 ms after stimuli, measured as r^2 for stimuli including vs. not including the desired character. Center: time courses at the vertex of the voltages for stimuli including (solid) or not including (dashed) the desired character. Right: corresponding r^2 time course [55]. (*Source:* From Leuthardt, E.C., Schalk, G., Moran, D., and Ojemann, G. The emerging world of motor neuroprosthetics: a neurosurgical perspective. *Neurosurgery*, 59:1–14; discussion 11–14, 2006. With permission.)

The stimulus that the user wants produces a P300 potential. By detecting this P300, the BCI system learns the user's choice. With this method, people (including those with ALS) can use a simple word-processing program. This P300-based communication does not appear to require any neuromuscular control because the amplitude of the P300 evoked by a specific stimulus in the BCI protocol depends mainly on whether the user wishes to attend to it. Note, however, that it is not yet entirely certain whether P300 amplitude in this setting depends to some extent on the subject's ability to fixate gaze on the desired selection (this would require some level of neuromuscular action). Figure 12.3C shows representative results from a P300-based BCI.

The EEG-based paradigms, in sum, are noninvasive methods that have been the basis for numerous BCI studies in humans to date. Using EEG, these studies have shown that healthy and motor-impaired individuals can control devices without a significant neuromuscular requirement. To date, the National Institute of Neurological Disorders and Stroke (NINDS) is sponsoring the only EEG BCI study, entitled

"Moving a Paralyzed Hand through Use of a Brain-Computer Interface." They are attempting to enroll thirty patients who are either healthy or have a chronic stroke history with residual severe unilateral paresis. The aim of this investigation is to attempt to utilize an EEG-based BCI system to control a hand orthosis [23]. There are, however, no companies that are currently attempting to market an EEG-based BCI. There are some practical considerations that should be made regarding clinical application of BCIs with an EEG platform. Because the electrodes are on the skin, the external nature of these brain signals make them susceptible to external forces (i.e., electrode movement) and signal contamination (i.e., interference generated by muscle movements or the electrical environment). Additionally, because signals are distant from the sources within the brain (greater than a centimeter), EEG signals have reduced fidelity, diminished spatial specificity, and a limited frequency detection (less than 40 Hz). These factors all coalesce to result in more prolonged requisite user training time to achieve higher levels of control. Furthermore, it is possible that these spatial and frequency limitations also prohibit the complexity of movements that can be supported by EEG. From a practical standpoint, external EEG electrodes placed in a cap or fixed to the skin are unlikely to provide long-term solutions for individuals who need to be continuously monitored or are significantly impaired such that they cannot manipulate their electrodes should they be displaced. Given the limitations of this signal platform, its clinical impact seems to be restricted to short-term applications to those individuals who are totally paralyzed and who require very basic levels of communication.

12.4.2 ECoG-Based Systems

Electrocorticography (ECoG) as a signal for a practical BCI platform has recently emerged over the past several years. ECoG is a measure of the electrical activity of the brain taken from beneath the skull. Either subdural or epidural in location, it is important to note that the signal is not taken from within the brain parenchyma itself. Due to the limited access of subjects, it has not been studied extensively until recently. The only manner to acquire the signal for evaluation to date is through the use of patients requiring invasive monitoring for localization of an epileptogenic focus.

The use of ECoG as an experimental platform was based on the insights garnered from previous EEG-based BCIs and the associated understanding of sensorimotor rhythms [75]. As mentioned above, sensorimotor rhythms comprise mu (8 to 12 Hz), beta (18 to 26 Hz), and gamma (greater than 30 Hz) oscillations [27, 50, 74]. The lower frequencies of mu and beta are thought to be produced by thalamo-cortical circuits [24, 35, 47, 53], while the higher frequencies (greater than 30 Hz), or *gamma rhythms*, are thought to be produced by smaller cortical assemblies [38]. BCIs based on EEG oscillations have focused exclusively on mu and beta rhythms because gamma rhythms are inconspicuous at the scalp [45]. In contrast, gamma rhythms as well as mu and beta rhythms are prominent in ECoG during movements [24, 38, 45, 53].

Until the past several years, the signal was assumed to be very similar to that of EEG with regard to the amount and type of discernable information. As recent studies have demonstrated, however, this was not true; the signal itself is substantively different [34]. The ECoG signal is much more robust compared to EEG signal: its magnitude is typically five times larger (0.05 to 1.0 mV vs. 0.01 to 0.2 mV for EEG) [5]; its spatial resolution as it relates to electrode spacing is much finer (0.125 vs. 3.0 cm for EEG) [19, 61]; and its frequency bandwidth is significantly higher (0 to 1000 Hz vs. 0 to 40 Hz for EEG). On a functional level, many studies have demonstrated that higher frequency bandwidths, unavailable to EEG methods, carry highly specific and anatomically focal information about cortical processing [9–11, 60]. Figure 12.4 shows a representative example of the focal nature of gamma frequency changes. ECoG's superior frequency range is attributable to several factors. First, the capacitance of cell membranes of the overlying tissue, combined with the intrinsic electrical resistance of the skull, constitutes a low-pass filter that largely eliminates higher frequencies from the EEG [61]. Second, higher frequencies tend to be produced by smaller cortical assemblies [38]. As a result, these high-frequency signals are more prominent when recorded by sub/epidural electrodes that are closer to cortex, and thereby achieve higher spatial resolution than EEG electrodes [61].

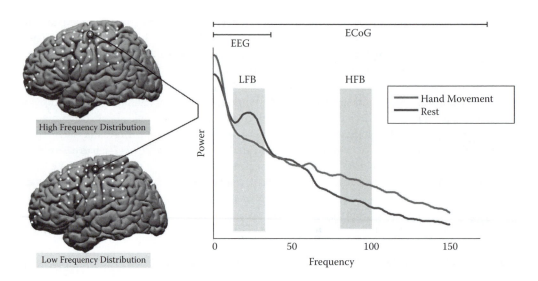

FIGURE 12.4 (See color insert following page 15-4). The difference between anatomic distribution of the lower frequencies (alpha and beta rhythms) compared to that of higher frequencies (gamma rhythms). On the left are two brains mapped in standardized Talairach space. The top brain represents the high-frequency distribution (High Frequency Band, 75 to –100 Hz, HFB) of power changes associated with hand movement when compared against rest. The bottom brain represents the low-frequency distribution (Low Frequency Band, 8 to –30 Hz, LFB) of power changes associated with hand movement in the same patient. The white dots represent subdural electrodes. On the right is frequency power spectra plot depicting the power change vs. frequency for the two conditions of hand movement vs. rest taken from the same electrode. In the LFB, there is a drop in power for hand movement when compared against rest, which is broadly distributed. In the HFB, there is an increase in power with hand movement when compared to rest, which is much more focal in nature. The horizontal bars represent the range of frequency detectable by the EEG and ECoG. The LFB is detectable by both EEG and ECoG, while the HFB, which is much more focal, can only be detected with ECoG. This has important implications when looking for independent signals to utilize for device control. Gamma rhythms are more easily separable than those of the lower frequencies. (*Source*: Adapted from Leuthardt, E.C., Schalk, G., Moran, D., and Ojemann, G. The emerging world of motor neuro-prosthetics: a neurosurgical perspective. *Neurosurgery*, 59:1–14; discussion 11–14, 2006. With permission.)

As a signal in BCI application, recent studies have cogently established its efficacy and speed of use in human subjects. Leuthardt et al. in 2004 demonstrated the first use of ECoG in closed-loop control. Over brief training periods of 3 to 24 min, four patients mastered control and achieved success rates of 74 to 100% in one-dimensional closed-loop control tasks. In additional experiments, the same group found that ECoG signals at frequencies up to 180 Hz accurately reflected the direction of two-dimensional joystick movements [34]. Subsequently, Schalk et al. in 2004 showed two-dimensional online control using independent signals from gamma rhythms inconspicuous in signals detectable by electroenceph-alography [54]. Additionally, Leuthardt et al. in 2006 demonstrated preliminary evidence that ECoG control using signal from the epidural space was also possible [32]. All these studies combined show the ECoG signal to carry a substantial degree of specific cortical information that can allow the user to gain control very rapidly.

In addition to evidence of feasibility, there is some pathologic and clinical evidence to support the implant durability of subdural-based devices. There are a substantial number of studies investigating the tissue response to intraparenchymal cortical electrodes and their associated signal prohibitive reactive gliotic sheaths [64, 66]. The studies that have investigated nonpenetrating subdural-placed electrodes, although somewhat limited, have been more encouraging. In cat and dog models, long-term subdural implants revealed minimal cortical or leptomeningeal tissue reaction while maintaining extended elec-trophysiologic recording [8, 37, 39, 76]. In clinical studies, the use of subdural electrodes as implants for motor cortex stimulation have been shown to be stable and effective implants for the treatment of chronic pain [6, 7, 43]. Additionally, preliminary work using the implantable NeuroPace device (NeuroPace, Mountainview, California) for the purpose of long-term subdural electrode monitoring for seizure identification and abortion has also been shown to be stable [68].

Currently, electrocorticography is a very promising intermediate BCI modality because it has higher spatial resolution, better signal-to-noise ratio, wider frequency range, and lesser training requirements than scalp-recorded EEG, and at the same time has lower technical difficulty, lower clinical risk, and probably superior long-term stability than intracortical single-neuron recording. Thus far, a clinical trial has not been initiated. Recent evidence of the high level of control with minimal training requirements and the numerous features favoring implant durability, however, shows an exciting potential for future real-world application in patients with motor impairment.

12.4.3　Single-Unit Based Systems

Since as early as the 1970s, Fetz and Finnochio were able to demonstrate that animals could be operantly trained to modulate the activity of a single neuron in their brains [17]. These early studies were originally limited to one-dimensional control; however, in the 1980s, Georgopoulos and colleagues developed a method of decoding three-dimensional hand movement direction from a population of neurons in primary motor cortex of nonhuman primates [21]. An accurate prediction of average hand movement direction was made *post hoc* by serially recording the single-unit activity from 50 to 200 individual neurons during repeated reaching tasks. These techniques advanced such that in the early 1990s, neurophysiologists had refined and enhanced these neural decoding methods to include prediction of both three-dimensional direction and speed (i.e., hand velocity) [42, 56]. With the advent of less-expensive fast computing, real-time recording of multiple single units simultaneously was possible by the late 1990s. Several groups achieved success in recording chronic, single-unit action potentials from a number of neurons simultaneously, which culminated in a number of manuscripts in the early 2000s showing elegant multidimensional real-time BCI control [58, 65, 69].

To date, there have been several trials in which single-unit based systems have been applied to quadriplegic subjects to achieve some form of overt device control. Currently, the results are somewhat limited and preclude making any definitive conclusions. There have been two modalities tested. The earliest attempt was the use of "neutrophic electrode" by Kennedy and Bakay in 1998. The group attempted to monitor the firing of a several neurons in a terminal ALS patient through the use of an electrode construct in which neurites were induced to grow within a surgically implanted glass cone electrode that contained neurotrophic factors [26]. The neurotrophic electrode was then later implanted in 2005 into a motor-impaired subject's Broca's speech cortex for the purpose of enhancing speech by identifying the various electrical signal changes associated with various phonemes [25]. The current company involved in the development of this construct is Neural Signals Inc. (Atlanta, Georgia). The group thus far has implanted this device into eight subjects.

The second approach has been through the use of electrode arrays that simultaneously monitor tens to a hundred cortical motor neurons. This type of construct, referred to as the "Utah Array," was first implanted in a spinal-injured quadriplegic in 2004 [57]. Cyberkinetics (Foxborough, Massachusetts) is the company currently involved with the development of this signal platform. To date, they have currently implanted four patients and are open for further recruitment of subjects.

Currently, the best signal for BCI control, from a control systems point of view, has been achieved with multiple, single-unit action potentials recorded in parallel directly from cerebral cortex. No other experimental BCI modality, to date, has provided as good control in terms of accuracy, speed, and DOF than single-unit data. There are, however, issues with current microelectrode technology in obtaining signals chronically. Modern single-unit recording techniques require insertion of a recording electrode into the cortex. Given the highly vascular nature of the brain, it is impossible to implant such a device without severing blood vessels and hence inducing a reactive response around the implant site [64]. Astrocytes and other glial tissue begin to encapsulate the implanted microelectrode via a standard foreign body response. The microelectrode is essentially electrically insulated from the surrounding tissue over time and can no longer discriminate action potentials [66]. Unlike stimulating neuro-prosthetic electrodes (e.g., deep brain stimulator for Parkinson's disease), the insulating nature of the gliotic sheath cannot be overcome by increasing the stimulation current. Once encapsulated, single-unit isolation cannot be reversed on the implanted electrodes. This irreversible insulation currently is a significant clinical issue to overcome in surgically applying these microelectrodes into the brain. Implantation will not be practical if they only provide a year of BCI control and require regular replacement. Given that these constructs are prone to scarring and would be implanted in eloquent regions of cortex, repetitive procedures could have significant detrimental effects to the brain and the patient's long-term functional and cognitive status. Invasive BCI electrodes, therefore, need a prolonged life span to warrant the risks of a neurosurgical procedure.

Currently, single-unit microelectrodes have chronic implant biocompatibility issues leading to limited life spans. There are, however, efforts currently underway to create new biomaterials as well as slow-release drug delivery systems that could significantly decrease gliotic encapsulation of implanted microelectrodes and hence make single-unit recordings practical in the future. Some strategies currently include cross-linking an anti-inflammatory agent such as dexamethasone to a hydrogel coating on the microelectrode. This might theoretically reduce the initial injury response of implantation [59]. Likewise, microelectrodes with incorporated microfluidic channels could also allow for chronic drug delivery to the implant site, not only to control reactive responses in the surrounding tissue, but also to enhance neural growth around the electrode to increase information content. Unfortunately, these studies are just beginning and there will be significant period of testing (years) before this technology is suitable for the clinic. While single-unit recordings are ideal from a neural control point of view, the technology needed to obtain long-term recordings remains controversial; and until new technologies improve microelectrode durability, single-unit BCIs will remain largely in the research sphere for the near term.

12.5 Conclusion

The field of neuroprosthetics today is in its infancy. Currently, research is only beginning to crack the electrical information encoding the information in a human's thoughts. Despite the field's youth, early advances have already demonstrated that these platforms can be utilized to significantly enhance an impaired user's ability to interact with his/her environment. Each of the reviewed signal platforms has the potential to substantively improve the manner in which patients with spinal cord injury, stroke, cerebral palsy, and neuromuscular disorders communicate with their world. Each platform also has distinctive barriers that it will need to overcome. For the population signal platforms of EEG and ECoG, increasing the complexity of control is critical; while for single-unit platforms, demonstrating chronic implant durability is of central concern. Given the rapid progression of these technologies over the past five to seven years and the concomitant swift ascent of computer processing speeds, signal analysis techniques, and emerging ideas for novel biomaterials, these issues should not be viewed as obstacles, but rather as milestones that will be achieved. The order in which these milestones will be accomplished remains to be seen. As research in this field begins to transition from basic research to one of clinical application, it will herald in a new era of restorative neurosurgery. Through neurosurgical intervention, the ability will exist to restore function that today is unrecoverable. In the future, a neurosurgeon's capabilities will go beyond the ability to remove offending agents such as aneurysms, tumors, and hematomas to prevent the decrement of function. Rather, he or she will also have the skills and tech-nologies in their clinical armamentarium to engage the nervous system to restore abilities that have already been lost.

References

1. Andersen, R.A., Burdick, J.W., Musallam, S., Pesaran, B., and Cham, J.G. Cognitive neural prosthetics. *Trends Cogn. Sci.*, 8:486–493, 2004.

2. Birbaumer, N., Ed. *Slow Cortical Potentials: Their Origin, Meaning, and Clinical Use.* Tilburg: Tilburg University Press, 1997.

3. Birbaumer, N., Ghanayim, N., Hinterberger, T., Iversen, I., Kotchoubey, B., Kubler, A., Perelmouter, J., Taub, E., and Flor, H. A spelling device for the paralysed. *Nature*, 398:297–298, 1999.

4. Birbaumer, N., Kubler, A., Ghanayim, N., Hinterberger, T., Perelmouter, J., Kaiser, J., Iversen, I., Kotchoubey, B., Neumann, N., and Flor, H. The thought translation device (TTD) for completely paralyzed patients. *IEEE Trans. Rehabil. Eng.*, 8:190–193, 2000.

5. Boulton, A.A., Baker, G.B., and Vanderwolf, C.H., Eds. *Neurophysiological Techniques: Applications to Neural Systems.* Totowa: Humana Press, 1990.

6. Brown, J.A. and Barbaro, N.M. Motor cortex stimulation for central and neuropathic pain: current status. *Pain*, 104:431–435, 2003.

7. Brown, J.A. and Pilitsis, J.G. Motor cortex stimulation for central and neuropathic facial pain: a prospective study of 10 patients and observations of enhanced sensory and motor function during stimulation. *Neurosurgery*, 56:290–297; discussion 290–297, 2005.

8. Bullara, L.A., Agnew, W.F., Yuen, T.G., Jacques, S., and Pudenz, R.H. Evaluation of electrode array material for neural prostheses. *Neurosurgery*, 5:681–686, 1979.

9. Crone, N.E., Boatman, D., Gordon, B., and Hao, L. Induced electrocorticographic gamma activity during auditory perception. Brazier Award-winning article, 2001. *Clin. Neurophysiol.*, 112:565–582, 2001.

10. Crone, N.E. Hao, L., Hart, J., Jr., Boatman, D., Lesser, R.P., Irizarry, R., and Gordon, B. Electro-corticographic gamma activity during word production in spoken and sign language. *Neurology*, 57:2045–2053, 2001.

11. Crone, N.E., Miglioretti, D.L., Gordon, B., and Lesser, R.P. Functional mapping of human sensorimotor cortex with electrocorticographic spectral analysis. II. Event-related synchronization in the gamma band. *Brain*, 121(Pt. 12):2301–2315, 1998.

12. Donchin, E. and Smith, D.B. The contingent negative variation and the late positive wave of the average evoked potential. *Electroencephalogr. Clin. Neurophysiol.*, 29:201–203, 1970.

13. Donchin, E., Spencer, K.M., and Wijesinghe, R. The mental prosthesis: assessing the speed of a P300-based brain-computer interface. *IEEE Trans. Rehabil. Eng.*, 8:174–179, 2000.

14. Duncan, D.E. Implanting Hope. *Technol. Rev.*, February 2005.

15. Elbert, T., Rockstroh, B., Lutzenberger, W., and Birbaumer, N. Biofeedback of slow cortical potentials. I. *Electroencephalogr. Clin. Neurophysiol.*, 48:293–301, 1980.

16. Farwell, L.A. and Donchin, E. Talking off the top of your head: toward a mental prosthesis utilizing event-related brain potentials. *Electroencephalogr. Clin. Neurophysiol.*, 70:510–523, 1988.

17. Fetz, E.E. and Finocchio, D.V. Operant conditioning of specific patterns of neural and muscular activity. *Science*, 174:431–435, 1971.

18. Fisch, B.J., Pedley, T.A., and Keller, D.L. A topographic background symmetry display for comparison with routine EEG. *Electroencephalogr. Clin. Neurophysiol.*, 69:491–494, 1988.

19. Freeman, W.J., Holmes, M.D., Burke, B.C., and Vanhatalo, S. Spatial spectra of scalp EEG and EMG from awake humans. *Clin. Neurophysiol.*, 114:1053–1068, 2003.

20. Gastaut, H. [Electrocorticographic study of the reactivity of rolandic rhythm]. *Rev. Neurol. (Paris)*, 87:176–182, 1952.

21. Georgopoulos, A.P., Schwartz, A.B., and Kettner, R.E. Neuronal population coding of movement direction. *Science*, 233:1416–1419, 1986.

22. Hochberg, L.R., Serruya, M.D., Friehs, G.M., Mukand, J.A., Saleh, M., Caplan, A.H., Branner, A., Chen, D., Penn, R.D., and Donoghue, J.P. Neuronal ensemble control of prosthetic devices by a human with tetraplegia. *Nature*, 442:164–171, 2006.

23. http://clinicalstudies.info.nih.gov/detail/A_2006-N-0012.html.

24. Huggins, J.E., Levine, S.P., BeMent, S.L., Kushwaha, R.K., Schuh, L.A., Passaro, E.A., Rohde, M.M., Ross, D.A., Elisevich, K.V., and Smith, B.J. Detection of event-related potentials for development of a direct brain interface. *J. Clin. Neurophysiol.*, 16:448–455, 1999.

25. Kennedy, P.R., Andeasen, D., Wright, E.J., Mao, H., and Ehirim, P. Towards conversational speech restoration in a locked-in patient recording from Brocas area with the neurotrophic electrode. Presented at *Society for Neuroscience,* Washington, D.C., 2005.

26. Kennedy, P.R. and Bakay, R.A. Restoration of neural output from a paralyzed patient by a direct brain connection. *Neuroreport,* 9:1707–1711, 1998.

27. Kostov, A. and Polak, M. Parallel man-machine training in development of EEG-based cursor control. *IEEE Trans. Rehabil. Eng.,* 8:203–205, 2000.

28. Kozelka, J.W. and Pedley, T.A. Beta and mu rhythms. *J. Clin. Neurophysiol.,* 7:191–207, 1990.

29. Kubler, A., Neumann, N., Kaiser, J., Kotchoubey, B., Hinterberger, T., and Birbaumer, N.P. Brain-computer communication: self-regulation of slow cortical potentials for verbal communication. *Arch. Phys. Med. Rehabil.,* 82:1533–1539, 2001.

30. Kubler, A., Nijboer, F., Mellinger, J., Vaughan, T.M., Pawelzik, H., Schalk, G., McFarland, D.J., Birbaumer, N., and Wolpaw, J.R. Patients with ALS can use sensorimotor rhythms to operate a brain-computer interface. *Neurology,* 64:1775–1777, 2005.

31. Laubach, M., Wessberg, J., and Nicolelis, M.A. Cortical ensemble activity increasingly predicts behaviour outcomes during learning of a motor task. *Nature,* 405:567–571, 2000.

32. Leuthardt, E.C., Miller, K.J., Schalk, G., Rao, R.N., and Ojemann, J.G. Electrocorticography-Based Brain Computer Interface - The Seattle Experience. *IEEE - Neural Syst. Rehabil. Eng.,* 2005.

33. Leuthardt, E.C., Schalk, G., Moran, D., and Ojemann, J.G. The emerging world of motor neuro-prosthetics: a neurosurgical perspective. *Neurosurgery,* 59:1–14; discussion 11–14, 2006.

34. Leuthardt, E.C., Schalk, G., Wolpaw, J.R., Ojemann, J.G. and Moran, D.W. A brain–computer interface using electrocorticographic signals in humans. *J. Neural. Eng.,* 1:63–71, 2004.

35. Levine, S.P., Huggins, J.E., BeMent, S.L., Kushwaha, R.K., Schuh, L.A., Passaro, E.A., Rohde, M.M., and Ross, D.A., Identification of electrocorticogram patterns as the basis for a direct brain interface. *J. Clin. Neurophysiol.,* 16:439–447, 1999.

36. Levine, S.P., Huggins, J.E., BeMent, S.L., Kushwaha, R.K., Schuh, L.A., Rohde, M.M., Passaro, E.A., Ross, D.A., Elisevich, K.V., and Smith, B.J., A direct brain interface based on event-related potentials. *IEEE Trans. Rehabil. Eng.,* 8:180–185, 2000.

37. Loeb, G.E., Walker, A.E., Uematsu, S., and Konigsmark, B.W., Histological reaction to various conductive and dielectric films chronically implanted in the subdural space. *J. Biomed. Mater. Res.,* 11:195–210, 1977.

38. Lopes da Silva, F.H. and Pfurtscheller, G., Eds. *Event-Related Desynchronization. Handbook of Electroencephalography and Clinical Neurophysiology.* Elsevier, Amsterdam, Elsevier, 1999.

39. Margalit, E., Weiland, J.D., Clatterbuck, R.E., Fujii, G.Y., Maia, M., Tameesh, M., Torres, G., D'Anna, S.A., Desai, S., Piyathaisere, D.V., Olivi, A., de Juan, E., Jr., and Humayun, M.S. Visual and electrical evoked response recorded from subdural electrodes implanted above the visual cortex in normal dogs under two methods of anesthesia. *J. Neurosci. Methods,* 123:129–137, 2003.

40. McFarland, D.J., Miner, L.A., Vaughan, T.M., and Wolpaw, J.R. Mu and beta rhythm topographies during motor imagery and actual movements. *Brain Topogr.,* 12:177–186, 2000.

41. McFarland, D.J., Sarnacki, W.A., and Wolpaw, J.R. EEG-based two-dimensional movement and target selection by a non-invasive brain-computer interface in humans: emulating full mouse control. Presented at *Society for Neuroscience,* Washington D.C., November 12, 2005.

42. Moran, D.W. and Schwartz, A.B. Motor cortical representation of speed and direction during reaching. *J. Neurophysiol.,* 82:2676–2692, 1999.

43. Nguyen, J.P., Lefaucher, J.P., Le Guerinel, C., Eizenbaum, J.F., Nakano, N., Carpentier, A., Brugieres, P., Pollin, B., Rostaing, S., and Keravel, Y. Motor cortex stimulation in the treatment of central and neuropathic pain. *Arch. Med. Res.,* 31:263–265, 2000.

44. Niedermeyer, E. and Lopes da Silva, F.H. *The Normal EEG of the Waking Adult.* Baltimore: Williams and Wilkins, 1999.

45. Pfurtscheller, G. and Cooper, R. Frequency dependence of the transmission of the EEG from cortex to scalp. *Electroencephalogr. Clin. Neurophysiol.,* 38:93–96, 1975.

46. Pfurtscheller, G., Flotzinger, D., and Kalcher, J. Brain-computer interface — a new communication device for handicapped persons. *J. Microcomp. App.*, 16:293–299, 1993.

47. Pfurtscheller, G., Graimann, B., Huggins, J.E., Levine, S.P., and Schuh, L.A. Spatiotemporal patterns of beta desynchronization and gamma synchronization in corticographic data during self-paced movement. *Clin. Neurophysiol.*, 114:1226–1236, 2003.

48. Pfurtscheller, G., Guger, C., Muller, G., Krausz, G., and Neuper, C. Brain oscillations control hand orthosis in a tetraplegic. *Neurosci. Lett.*, 292:211–214, 2000.

49. Pfurtscheller, G. and Neuper, C. Motor imagery activates primary sensorimotor area in humans. *Neurosci. Lett.*, 239:65–68, 1997.

50. Pfurtscheller, G., Neuper, C., Guger, C., Harkam, W., Ramoser, H., Schlogl, A., Obermaier, B., and Pregenzer, M. Current trends in Graz Brain-Computer Interface (BCI) research. *IEEE Trans. Rehabil. Eng.*, 8:216–219, 2000.

51. Pierce, J.R. *An Introduction to Information Theory.* New York: Dover Press, 1980.

52. Rockstroh, B., Elbert, T., and Canavan, A., Lutzenberger, W., Birbaumer, N. *Slow Cortical Potentials and Behavior.* Baltimore: Urban & Schwarzenberg, 1989.

53. Rohde, M.M., BeMent, S.L., Huggins, J.E., Levine, S.P., Kushwaha, R.K., and Schuh, L.A. Quality estimation of subdurally recorded, event-related potentials based on signal-to-noise ratio. *IEEE Trans. Biomed. Eng.*, 49:31–40, 2002.

54. Schalk, G., Leuthardt, E.C., Moran, D., Ojemann, J., and Wolpaw, J.R. Two-dimensional cursor control using electrocorticographic signals in humans. Presented at *Society for Neuroscience*, San Diego, October 23, 2004.

55. Schalk, G., McFarland, D.J., Hinterberger, T., Birbaumer, N., and Wolpaw, J.R. BCI2000: a general-purpose brain-computer interface (BCI) system. *IEEE Trans. Biomed. Eng.*, 51:1034–1043, 2004.

56. Schwartz, A.B. Direct cortical representation of drawing. *Science*, 265:540–542, 1994.

57. Serruya, M.D., Caplan, A.H., Saleh, M., Morris, D.S., and Donoghue, J.P. The BrainGate pilot trial: building and testing a novel direct neural output for patients with severe motor impairments. Presented at *Society for Neuroscience*, San Diego, CA, 2004.

58. Serruya, M.D., Hatsopoulos, N.G., Paninski, L., Fellows, M.R., and Donoghue, J.P. Instant neural control of a movement signal. *Nature*, 416:141–142, 2002.

59. Shain, W., Spataro, L., Dilgen, J., Haverstick, K., Retterer, S., Isaacson, M., Saltzman, M., and Turner, J.N. Controlling cellular reactive responses around neural prosthetic devices using peripheral and local intervention strategies. *IEEE Trans. Neural Syst. Rehabil. Eng.*, 11:186–188, 2003.

60. Sinai, A., Bowers, C.W., Crainiceanu, C.M., Boatman, D., Gordon, B., Lesser, R.P., Lenz, F.A., and Crone, N.E. Electrocorticographic high gamma activity versus electrical cortical stimulation mapping of naming. *Brain*, 128:1556–1570, 2005.

61. Srinivasan, R., Nunez, P.L., and Silberstein, R.B. Spatial filtering and neocortical dynamics: estimates of EEG coherence. *IEEE Trans. Biomed. Eng.*, 45:814–826, 1998.

62. Sutter, E.E. The brain response interface: communication through visually-induced electrical brain responses. *J. Microcomp. App.*, 15:31–45, 1992.

63. Sutton, S., Braren, M., Zubin, J., and John, E.R. Evoked-potential correlates of stimulus uncertainty. *Science*, 150:1187–1188, 1965.

64. Szarowski, D.H., Andersen, M.D., Retterer, S., Spence, A.J., Isaacson, M., Craighead, H.G., Turner, J.N., and Shain, W. Brain responses to micro-machined silicon devices. *Brain Res.*, 983:23–35, 2003.

65. Taylor, D.M., Tillery, S.I., and Schwartz, A.B. Direct cortical control of 3D neuroprosthetic devices. *Science*, 296:1829–1832, 2002.

66. Vetter, R.J., Williams, J.C., Hetke, J.F., Nunamaker, E.A., and Kipke, D.R. Chronic neural recording using silicon-substrate microelectrode arrays implanted in cerebral cortex. *IEEE Trans. Biomed. Eng.*, 51:896–904, 2004.

67. Vidal, J.J. Real-time detection of brain events in EEG. *IEEE Proc. Spec. Issue Biological Signal Process. Anal.*, 65:633–664, 1977.

68. Vossler, D.D., Goodman, R., Hirsch, L., Young, J., and Kraemer, D. Early safety experience with a fully implanted intracranial responsive neurostimulator for epilepsy. Presented at *Annual Meeting of the American Epilepsy Society (AES)*, New Orleans, LA, December 2004.

69. Wessberg, J., Stambaugh, C.R., Kralik, J.D., Beck, P.D., Laubach, M., Chapin, J.K., Kim, J., Biggs, S.J., Srinivasan, M.A., and Nicolelis, M.A. Real-time prediction of hand trajectory by ensembles of cortical neurons in primates. *Nature*, 408:361–365, 2000.

70. Wolpaw, J.R., Birbaumer, N., Heetderks, W.J., McFarland, D.J., Peckham, P.H. Schalk, G., Donchin, E, Quatrano, L.A., Robinson, C.J., and Vaughan, T.M. Brain-computer interface technology: a review of the first international meeting. *IEEE Trans. Rehabil. Eng.*, 8:164–173, 2000.

71. Wolpaw, J.R., Birbaumer, N., McFarland, D.J., Pfurtscheller, G., and Vaughan, T.M. Brain-computer interfaces for communication and control. *Clin. Neurophysiol.*, 113:767–791, 2002.

72. Wolpaw, J.R. and McFarland, D.J. Multichannel EEG-based brain-computer communication. *Electroencephalogr. Clin. Neurophysiol.*, 90:444–449, 1994.

73. Wolpaw, J.R. and McFarland, D.J. Control of a two-dimensional movement signal by a noninvasive brain-computer interface in humans. *Proc. Natl. Acad. Sci., U.S.A.*, 101:17849–17854, 2004.

74. Wolpaw, J.R., McFarland, D.J., Neat, G.W., and Forneris, C.A. An EEG-based brain-computer interface for cursor control. *Electroencephalogr. Clin. Neurophysiol.*, 78:252–259, 1991.

75. Wolpaw, J.R., McFarland, D.J., Vaughan, T.M., and Schalk, G. The Wadsworth Center Brain-Computer Interface (BCI) Research and Development Program. *IEEE Trans. Neural. Syst. Rehabil. Eng.*, 11:204–207, 2003.

76. Yuen, T.G., Agnew, W.F., and Bullara, L.A. Tissue response to potential neuroprosthetic materials implanted subdurally. *Biomaterials*, 8:138–141, 1987.

13

Implantable Brain–Computer Interfaces: Neurosurgical Experience and Perspective

Roy A.E. Bakay

13.1 Introduction

We perceive thoughts as powerful forces. It is often said, "They won by sheer force of will." We think of movement and our body performs accordingly. Why then can we not simply think of bending a spoon and have it bend? Myths, legends, and science fiction are full of thought as a power able to move objects and submit the physical world to our will. It is possible that the brain–computer interface (BCI) has origins in this type of thinking. More likely, it is the logical extension of an electrical organ to control electrical devices. We know the brain's electrical activities are capable of pattern changes that can be conditioned. A simple example is alpha wave activity with meditation. The key problem is and remains the interface. Neural signals need to be accurately detected and translated into useful command signals to effect control over computers or prostheses. The question is no longer can these changes be used even in a damaged or degenerative brain to provide a basis for brain–computer interface as this is now established in principle [19, 26, 27, 44].

The first direct human brain–machine interface (Figure 13.1) was an outgrowth of a great deal of work on primates over a long period of time; that work demonstrated the ability of single neurons to change their firing pattern over time in a plastic manner [25, 28, 30]. The ability to record signals over a long period of time from the same neuron suggested that the neuronal activity of an individual could be reliably used to control machines. The obvious machine to control was a computer. The obvious first approach would be to restore communication to the patient who has lost that ability. The first paradigm was an on–off binary response [26] that subsequently became more complex [27]. The intention was to communicate, and that is what out implanted patient J.R. was able to do by thought alone. The first step

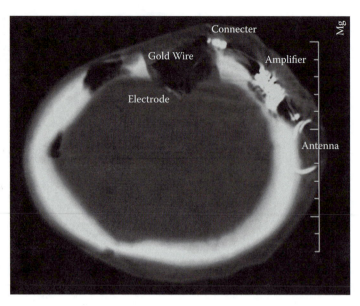

FIGURE 13.1 CT scan demonstrating the location of the Neutrophic electrode and its cortical wires atop the motor cortex. The air distorts the craniotomy site as well as the frontal lobes, but the post-op course was uneventful. The remainder of the electronics is not evident on this tomographic image.

was simple activation of computer icons to produce computer-generated phrases such as, "Hello, my name is J.R." By learning to control unit signals, he was able to move a cursor across a computer screen and stop at the appropriate location. The next step was to move in two directions and use the computer as a virtual typewriter to produce short responses by stopping on letters or punctuation. Following this success, a company, Neural Signal, Inc. (Atlanta, Georgia), was created.

The second major clinical long-term study of patients with brain–computer interfaces are being conducted by Cyberkinetics Neurotechnology Systems, Inc. (Foxborough, Massachusetts). Similar to the Neural Signal background, this work grew out of primate studies [8, 12, 50]. The procedure is very similar to the Neural Signal procedure, although the electrode is very different [19, 44]. The electrode uses 100 tines, and because of that, the amount of information needed to be transferred is entirely too large to go through a radio-frequency transmitter. As a result, the information is transferred directly to a plug, which is then connected to the electronics. While there is a lot more data to analyze than in the Neural Signal system, it is essentially the same type of output. There are a variety of computer games that can be played and even simple demonstration of robotic movements. Although the signals are reportedly stable, the initial three patients have failed to demonstrate that a large number of signals can be maintained for longer than six months without requiring significant intervention. In addition, there is a set-up time required each day to evaluate which units are still active and able to engage. There is then an additional retraining time required. Although an extremely sophisticated system, the degree of utility is still limited.

Brain-computer interfaces have great potential to allow patients with severe neurological disabilities to return to interaction with society through communication devices, environmental controllers, and movement devices. Interest in this field has dramatically increased. At the end of the last century, there were but a handful of centers investigating brain–computer interfaces. There are now multiple centers throughout the world with considerable intranational interest in resolving the issues of how to use the brain–computer interface to affect communication and mobility problems. The problems are equally biological as well as with computer and engineering problems. This book is a manifestation of much of this effort. In this particular chapter specifically, the problems are discussed from a neurosurgical perspective: (1) patient selection, (2) lead configuration, (3) lead location, (4) housing for the electronic components, and (5) device maintenance.

13.2   Patient Selection

Because of the experimental nature of the procedure, the patients must have severe fixed and/or progressive deficits. This has included patients with high cervical injury, stroke (especially brainstem stroke), cerebral palsy, or amyotrophic lateral sclerosis (ALS); but other diseases may qualify. One of the keys to the success of any surgical procedure is to carefully design the inclusion and exclusion criteria. There is always a temptation to include exception patients. This should be avoided. Reasons for exclusion should be carefully considered. These need to reflect the limits of both the device as well as the limits of surgery.

Inclusion criteria should not stretch the limits of what is possible. It is clear that any current treatment is for patients with severe disability. Nevertheless, one would like to obtain patients who are as intact as possible. In the case of ALS, the history is well known; and to obtain permission to perform surgery before severe deficits occur would be an ideal situation to advance the understanding of what is possible for these patients and what may not be possible after severe deficits ensue. Similarly, although the stroke or spinal cord injury patient may need to have a high degree of disability, some residual motor activity, even if nonfunctional, could be extremely useful and might be essential for optimal function of a brain–computer interface. Plasticity is necessary, and any residual motor function may suggest that the circuitry is still readily available for the brain–computer interface. Again, even minimal residual sensory feedback may be ideal for pushing forward the learning curve. Obviously, the brain must have some functional activity for it to be useful, and a functional MRI (fMRI) is necessary to identify functional activity. These patients are generally quite sick and may require interventions of a variety of sorts unrelated to the brain–computer interface. These setbacks can be quite serious, and the selection process should avoid patients who are metabolically unstable and, in general, are marginally fit for surgical intervention. Communication with these patients is essential, and some form of communication would have to be demonstrable. The more robust the communication, the better off the investigator and the patients are as they proceed in the study.

Exclusion criteria are those that have already been discussed in a negative way in the inclusion criteria. Patients who are (1) medically unstable and would have difficulty undergoing surgical procedures, (2) are cognitively impaired — unable to learn the protocol, or (3) unable to adequately communicate should be avoided. Furthermore, patients who are angry about their injury or emotionally unstable or depressed are patients who should be avoided. Litigation in the background certainly should raise a red flag. Similarly, patients with recent injuries should be avoided until it is clear that recovery will not occur. On the other hand, it is probably advisable to proceed with the implantation as soon as that evaluation is made because patients with plasticity change still needs to be possible. However, studies from stroke patients demonstrate the potential for plasticity changes even months or years following stroke, so timing may be less critical than originally thought [5, 17]. History suggesting poor wound healing, chronic infections, visual loss, cancer, etc., all should be exclusion criteria as these will impact the ability to evaluate and use any type of interface.

And then there are those inexplicable aspects that must be considered. Patients with positive mental attitudes are going to do the best and are the ideal patients to work with before and after surgery. Another ideal aspect would be someone with computer familiarity, or at least someone with a good deal of cognitive understanding of what is being attempted. Expectations must be reasonable. There are some patients who realize the severity of their disease but still want to make a contribution and understand that the benefit would be far greater for those who come after their participation than to themselves. Family support is critical. These patients are in need of continuous care, and a supportive family is a huge asset in this regard. Like all clinical medical research, it is the families and the patient who are the heroes and who are essential to push this work forward.

13.3   Lead Configuration

There are basically three types of electrodes that are useful for recording cortical or subcortical activity. The characteristics of the electrode will determine which localizations are possible and the electronics will follow. There are surface recording electrodes, either from the scalp electroencephalograph (EEG)

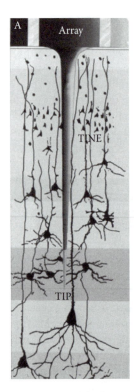
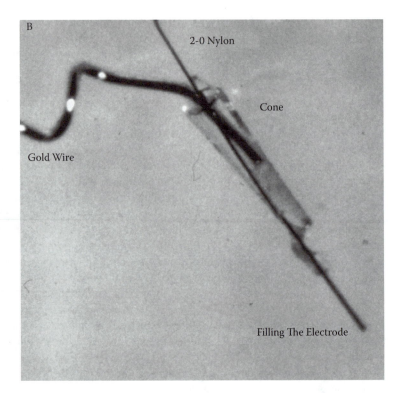

FIGURE 13.2 (A) The tine from an electrode array. The cortex is illustrated and surrounding this electrode will be gliosis and edema. Exactly which neurons are recorded from is unclear. (B) The loading of a Neutrophic electrode. The glass cone is connected to a gold wire and is being filled with peripheral nerve fascicles using a nylon suture to force the fascicles into the wide end of the cone and out the narrow end.

[18, 52, 57, 58] or directly from the pial surface electrocorticogram (ECoG) [35]. The advantages of both are the ease of application, the absence of brain tissue damage, and the ease with which areas of interest can be sampled. Signaling requires the detection of frequency changes, which require averaging that slows detection and data transfer. Communication devices based on EEG are commercially available now. The EEG activity is in the microvolt (μV) range and occasionally can be difficult to separate from electrical noise in the environment. Other disadvantages are the low spatial resolution and extensive training to establish the two independent control signals that are required for effective use. Advantages for the location closer to the electrical activity (ECoG) are higher spatial resolution, better signal-to-noise ratio, wider frequency range, and better long-term stability. The use of ECoG activity in closed-loop, real-time control of a cursor to increase flexibility and potential utility of such recordings has been demonstrated, in association with motor or speech imagery [35]. The basic technique for electrode placement is well established in the treatment of epilepsy. Currently, no one has used specifically designed cortical recording electrodes to provide signals for brain–computer interactions. The disadvantage is that these are large regions that may or may not be able to be specifically and separately controlled for three-dimensional movements. The activity is still delayed, albeit only slightly, from direct recordings. Because they are on the surface, they can move and lose specific registration. The greatest disadvantage is the apparent low rate of information transfer.

The other types of electrodes are those that are penetrating (Figure 13.2). This is the future of brain–computer interface as the time scale shifts from 1 to 2 s to 2 to 400 ms. These are either in the cortical or subcortical areas. They may be individual or multiple. They may record field potentials or individual units. High-amplitude, local-field potential (LFP) recordings have the advantages of not requiring that specific units be identified and theoretically can be maintained indefinitely by simply recording activity over a discrete area. In motor cortex, these represent neural activity from widespread neurons that are temporally coupled and related to motor planning and preparatory activity rather than precise motor encoding [12]. Elsewhere, LFP may be useful to recognize intention and planning [1]. Intracortical LFP signals were able to be used in a crude manner to control digital movements of a cyber hand [29]. This system requires threshold data, and therefore there is only a very short detection delay. Extracranial-obtained LFP can be used but appears to be less useful due to limited information content. The disadvantages of LFP recording are that the field must respond to conditioning, which is in some ways more difficult than individual units, and control of multiple degrees of freedom has not been demonstrated. These appear quite capable of providing binary or switching function in realtime, but lack the robustness of proportional responses for fine prosthesis movements.

The most commonly used invasive electrodes have multiple prongs (Figure 13.2A), such as the Utah or Michigan electrode arrays [37, 53]. Various versions of these arrays all suffer from the basic problem: it is not possible to record from all of the electrodes. Over time, units are lost. One has difficulty determining whether the same unit can be recorded from continuously over time; therefore, currently reregistration of the units is performed on a daily basis. This provides a significant disadvantage for long-term programming and conditioning. An alternative is an array of microwires implanted in loose bundles [31] or fixed grids [8, 42]. It has yet to be determined whether these behave differently in subcortical locations than cortical, but it is unlikely that they would. The multitude of leads presents multiple problems down the line in terms of transmitting the signal, as we discuss later. In addition, the wires coming from such an electrode are heavy and can cause problems in terms of pulling or tugging in response to Valsalva maneuvers or any brain movement (such as from minor trauma). The results are micromovements, signal instability, and loss of units. This has been frequently seen in monkey studies.

Glial scarring around the electrodes may result in decreased activity, loss of signal intensity, etc. Edema and protein coating of the electrode all contribute to signal loss. Although the Cyberkinetic electrode is not supposed to have this problem, the first three patients all have had loss or diminished signal at relatively short times postimplant. Newer nanoporous silicon surface electrodes may be more biocompatible [39], but the problem is not solved. A very different recording device is the microelectromechanical system (MEMS) that allows recording from multiple cortical layers and as far away as 140 μm [7]. The MEMS uses a tetrode to separate action potentials from more than 1000 neurons in three-dimensional space. It has not yet been tried for this purpose.

Finally, there is the bioelectrode. Electrodes such as the Neurotrophic electrode integrate the regenerative biology with recording techniques. Briefly, the Neurotrophic electrode consists of a chronically implanted glass cone through which neuronal processes can grow (Figure 13.2B). The cone contains gold contact recording wires inserted through the wide end. These wires differentially record the electrical activity of the ingrown neuronal processes (axonal and dendritic) [30]. Prior to implantation, neurotrophic substances or peripheral nerve fascicles are placed inside the cone. Autologous sciatic nerve was first used as the attractant for axonal growth following the work of Benfey and Aguayo [4], which showed that Schwann cells from the sciatic nerve when placed in rat cortex would induce axonal sprouting from the underlying neurons. But Schwann cells from any nerve source will provide the same effect. Following injury, Schwann cells will activate more than sixty genes and produce growth factors, cytokines, neuropeptides, cytoskeleton proteins, and extracellular matrix molecules [22]. Other sources of neurotrophic factors and matrix molecules may also prove effective [25, 28].

The key feature of this electrode compared with any other is that instead of bringing the electrode's metal recording tip to the neurons, the neuronal processes grow against the electrode tip and are held there as a bridge of tissue isolated within the glass cone. After a few weeks, the processes that grew into the core from surrounding neurons become myelinated. Because the tissue grows through the cone and anchors it in the cortex, the recording wires in the cone move with the recorded tissue during normal movements of the brain, thus providing both mechanical and signal stability. The actual duration for which recordings can be made has not been yet been determined but individual units have been recorded for up to nineteen months in monkeys and four years in humans [25, 27]. This type of integration provides a potentially more stable environment, and the same single units can be recorded for a long period of time. A disadvantage is that only a small number of units (usually five to ten) can be recorded from each electrode. Not all these will be adequately isolated for use in conditioning. Although multiple biologic electrodes can be placed and appear not to significantly damage the adjacent cortex, the increased number of electrodes will again produce problems in transmitting multiple signals for analysis. Other disadvantages are the irregular manufacturing of the electrode, the delay of months for the electrode to mature after implantation, and the very extensive and intensive training required to gain control of a cursor. Finally, this approach also appears to be limited to cortical activity rather than subcortical activity.

Single-unit conditioning is believed to be the key to implementing proportional signals for real-time brain–computer interfaces for prosthesis. Fetz and Finocchio [14] showed that a cortical motor neuron could be used to control movement in a single direction. In 1973, Fetz and Baker [13] demonstrated in monkeys that one neuron could be conditioned to fire and another to suppress firing. This suggests that not only can the desired neurons be conditioned to fire at specific rates, but also that undesirable neurons can be suppressed if contributing to unwanted background activity, thus physiologically improving the signal-to-noise ratio. Burnod et al. [6] confirmed that conditioning of motor neuronal firings in monkeys is possible during task-specific conditioning.

Wyler et al. [60] confirmed that monkeys could control firing rates within predetermined firing ranges or levels (expressed as the modal interspike intervals of unit firings). Wyler [59] also demonstrated that when pairs of units were recorded and reward was contingent upon the firing rate of one of the pair, the other unit's firing rate did not co-vary with the firing rate of the conditioned unit. This result, in addition to the results of Fetz et al. [15], suggests that two units can be conditioned separately, especially if they are related to different movements or separate aspects of one movement (for example, agonist/antagonist pairs). The Kennedy and Bakay [25, 28] monkey experiments provide evidence that recordings can be made of units that fire reciprocally. In addition, longer recording times are available in patients than were available to the above workers, and this persistence should provide the time needed to thoroughly test all recorded units. Multiple precise control functions in three-dimensional arm movement can be achieved by isolating independent single units and developing an algorithm to predict movement from a population of neurons [21]. But not until 2002 was it possible to demonstrate three-dimensional, closed-loop, real-time control of a cursor. There are now a number of primate studies demonstrating that neural output from motor cortex can be used to control computer cursors or prostheses as effectively as natural hand movement [8, 12, 34, 40, 42, 50, 55].

The work of Wyler and colleagues [62] also suggested that closed-loop control was required for operant conditioning to occur: If the spinal cord dorsal columns were transected at C1-2, operant conditioning was diminished but not lost. If the monkey's contralateral ventral roots were sectioned, however, operant control was totally lost [61]. This provides evidence against open-loop control. However, the role of vision and auditory as compensatory means of closing the loop were not investigated systematically, and other data suggest visual and auditory feedback may be sufficient [49]. But control of grip force in the absence of somatosensory feedback may be very inaccurate [43]. These results have important implications for patients because their lesion would leave an open feedback loop. In fact, there are differences in imagined movements between controls, paraplegics, and quadriplegics [32]. However, it has been shown that a well-motivated paralyzed human can condition units using auditory and visual feedback via internal hemispheric, cerebellar, and/or brainstem loops. However, it should also be noted that the best results were obtained when some residual function was present.

Table 13.1 provides a summary of the potential of the clinically used devices. For simple tasks such as a binary switch, all the leads will provide effective signals. Less invasive and robust recording of LFP could be favored for this binary function. Dual switches provide information for communication and simple environmental controls such as turning on or off the lights. But for complex communication and prosthetic control, there will be a need for multiple, functionally independent real-time single-unit potentials. The number of units is still being determined [41, 45] but the clear advantage here is that the signals can be proportioned, and that has far greater impact than a binary switch. Thus, the emphasis of many groups is not on the individual units but rather on the population to predict motor action and use this to interface with a prosthesis. Clearly, conditioning and plasticity still play a role, and the model cannot be simple "wire tapping" to reconstruct the motor system.

TABLE 13.1   Characteristics of Brain-Computer Interface Electrodes Used in Clinical Studies

	Tines[*]	Wires[†]	Neurotrophic[=]
Active Contacts	100 per grid	32 microwires	1 per cone
Units/Contact	1–2	1–2	5–10
Units recorded	7–130	28–119	5–10
Stability of Signal	Unknown	Not tested	Years
Longevity of recording	Problems with maintaining signals	Only acutely tested	>4 years
Directionality	2D	2D	2D

[*] Park et al. (44)
[†] Patel et al. (45)
[=] Kennedy et al. (27)

The other aspect of electrodes that has not been addressed in any of these studies to date is the potential for stimulation to augment or control activities [3]. Such augmentation clearly has been demonstrated in the past and certain types of behaviors can be initiated [10]. Subthreshold stimulation could enhance performance of the recording neurons as has been done for stroke or head injury [5, 17]. Alternatively, stimulation could arrest unwanted activity such as seizures [16, 38]. There is also the potential to put in sensory feedback signals by converting mechanical position, and directional and force information into microstimulation of the sensory cortex. The experimental basis to provide this type of substitute feedback already exists [46, 56] and could use technology that is already being developed to put sensory prosthesis into cortical interfaces [36]. The possibilities are endless, but so may be the ethical problems as now the question becomes, "Who is controlling whom?" However, the possibility of true mind control is extremely remote [24]. Whether this type of stimulation would be helpful for prosthetics or communication devices remains to be determined. With regard to stimulation, the bioelectrode is least useful and the penetrating metal electrodes are most useful.

Thus, the type of electrode will determine where exactly the opening will be made, how the lead will be inserted, what type of support devices will be needed, and finally, how to protect the electrode and its wires from damage. Obviously, the CSF space will be violated with anything other than superficial electrodes and repair of that space will be essential. Preventing migration of the electrodes, allowing flexibility of the leads so that pulling and local trauma does not occur will be important.

13.4   Lead Location

Location of the electrode will dramatically depend on the type of electrode. Most electrodes record cortical activity and require only a surface approach. There is the possibility that interhemispheric, subfrontal, or even subtemporal may be sites that would be useful. Initial work focused on placing electrodes in specific sites such as motor cortex or speech cortex in an effort to use the highly segmented activity to generate specific impulses. This type of work does not require that motor activity necessarily be performed by motor neurons. Thus, the movement of a cursor across the screen of a computer can serve the function of positive feedback for cortical activity. Motor cortex is useful in this regard only if it has controllable firing patterns (functional activity). The use of fMRI to document changes in cortical activity with imagined movement is essential (Figure 13.3). With neurosurgical image-guidance techniques, it is relatively straightforward to identify the exact area of the cortex for implantation (Figure 13.4). At the time of craniotomy, macroscopic optical imaging of blood flow could further refine localization for implants. While these areas may be useful in generating potentials for prosthesis, it is also quite possible that premotor, prefrontal, supplementary motor, or even parietal areas may be as useful in defining potentials for the movement [4, 40]. Likewise, the use of the Broca's motor speech area may be very useful for developing the use of individual phonemes to generate words and speech; but other areas such as Wernicke's area may be equally useful. It is simply unknown at this time.

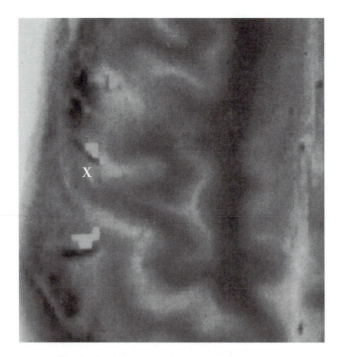

FIGURE 13.3 A functional MRI sectioned through the area of interest and demonstrating the area of hand activity at **X**. There are activation changes indicated on either side of the **X**.

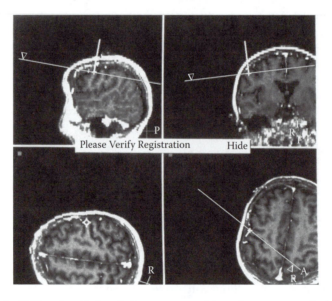

FIGURE 13.4 The Stealth™ station Medtronic image reconstructions demonstrate sagittal, coronal, axial, and trajectory approach views of the hand area of the motor cortex target (**X**) as in Figure 13.3.

Electrical activity, especially motor activity, can be highly concentrated in subcortical sites; and this has been used in at least a test of principle [45]. However, these were on awake, intact patients; and following injury, the results may be less robust. We know in disease states that units no longer respond to a single joint movement but more commonly respond to multiple units [54]. There is growing evidence that rhythmic movements are not part of a more generalized discrete movement system but are separate neurophysiologically [48]. One could question — at least on a theoretical basis — whether or not extrapyramidal sites would be as useful as pyramidal sites in motor control. Clearly, they could be used for other types of control. There is, of course, the potential for multiple sites. This may be especially important for positive and negative feedback controls, as well as other safety issues. The minimum requirement would be the functional activity in the target area of brain. There is at least one instance where inadequate cortical activity failed to produce anticipated results, despite fMRI indications of activity [2]. Not only must there be activity, but there also must be functional activity that can be regulated. Another major unknown is the effect of sensorimotor reorganization.

13.5   Implantation of Electronics

Once the wires exit the calvarium, they are then connected to the electronics. The electronics may simply exit to a plug on the skull. This type of device has been used by Cyberkinetics [44]. The advantage is it is the simplest means of getting the data from the 100 electrodes. A disadvantage is the risk for infection, which may pose challenges for these plugs, including the weakening of supports due to osteomyelitis. Trauma can dislodge the device, and mechanical connections mandate support personnel and thus make stand-alone capabilities impossible. Thus, this technology must overcome significant challenges if it is to be used in the future. Since Delgado's work in 1969, the use of totally subcuticular electronics has been employed in both monkeys and humans [10]. Future surgery must include total internalization of wireless devices for real-time biotelemetry. Power for the internal electronics can come from an induction source, which requires the placement of an external device over top the induction coil to transmit the energy through the skin. "Smart caps" and other types of devices have been used for this purpose. However, in the future, this also restricts the mobility and utility of the device. With the potential for the device to be "on" 24/7, there will be the need to develop a control "switch" to activate or inactivate the device as needed and save energy. Ultimately, some type of internal battery, preferably rechargeable through induction, should provide the energy, mobility, and independence necessary for optimal function. While not yet available, one can conceive of batteries that draw on energy sources from the patients themselves and thus may be self-sustaining.

The electronic component must be placed in the subgaleal compartment. In the past, this was a simple preamplifier. In the future, more and more electronic components will be added but microengineering is essential. The initial implants for the Neural Signal patients required neurosurgical creativity [25, 27]. The devices were too large, even with wide scalp elevation, to easily close the skin. Skin closure is absolutely essential to prevent infection. What was done was to outline the devices on the patient's calvarium, drill down the outer cortical bone into the diploic space so that the inner cortical bone would support the electronic device and allow placement of the device without tension on the suture line (Figure 13.5). Protection of the device is essential, and initially covering with cranioplasty provided a degree of protection (Figure 13.6). In the future, miniaturization will be necessary to decrease the volume of the implants and development of new material will be necessary to increase the effectiveness of the protection. As in all situations, good neurosurgical closure is essential. This involves decreasing any stress on the incision line by placing the devices as far as possible from the incision line.

13.6   Device Maintenance

Maintenance of the device must also be simplified. Upgrades will be coming in rapid succession, at least early on, and the device should have components that can be interchangeable when upgraded. Power sources should be remote so as to diminish the potential for infection and streamline the replacement process. In addition, the electronic components must have internal sensing devices and fail-safe type capability. Currently, none of the devices have such protection for patients. Seizures can easily disrupt such a system — and with a damaged brain, this may be highly likely. Similarly, sleep states could dramatically affect the firing patterns, and connection to a computer 24/7 would require additional safety considerations. The potential for short circuits, power surges, etc. will further require protection for the

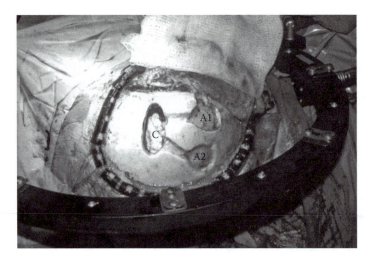

FIGURE 13.5 The photograph demonstrates the craniectomy (**C**) for approach to the hand motor area as determined by Stealth™ neuronavigation. To accommodate the amplifiers (**A1** and **A2**), the outer cortical and spongiform bone were removed along with troughs from the craniectomy for the cables.

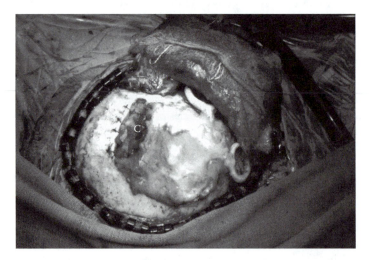

FIGURE 13.6 The craniectomy (**C**), which is covered by gel foam and tack-up stitches placed anteriorly. The amplifiers are in place and covered with crainoplasty for protection. In addition, the two loops posteriorly are the antenna for induction and radiosignals.

patient. Materials will have to be nontoxic, capable of withstanding minor trauma, waterproof, and not subject to breakdown over the duration of need for the implant. Microelectromechanical systems are undergoing revolutionary advances in performance and capabilities [46]. These will be needed to reach the goals of the ideal device (Table 13.2).

One of the major problems is getting the information from the electrodes to the computer for analysis. With a great number of potential signals, there is a great deal of data that has to be rapidly transferred. Increasingly, more and more data processing will need to be performed closer and closer to the electrodes. Information sent from this source to the receiving unit must be processed to a high degree to obtain mobility. Single one-to-one correspondence to 100 electrodes is simply not possible. If only 20 to 30 of those have information that is of value, then restricting transport to those electrodes decreases some of the tremendous amount of information that must be transferred. There is a tremendous amount of work trying to define specific ensembles of neurons with weighted values that are greater than the population

TABLE 13.2 The Ideal Brain-Computer Interface

Safe	No harmful effects from internal or external electrical or magnetic interference
Dependable	Available 24/7 on demand without the need for care givers to activate or interface
Accurate	The information transfer must have a very high level of fidelity to be effective
Real-Time	Able to perform immediately and appropriately at the speed of normal speech or movement
Reliable	Durable construction and long-term performance without the need for frequent maintenance adjustments or re-settings
Interactive	Develop analogs to anticipate and simplify activity so that fatigue does not become a limiting factor
Available	Off the shelf products at reasonable prices
Aesthetic	Minimal obtrusion and maximum acceptable for the patient

as a whole [20, 23, 34, 50, 51]. Further development of robust decoding algorithms to decrease this to four or five channels of information will be essential to make a workable real-time system. Operating at low power will be essential for long-term energy conservation and heat dissipation of an implanted device. The key will be to accurately and rapidly transmit highly enriched information.

Most of the information that comes from the patient immediately goes to a series of computers. The signal obviously must be amplified, processed, analyzed, and various analogs used to then convert the information to a practical function. To date, this has required a very large amount of equipment and hands-on technicians. This then restricts the amount of time that the patient can engage in learning or utilizing the device. The BrainGate device has streamlined this but remains a very bulky series of equipment that must accompany the patient in order to have it function. The same is true for Neural Signal patients. During this early learning phase, such equipment may be necessary, but obviously in the future this would be unacceptable and create difficulties in utility and mobility. None of the systems are stand-alone, and we are a long way from having any system that could be useful in that regard. Patients may want the advantage of robotics, but they do not necessarily want to look like a robot; so aesthetics also must be addressed.

The growth in this field will continue to be slow as long as the work is supported by small grants and small companies. Too much equipment is off-the-shelf and not specifically designed for this work. Too much computer programming simply has been transferred by cutting and pasting from old programs rather than the development of new types of programming necessary to advance the field more rapidly. The electrodes are decades old, and new innovations are not rapidly forthcoming. While it is indeed true that we do not know enough about the electrophysiology of the nervous system we are trying to tap into, the main problem behind slow growth is lack of resources. This all may change in the near future as countries such as Japan and France make brain–computer interfaces part of their national scientific goals.

13.7 Conclusion

A functional brain–computer interface has been demonstrated in a proof-of-principle study and confirmed in a similar design study using different electrodes and recording techniques. Nevertheless, there is considerable work that that must be done to bring this type of technology to a useful state. Currently, far too much energy is required to maintain the system, and nothing yet even approaches stand-alone technology. There is the need to improve all aspects of the technology. There is excessive overlap and redundancy among current research and development efforts. This is not the time to standardize or obstruct the development of any model. The ultimate design may well be a hybridized combination of technologies and may include repair and regenerative therapeutics of neural transplantation and gene transfer. What is needed are more innovative approaches, and in this a neurosurgeon can play a significant role in helping the bioengineers understand the limitations and possibilities of surgical manipulation of the brain. Likewise, intimate knowledge of bioengineering certainly will help neurosurgeons understand where they can be helpful in improving the procedures.

References

1. Andersen, R.A. and Buneo, C.A. Intentional maps in posterior parietal cortex. *Annu. Rev. Neurosci.*, 25:189–220, 2002.
2. Bakay, R.A.E. Limits of brain-computer interface. *Neurosurg. Focus*, 20(5):e6, 2006.
3. Benabid, A.L., Wallace, B., and Mitrofanis, J. Therapeutic electrical stimulation of the central nervous system. *C. R. Biol.*, 328(2):177–186, 2005.
4. Benfey, M. and Aguayo, A.J. Extensive elongation of axons from rat brain into peripheral nerve grafts. *Nature*, 296:105–107, 1982.
5. Brown, J.A., Lutsep, H.L., Weinand, M., and Cramer, S.C. Motor cortex stimulation for the enhancement of recovery from stroke: a prospective, multicenter safety study. *Neurosurgery*, 58(3):464–473, 2006.
6. Burnod, Y., Maton, B., and Calvet, J. Short-term changes in cell activity of areas 4 and 5 during operant conditioning. *Exp. Neurol.*, 78:227–240, 1982.
7. Buzsaki, G. Large-scale recordings of neuronal ensembles. *Nat. Neurosci.*, 7:446–451, 2004.
8. Carmena, J.M., Lebebev, M.A., Crist, R.E., O'Doherty, J.E., Santucci, D.M., Dimitrov, D.F., Patil, P.G., Henriquez, C.S., and Nicholelis, M.A.L. Learning to control a brain-machine interface for reaching and grasping by primates. *PloS Biol.*, 1:2, E42, 2003.
9. Cheung, G., Gawel, M.J., Cooper, P.W., Farb, R.I., Ang, L.C., and Gawal, M.J. Amyotrophic lateral sclerosis: correlation of clinical and MR imaging findings. *Radiology*, 194(1) 263–270, 1995.
10. Delgado, J.M. Radio-controlled behavior. *NY State J. Med.*, 69(3):413–417, 1969.
11. Donoghue, J.P. Connecting cortex to machines: recent advances in brain interfaces. *Nat. Neurosci.*, 5(Suppl):1085–1088, 2002.
12. Donoghue, J.P., Sanes, J.N., Hatsopoulos, N.G., Gaal, G. Neural discharge and local field potential oscillations in primate motor cortex during voluntary movements. *J. Neurophysiol.*, 79(1):159–173, 1998.
13. Fetz, E.E. and Baker, M.A. Operantly conditioned patterns of precentral unit activity and correlated responses in adjacent cells and contralateral muscles. *J. Neurophysiol.*, 36:179–204, 1973.
14. Fetz, E.E. and Finnocchio, D.V. Operant conditioning of specific patterns of neural and muscular activity. *Science*, 174:431–435, 1971.
15. Fetz, E.E., Finnochio, D.V., Baker, M.A., and Soso, M.J. Sensory and motor responses of precentral cortex cells during comparable passive and active joint movements. *J. Neurophysiol.*, 43(4):1070–1089, 1980.
16. Fountas, K.N., Smith, J.R., Murro, A.M., Politsky, J., Park, Y.D., and Jenkins, P.D. Implantation of a closed-loop stimulation in the management of medically refractory focal epilepsy: a technical note. *Stereotact. Funct. Neurosurg.*, 83(4):153–158, 2005.
17. Frazer, C., Power, M., Hamdy, S., Rothwell, J., Hobday, D., Hollander, I., Tyrell, P., Hobson, A., Williams, S., and Thompson, D. Driving plasticity in human adult motor cortex is associated with improved motor function after brain injury. *Neuron*, 34:831–840, 2002.
18. Freeman, W.J., Holmes, M.D., Burke, B.C., and Vanhatolo, S. Spatial spectra of scalp EEG and EMG from awake humans. *Clin. Neurophysiol.*, 114:1053–1058, 2003.
19. Friehs, G. and Hochberg, L. BrainGait, a human brain-machine interface: initial experience in an ongoing pilot clinical trial. American Association of Stereotactic and Functional Neurosurgery meeting abstracts June 1–4, 2006.
20. Gao, Y., Black, M.J., Bienenstock, E., Shoham, S., and Donoghue, J. Probabilistic inference of hand motion from neural activity in motor cortex. *Proc. Adv. Neural Info. Process. Syst.*, The MIT Press, 2002.
21. Georgopoulos, A.P., Schwartz, A.B., and Kettner, R.E. Neuronal population coding of movement direction. *Science*, 233:1416–1419, 1986.
22. Gillen, C., Korfhage, C., and Muller, H.W. Gene expression in nerve regeneration. *The Neuro-scientist*, 3:112–122, 1997.

23. Helms Tillery, S.I., Taylor, D.M., and Schwartz, A.B. Training in cortical of neuroprosthetic devices improves signal extraction from small neuronal ensembles. *Rev. Neurosci.*, 14:107–119, 2003.

24. Horgan, J. The myth of mind control. *Discover*, 28(10):40–47, 2004.

25. Kennedy, P.R. and Bakay, R.A. Activity of single action potentials in monkey motor cortex during long-term learning. *Brain Res.*, 760:251–254, 1997.

26. Kennedy, P.R. and Bakay, R.A. Restoration of neural output from a paralyzed patient by a direct brain connection. *NeuroReport*, 9:1707–1711, 1998.

27. Kennedy, P.R., Bakay, R.A., Moore, M.M., Adams, K., and Goldwaithe, J. Direct control of a computer from the human central nervous system. *IEEE Trans. Rehab. Eng.*, 8198—202, 2000.

28. Kennedy, P.R., Bakay, R.A., and Sharpe, S.M. Behavioral correlates of action potentials recorded chronically in the cone electrode. *NeuroReport*, 3:605–608, 1992.

29. Kennedy, P.R., Kirby, M.T., King, B., Mallory, A., Adams, K., and Moore, M.M. Computer control using human cortical local field potentials. *IEEE Trans. Neural Syst. Rehab. Eng.*,12(3):339–344, 2004.

30. Kennedy, P.R., Mirra, S., and Bakay, R.A.E. The cone electrode: ultrastructural studies following long-term recording. *Neurosci. Lett.*, 142:89–94, 1992.

31. Kreiman, G., Koch, C., and Fried, I. Category specific visual responses of single neurons in the human medial temporal lobe. *Nat. Neurosci.*, 3:946–953, 2000.

32. Lacourse, M.G., Cohen, M.J., Lawrence, K.E., and Romero, D.H. Cortical potentials during imagined movements in individuals with chronic spinal cord injuries. *Behav. Brain Res.*, 104(1–2):73–88, 1999.

33. Lauer, R.T., Peckham, P.H., Kilgore, K.L., and Heetderks, W.J. Applications of cortical signals to neuroprosthetic control: a critical review. *IEEE Trans. Rehabil. Eng.*, 8:205–208, 2000.

34. Lebedev, M.A., Carmena, J.M., O'Doherty, J.E., Zacksenhouse, M., Henriquez, C.S., Principe, J.C., and Nicolelis, M.A. Cortical ensemble adaptation to represent velocity of an artificial actuator controlled by a brain-machine interface. *J. Neurosci.*, 25(9):4681–4893, 2005.

35. Leuthardt, E.C., Schalk, G., Wolpaw, J.R., Ojemann, J.G., and Moran, D.W. A brain-computer interface using electrocorticographic signals in humans. *J. Neural. Eng.*, 1:63–71, 2004.

36. Loeb, G.E. Neural prosthetic interfaces with the nervous system. *Trends Neurosci.*, 12(5):195–201, 1989.

37. Maynard, E.M., Nordhausen, C.T., and Normann, R.A. The Utah intercortical electrode array: a recording structure for potential brain-computer interfaces. *Electroencephalogr. Clin. Neurophysiol.*, 102:228–239, 1997.

38. Morrell, M. Brain stimulation for epilepsy: can scheduled or responsive neurostimulation stop seizures? *Curr. Opin. Neurol.*, 19(2):164–168, 2006.

39. Moxon, K.A., Kalkhoran, N.M., Markert, M. Sambito, M.A., McKenzie, J.L., and Webster, J.T. Nanostructured surface modification of ceramic-based microelectrodes to enhance biocompatibility for a direct brain-machine interface. *IEEE Trans. Biomed. Eng.*, 51(6):881–889, 2004.

40. Musallam, S., Corneil, B.D., Greger, B., Scherberger, H., and Andersen, R.A. Cognitive control signals for neural prosthetics. *Neuroscience*, 305:259–262, 2004.

41. Nicolelis, M.A. Actions from thoughts. *Nature*, 2003; 409:203–407, 2003.

42. Nicolelis, M.A., Dimitrov, D., Carmena, J.M., Crist, R., Lehew, G., Kralik, J.D., and Wise, S.P. Chronic, multisite, multielectrode recordings in macaque monkeys. *Proc. Natl. Acad. Sci., U.S.A.*, 100:11041–11046, 2003.

43. Nowak, D., Glasauer, S., and Hermsdorfer, J. How predictive is grip force control in the complete absence of somatosensory feedback?. *Brain*, 127:182–192, 2003.

44. Park, M.C., Zerris, V., Hochberg, L., Donoghue, J., Mukand, J.A., and Friehs, G. BrainGate — First experience with an implantable human neuromotor prosthesis. *Neuromodulation*, 9(1):19–20, 2006.

45. Patil, P.G., Carmena, J.M., Nicolelis, M.A.L., and Turner, D.A. Ensemble recordings of human subcortical neurons as a source of motor control signals for a brain-machine interface. *Neurosurgery*, 55(1):27–35, 2003.

46. Romo, R., Hernandez, A., Zainos, A., Brody, C.D., and Lemus, L. Sensing without touching: psychophysical performance based on cortical microstimulation. *Neuron*, 26:273–278, 2000.
47. Roy, S., Ferrara, L.A., Fleischman, A.J., and Benzen, E.C. Microelectromechanical systems and neurosurgery: a new era in a new millennium. *Neurosurgery*, 49(4):779–797, 2001.
48. Schall, S., Sternad, D., Osu, R., and Kawato, M. Rhythmic arm movement is not discrete. *Nat. Neurosci.*, 7(10):1136, 2004.
49. Schmidt, E.M. Single neuron recording from motor cortex as a possible source of signals for control of external devives. *Ann. Biomed. Eng.*, (8):339–349, 1980.
50. Serruya, M.D., Hatsopoulos, N.G., Paninski, L., Fellows, M.R., and Donoghue, J.P. Instant neural control of a movement signal. *Nature*, 416:141, 2002.
51. Serruya, M.D., Hatsopoulos, N.G., Fellows, M.R., Paninski, L., and Donoghue, J.P. Robustness of neuroprosthetic decoding algorithms. *Biol. Cybern.*, 88(3):219–228, 2003.
52. Srinivasan, R., Nunez, P.L., and Silberstein, R.B. Spatial filtering and neocortical dynamics: estimates of EEG coherence. *IEEE Trans. Biomed. Eng.*, 45:814–826, 1998.
53. Taylor, D.M., Tillery, S.I., and Schwartz, A.B. Direct cortical control of 3-D neuroprosthetic devices. *Science*, 7:1829–1832, 2002.
54. Vitek, J.L., Bakay, R.A.E., Hashimoto, T., Kaneoke, T., Mewes, K., Zhang, J.Y., Rye, D., Starr, P., Baron, M., Turner, R., and DeLong, M.R. Microelectrode-guided pallidotomy: technical approach and its application in medically intractable Parkinson's disease. *J. Neurosurg.*, 88(6):1027–1043, 1998.
55. Wessberg, J., Stambaugh, C.R., Kralik, J.D., Beck, P.D., Laubach, M., Chapin, J.K., Kim, J., Biggs, S.J., Srinivasan, M.A., and Nicolells, M.A.L. Real-time prediction of hand trajectory by ensembles of cortical neurons in primates. *Nature*, 2000; 408:361–365, 2000
56. Wickersham, I. and Groh, J.M. Neurophysiology: electrically evoking sensory experience. *Curr. Biol.*, 8:R412–R414, 1998.
57. Wolpaw, J.R., Birbaumer, N., McFarland, D.J., Pfurtscheller, G., and Vaughan, T.M. Brain-computer interfaces for communication and control. *Clin. Neurophysiol.*, 113:767–791, 2002.
58. Wolpaw, J.R., McFarland, D.J., Neat, G.W., and Forneris, C.A. An EEG-based brain-computer interface for cursor control. *Electroencephalogr. Clin. Neurophysiol.*, 78:252–259, 1991.
59. Wyler, A.R. Interneuronal synchrony in precentral cortex of monkeys during operant conditioning. *Exp. Neurol.*, 80:697–707, 1983.
60. Wyler, A.R., Lange, S.C., Neafsey, E.J., and Robbins, C.A. Operant control of precentral neurons: control of model interspike intervals. *Brain Res.*, 190:29–38, 1980.
61. Wyler, A.R. and Burchiel, K.J. Operant control of pyramidal tract neurons: role of spinal dorsal. *Brain Res.*, 157(2):257–265, 1978.
62. Wyler, A.R., Burchiel, K.J., and Robbins, C.A. Operant control of precentral neurons: evidence against open loop control. *Brain Res.*, 171:29–39, 1979.

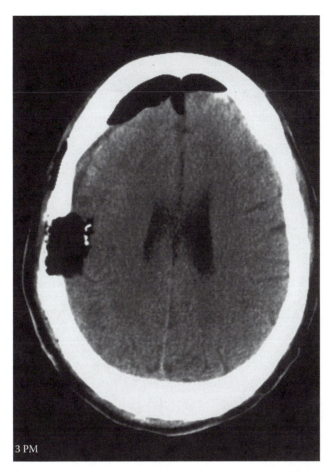

unlabeled013x001.eps

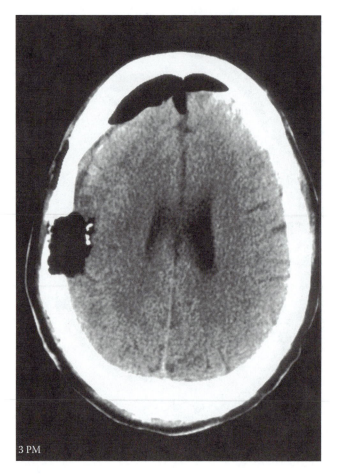

3 PM

unlabeled013x002.eps

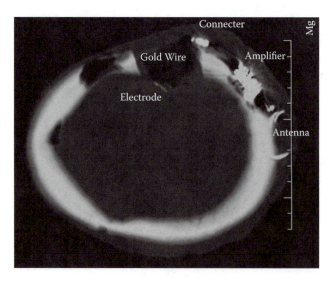

unlabeled013x003.eps

14

Comparing Electrodes for Use as Cortical Control Signals: Tiny Tines, Tiny Wires, or Tiny Cones on Wires: Which Is Best?

Philip R. Kennedy

14.1 Introduction and Background

In the fields of neural prosthetics and neural engineering there are several viable contenders for the prize of best long-term electrode to access cortical control signals for restoration of communication and movement in humans. These contenders can be classified into three main groups. The first group includes those who have developed tiny tines that are driven into the cortex and provide signals for months and sometimes years [1, 10, 11]. The second group produces flexible wires that are inserted into the cortex and also provide signals for months and sometimes years [2]. The third type of electrode is also a wire configuration but allows for growth of the brain's neuropil into the hollow glass tip of the electrode. Robust signals have been recorded for years from this neurotrophic electrode [5]. Thus, these electrodes can be classified into (1) those that protrude towards neurons and (2) the neurotrophic electrode that welcomes the neurites into its tip and thus fuses with the neuropil.

The "holy grail" of all these efforts is the restoration of movement to the paralyzed. For example, quadriplegics need to use their hands, and cortical control of functional neuromuscular stimulation devices would appear to be the answer to this need. These systems, although they will provide access to cortical control signals, may not alone restore movement to those with spinal cord injury. Instead, this author suspects that these recording technologies will be hybridized with spinal cord regeneration efforts

TABLE 14.1 Characteristics of Electrodes

Investigators	Tines Anderson	Tines Donoghue	Tines Schwartz	Wires Nicolelis	Neurotrophic Kennedy
1-Longevity	1+ yrs	4+ yrs	1+ yrs	1+ yrs	4+ yrs human
2-Stability	No	No	No	No	Yes
3-Plasticity	Yes	Yes	Yes	Yes	Yes
4-Directionality	2D	2D	3D	2D	2D
5-Force	N/T	N/T	N/T	Yes	N/T
6-LFPs	Yes	N/T	N/T	N/T	Yes
7-Units/contact	1–2	1–2	1–2	1–2	19
8-Units/electrode	40	40	40	100++	19
9-EMG related	N/T	N/T	N/T	N/T	Yes
10-Stimulation	Yes	Yes	Yes	Yes	No

Note: N/T = not tested; yrs = years.

to restore movement to those paralyzed by spinal cord injuries. Similarly for amputees, robotic technologies will wed with cortical control signal technologies for successful control of artificial limbs.

Herculean efforts have been expended by all workers in this field to provide single-unit recordings in the belief that the precision needed for control of digits is found only in the firing patterns of cortical single units recorded from primates. While not doubting this conclusion, less precise control may be sufficient for some prosthetic applications. An example of less precision is found in local field potential recordings (LFPs), which are simply an aggregate of single-unit recordings. These LFPs may prove very useful as prosthetic controllers [1, 7–9]. Furthermore, this chapter concludes with the surmise that not only precision, but plasticity also, may be the unique (and a very necessary) feature available from single-unit recordings and not from LFPs. Yet LFPs may provide some degree of prosthetic control that, although less precise, may be useful.

14.2 Electrodes

First, let's look more closely at these three electrode categories and discuss the pros and cons. Conflict of interest statement: Yes, it is true that this author is the developer of the neurotrophic electrode (NE). Nevertheless, this author will try to be as impartial as possible and assess the facts as published and known to him. Table 14.1 summarizes electrode similarities and differences. [The author apologizes in advance if any worker in this field is under-represented.]

The first three investigators use tine type of electrodes that are either the 100-pin array (devised by Dick Norman), and used by Donoghue and colleagues [10], or the Michigan probe used by Schwartz et al. [11]. Andersen and colleagues use their version of the array [1]. Nicolelis et al. use microwires that they have devised [2]. Let us take these characteristics point by point. Kennedy is the only one to use the neurotrophic electrode (NE) thus far [3–5].

14.2.1 Longevity

Longevity is of prime importance in a chronic electrode that will be implanted in a young adult human for a lifetime that can extend beyond fifty years. Clearly, the NE is ahead in this respect so far [5], although recent reports from Donoghue's lab show a few signals enduring for years [10]. In all animal and human subjects, the signals from the NE continued until the preparation was destroyed or the subject died. The NE has endured in two humans for more than four years [Kennedy et al., manuscript in preparation].

14.2.2 Stability

The *stability* of the signal is also of great importance. This is difficult to assess over long time periods, especially if the subject is totally paralyzed — because in that case there is no behavior or EMG (electromyographic) activity available for correlation. Stability has been shown in the NE [3–6] and to some extent in Donoghue's recordings [10]. One can argue, however, that stable single units may not be of great importance now with the advent of LFP recordings. Because the LFP signals are inherently more stable, the loss of a few single units may not matter to the overall quality of the LFP. Nevertheless, there is as yet no firm evidence of that, so stability still remains important.

14.2.3 Plasticity

Plasticity is an important feature of single units and one that is probably not available from LFPs. It ought to be easier to train one unit than an unruly classroom of poorly correlated units. The expectation is that perhaps any unit recorded from anywhere in the brain can be trained to control a specific output. If this plasticity expectation is to be fulfilled, however, there must be *a priori* stability of the unit. Without stability and longevity, repeated sessions of retraining would be required. Thus far, the NE is unique in achieving the goals of plasticity, stability, and longevity [4, 5, manuscript in preparation].

14.2.4 Directionality

Directionality has been tested by all investigators and found to be present. Schwartz and colleagues are the only ones to have shown directionality in three dimensions using a virtual reality environment [11]. In these trials, the cursor was driven under the influence of the single units into all eight corners of the virtual cube space. There does not seem to be any overwhelming reason why other electrode configurations should be deficient in this task. Directionality with the NE was achieved without resort to the firing rate, but was deduced from the initial direction of depolarization of the action potentials [6].

14.2.5 Force

Force has been tested by the Nicolelis group, who found that force and direction can be controlled by the monkey [2]. This feature has not been tested by others. Again, there is no overwhelming reason why force relationships could not be found with other electrode configurations.

14.2.6 Local Field Potentials

Local field potentials have been studied by Andersen et al, [1], Kennedy et al, [7, 8], and Leuthardt [9] and found to be useful. Andersen's group found that LFPs indicate the cognitive state of the monkey and has also indicated directionality. The Kennedy group also found directionality within the LFPs [5] and, in addition, used them to control a cursor and a cyber hand on the computer monitor. The subject was able to flex the cyber digits under control of the LFPs with reasonable speed [5]. Again, there is no overwhelming reason why other electrode configurations should be deficient in recording LFPs. In fact, it may minimize the impact of unit instability inherent in other electrodes and thus improve their functional longevity.

14.2.7 Units per Contact and per Electrode

Units per contact and per electrode have been presented by all authors. Tines and wires usually have only a few units, but have many tines or wires, thus providing a large number of units overall. This makes processing of the signal outputs per electrode relatively straightforward, but implies that a large number of electrodes are needed to provide many signals. On the other hand, the NE has five to ten signals per electrode and up to nineteen signals in one subject [manuscript in preparation], which means that fewer electrodes are required. Processing of these signals using spike sorting technologies is complicated but very achievable with today's systems [10, manuscript in preparation]. The NE advantage is that fewer electrodes, and hence implantable amplifiers, are needed.

14.2.8   Relatedness to EMG Activity

Relatedness to electromyographic (EMG) activity has been studied by the Kennedy group in one almost-paralyzed subject. They found that EMG was related to movement onsets, and poorly related to single-unit recording, although fairly well related to LFP onsets [manuscript in preparation]. Although interesting, these results are not essential to the success of any of the electrodes. After all, the subjects to be implanted will be paralyzed *a priori.* Furthermore, there is no overwhelming reason why other electrode configurations should be deficient in this task.

14.2.9   Stimulation

Stimulation of the underlying cortex can be achieved by all electrodes except the NE. The NE is not designed for stimulation. However, its design constraints do not preclude it from being used as a stimulation electrode. Tests cannot be carried out in the human for technical reasons (no implantable stimulating electronics, subjects are paralyzed) and for ethical reasons (implants are allowed so as to provide communication with the external world). In animal studies, stimulation was attempted and no response, such as limb movement, was observed. It would have been surprising if a response was observed because the NE contains a limited number of axons within its tip (see below), and stimulation with conventional electrodes affects a large number of neurons (with or without passing fibers) to produce a measurable response. If it can never be shown to produce a response with stimulation, then it has the disadvantage of being used only for recording.

14.3   Neurotrophic Electrode in Humans

Now take a closer look at the evidence for some functional advantages of the neurotrophic electrode in humans. Disadvantages are enumerated and discussed at the end of this section. Published data have shown the following:

1. Subjects drove a cursor in two dimensions for communication using single action potentials and LFPs [5, 6].
2. Plasticity was demonstrated for single APs, but not attempted with LFPs [5, 6].
3. Directionality was detected in single APs [5, 6] and LFPs [7].
4. LFPs were used to drive simulated cyber digits [7].
5. Binary switch control was obtained from single APs [4] and LFPs [8].

14.3.1   Brief Description of the Neurotrophic Electrode

First, one should understand the unusual configuration of the NE. Figure 14.1 illustrates the salient features of the NE. The glass tip is 1 to 2 mm in length, with a 50-μm diameter at the lower end where the neurites enter. The upper end is about 300 μm wide to allow at least two wires to enter. They are held in place by methylmethacrylate glue. The wire ends are usually 500 μm from each other and from each end of the glass. The amplifier is connected to the wires. Because the ingrown neurites become myelinated, the neural signals recorded must be considered as action potentials (see histology discussion below). The action potentials shown have opposite initial depolarization directions. This is because the amplifier has fixed polarity with positive and negative wires. Thus, axons close to one wire will have initial action potential depolarizations opposite in direction to the action potential depolarizations recorded at the other wire. This has important implications for directionality, as discussed later.

A brief description of implantation techniques is required. Prior to surgery, localization of an active cortical site is essential and this is achieved using functional MRI. An example of active areas is shown in red in Figure 14.2. The subject imagines movement of digits, for example; and the scanner detects this as blood flow changes. We describe this in detail in a future publication [manuscript in preparation]. All subjects undergo fMRI prior to implantation.

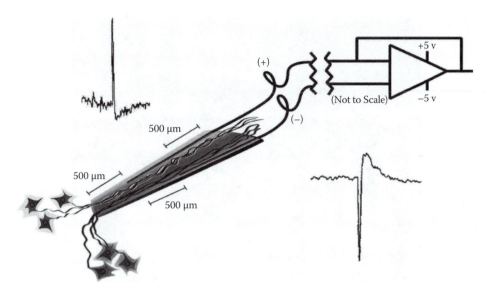

FIGURE 14.1 (**See color insert following page 15-4**). Neurotrophic electrode configuration. (See text for details.)

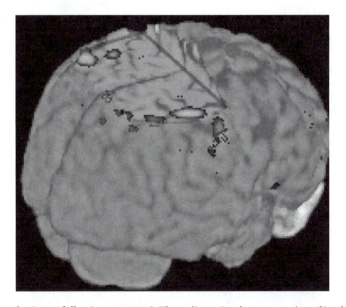

FIGURE 14.2 (**See color insert following page 15-4**). Three-dimensional reconstruction of implantation target site.

The implantation site is chosen based on stereotaxic three-dimensional guidance using a system such as Stealth™ (Medtronic Inc. Minneapolis, MN). Figure 14.3 shows the white pointer wand that is registered in three dimensions with the computer that contains the subject's MRI. The pointer indicates the active area by moving it over the active area of the fMRI.

Histological processing has shown that there are myelinated neurites inside the cone tip of the electrode, as shown in the electron microscopic image in Figure 14.4. The tissue contains normal neuropil except for the lack of neurons [4]. There are myelinated neurites, axo-dendritic synapses, blood vessels, and dendroglial cells, but no microglial scavenger cells, no gliosis, and no neurons. Our interpretation is that the neurons sprouted neurites that grew into the cone tip and became myelinated. Thus, the NE records action potentials from axons; and because there are usually many axons close together, we appear to be recording compound action potentials when the firing rates are high. Stable [1]

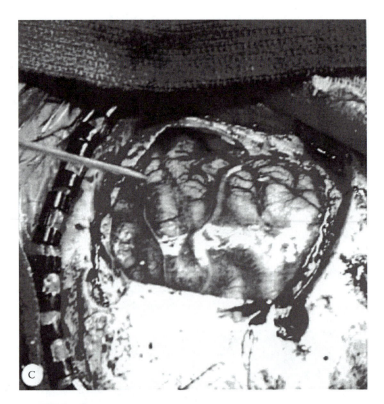

FIGURE 14.3   (See color insert following page 15-4). Surgical implantation site.

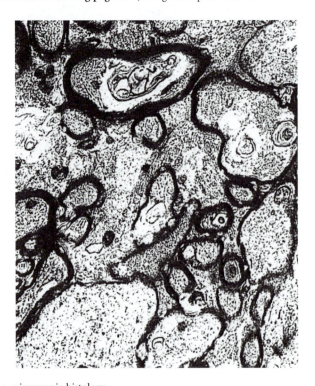

FIGURE 14.4   Electron microscopic histology.

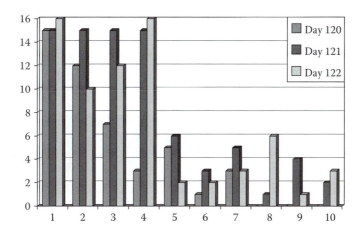

FIGURE 14.5 Learning curve for cursor control. (See text for details.)

and long-lasting [2] signals have persisted in two subjects for more than four years, when the subjects died from their underlying diseases.

14.3.2 Functionality of the Action Potentials Recorded from the NE

14.3.2.1 Cursor Control

Subject JR was the world's first cyborg because he was the first to control a computer directly from his brain. Our first subject (MH) only provided binary signals and did not demonstrate control of a computer. Subject JR was implanted on March 24, 1998, and by midsummer was controlling the cursor [4]. We thresholded his signals and separated them into large- and small-amplitude action potentials (APs). One set drove the cursor in the horizontal direction, and the large units drove the cursor in the vertical direction. The firing rate (or the AP) was directly proportional to the cursor movement above a user-determined threshold firing. Gain of firing rate to cursor velocity was also user determined but held fixed during trials. Results of testing on days 120, 121, and 122 after implantation demonstrated that within five trials, he could control the cursor as shown in Figure 14.5. The cursor was placed at the top left of the screen and he had to move it across and down the screen as quickly as possible, thus forcing him to fire the large units that drove the cursor vertically downward.

When asked to drive the cursor to a particular icon about halfway across the screen, the subject succeeded quite well, as shown in Figure 14.6 for day 243. Target 4 (on the ordinate) was the requested target, which he hit repeatedly after initial inaccurate hitting of target 3 for five trials. Gaps between bars indicate rest periods. Thus, subject JR demonstrated cursor control in two dimensions even before it was demonstrated in monkeys [5].

- *Plasticity.* In subject JR, we implanted area 4, hand representation, as determined by the functional MRI. We realized that he was moving facial muscles, specifically eyebrow movements, to produce neural activations. We preferred, of course, that he use neural activity that was not related to face or other residual movements. We asked JR not to use face movements of any kind during cursor driving. He appeared to comply with this request. To ensure that he did not move, we placed electrodes over his eyebrows to measure electromyographic (EMG) activity. This activity would have driven the cursor in the vertical direction. This would have upset his performance in a task that required him to move horizontally to hit icons. Thus, he had to maintain relaxation of his eyebrow muscles. The neural activity that drove the cursor is shown in Figure 14.7 with the target icon entry point on the right above, and the EMG activity shown below over a 10-s timebase. Note the neural bursts that are not accompanied by EMG activity.

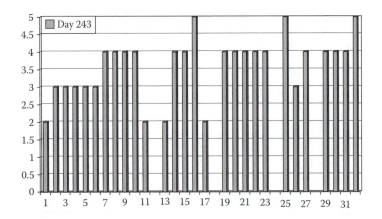

FIGURE 14.6 Learning curve for two-dimensional cursor control. (See text for details.)

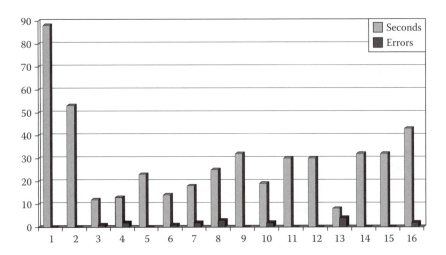

FIGURE 14.7 Neural signal activity during cursor movement with no EMG phasic activity.

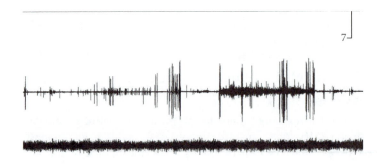

FIGURE 14.8 Learning curve while using neural signals alone.

His performance during this task is shown in Figure 14.8. When he tried to perform too quickly (less than 20 s), he produced errors, as shown by the gray bars. Trial 5 was error-free with a time of 22 s, whereas trials 3 and 4 performed in 11 or 12 s produced errors. Thus, he could perform without activating his face muscles.

When asked what he was thinking during these trials, he spelled out "n-o-t-h-i-n-g" [5]. However, the next day he admitted he was thinking of the cursor. He was focused on the cursor alone.

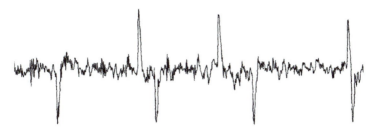

FIGURE 14.9 Action potentials.

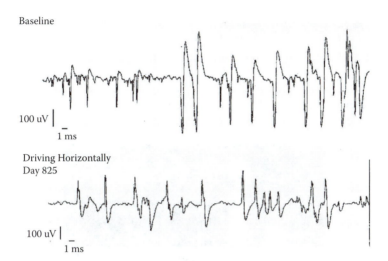

FIGURE 14.10 Directionality detected by phase relationships.

He was driving the cursor simply by thinking about it. Thus, we concluded that what had once been hand-related cortex was now cursor-related cortex. This is the first demonstration of plasticity in human cortical recording.

- *Directionality.* In other electrode configurations, directionality is detected by the firing rate of a neuron in a specific direction [11]. With the NE, however, directionality is independent of firing rate. Instead, it is determined by the initial depolarization direction of the action potential, as shown in Figure 14.9. We noted that individual action potentials can be discrete depolarizations in the positive or negative direction, or can appear to be a single biphasic unit that is, in reality, the overlying of positive and negative depolarizing directions.

Deflections in one direction are shown in the upper panel of Figure 14.10 where most action potentials have initial deflections in the negative or downward direction at rest. Subject JR was then requested to think of moving the cursor in the horizontal direction, and all action potentials were thresholded as a group to drive the cursor. Subject JR saw the cursor moving horizontally. The lower panel of Figure 14.10 shows the resulting action potential directions, namely upward (or positive). Later recordings, when all the action potentials were used to drive the cursor vertically downward, resulted in action potential deflections in the opposite direction. Thus, the initial direction of depolarization allows detection of directionality.

Directionality was also detected during LFP recordings, as shown in Figure 14.11. In these recordings, LFPs were recorded separately from each wire inside the cone tip of the electrode. The wires are in the rows with 5 s of recording in each panel. In the left column, the subject JR was moving the cursor horizontally; and in the right column, he was moving it vertically. Note the clearly distinct separation of activity for each wire that depended on the direction of cursor movement.

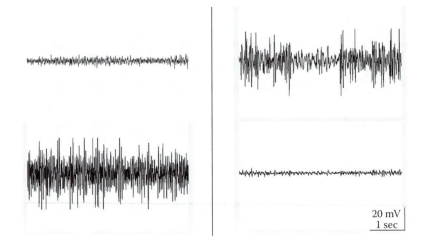

20 mV
1 sec

FIGURE 14.11 Directionality detected in LFPs. Neural signals from subject directing cursor movement horizontally (left) and vertically (right).

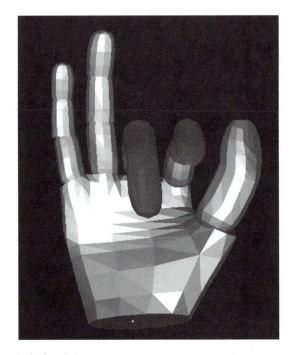

FIGURE 14.12 LFP control of cyber digit movements.

- *Force.* Clearly, we cannot test force relationships in paralyzed people. Force relationships were not tested in animals.
- *Local field potentials (LFPs).* Intracortical LFPs were tested in two subjects (JR and TT), as published [7]. Intracortical LFPs have large amplitudes, as can be seen in Figure 14.11. In one subject (JR), a cyber hand was developed with digits that moved with the firing rate as shown in Figure 14.12. The subject received feedback visually, auditorially, and sensorially (with a finger tap) with each AP firing. He was instructed to move a digit on receiving a verbal "go" signal. The latency between the "go" signal and movement onset is plotted in Figure 14.13. After six trials, he usually performed this task in less than 5 s. This performance improvement indicated that LFPs could be used in this crude manner for movement control.

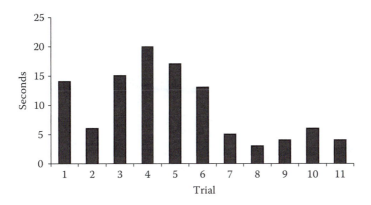

FIGURE 14.13 Learning curve for LFPs.

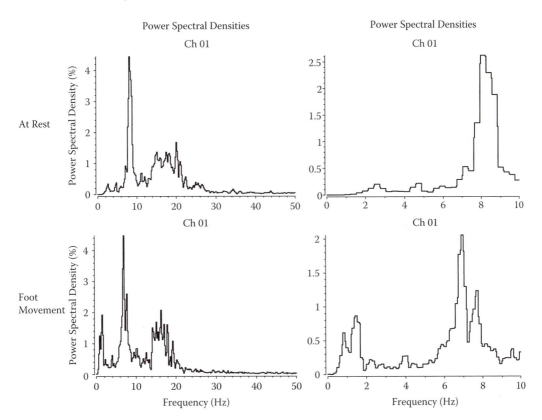

FIGURE 14.14 Differences in LFP spectrum during movement and rest.

Extracortical LFPs were recorded in two subjects (RR and GT) [8]. The signal amplitudes were low in this mode and thus detection methods other than voltage thresholds were sought. Thus, we analyzed the data using the frequency domain as shown in Figure 14.14. In this subject, a stainless steel skull screw was implanted over the leg area of motor cortex. At rest, the dominant frequency was the resting 8-Hz (alpha waves) and a 16- to 20-Hz signal. During attempted (and very slight) foot movements, the 8-Hz signal shifted lower and the lowest frequency increased in amplitude near 2 Hz. Changes such as these can provide a binary switch signal. In this subject, this binary output was used to operate a light switch, as recently published [8]. Thus, LFPs can be used as binary outputs and for controlling crude movements.

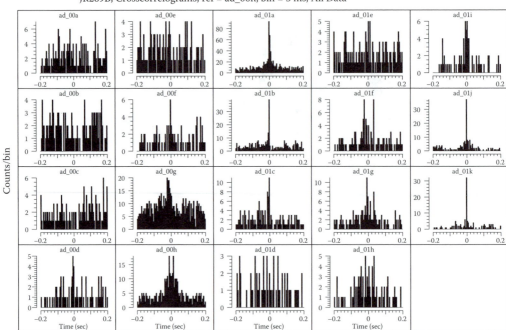

FIGURE 14.15 Cross-correlograms of single action potentials. (See text for details.)

- *Units per contact and per electrode.* One of the misconceptions about the NE is that only one or two can be implanted, and thus only one or two signals can be obtained. It is true that only a few can be implanted, and that is due to the bulk of the implanted electronics. Once the electronics can be further miniaturized, several electrodes can be implanted. Even with one electrode, however, recent offline analysis of human recorded data has revealed as many as nineteen units from one electrode in subject JR, as shown in Figure 14.15. About half are not correlated and thus are expected to be independent channels of communication [manuscript in preparation]. Therefore, it is not unrealistic to expect that each electrode will yield between five and ten useful signals that the patient can control. Such control remains to be tested in subsequent subjects. These large numbers of units per electrode are in sharp contrast to other electrodes where about one, or maybe two units are recorded at each electrode tip. With 100 tines on the Cyberkinetic's probe, this is indeed impressive [10]. In fact, over time, this number drops to 40% or less of tines that continue to record unit activity. Furthermore, each unit varies due to micromovements, and thus the stability of units is questionable.
- *Electromyographic (EMG) relatedness.* In one subject (DJ), we were able to record EMG activity from the forearm as he made minute movements. The EMG was correlated with LFPs and action potentials (APs) recorded from the electrode implanted in the contralateral area 4 motor cortex [manuscript in preparation]. The LFPs were correlated, and the APs had weak correlations, as expected. Recording EMG in near-paralyzed subjects is unique to our project but would be feasible with any of the electrodes.
- *Stimulation.* Attempts at microstimulation in rats through the NE implanted in the leg area did not produce movement. Evoked movements would have been surprising because only a few tens of axons could have been stimulated, and usually an observable leg movement in response to stimulation would be produced in response to a large area of stimulated neurons and passing axons. Increasing the stimulating current was not an option, due to the danger of electrolytically destroying the axons inside the electrode tip. All other electrodes, whether wires or tines, have safely provided stimulation. Thus, for stimulation, the other electrodes are preferred.

14.4   Conclusions

In nine of the ten characteristics discussed above, the NE shines. The one exception is *stimulation*, as discussed above. Thus, it could be the outright winner in this race except that it has some unique problematic characteristics:

1. Accurate histological reconstruction of recorded units is not possible due to the destruction inherent in placing it in the cortex and the trophic changes that take place as the tissue grows into the cone tip. This is not, however, of importance in neural prosthetics where functionality is paramount.
2. Manufacturing the electrode at present is difficult and requires many months of practice.
3. Implantation of the electrode is also difficult and requires much practice by someone with micro-surgical skills.
4. There is a delay of three or four months before the tissue has grown in and signals stabilize.
5. Replacement of the electrode in the exact same area would hardly produce the same signals unless many months passed to allow healing and reconstitution of the tissue.
6. Training time may be prolonged because plasticity changes may be needed to produce useful function.

Thus, for successful usage, skill in the manufacture and implantation is required. The delay in signal acquisition of three months is surely tolerable for someone who requires a lifetime of use. Replacement is hardly going to be needed if it continues recording, as studies strongly suggest. Training time again is hardly a problem when used for the lifetime of the subject.

We will allow readers to judge for themselves as to who the winner will eventually be because this author cannot be considered unbiased. The LFPs discussed above may allow other electrodes to produce useful signals although they may not be able to retain single units over the required lifetime of the subject. LFPs may prove adequate when crude control is needed, but will hardly prove adequate for precise control. Only time and effort will tell which electrode and recording technique will be instrumental in providing cortical control of prosthetic devices.

Acknowledgments

Support from NIH, NINDS, Neural Prostheses Program, Grant No. 2 R44 NS36913-02, is gratefully acknowledged. Support from Neural Signals Inc. internal funds is also acknowledged.

Financial Disclosure: The author may derive some financial gain from the sale of the Neurotrophic Electrode, U.S. Patent 4,852,573.

References

1. Andersen, R.A. and Buneo, C.A. 2002. Intentional maps in posterior parietal cortex. *Annu. Rev. Neurosci.*, 25:189–220.
2. Carmena, J.M., Lebedev, M.A., Crist, R.E., O'Doherty, J.E., Sci, D.M., Dimitrov, D.F., Patil, P.G., Henriquez, C.S., and Nicolelis, M.A.L. 2003. Learning to control a brain-machine interface for reaching and grasping by primates. *PloS Biol.*, 1(2): E42.
3. Kennedy, P.R., Mirra, S., and Bakay, R.A.E. 1992. The cone electrode: ultrastructural studies following long-term recording. *Neurosci. Lett.*, 142:89–94.
4. Kennedy, P. 1989. A long-term electrode that records from neurites grown onto its recording surface. *J. Neurosci. Methods*, 29:181–193.
5. Kennedy, P.R., Bakay, R.A., Moore, M.M., Adams, K., and Goldwaithe, J. 2000. Direct control of a computer from the human central nervous system. *IEEE Trans. Rehabil. Eng.*, 8(2):198.
6. Kennedy, P.R. and King, B. 2001. Dynamic interplay of neural signals during the emergence of cursor related cursor in a human implanted with the Neurotrophic electrode. In *Neural Prostheses for Restoration of Sensory and Motor Function*, Chapin, J. and Moxon, K., Eds. Boca Raton, FL: CRC Press, chap. 7.

7. Kennedy, P.R., Kirby, M.T., King, B., Mallory, A., Adams, K., and Moore, M.M. 2004. Computer Control Using Human Cortical Local Field Potentials. *IEEE Trans. Neural Syst. Rehab. Eng.*, 12(3):339–344.

8. Kennedy, P., Andreasen, D., Ehirim, P., King, B., Kirby, T., Mao H., and Moore, M.M. 2004. Using human extra-cortical local field potentials to control a switch. *J. Neural Eng.*, 1:63–71. FDA approval number: G960032/S10, Brain to computer interfacing device.

9. Leuthardt, E.C., Schalk, G., Wolpaw, J.R., Ojemann, J.G., and Moran, D.W. 2004. A brain-computer interface using electrocorticographic signals in humans. *J. Neural. Eng.*, 1:63–71.

10. Serruya, M.D., Hatsopoulos, N.G., Paninski, L., Fellows, M.R., and Donoghue, J.P. 2002. Instant neural control of a movement signal. *Nature*, 416:141.

11. Taylor, D.M., Tillery, S.I., and Schwartz, A.B. 2002. Direct cortical control of 3D neuroprosthetic devices. *Science*, 7(296):1829–1832.

12. Plexon Inc., Dallas TX; Neuralynx Inc., Denver, CO.

15

Neuromotor Prosthetics: Design and Future Directions

Mijail D. Serruya,
Sung-Phil Kim, and
John P. Donoghue

15.1 Overview of Neuromotor Prosthetics

Brain–computer interfaces (BCIs) include a broad range of systems that translate recorded brain activity into a signal that provides a coupling of physical sensor to devices to record and interpret brain activity. The purpose of these interfaces can range from diagnostic signals to those that provide a replacement output for lost function, such as voluntary arm or leg movement or speech. Neuromotor prosthetics (NMPs) are a subset of BCIs in which signals are recorded from movement-related areas of the brain that give rise to voluntary movement in healthy, able-bodied adults. NMPs can be considered an alternative and complementary approach to therapies based in molecular biology. In the case of spinal cord injury, for example, molecular approaches might promote the regeneration of corticospinal axons across damaged areas of the cord to synapse onto motor neuron pools below, whereas an NMP would record directly from the primary motor cortex where the cell bodies of those axons are located, decode ensemble activity, and then map the resulting control signals onto electrical stimulators implanted in muscles, thereby bypassing the spinal cord injury. One could envision two such devices working in tandem to promote and restore function.

NMPs comprise three basic components: (1) a sensor to record neural activity, (2) a decoder to map the recorded activity into a kinematic variable that can be used as a control signal, and (3) an output effector, such as a computer cursor, wheelchair, or muscle stimulators, which can restore communication or voluntary movement to a user.

BCIs can be divided into direct and indirect systems. The NMP is a direct system because it relies on signals from the source of movement information in the brain. By contrast, an indirect system uses an unrelated or tangentially related brain signal (such as changes in overall brain electrical activity during relaxation) to substitute for movement commands. This chapter provides an overview of neuromotor prosthetics and gives particular attention to how each component has been designed and implemented both in the Cyberkinetics' BrainGate™ system, currently being assessed in a multicenter FDA-regulated investigational device exemption pilot clinical trial, and in prototype and algorithm development at Brown University.

15.1.1   Indirect NMPs

Indirect neuromotor prosthetic systems are based on a brain-derived signal that can be used to substitute for the normal "direct" channel of voluntary movement, which emanates from the spiking patterns of large numbers of neurons in motor cortex. This indirect signal is based on ongoing summed electrical current flowing in an area or from time-averaged mass potentials evoked by a triggering stimulus. These signals are called EEG when recorded on the scalp, electrocorticogram (ECoG) when recorded on the cortical surface, and the local field potential (LFP) when recorded intracortically. EEG has had appeal as a signal source because it can be recorded from the scalp using fairly straightforward recording equipment. However, these signals, which are believed to emerge from the sum of the local currents flowing in an area, have a complex relationship to movement-related spiking (Murthy and Fetz, 1996; Donoghue et al., 1998) and to brain state. Thus, the EEG signal may be considered a surrogate for the actual neural signal that encodes movement; it is an arbitrary signal requiring the user to learn an association between a desired action and some externally or mentally generated cue. For example, one might generate an EEG suppression by imagining kicking with the leg after being in a relaxed state. Similarly, evoked potentials (EPs) such as the P300 wave are indirect because they exploit one system — the brain state change to an attended stimulus — to drive another. EEG and EP signals are linked to brain state and their form can be altered by distractors. Scalp EEG electrodes provide a very weak signal that is highly prone to noise, for example, from facial muscle contractions, eye movements, or external electrical fields. A long process of electrode application is required for each session; the process is time consuming and there is concern that repeated application could potentially be a source of skin damage and possible increased risk of infection. Finally, daily application would require a caretaker or technician for those unable to move their arms. The development of noncontact electrodes might resolve some of these issues but these are not currently available. Finally, externally placed electrodes present substantial cosmesis issues, which might not be acceptable to users. Subdurally placed grids, which are used diagnostically prior to epilepsy surgery, greatly enhance the EEG signal strength but the typical grid placement involves a very large craniotomy. Such grids are not permitted as a chronic implant in the United States. Smaller grids are less concerning and are being tested, but they would provide EEG from more limited areas, which might restrict their utility.

Despite these signal acquisition and processing challenges, humans have learned to use various components of the EEG, such as slow cortical potentials, mu or beta rhythms, or event-related potentials such as the P300 as a signaling source. Study participants have learned to move a computer cursor in two dimensions or to indicate discrete choices (Wolpaw and McFarland, 2004; Blankertz et al., 2004), although learning times were long and no patients with tetraplegia have been tested with this device at this time. Indirect NMP systems, while providing a valuable new output channel for paralyzed individuals, retain several shortcomings as a system. The learned and arbitrary mapping of a signal to desired action and limits to timing or flexibility compared to the system it is trying to emulate are further reasons that these systems are considered indirect. EP-based systems require averaging of multiple trials, which slows system response. The global nature of the systems' signal source makes it difficult to see how they could be scaled to emulate the functions of the arms, hands, fingers, and leg at the speed of natural actions, which would be one lofty goal of an NMP. Such systems could provide a useful control signal for those lacking any communication alternative, although there are no systems available commercially for patients at the present time.

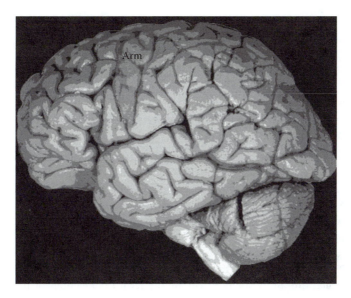

FIGURE 15.1 (See color insert following page 15-4). The arm representation in primary motor cortex. The precentral gyrus is indicated in pink; the arm representation is shown midway in the gyrus.

15.1.2 Direct NMPs

Direct NMPs extract signals directly from ensembles of neurons that are engaged during the function that the system is attempting to recreate. By recording directly from the actual substrate for voluntary movement, intracortical NMPs may ultimately promise to restore to a user the widest range of high degree-of-freedom control, and to do so in a manner as effortless, and not mutually exclusive of other activities, as movement is in an able-bodied person.

Direct NMPs are based on a sensor designed to capture the spiking activity for many neurons engaged in movement. The only method to record extracellular spiking is via a microelectrode; recording an assembly of neurons requires a multielectrode array. The necessity that the recording surface be very near neurons requires that it be placed into the brain, and thus involves a surgical procedure. Signal processing demands involve sophisticated signal detection and processing features. Further, for the sensor to obtain a successful signal, electrodes must remain reliable for long periods — the sensor must be immune to degradation by the body and biological responses to the implant must not limit recording lifetime (Suner et al., 2005). Further, there must be a way to locate the cortical motor areas of interest reliably and in a repeatable manner. The system must have the potential to be nonobtrusive and automated. Finally, the signals must be rich and sufficiently controllable to provide device control that can restore independence through the control of useful devices that are meaningful for those unable to move. Below we discuss each of the issues of sensor, decoder, and user interface, and comment on recent advances in humans from our own human pilot studies that have provided evidence for the feasibility of a direct interface in humans with tetraplegia.

15.2 Sensor and Site

15.2.1 Implantation Sites

Movement-related neuronal activity is present in a large number of cortical areas. In the skeletomotor system, most knowledge has derived from studies of arm control in nonhuman primates. Of these many cortical arm areas, the primary motor cortex (MI) is thought to be the most directly related to the spinal motor apparatus and is thus a prime target to derive motor signals (Figure 15.1). We know already that neurons in MI fire in a way that suggests that they participate in the initiation of movement and in directing arm actions. Areas outside the MI, nonprimary motor areas, carry some of the same information but also appear to have a role in planning actions that may or may not be performed. These may also be useful sites to obtain human NMP control signals, as suggested by recent studies in able-bodied monkeys (Santhanam et al., 2006). Localizing the human MI arm area has a special advantage in that it has a distinct twist in the precentral gyrus, termed a "knob," that is a gross landmark for the arm MI area (Yousry et al., 1997).

15.2.2 Sensor Design

All neuromotor prosthetic devices begin with a sensor that is chronically implanted into the central nervous system and has a means to deliver the recorded signals outside the body. The multielectrode array currently used in the BrainGate™ system is tethered to external processors through a percutaneous port and a cable. A wireless telemetry system is ultimately required, at a minimum to remove this bulky external connection. In the early 1990s, the Donoghue laboratory at Brown University began testing the ability of the "Utah" multielectrode array, invented by Richard Normann and colleagues at the University of Utah (Jones et al., 1992), to chronically record the activity of neural populations in monkey motor cortex (Suner, 2005). This silicon-based monolithic unit with up to 100 Pt/PtSi-coated electrodes has been shown to be successful in its ability to chronically record populations of individual neurons in primary motor cortex in monkeys (Maynard et al., 1999; Suner et al., 2005). This Utah-array was eventually sold as an animal research tool by Bionic Technologies, and continues to be sold as the Bionic Array for this purpose by Cyberkinetics Neurotechnology Systems (which was cofounded by two of the authors of this chapter). Over the past fifteen years, the Donoghue Laboratory at Brown University developed the bionic array technology into a chronic interface system; this longitudinal development included more than forty-one Bionic/Utah arrays in eighteen macaque monkeys. It has been shown to continue to record in monkeys as long as it was tested (more than one year; Suner et al., 2005). On the basis of this data, the technology was modified and adapted for an FDA-approved human pilot trial with successful multineuron recording demonstrated in humans.

Other electrode arrays are in development. One novel design allows for multiple recording conductor ports to populate individual shanks, while the Bionic array has only a single recording site at the tip of each electrode, each of which lies in a single plane. Multiport electrodes enable one to sample a larger number of sites per electrode along the depth of cortex. However, there appears to be a significant challenge in obtaining long-term recording with these side ports. They appear to lose recordings over time; efforts are underway to develop these systems to allow them to record chronically, but these are not yet validated in primates (Otto et al., 2006). The most well-developed of these silicon electrodes was created at the University of Michigan. These probes are fabricated using semiconductor manu-facturing (planar photolithography) techniques, which provide for wide design flexibility (Bai and Wise, 2001; Csicsvari et al., 2003). New polyamide and ceramic electrode arrays, which provide flexibility with a biocompatible material, are currently in development (Moxon, 1999; Rousche et al., 2001). These electrodes can conform to different shaped surfaces but their lack of stiffness can make insertion difficult. Trophic factors are being developed to attract neurites into the electrodes in an effort to integrate brain and sensor into a stable, long-lasting unit. By integrating neural tissue into the recording device, it is possible that such sensors could provide improved or longer-lasting signals; however, much more preclinical investigation is required before human application, especially because they involve the growth of tissues that have not been proven to produce better, long-term recording interfaces. Furthermore, neurite attachment may be undesirable if one wants to be able to remove and replace sensors, as has been successfully achieved in a human (Hochberg et al., 2006) as well as in many monkeys in our laboratory.

The "cone electrode" developed by Kennedy et al. (1989; Kennedy and Bakay, 1998; Neural Signals, Inc.) is not an array, but is noteworthy because it has been used in humans as a way to obtain neural spiking signals. This electrode, which consists of a glass pipette tip-electrode containing growth factors,

FIGURE 1.1

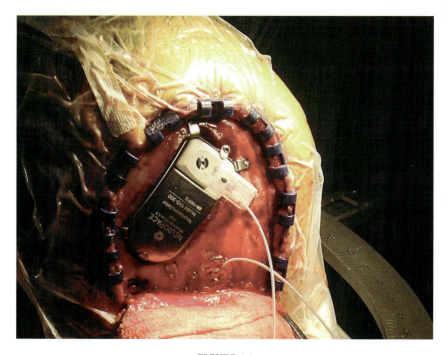

FIGURE 4.6

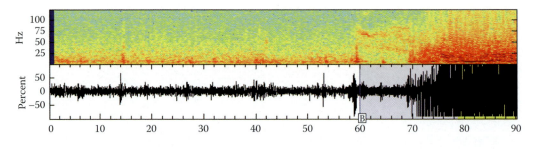

FIGURE 5.3

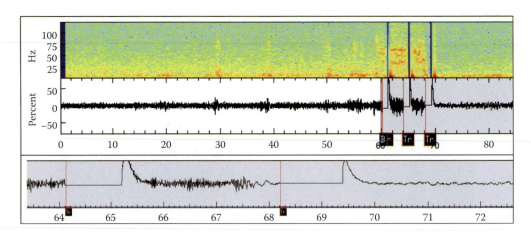

FIGURE 5.4

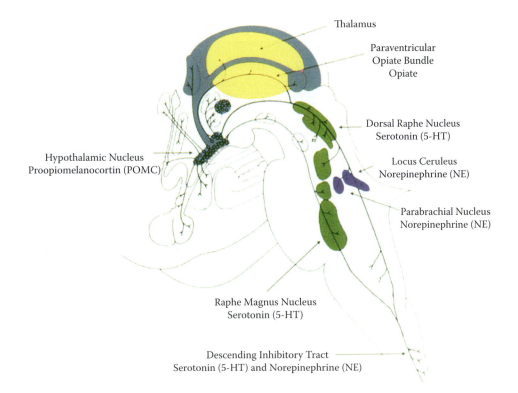

Thalamus

Paraventricular
Opiate Bundle
Opiate

Dorsal Raphe Nucleus
Serotonin (5-HT)

Locus Ceruleus
Norepinephrine (NE)

Hypothalamic Nucleus
Proopiomelanocortin (POMC)

Parabrachial Nucleus
Norepinephrine (NE)

Raphe Magnus Nucleus
Serotonin (5-HT)

Descending Inhibitory Tract
Serotonin (5-HT) and Norepinephrine (NE)

FIGURE 6.3

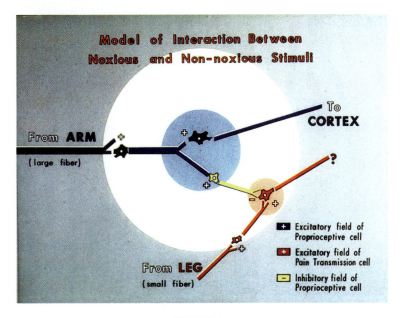

FIGURE 6.4

FIGURE 7.2a

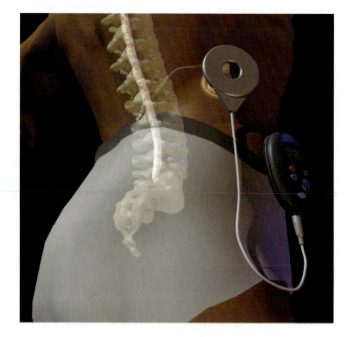

FIGURE 7.2b

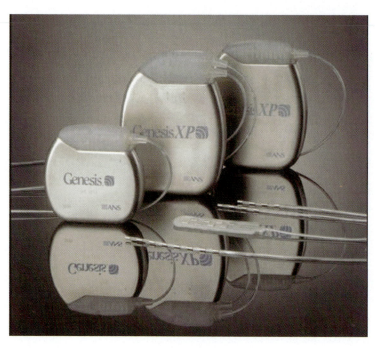

FIGURE 7.2c

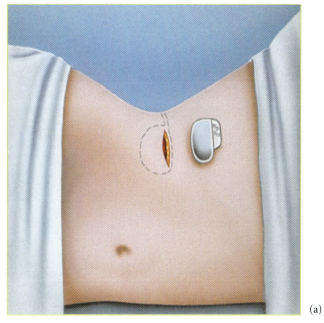

(a)

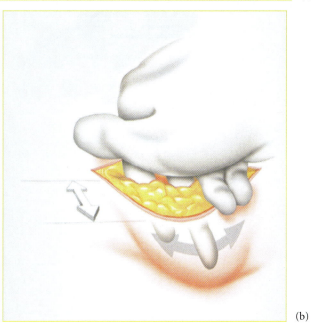

(b)

FIGURE 7.5

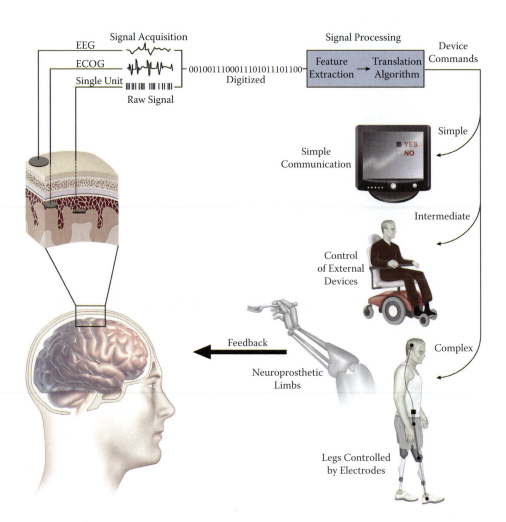

FIGURE 12.1

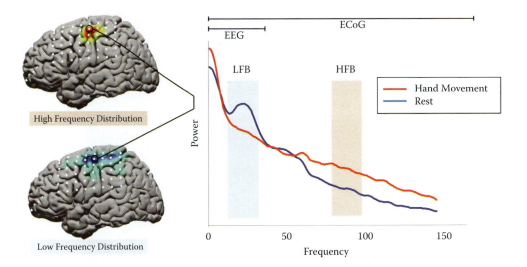

High Frequency Distribution

Low Frequency Distribution

FIGURE 12.4

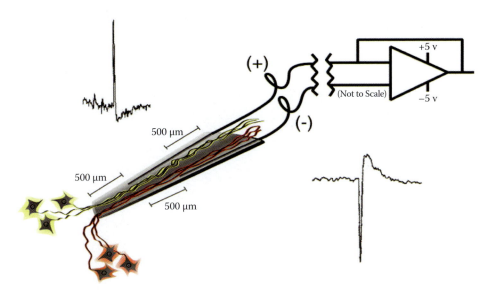

FIGURE 14.1

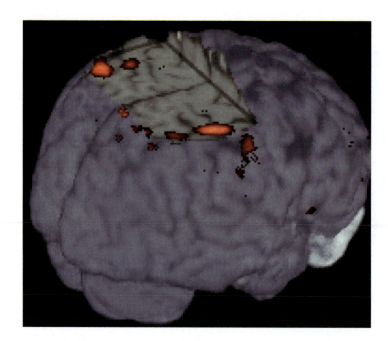

FIGURE 14.2

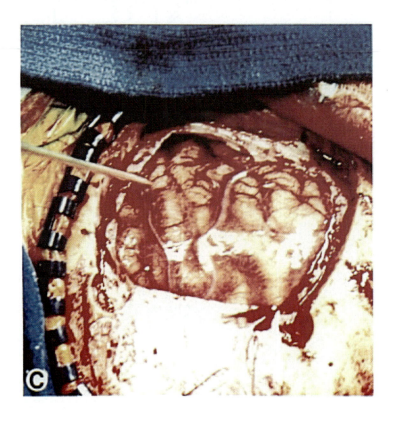

FIGURE 14.3

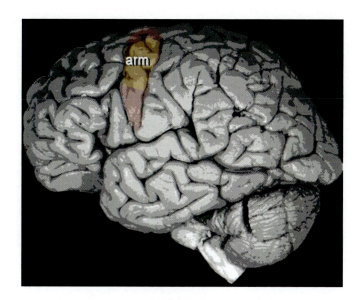

FIGURE 15.1

FIGURE 15.2

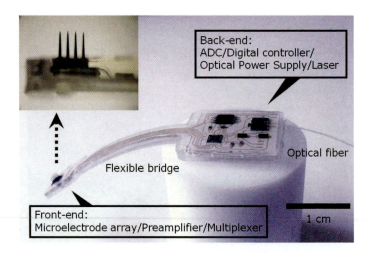

FIGURE 15.4

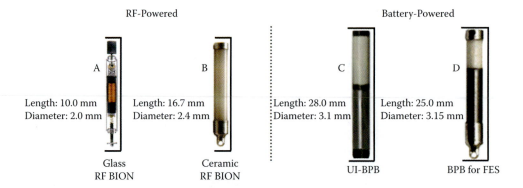

RF-Powered

Battery-Powered

A

Length: 10.0 mm
Diameter: 2.0 mm

Glass
RF BION

B

Length: 16.7 mm
Diameter: 2.4 mm

Ceramic
RF BION

C

Length: 28.0 mm
Diameter: 3.1 mm

UI-BPB

D

Length: 25.0 mm
Diameter: 3.15 mm

BPB for FES

FIGURE 18.1

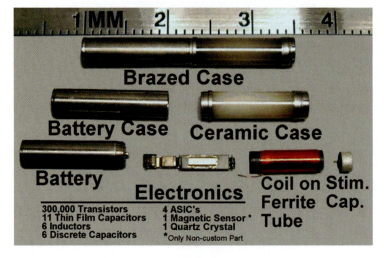

FIGURE 18.4

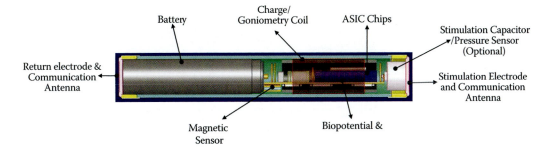

FIGURE 18.5

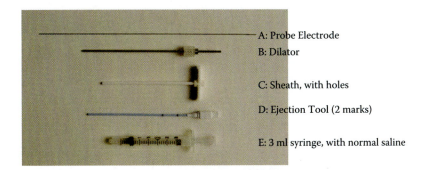

FIGURE 18.10

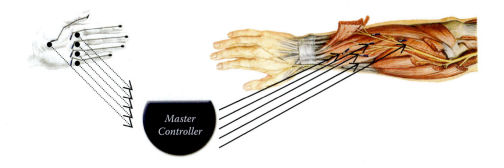

FIGURE 18.13

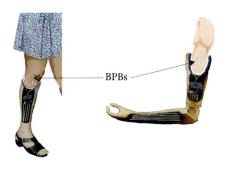

FIGURE 18.14

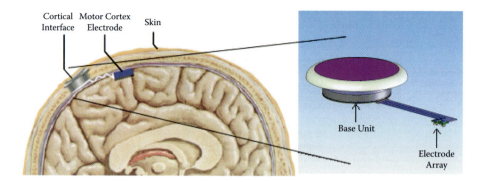

FIGURE 18.15

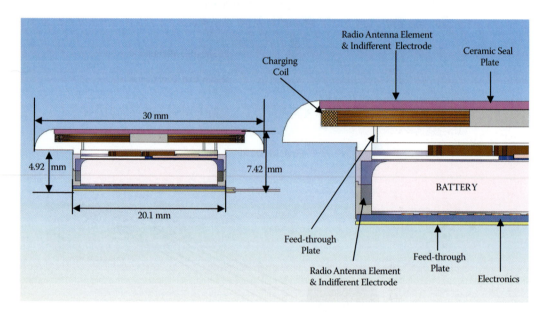

FIGURE 18.16

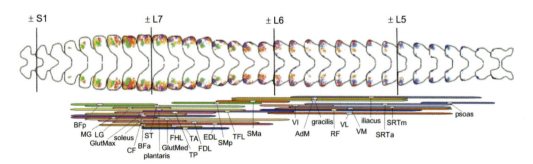

FIGURE 19.1

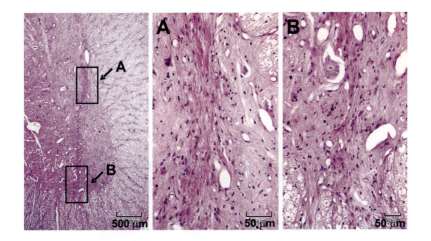

FIGURE 19.3

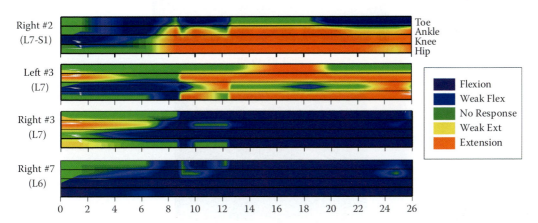

Right #2
(L7-S1)

Toe
Ankle
Knee
Hip

Left #3
(L7)

Right #3
(L7)

Right #7
(L6)

0   2   4   6   8   10   12   14   16   18   20   22   24   26

	Flexion
	Weak Flex
	No Response
	Weak Ext
	Extension

FIGURE 19.5

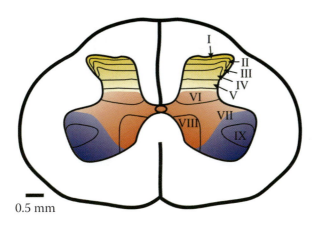

I
II
III
IV
V
VI
VII
VIII
IX

0.5 mm

FIGURE 19.7

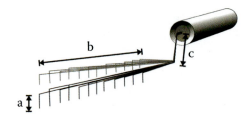

FIGURE 19.9A

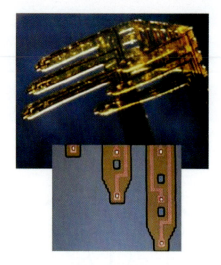

FIGURE 19.9B

FIGURE 19.12B

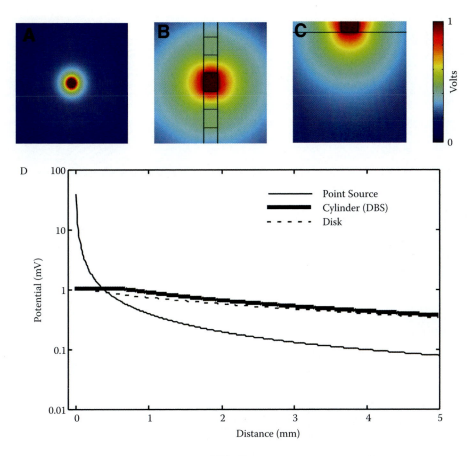

FIGURE 20.2

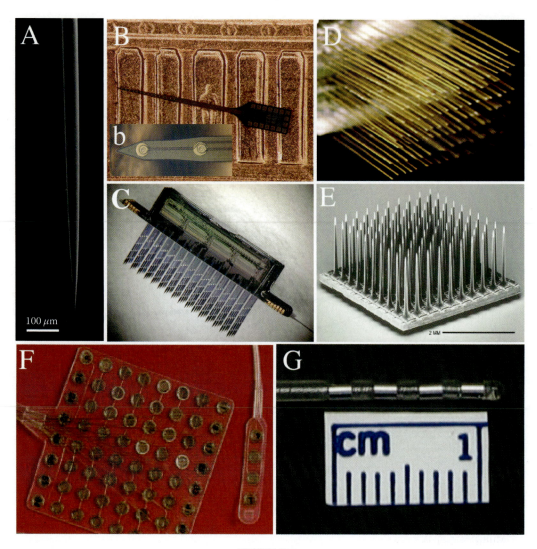

FIGURE 20.3

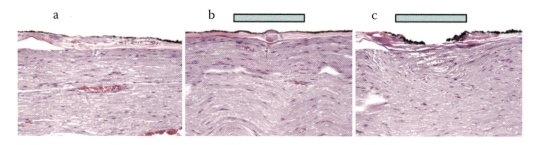

FIGURE 21.5

FIGURE 15.2 The BrainGate™ sensor.

has been inserted into cortex of paralyzed patients. Cone electrodes are capable of providing neural signals for prolonged periods by promoting the growth of neuritic processes from surrounding parenchyma into the interior of the glass cone. However, it is not known what types of neuronal processes grow into a cone electrode — sprouted fibers could arise from any of the inputs or local collaterals of neurons in an area. Standard extracellular recording sensors are strongly biased toward pyramidal neurons, which are a class of very well characterized projection neurons in cortex with specific movement-related properties that have been extensively defined in able-bodied monkeys. Population recordings would require a very large number of individual cone insertions, which has not been attempted. Furthermore, integration of electrodes in a large area of cortex may by undesirable if the ability to remove these sensors for medical reasons or to upgrade technology is an important design feature. An estimate derived from the surgical literature is that roughly 5% of all implanted devices, which now include more than 30,000 implant procedures (largely deep brain stimulating electrodes; Lyons et al., 2004; Beric et al., 2001), fail in some way (e.g., infection, physical breakage) and require removal. However, smaller and more superficial cortical electrodes could potentially be less prone to these issues.

Multielectrode sensors presently require complex, cumbersome cabling and a significant amount of physical wiring to transmit the recorded neural activity to an extracranial amplifier and signal processor. Amplifiers are mounted on headstages that are screwed to the skull and connected by a cable to electronics for capture and storage. Recent technological advances are making it possible to miniaturize signal processing components and integrate them into an implant that can be contained within the body.

15.2.2.1 The BrainGate™ Sensor

The BrainGate™ system is based on the Bionic sensor component of the system. We discuss the design of this sensor in greater detail because it is currently part of the only direct neural interface in human pilot trials, and it has been FDA approved for short-duration use in human investigations, such as for epilepsy monitoring. The sensor comprises a 10×10 grid of microelectrodes, with a 4×4-mm footprint and a 400-μm inter-electrode spacing (Figure 15.2). Electrodes are uniformly the same length, typically 1 or 1.5 mm. The remainder of the implant consists of a cable from the array to a percutaneous Ti pedestal, similar to ones used in other early-stage neural interfaces (Downing et al., 1997).

In the first Investigation Device Exemption (IDE) pilot clinical trial, the array is attached to a connector pedestal that penetrates the skin. When not in use, a protective plastic cap is used to cover the connector. When in use, the connector is attached to a series of amplifiers and external computers that process the signals and decode them according to mathematical algorithms.

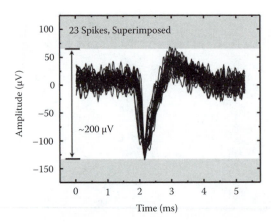

FIGURE 15.3 Example action potential waveforms recorded in a rat brain (acute *in vivo*) by the ultra-low-power analog CMOS chip-scale amplifier prototype.

15.2.2.2 The Brown University Brain-Implantable Chip

A significant step in the development of neuromotor prosthetics is to create a fully implantable, portable system. Numerous groups are developing methods to miniaturize electronics to render multielectrode recording systems fully implantable (Guillory and Normann, 1999; Kipke et al., 2003; Obeid et al., 2003). There are no active multichannel primate brain interfaces currently available, and there are many challenges to creating a system beyond miniaturizing electronics, including power and heat dissipation. To our knowledge, the first successful effort to integrate amplifiers with a sensor probe was accomplished by Najafi and Wise in 1986 (see Najafi and Wise, 2004), and this effort has continued to the present (Olsson et al., 2002; Kipke et al., 2003). This system is based on silicon blades with side contacts built using anisotropic etching, with CMOS (complementary metal oxide semiconductor) circuitry at one end of the assembly. Several sites along each blade are multiplexed into a smaller set of amplifiers, which in turn multiplex to a single output. They report noise levels down to 11 µV root-mean-square referred-to-input with preamplifier power levels down to 100 µW. These have not been evaluated in primates.

Recently, Nurmikko and collaborators have been developing active electronics that integrate with the Bionic array to move toward a fully implantable system (Serruya et al., 2004; Hochberg et al., 2005). The choice of array was made based on the established track record of the sensor in nonhuman and human pilot studies. The plan for this system is to create a fully implantable multielectrode sensor that captures, amplifies, digitizes, and then wirelessly transmits up to one hundred channels at a sampling rate of 30 kHz. Prototype systems have been developed and have demonstrated the successful integration and operation of a chip-scale unit. The system integrates an ultra-low-power, high-performance silicon microelectronic integrated circuit onto the array (Patterson et al., 2004; Song et al., 2005). Further, prototypes of this system at a sixteen-channel scale have been tested in benchtop saline tests, *in vitro* test beds based on activity measurements from acutely prepared rat brain slices, and *in vivo* tests based on acute rat brain recordings (Figure 15.3).

Powering and transmissions of signals represent formidable challenges for a fully implantable sensor that must convey very large amounts of data required when sampling ninety-six channels of neuronal activity. We have adopted a unique perspective that exploits the high bandwidth, robustness, and small scale of optical systems. We are developing an optical power system at chip scale that can fully power the active components in our implant system. In addition, optical transmission carries the signals from the amplifier to other components. This effectively eliminates bandwidth issues. In addition, optical fibers may provide a convenient means to route information from the sensor to remote transmitters, or even to interfaces that control muscle stimulators used to reanimate paralyzed muscles. In this way, the system could become a physical replacement of what cannot be repaired biologically. A model of this prototype of the complete, implantable neural recording system is shown in Figure 15.4.

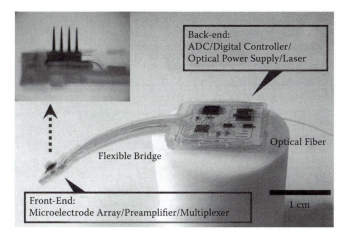

FIGURE 15.4 (See color insert following page 15-4). Model of the fully implantable neural recording system in development at the Nurmikko Laboratory at Brown University.

15.3 Decoders

Once neural activity has been amplified and digitized, algorithms must translate the activity into useful control signals. The decoded signals must be capable of achieving useful control of real-world assistive devices for patients. The decoding can be thought of as a mathematical mapping between the activity patterns of neural populations and movement.

Direct NMPs have grown out of chronic microelectrode recording in animals; hence, clinical applications are based on mathematical models originally designed to predict a variety of movement parameters from the firing rates of ensembles of neurons in motor areas of cortex (Serruya et al., 2003; Wu et al., 2006). Before decoders can be applied to firing rates, action potentials must be extracted from the ongoing electrical signal recorded at a given channel. The sharp, line-like appearance of these action potential waveforms, especially on compressed time scales, has given rise to the term "spike." Trained technicians can use window discriminators to separate spikes from different neurons.

Decoding can be considered a problem of classification or regression of spike activity patterns. Spikes are usually counted in small time bins (~50 ms) across many neurons and are first compared to a desired or real output (e.g., hand trajectory), and then a function is created to relate the pattern of spike counts across the population of cells to that output. Linear filters, maximum likelihood discrete classifiers, and neural networks and other models have all been successful in decoding neural activity into reasonable estimates of hand motion or selection of a switch (see Serruya et al., 2003 for an overview with references). Which algorithms perform best continues to be an area of active investigation.

Several groups have recently demonstrated that monkeys are able to achieve behaviorally useful real-time control of cursors and robotic arms using neural activity alone (Serruya et al., 2002; Taylor et al., 2002; Carmena et al., 2003; Shenoy et al., 2003). The fact that they used a variety of sensors to record the activity, and a variety of mathematical algorithms to decode that activity, may indicate an inherent redundancy and robustness of the neuronal ensemble representation of movement.

15.3.1 Linear Filters

A simple linear model, adapted from Warland et al. (1997), incorporating the activity of between seven and thirty neurons, is sufficient to enable macaques to move a neurally driven cursor to hit targets with a success rate and time-to-target distribution significantly similar to the performance achieved using the hand and a joystick-manipulandum (Serruya et al., 2002, 2003).

In animal studies, decoding algorithms are built by first observing the relationship between hand motion and neural activity. The resulting model then uses this relationship to make predictions about movement from the neural activity alone. However, hand motion is not available in paralyzed patients. Functional MRI studies had shown that paralyzed humans activate the hand area of M1 when they imagine moving (Shoham et al., 2001; Turner et al., 2001). In addition, Kennedy and colleagues demonstrated that a paralyzed patient was able to activate neurons recorded by a single cone electrode by imagining movement (Kennedy and Bakay, 1998; Kennedy et al., 2000). Other groups have demonstrated that kinematic data is not necessary to build certain velocity decoders in animals whose movements were constrained and myographic activity monitored to ensure little or no movements (Taylor et al., 2002).

Preliminary results from the first human patient implanted with the Cyberkinetics' BrainGate™ system, using a variation on the same linear filter model investigated in the animal work at Brown University, indicate that this decoding approach is sufficient to enable a paralyzed human to control a two-dimensional cursor on a range of navigation and target selection tasks (Serruya et al., 2004; Hochberg et al., 2005). The patient was asked to imagine moving his hand in a trajectory indicated by a stimulus cursor. This stimulus cursor position was thus used as a proxy variable for the unavailable actual hand position in order to build the decoding model. To our knowledge, these reports are the first evidence that linear filters can be implemented in paralyzed humans to translate the ensemble activity of several primary motor cortex neurons to generate real-time, behaviorally useful neural cursor control.

15.3.2 New Approaches

Previous work has shown that numerous features of intended movement, including hand trajectory and muscle contraction states, can be decoded from the activity patterns of ensembles of motor cortex neurons (Georgopoulos et al., 1986; Schieber et al., 2002). A statistical, probabilistic approach gives a practical framework to continue the improvement of future algorithms. Decoding models can be categorized into two groups: (1) direct models and (2) generative models. The linear filters belong to the first group, where decoding models estimate the direct input–output mapping from neural activity to movements. The generative models view the decoding problem as inference over time from sparse and uncertain data. This group of models incorporates broader aspects of neural population coding, modeling encoding, and decoding of movements through neural activity. Hence, the direct models only concern the decoding problem, while the generative models estimate all the parameters.

The literature on control theory and probabilistic inference addresses this problem and provides such a framework, which can be presented as follows. At a given instant in time, we seek an estimate of the intended movement given the history of neural activity up to that time, the history of previous movements, and any contextual information we might know about the task. When we try to reconstruct movement, we use our knowledge about how the system performs naturally to help estimate the desired output. For example, if past predictions of arm motion indicate that the arm is moving smoothly to the left, a decoded sample that predicts the arm should suddenly be at a distant location to the right would be unexpected. Such improbable estimates could be eliminated by gathering more information about what the neuron output specifies. The probabilistic approach supplies a framework to approach such situations. In this case, Bayes' rule inverts what we know to what we predict: given measurements of the probability of an observed firing pattern when we know the hand is moving left (written as p(CurrentFiring|Movement)), we then can predict how a firing pattern predicts a particular movement. The problem can be described as one of modeling the probability of movement in terms of observations and history:

$$p(\text{Movement}|\text{CurrentFiring},\text{FiringHistory},\text{MovementHistory}) =$$

$$\text{constant } p(\text{CurrentFiring}|\text{Movement}) \times p(\text{Movement}|\text{FiringHistory},\text{MovementHistory})$$

where we have exploited Bayes' rule (and a first-order Markov assumption) to rewrite the *a posteriori* probability in terms of two simpler probabilities. The first term, p(CurrentFiring|Movement), signifies the *encoding* model of motor cortical activity. This term quantifies the "likelihood" of observing a particular pattern of neural activity given a particular movement. The second term on the right-hand

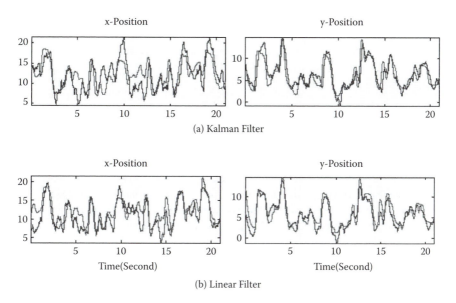

FIGURE 15.5 Examples of reconstructed hand position trajectories by (a) the Kalman filter and (b) the linear filter: actual target motion (red) and reconstruction by the filters (blue). (Source: From Wu et al., 2003.)

side signifies the *a priori* probability of the movement. The "prior" captures the continuity and physical limits of hand motion. This prior also specifies how the history of preceding measurements is incorporated into the current estimate of the movement. We take movement to be described by the position, velocity, and acceleration of the hand. These properties have been shown to be roughly linearly related to the firing rate of M1 cells (see Georgopoulos et al., 1982; Serruya et al., 2003). We next assume that the observed firing rates are distorted by Gaussian noise. This leads to approximation of the likelihood as a simple linear Gaussian model, which is a straightforward technique to model a noise process. We further assume that the hand motion at one time instant is linearly related to the motion at the previous time instant and is likewise distorted by Gaussian noise. When these assumptions are met, the *a posteriori* probability is also linear and Gaussian, and can be optimally estimated using a Kalman filter, which is also a well defined and widely employed method for real-time inference from noisy data. Even with these simplifying assumptions, we have found that this Kalman filter is more accurate than linear filter or population vector methods (Wu et al., 2003a,b, 2005). The Bayesian formulation makes explicit modeling assumptions. Without altering the basic framework, these assumptions can be relaxed and improved in principled ways. For example, we have been developing nonlinear and non-Gaussian encoding models that may capture additional complexity of neural activity (Gao et al., 2003). As NMPs develop, they will be able to leverage decoding algorithms developed from current computer science research in probabilistic modeling and machine learning. We present an illustrative example of decoding outputs by the linear filter and the Kalman filter for two-dimensional target hitting tasks in which monkeys move a feedback cursor to hit a randomly placed target cursor (see Figure 15.5).

In parallel to the Bayesian generative decoding approach, there also have been a number of studies to find better models than the linear filter to directly approximate a transfer function that maps neural activity into movements. The first class of model concerns a nonlinearity in the transfer function. Time-delay neural networks (TDNNs) are a straightforward extension of the linear filter in that, like linear filters, they take into account the firing rate of neurons at different delays relative to a given hand position, yet unlike linear filters they insert what is called a hidden layer between the input and output. This new hidden layer includes a number of nonlinear (usually nonlinear with a sigmoid shape) nodes that process the time-delayed neural inputs to estimate the hand position (Wessberg et al., 2000). Despite the possible advantages of adding this nonlinear feature, these models suffer from having too many degrees of freedom: if we are given 100 neurons with 10 delays, and TDNN has 30 hidden nodes, then the degrees of freedom will exceed 30,000. It is not feasible to learn such a large number of parameters with the finite number of data samples that might be gathered at a typical clinical calibration session. To address this issue, a different nonlinear modeling technique has been employed called recurrent neural networks (RNNs) (Chapin et al., 1999; Sanchez et al., 2005). Like the TDNN, the RNN is a two-layer network but has recurrent connections in its hidden layer. The recurrent structure gives this model special properties: (1) it allows the model to approximate a function with an infinite input history; and (2) by eliminating the time delays in the input layer, it ends up with fewer degrees of freedom. However, training of RNN, which requires more complicated computations than TDNN, may not be feasible for real-time applications. The second class of model takes a *divide-and-conquer* approach in which the transfer function is approximated by a mixture of simple filters. As the simplest case, a special decoder called a discrete state decoder could be used to switch between decoding done by two linear filters, one to decode changing movement and one to decode stationary states, respectively. For example, two linear filters were switched by a type of discrete decoder called the hidden Markov model (HMM) for approximating movement and stationary kinematics (Darmajan et al., 2003), and another alternative of the linear filter called particle filters were switched by a discrete decoder called the linear classifier to decode when the subject was actively moving and when the subject was not active (Wood et al., 2005). This feature of being able to determine when a group of neurons is representing something useful that needs to be decoded to control the device output (such as a cursor) is important not just for neuromotor prosthetics, but perhaps all neurotechnology devices. In devices used to treat epilepsy and Parkinson's disease, for example, a *divide-and-conquer* model could help determine whether neural activity is representing something relevant to predicting disease state (which might be used to drive a stimulator) or an idling state not relevant to device function.

Decoding models that predict not just a movement parameter (such as hand position) or just two states (active vs. inactive), but additional, multiple states have also been proposed for *divide-and-conquer* models. One example is a mixture of multiple linear filters, each specialized in a certain aspect of reaching movements, that are switched by a nonlinear filter (Kim et al., 2003). Another example used multiple versions of a broader type of linear filter called a Kalman filter, which in turn were each selected based on Gaussian statistical patterns found in the data (Wu et al., 2004). Despite this plethora of choices for the *divide-and-conquer* approach, however, there does not yet exist an established way to determine what the states should be (e.g., "about to seize" vs. "not about to seize") and how to switch between them.

The third class of model adopts what is called a kernel method to map a neural firing rate space into some feature space (such as movement parameters for a neuromotor prosthetic, or arousal and other global and local cortical states for seizure prediction). When one is able to define an appropriate mapping to a feature space on which the target parameters can be well regressed in a linear fashion, this approach may be able to decode with a better performance (e.g., Spikernels as described in Shpigelman et al., 2005). Similar to the previous models, however, high computational complexity remains a bottleneck for building and implementing the models in real time for clinical neuroprosthetic devices.

Finally, we would like to remark that there have also been attempts to further enhance the original, humble linear filters. One approach is to regularize the linear filters with various statistical methods that are called subspace projection or penalizing parameters (Kim et al., 2006). Other approaches include the autoregressive moving average (ARMA) model, which incorporates simple dynamics of the hand kinematics into decoding (Fisher et al., 2005).

15.4   Output Devices

NMPs generate output signals to replace those compromised or lost by disease or injury. These output signals must be connected to effectors that restore lost functions. These effectors might be in the form of a computer interface, a switch controller to activate appliances, a robotic limb, or a stimulator to activate the paralyzed muscles themselves. As systems that produce control signals become practical, it is essential to consider the types of technology that will benefit and be acceptable to users. For example, it may be a goal to reanimate muscles in those with spinal cord injury, but a robotic agent may be essential for those with muscular dystrophy or ALS, where muscle use is not feasible. Below we consider computer, robotic, and functional electrical stimulation interfaces.

15.4.1 Computers

Computers are themselves ideal effectors because they are simple for a human to control and yet are highly developed to facilitate a wide range of uses. They are both a device that can carry out assistive functions and a requisite gateway to other assistive technologies. The capabilities of the modern computer are vast. Many able-bodied people spend most of their waking hours communicating, learning, creating, working, and being entertained by pointing, clicking, and typing on a computer. Many movement disorders compromise the ability to use the mouse and keyboard; although several PC-based assistive products have been developed to enable people with such impairments to access the computer using alternative input devices and control methods that reduce or adapt to the movements required to operate a computer. Flexible assistive software is available (e.g., EZKeys, Words+, Lancaster, California; Roll Talk). Such products are often designed around a simple switch or choice, such that the computer automatically scans the cursor over the screen and a switch press is required to make the choice when the cursor moves over the desired target. Although such systems are useful when no other effectors are available, the rates for typing are so slow that even very disabled users may prefer alternative, simpler methods (e.g., alphabet boards) to communicate. Speech recognition software (e.g., Dragon Dictate) offers another method of computer control, both for dictation-based typing and for word-based Web navigation. Single-switch devices, speech recognition, and other options such as eye trackers may all be very useful assistive systems for paralyzed individuals. Still, each of these hand-movement surrogates are ultimately limited in that they substitute another type of action for real, natural, complex hand movement to mouse and keyboard. One hallmark of a very successful interface (like a mouse) will be reflected in its widespread use. This highlights one of the primary goals of an NMP, namely that it must restore communication and movement without co-opting remaining abilities (such as speech); and likewise, they must, as in able-bodied people, be usable simultaneously with those abilities, such as moving a cursor while carrying on a conversation with a person in the room.

Simple interfaces through a computer can be designed to use the cursor to type, Web surf, and drive external functions. Inexpensive, commercially produced "X10" modules can operate through a simple switch interface on the computer to perform such commands as turning on and off the lights, adjusting the room temperature, and performing other actions otherwise lost to individuals with motor impairment. Indeed, the first human user of the BrainGate™ demonstrated the ability to use the cursor to use an X10 system to turn on and off a television, as well as to change channels and the volume (Hochberg et al., 2006).

15.4.2 Robotics

Assistive robots are another class of output devices that could potentially be driven by NMPs and be useful to movement-impaired individuals. Robots can be broadly defined: they include dexterous limbs, mobile transports, exoskeletons, and even autonomous vacuum cleaners. Autonomous complex functions with minimal commands seem to define a class of robots that can aid those with limited movement capabilities. By extension, a wheelchair might be considered a type of assistive robot. An NMP-controlled wheelchair could free "sip and puff" users of the need to use their mouth to control their wheelchairs, thus allowing them to concurrently engage in conversation, eating, or other orofacial actions currently obstructed by this technology. Autonomous mobile robots use various sensors (sonar, infrared, laser, and video) to perform tasks such as obstacle avoidance, path planning, map learning, and navigation through rooms and doorways. Likewise, current robotic arms use various end effectors (grippers or multijoint hands) and exploit visual cues from video camera feeds to perform automatic hand–eye coordination to grasp or manipulate objects. Neural signals could eventually be leveraged to control robots but considerable development and testing will be required to determine how much of the control burden must be assumed by the robot.

Robotic prosthetic limbs comprise another example of an NMP output. Such prosthetics can serve as a training system for paralyzed patients to master multidimensional control or as an actual restoration system for amputees. The first BrainGate™ participant demonstrated the ability to voluntarily open and close a prosthetic hand connected to the computer (Hochberg et al., 2006). Interestingly, although initial control of the prosthetic hand was achieved through cursor movement on the screen (which was in turn mapped to a prosthetic hand switch), the patient reportedly controlled the hand even after the monitor was moved out of visual range, implying the patient's ability to generalize the control signal without detailed visual feedback (Serruya et al., 2004). Thus, direct interaction with robots without a visual, control signal intermediary seems feasible. In addition to the prosthetic hand, the first human user demonstrated the ability to drive a three-degree-of-freedom robotic arm (shoulder, elbow, claw) to grab, move, and deposit an object using a simple cursor–target interface. These simple and early demonstrations suggest that paralyzed humans might readily establish control over a range of complex agents that can restore the ability to interact with the world. Although these initial demonstrations are promising, much more work is necessary to improve the range and accuracy of NMP output signals.

15.4.3   FES

By reconnecting the motor control centers of the brain to skeletal muscles, NMPs could restore voluntary movement to patients paralyzed by spinal cord injury or brainstem stroke. That is, in these conditions, the muscles and their connections to the spinal cord are preserved and available to be reanimated. To achieve this, NMP outputs would be coupled to functional electrical stimulation (FES) systems that provide electrical stimulation to the muscles or their innervations. Fully implantable FES systems being developed at Case Western (Mauritz and Peckham, 1987) and the BION system (Loeb et al., 2001) comprise the two most advanced of these technologies. FES systems are triggered by available, non-paralyzed muscle groups that activate a switch; for example, a patient with spinal cord transection at the C5 level can use shoulder movements or an EMG sensor to trigger an FES system implanted in the forearm, in order to close the hand. A cortical control signal could provide a direct and more natural input to the paralyzed muscles (Lauer et al., 1999). In 2005, Cyberkinetics and investigators at Case Western Reserve University and the Cleveland FES Center received federal funding to develop a neuro-prosthetic system to restore voluntary arm and hand function (Peckham, 2007). Although restoration of complex, voluntary hand and arm movement embody the ultimate goal, intermediate steps, in which neural control would be coupled to simpler, preprogrammed FES routines, could also serve a useful rehabilitation purpose by improving range of movement, vascular flow, and skin integrity and simple volitional actions in otherwise immobile limbs.

15.5   Future Directions

15.5.1   Noninvasive Sensors

As the functional capabilities of neurotechnology develop, there will be a growing drive to create non-invasive sensors. Accessing useful signals without an intracranial implant poses significant challenges. Spikes (or, more correctly, action potentials) form the fundamental basis for information coding in the nervous system; these tiny impulses are simply inaccessible at a distance. Imaging systems, such as MRI or MEG, yield filtered, indirect, or more global signals, and the machines are very large and expensive in their current form. Methods to amplify and record spikes noninvasively, in a manner that retains the spatial and temporal resolution possible with microelectrodes, do not yet exist. One noninvasive strategy would be to deliver molecular tags to a subset of neurons, which in turn could convert neural electrical signals to light of an appropriate intensity and tissue transparent wavelength such that an external optical sensor could detect spiking. Fluorescent semiconductor quantum dots fulfill some of these requirements (Shapiro et al., 2003), but the approach presents formidable challenges to become useful. Nevertheless, if these or other noninvasive approaches to recording spikes ever become feasible, they can leverage the decoder and output device advances already made for the invasive sensors, which will be based on the same fundamental code.

15.5.2 New Applications

Successful NMPs will deliver a new generation of neural interfaces with intelligent components and greater capabilities to sense the brain's activity. Among these applications of such neural sensors might be diagnostic aids in which sensors detect abnormal brain states and report them back to patients or medical professionals. Therapeutic systems could both diagnose and treat disease, especially if neural interfaces could interact bidirectionally. These devices can include the ability to stimulate or to deliver pharmaceuticals. Closed-loop systems to sense and block seizures are already being investigated in pilot clinical trials, although the current systems use only gross brain potentials and macrostimulation. NMPs based on ensemble spiking activity would provide a more detailed description of neural state. Although currently not known, these higher resolution signals could possibly predict the future appearance of a clinical seizure or report changes in brain state in response to therapy after traumatic brain injury. Other applications might include closed-loop systems to detect and treat developing stroke, or treatments for neuropsychiatric conditions such as intractable depression.

15.5.3 Sensory Feedback

The NMPs described above detect and decode signals and map outputs onto effectors such as cursors or FES systems. Although vision is sufficient to guide cursor movement, manipulation of a robotic prosthesis or an anaesthetic limb would likely benefit greatly from some form of haptic feedback to allow users to judge grip force, object contact, or limb position (Chapin, 2000). While external vibrotactile feedback onto sensate skin may serve as a useful development tool, ultimately tactile feedback could be provided by direct electrical stimulation of sensory areas onto peripheral nerves or the brain itself. Electrical stimulation is highly artificial in the way it activates neurons, but may be able to reproduce certain aspects of sensation. Dhillon and Horch (2005) recently demonstrated that stimulation of electrodes implanted into the individual fascicles of peripheral nerve stumps in human amputees was able to evoke graded sensations of touch and movement such that these cues could be used to judge grip force and set joint position in artificial arms. Electrical stimulation has also been used to map eloquent brain areas as part of functional neurosurgery since the time of Penfield, and has been found to produce a variety of tactile percepts, including pressure and vibration, but also dysesthesias (Ohara et al., 2004). Quantitative analyses in monkeys indicate that sensory cortex stimulation can be used about as effectively as actual touch (Romo et al., 1998). Several groups have used microstimulation of primary sensory cortex (S1) in animals as a means of delivering target information (Miller et al., 2005; Oliviera et al., 2005). Although these groups have established that animals can be conditioned to interpret S1 stimulation to guide gross behavioral decision-making (e.g., left-or-right target selection), the systems remain primitive and the percept of S1 microstimulation in the brain by these groups may more akin to the tactile sensation experienced by horse with the bit tugged in its mouth to signal direction rather than a continuous stream of tactile and proprioceptive feedback that accompanies natural action. The potential benefits of restoring paralyzed users' tactile sensation and proprioception ensure that research in this area will likely continue.

15.5.4 Demands on the Medical System

Just as neurotechnology may relieve the medical system by one day rendering paralyzed individuals more independent of caregivers and provide rehabilitation health benefits to prevent hospitalizations, it will likely place new demands on a variety of medical services to sustain and promote this transition toward greater patient independence. If clinical trial data establishes the safety and efficacy of NMPs in patients with severe paralysis due to spinal cord injury, muscular dystrophy, brainstem stroke, and ALS, this could facilitate the application to an even wider range of movement disorder patients such as cerebral palsy, unresolved Guillian–Barre syndrome, and hemiparetic strokes. If demand increases, a commensurate increase in the number of functional neurosurgeons will be necessary.

The advent of DBS (deep brain stimulation) and other FDA-approved neurotechnologies, which now includes tens of thousands of human recipients, is already stimulating growth in this area, and is bringing the experience of neurosurgeons, neurologists, and other health-care providers toward the development of NMPs. There will likely be a need for additional training programs, perhaps as subspecialty fellowships in neurotechnology, to meet the special multidisciplinary demands of these patients. Physicians will be required to track the function of these devices and adjust them when necessary; such functions have already emerged for DBS and vagal nerve stimulators, as well as other stimulators and drug pumps. The unique integration of brain and machine in an NMP, coupled with the ongoing medical needs of patients with movement disorders, will require specially trained physicians and health-care workers. As assistive technology becomes more complex, it may be necessary to train a new generation of rehabilitation therapists and clinicians in computer science, robotics, and programming. NMPs will elicit many important questions that will hopefully be addressed in the coming decades: What will be the optimal form of output signals? Will functional abilities improve over time? How will rehabilitation professionals best teach patients to use NMPs, select the best assistive devices, and put it all together to restore a productive life? In addition to training providers to be skilled in the selection and therapeutic use of NMPs, it is likely that specialized neurotechnology neurorehabilitation clinics will emerge within major academic medical centers (Okun et al., 2005). These translational clinics will serve multiple functions, including training next-generation neurorehabilitation health-care workers, providing a central location for patients to receive customized care, and serve as an interface to neuroscience and engineering collaboration to promote ongoing improvement and development of the devices.

Neurotechnology also poses new questions to clinicians and medical ethicists. The importance of building trusting provider–patient relationships will be paramount as such devices record and stimulate brain areas that may give rise to conscious self-awareness. Questions about how and when behavior could or should be manipulated by such devices will arise. The possibility of enhancing intelligence or modifying deviant behavior raises a host of concerns that will need to be addressed. It is important to note that although NMPs represent a novel technology, precedent exists for grappling with such ethical dilemmas as the use and abuse of psychoactive drugs, psychosurgery, electroconvulsive shock, and even earlier versions of brain implants have been previously debated (Fins, 2003). It is essential to identify ethical issues and develop rational policies now rather than wait for some ignominious application to impede the meaningful uses of neurotechnology.

15.5.5 Conclusions

The arrival of NMPs and other neurotechnologies offer the promise of a new way for physicians to treat their patients. NMPs may provide humans with a variety of movement disorders a way to interact more effectively with the world. These devices will likely retrace the development process of other implantable systems; while first tied by transcutaneous wires to large external systems, their success would promote miniaturization into fully implantable and portable systems. The next-generation systems could be coupled to a growing array of computer functions, assistive robots, and FES systems to restore muscle movement. These miniaturized sensors and stimulators could, in turn, find new applications in the diagnosis and treatment of a wide range of neurological and psychiatric disorders. Such new devices will place demands on the medical system for new skills and carefully measured ethical policies. The opportunity for improved health care and quality of life for many people with disabilities may be great.

Acknowledgments

The authors would like to thank the following for support of work described in this chapter: NIH/ NINDS grant NS25074 and contract N01-NS-2-2345, DARPA BioInfoMicro Program, Keck Foundation, ONR grant NRD-386. We would like to thank Yoon-Kyu Song and Arto Nurmikko for providing Figures 15.2 and 15.3.

Conflict of Interest Disclosure

Mijail Serruya and John Donoghue are co-founders and shareholders in Cyberkinetics Neurotechnology Systems, Inc., a neurotechnology company developing medical devices to restore movement and community to patients with motor impairment.

References

Beric, A., Kelly, P.J., Rezai, A., Sterio, D., Mogilner, A., Zonenshayn, M., and Kopell, B. Complications of deep brain stimulation surgery. *Stereotact. Funct. Neurosurg.*, 77(1–4):73–78, 2001.

Blankertz, B., Muller, K.R., Curio, G., Vaughan, T.M., Schalk, G., Wolpaw, J.R., Schlogl, A., Neuper, C., Pfurtscheller, G., Hinterberger, T., Schroder, M., and Birbaumer, N. The BCI Competition 2003: progress and perspectives in detection and discrimination of EEG single trials. *IEEE Trans. Biomed. Eng.*, 51(6):1044–1051, 2004.

Carmena, J.M., Lebedev, M.A., Crist, R.E., O'Doherty, J.E., Santucci, D.M., Dimitrov, D.F., Patil, P.G., Henriquez, C.S., and Nicolelis, M.A. Learning to control a brain-machine interface for reaching and grasping by primates. *PLoS Biol.*, 1(2):E42, 2003. Epub 2003 Oct. 13.

Donoghue, J.P., Sanes, J.N., Hatsopoulos, N.G., and Gaal, G. Neural discharge and local field potential oscillations in primate motor cortex during voluntary movements. *J. Neurophysiol.*, 79(1):159–173, 1998.

Downing, M., Johansson, U., Carlsson, L., Walliker, J.R., Spraggs, P.D., Dodson, H., Hochmair-Desoyer, I.J., and Albrektsson, T. A bone-anchored percutaneous connector system for neural prosthetic applications. *Ear Nose Throat J.*, 76(5):328–332, 1997.

Fins, J.J. From psychosurgery to neuromodulation and palliation: history's lessons for the ethical conduct and regulation of neuropsychiatric research. *Neurosurg. Clin. N. Am.*, 14(2):303–319, ix–x. 2003.

Guillory, K.S. and Normann, R.A. A 100-channel system for real time detection and storage of extracellular spike waveforms, *J. Neurosci. Meth.*, 91:21–29, 1999.

Hochberg, L.R., Serruya, M.D., Mukand, J., Polykoff, G., Friehs, G.M., and Donoghue, J.P. BrainGate neurmotor prosthesis: nature and use of neural control signals. Society for Neuroscience Abstracts, *35th Annual Meeting of the Society for Neuroscience*, Washington, D.C., October 2005.

Hochberg, L.R., Serruya, M.D., Friehs, G.M., Mukand, J.A., Saleh, M., Caplan, A.H., Branner, A., Chen, D., Penn, R.D., and Donoghue, J.P. Neuronal ensemble control of prosthetic devices by a human with tetraplegia. *Nature*, 442(7099):164–171, 2006.

Jones, K.E., Campbell, P.K., and Normann, R.A. A glass/silicon composite intracortical electrode array. *Ann. Biomed. Eng.*, 20:423–437, 1992.

Kipke, D.R., Vetter, R.J., Williams, J.C., and Hetke, J.F. Silicon-substrate intracortical microelectrode arrays for long-term recording of neuronal spike activity in cerebral cortex. *IEEE Trans. Neural Syst. Rehab. Eng.*, 11:151–155, June 2003.

Lyons, K.E., Wilkinson, S.B., Overman, J., and Pahwa, R. Surgical and hardware complications of subthalamic stimulation: a series of 160 procedures. *Neurology*, 63(4):612–616, 2004.

Mauritz, K.H. and Peckham, H.P. Restoration of grasping functions in quadriplegic patients by functional electrical stimulation (FES). *Int. J. Rehabil. Res.* 10(4 Suppl. 5) 57–61, 1987.

Maynard, E.M., Hatsopoulos, N.G., Ojakangas, C.L., Acuna, B.D., Sanes, J.N., Normann, R.A., and Donoghue, J.P. Neuronal interactions improve cortical population coding of movement direction. *J. Neurosci.*, 19:8083–8093, 1999.

Mehring, C., Rickert, J., Vaadia, E., Cardosa de Oliveira, S., Aertsen, A., and Rotter, S. Inference of hand movements from local field potentials in monkey motor cortex. *Nat. Neurosci.*, 6(12):1253–1254, 2003.

Murthy, V.N. and Fetz, E.E. Oscillatory activity in sensorimotor cortex of awake monkeys: synchronization of local field potentials and relation to behavior. *J. Neurophysiol.*, 76(6):3949–3967, 1996.

Najafi, K. and Wise, K.D. An implantable multielectrode array with on-chip signal processing, *IEEE J. Solid-State Circuits*, 21:1035–1044, 1986. *IEEE Trans. Biomed. Eng.*, 51(10), 2004.

Obeid, I., Morizio, J.C., Moxon, K.A., Nicolelis, M.A.L., and Wolf, P.D. Two multichannel integrated circuits for neural recording and signal processing. *IEEE Trans. Biomed. Eng.*, 50:255–258, 2003.

Ohara, S., Weiss, N., and Lenz, F.A. Microstimulation in the region of the human thalamic principal somatic sensory nucleus evokes sensations like those of mechanical stimulation and movement. *J. Neurophysiol.*, 91(2):736–745, 2004.

Okun, M.S., Tagliati, M., Pourfar, M., Fernandez, H.H., Rodriguez,R.L., Alterman, R.L., and Foote, K.D. Management of referred deep brain stimulation failures: a retrospective analysis from 2 movement disorders centers. *Arch. Neurol.*, 62(8):1250–1255, 2005.

Olsson, III, R.H., Gulari, M.N., and Wise, K.D. Silicon neural recording arrays with on-chip electronics for *in vivo* data acquisition, in *Proc. 2nd Annu. Int. IEEE-EMB Special Topic Conf. Microtechnologies in Medicine and Biology,* May 2–4, 2002, p. 237–240.

Otto, K.J., Johnson, M.D., and Kipke, D.R. Voltage pulses change neural interface properties and improve unit recordings with chronically implanted microelectrodes. *IEEE Trans. Biomed. Eng.*, 53(2):333–340, 2006.

Patterson, W.R., Song, Y.K., Bull, C.W., Ozden, I., Deangellis, A.P., Lay, C., McKay, J.L., Nurmikko, A.V., Donoghue, J.D., and Connors, B.W. A microelectrode/microelectronic hybrid device for brain implantable neuroprosthesis applications. *IEEE Trans. Biomed. Eng.*, 51(10):1845–1853, 2004.

Peckham, H. Smart prosthetics: interfaces to the nervous system help restore independence, American Association for the Advancement of Science, San Francisco, CA, 2007 session, 180–181.

Santhanam, G., Ryu, S.I., Yu, B.M., Afshar, A., and Shenoy, K.V. A high-performance brain-computer interface. *Nature*, 442(7099):195–198, 2006.

Serruya, M., Hatsopoulos, N., Fellows, M., Paninski, L., and Donoghue, J. Robustness of neuroprosthetic decoding algorithms. *Biol. Cybern.*, 88(3):219–228, 2003.

Serruya, M.D., Hatsopoulos, N.G., Paninski, L., Fellows, M.R., and Donoghue, J.P. Instant neural control of a movement signal. *Nature*, 416(6877):141–142, 2002.

Serruya, M., Caplan, A., Saleh, M., Morris, D., and Donoghue, J. The BrainGate Pilot Trial: Building and Testing a Novel Direct Neural Output for Patients with Severe Motor Impairment. Society for Neuroscience Abstracts, *34th Annual Meeting of the Society for Neuroscience*, San Diego, CA, October 2004. (http://techhouse.brown.edu/~dmorris/publications/SFN2004.ck_poster.pdf).

Shenoy, K.V., Meeker, D., Cao, S., Kureshi, S.A., Pesaran, B., Buneo, C.A., Batista, A.P., Mitra, P.P., Burdick, J.W., and Andersen, R.A. Neural prosthetic control signals from plan activity. *Neuroreport*, 24;14(4):591–596, 2003.

Shoham, S., Halgren, E., Maynard, E.M., and Normann, R.A. Motor-cortical activity in tetraplegics. *Nature,* 413(6858):793, 2001.

Song, Y.K., Patterson, W.R., Bull, C.W., Beals, J., Hwang, N., Deangelis, A.P., Lay, C., McKay, J.L., Nurmikko, A.V., Fellows, M.R., Simeral, J.D., Donoghue, J.P., and Connors, B.W. Development of a chipscale integrated microelectrode/microelectronic device for brain implantable neuroengineering applications. *IEEE Trans. Neural Syst. Rehabil. Eng.*, 13(2):220–226, 2005.

Suner, S., Fellows, M.R., Vargas-Irwin, C., Nakata, K., and Donoghue, J.P. Reliability of signals from chronically implanted, silicon-based electrode array in non-human primate primary motor cortex. *IEEE Trans. Neural Syst. Rehabil. Eng.*, 13, 524–541, 2005.

Taylor, D.M., Tillery, S.I., and Schwartz, A.B. Direct cortical control of 3D neuroprosthetic devices. *Science,* 7;296(5574):1829–1832, 2002.

Turner, J.A., Lee, J.S., Martinez, O., Medlin, A.L., Schandler, S.L., and Cohen, M.J. Somatotopy of the motor cortex after long-term spinal cord injury or amputation. *IEEE Trans. Neural Syst. Rehabil. Eng.*, 9(2):154–160, 2001.

Warland, D.K., Reinagel, P., and Meister, M. Decoding visual information from a population of retinal ganglion cells. *J. Neurophysiol.*, 78(5):2336–2350, 1997.

Wolpaw, J.R. and McFarland, D.J. Control of a two-dimensional movement signal by a noninvasive brain-computer interface in humans. *Proc. Natl. Acad. Sci., U.S.A.*, 101(51):17849–17854, 2004. Epub 2004 December 7.

Wu, W., Gao, Y., Bienenstock, E., Donoghue, J.P., and Black, M.J. Bayesian population decoding of motor cortical activity using a Kalman filter. *Neural Comput.*, 18(1):80–118, 2006.

Yousry, T.A., Schmid, U.D., Alkadhi, H., Schmidt, D., Peraud, A., Buettner, A., and Winkler, P. Localization of the motor hand area to a knob on the precentral gyrus. A new landmark. *Brain,* 120(Pt. 1):141–157, 1997.

Motor Prostheses— Effector Subsystem Technologies

16

Transcutaneous FES for Ambulation: The Parastep System

Daniel Graupe

16.1 Introduction and Background

16.1.1 Historical Background

Functional electrical (neuromuscular) stimulation, denoted as FES (or FNS), has its origins in Luigi Galvani's experiment on electrically exciting a frog's leg in the 1780s as described by him in *De Viribus Electricitatis in Motu Muscular* (1791), which despite some faults due to the state of scientific knowledge at that time, can be shown to lay the foundation to two great disciplines: electrical engineering and neurophysiology [1].

The first demonstrated modern application of FNS to a human patient for functional movements of extremities was reported by Lieberson in 1960 [2] in the case of a hemiplegic patient, whereas the first application to a paraplegic patient was that by Kantrowitz [3].

Unbraced short-distance ambulation by transcutaneous FNS of a complete paraplegic was first described in 1980 by Kralj et al. [4].

In early 1982, the first patient-controlled ambulation for a complete paraplegic as necessary for independent ambulation was achieved by Graupe et al. [5–7] employing EMG (electromyographic) control. A manually controlled system known as the Parastep FNS system was tested from 1982 and received FDA approval in 1994 to become the first FNS ambulation system to be so approved and to be commercially available for use by individuals beyond research environments.

The systems of Graupe et al. [5, 7] employ a walker for balancing support. It was commercialized by Sigmedics Inc. (founded for this purpose in 1987) as the Parastep System (Parastep-1 System), and was the first (and still the only) FES ambulation system to have received FDA approval in 1994 [8] and approval by Medicare/Medicaid for reimbursement in 2003 [9, 10].

In parallel, work has been carried out since the early 1980s on percutaneous FNS, especially at Case Western Reserve University [11, 12]; in Vienna, Austria [13]; and in Augusta, Maine [14].

FNS as above is applicable for traumatic complete (or near-complete, as far as sensation, leg-extension, and hip-flexion are concerned) upper-motor-neuron thoracic-level paraplegics. To date, approximately 1000 such patients are or have been able to ambulate over short distances with the FDA-approved Parastep system. They have been trained in more than twenty hospitals or rehabilitation centers in the United States and in Europe, with no known detrimental effects. These patients can ambulate independently between twenty and several hundred meters (some up to one mile) without sitting down. The number of complete paraplegic patients who ambulate with implanted (percutaneous) electrodes is very small — about a dozen. These latter patients must undergo surgery (often of several hours), and they require occasional repeat surgery to correct electrode breakage or slippage, which are still unresolved problems. They also experience infections at the sites of electrode penetration through the skin. Consequently, at the present state of implantation, and noting the performance of the transcutaneous Parastep FNS users, it appears that for some time to come, the transcutaneous approach will be the more common one, and not just due to it being the only one available approach outside the research lab. It is for these reasons and because it is the only system for which there exists a body of independent-source, published material on clinical experience and data collection, that this chapter concentrates on the Parastep ambulation system. Still, for completeness of this chapter, a very brief discussion of the major other ambulation systems is given below.

16.1.2   Brief Review of FES Systems for Ambulation by Paraplegics

A very brief discussion of the other major FES ambulation systems is given below for completeness.

16.1.2.1   Noninvasive (Transcutaneous) FES Systems

The only other transcutaneous FES system (but for the Parastep) for both standing and ambulation that has been used outside their inventors' laboratory is the Ljubljana FES system, which is based on the work of Kralj et al. [15, 16]; it emanated from that group's earlier pioneering work [4] on FES (related to the still earlier work of Lieberson et al. [2] concerning hemiplegia). The bench model of the Ljubljana system was the first to achieve ambulation via FES by a complete thoracic-level paraplegic (in 1980; [4]). Its principles are similar to those of the Parastep system in their purpose and in their general function, which can, in part, already be found in the principles of the earlier Ljubljana work on hemiplegia and in Lieberson's work. It differs from the Parastep system in that it was not designed to maximize walking distance, in its control, and in its channel coordination (to result in a bulkier system than the Parastep system). Its patient-borne version is usually a four-channel system. Its signal generation is essentially a two-channel signal generator, such that the four-channel system is a double two-channel system. The Ljubljana system is not yet commercially available (at least not outside its use in research programs, mainly in Europe), and is presently not FDA-approved. No independent multipatient ambulation performance studies and statistics, and no multipatient medical evaluations or psychological evaluations, were published on that system.

Other noninvasive (transcutaneous) FES systems for standing and ambulation, apart from the Ljubljana system and the Parastep system, are essentially all bench devices, as developed in various research laboratories for the sole purpose of their own research (see [17–19]). Whereas all FES ambulation systems can be and are used for standing, there are several transcutaneous FES systems for standing alone (see [20–22]). These are obviously limited in scope and are not within the main focus of this chapter. None is commercially available.

16.1.2.2 Hybrid FES-Long-Leg Brace Ambulation Systems

Hybrid FES-long-leg-brace or FES-body-brace systems, which combine transcutaneous FES with long-leg braces or with a body brace for standing and ambulation by paraplegics, have been developed since the 1970s [23–25]. These systems are also intended for upper-motor-neuron (thoracic-level) spinal cord injury (SCI). They represent a regression from FES because they give up one major goal: of FES–ambulation — namely, the patient's independence. Because hybrid systems use a body brace or long-leg braces, they are far heavier and far more cumbersome than, say, the 10.5-ounce Parastep. They require 30 min to don and a long time to doff, requiring help from an able-bodied person in donning and in doffing the system. This also affects patient compliance and regular use of the system, while the system's weight reduces ambulation distances [25].

16.1.2.3 Implanted FES Ambulation Systems

As stated previously in this section, research on implanted FES for standing and ambulation has been carried out in parallel with the work on transcutaneous noninvasive FES, the latter being the subject of the present review. It is, however, important for the completeness of this review to comment briefly on implanted FES.

Work on both invasive FES and noninvasive FES started in the late 1970s and early 1980s. Also, the first applications to thoracic-level, complete traumatic paraplegics were reported for both approaches in the early 1980s [4–6, 11, 13]. Also, both approaches are based on the fundamentals used by Lieberson, et al. [2]. However, invasive methods, both *percutaneous* [11, 13] and of fully implanted systems [26], always involve major surgery, in contrast to the noninvasive transcutaneous methods on which this chapter concentrates. It is not only the surgery (and its cost). Furthermore, thus far all invasive methods encounter loss of contact of electrodes, wire breakage, and sometimes even tears of the nerve fibers. Such occurrences then require reoperation. Fully implanted FES [26] does not encounter infections at locations where wires penetrate the skin, as happens with percutaneous methods; in fully implanted systems, a radio-frequency (RF) receiver is implanted that receives RF signals through the skin, from a transmitter attached above the skin, and the received RF is rectified to provide electric power. All invasive systems require some kind of patient control from a nonimplanted device, as do noninvasive systems. Also, all invasive systems require similar patient training and muscle strengthening. Of course, an implanted device requires no electrode placement each morning and removal each evening. However, with the Parastep system, donning time is 5 to 8 min for a trained user and doffing time is 3 to 4 min. Connection and disconnection of the FES control device and, in percutaneous systems, connection of wires that to the implanted electrodes from outside, also still takes a few minutes. It is therefore not surprising that the Parastep system was the first and is still the only FES system for standing and ambulation that has received (1994) FDA approval and is commercially available. We note that there are presently some 600 Parastep users, and it is used both at home and at the workplace. In contrast, there are presently only a few (on the order of a dozen) users of even the most advanced percutaneous systems (based on Marsolais' work and that of his colleagues in Cleveland, as mentioned above), whereas the fully implanted system is not yet complete, to allow out-of-clinic ambulation.

We comment that the work on implanted FES has resulted in great advances in implantation techniques and materials that are of value in situations where there is no alternative to implantation (unlike the case of FES for standing and ambulation). However, the difficulties in implanted systems are still with us, and they will always require surgery.

16.2 The Parastep System

16.2.1 System's Electric Charge and Charge Density Parameters

FES consists of sequences (trains) of electrical impulses that are applied transcutaneously so that stimulation reaches peripheral motor neurons at selected sites. Stimulation serves only to trigger action potentials (APs) at these motor units. The resultant action potentials produced in the motor neurons

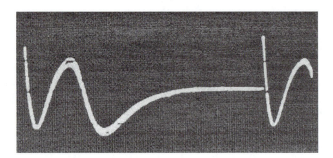

FIGURE 16.1 Action potential (AP) in response to stimulation at a quadriceps stimulated site. This AP is a summation of many synchronous action potentials produced in response to a stimulation signal. It is recoded by surface electrodes at the stimulation site. The sharp peak at the beginning of each AP is an artifact of the stimulus.

FIGURE 16.2 The Parastep unit.

concerned, in response to these triggers (stimulation impulses), subsequently cause contraction of muscle fibers that are associated with these motor neurons [27, 28] (Figure 16.1).

16.2.1.1 Parameters of Stimulation Signals and Safety Standard Constraints

The stimulation trains employed are trains of impulses of 120 to 150 μs in duration (width), and their rate is of 20 to 25 pps (pulses per second). The pulse duration was selected to be as low as possible while still allowing full contractions [27, 29]. This is necessitated by considerations of minimizing the electrical charge density applied to the stimulation site for the patient's safety: It therefore also minimizes battery power, resulting in a compact, lightweight portable system (Figure 16.2).

The system is powered by a 9.6-VDC battery pack consisting of 8 AA or AAA Ni-Cad rechargeable batteries, to power the stimulator and its computer.

The maximum current per pulse is limited [27] in our system to 0.3 A = Io. By the 1985 ANSI standard, Section 3.2.2.2 of the Association for Advancement of Medical Instrumentation and the American National Standard Institute [30], stimulation should be limited to below 10 mA average current. Hence, stimuli of $T = 150$-μs duration (pulse width) at $f = 24$ pps, result in average current $Iave$ of:

$$Iave = Io \times T \times f \qquad (16.1)$$

that is, $0.3 \times 24 \times 0.00015 = 0.00108$ A (or 1.08 mA), and is well below the ANSI limit. Another critical ANSI parameter is that of maximal electrical charge per pulse of 75 μC/pulse. The Parastep system's maximal output electrical charge value is given by:

$$Q = Io \times T \qquad (16.2)$$

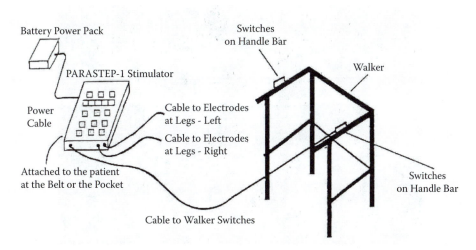

FIGURE 16.3 Parastep system with battery pack and walker.

or 0.3×0.00015 C = 45 µC = Q. Thus, the current density *Iave/S* is for electrodes of dimensions 1.75 in. $\times$ 3.75 in. (that is, $S = 40$ cm^2 = 4000 mm^2). The current density in the case of the Parastep system is therefore:

$$Iave/S = 0.00108/4000 \text{ A/mm}^2 = 0.25 \text{ µC/mm}^2 \qquad (16.3)$$

which is well below the ANSI limit of 10 µC /mm^2.

16.2.2 System Parameters and Design

The Parastep system is based on a single microprocessor [27, 29], which is its main component and in which the stimulation signals of all channels are shaped and controlled, and in which synchronization between channels is performed for the four different stimulation operational menus. The microprocessor generates and shapes trains of stimulation pulses that are multiplexed and directed by the algorithm embedded in that microcomputer to six output channels that are individually controlled by the micro-computer, in response to menu selection by the patient, to avoid robotic-like movements. Channel separation is performed by a timing program, which is passed from the microcomputer to an array of microcomputer-controlled opto-isolators and then appropriately amplified, thus providing the system's outputs to twelve surface electrodes attached to the skin at appropriate placements. These skin electrodes are self-adhesive and are reusable for fourteen days. They are to be attached by the patient himself in the morning and removed each evening or as desired, at locations that the patient has been taught to remember. The stimulator unit weighs 7.6 ounces (Figure 16.2), excluding a battery pack of six AA 1.5-V rechargeable alkaline (or eight rechargeable NiMH) batteries to allow at least 60 min of standing or walking [27]. The system is shown in Figure 16.3 and Figure 16.4.

16.2.2.1 Pulse Width and Pulse Repetition Rate (Frequency)

Pulse durations (widths) are set to 120 to 150 µs [27]. Higher durations are undesirable and unnecessary. Higher pulse width speeds up the rate of muscle fatigue and therefore reduces the maximal ambulation distance (see Table 16.1) and the maximal time a patient can stand or walk via FES. It also enters into the body more electrical charge than needed and requires higher battery power and hence higher battery weight.

The inter-pulse repetition rate (frequency) is set higher than the average pulse rate in the able-bodied individual, but is still kept as low as possible (22 to 24 pps) to reduce the rate of fatigue (see Table 16.1). It is determined with consideration for fatigue, tetanization, and force (note that while standing or

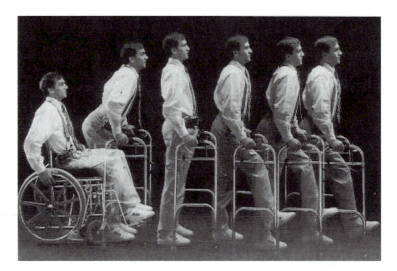

FIGURE 16.4 Parastep system in use during ambulation by a paraplegic patient. The same microchip also controls optical isolation chips to allow using a single power amplifier for all channels. This allows the Parastep to employ six stimulation channels (twelve electrodes) rather than the usual four stimulation channels and to integrate them to reduce the system's weight, while facilitating full patient control of all channels. Furthermore, it facilitates considerable battery power savings. The additional two stimulation channels (at the paraspinals, for trunk stability) play a major role in enhancing standing time, ambulation distances, and speeds as compared with four-channel systems.

TABLE 16.1 Stimulation Frequency vs. Rate of Muscle Fatigue

Stimulation frequecy (Hz)	% drop in isometric moment at ankle joint		
	After 10 min	After 20 min	After 30 min
20	less than 2%	less than 2%	3%
30	4%	14%	40%
50	17%	53%	71%

Pulse duration: 300 ms throughout.

walking, the body weight dampens vibrations considerably). At even lower frequencies, muscle vibrations are observed that are no longer dampened and this may affect the patient's balance when standing or walking. Higher frequencies also imply that a higher electrical charge enters the body, and requires higher battery power and heavier batteries. Furthermore, higher pulse rates speed up the rate of muscle fatigue to reduce duration and range of ambulation. Both pulse widths and rates, while constant, can be adjusted if necessary (see Figure 16.5).

Therefore, a combination of a short pulse duration, low stimulation levels, and low pulse rate is essential to reduce muscle fatigue, thus extending walking distance (walking time) per walk [5]. The consequences of lower battery power, of lower system weight, and of the resultant effect on compactness, are of course also significant for a patient-borne system and for user friendliness, especially in a body-borne system.

16.2.2.2 Stimulation Sites

The stimulation electrodes are self-adhesive electrodes (twelve in total), placed at six stimulation sites [27], two electrodes per site, as follows: two electrodes over the right quadriceps and two over the left quadriceps, to stimulate knee extension; two electrodes over the common peroneal nerve right and left (to activate dorsi-flexion and to elicit a hip flexion reflex via sensory neural feedback). Finally, two electrodes are placed over the right paraspinals at right and two at left, for upper trunk stability in patients with lesions at T7 or higher (to be placed approximately one inch below the level of the level of start of sensation, but not too close to the heart). Patients with SCI lesions at lower levels will have

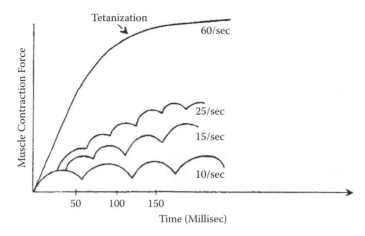

FIGURE 16.5 Muscle contraction force vs. stimulation rate and time.

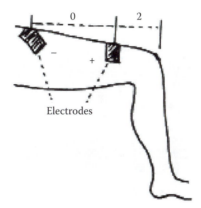

FIGURE 16.6 Placement of quadriceps stimulation electrodes (right leg, lateral side view).

electrodes placed over the gluteus medius and maximus for improved stability, whereas patients with lesions at T-10 or lower usually do not require paraspinal stimulation at all (Figures 16.6, 16.7, and 16.8). We comment that improved trunk stability affects not just patient safety, but also helps to reduce fatigue, thus improving ambulation performance and appearance (which is not just an aesthetic aspect but also a psychological one).

As shown in Figure 16.8, alternatives to the peroneal nerve placements are possible in some cases, (see Chapter 7 of [5]). These alternatives involve other branches of the sciatic nerve, which trigger the hip flexion reflex.

The number of channels (of electrode pairs) to be used is a matter of trade-off. Obviously, with more channels, more muscle groups (at below the SCI lesion) can contract. However, when increasing the number of channels, say from six to eight, the patient must place (every morning) sixteen electrodes instead of twelve; and for a paraplegic patient, this involves significant additional effort and time. Furthermore, the six channels stimulated by the Parastep system, as discussed above, are the ones that are the easiest to reach by the user and the ones where there is the greatest tolerance in terms of error in placement localization, while additional sites will require more care in exact placement. It is our experience that with more than six channels, most patients will soon stop using the system. Hence, human factor considerations imply that one should limit the system to the most important functions (channels), as far as performance is concerned. The resulting performance, as discussed later in this chapter, appears to justify this choice and to result in a rather smooth walk, as can be viewed in a 15-minute movie of a walk of a complete thoracic-level paraplegic using the Parastep system (see: http://www.ece.uic.edu/~graupe).

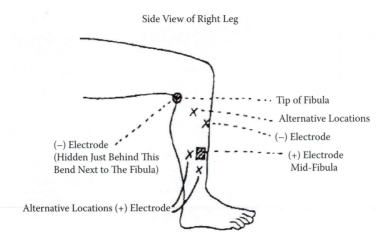

FIGURE 16.7   Placement of peroneal nerve electrodes (to elicit step via hip flexion reflex).

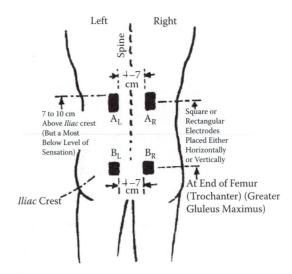

FIGURE 16.8   Placement of paraspinal electrodes on patient's back for truncal stability.

16.2.2.3   Sequencing Control Menus for the Stimulation Signals

Pulse-amplitude shaping is a major aspect of the pulse-shaping algorithm and it is the subject of four different menus within that algorithm. The menus are patient-selectable, through touch of finger-touch switches located on the Parastep's walker or on the Parastep's elbow-support canes. The menus are those for standing-up, for right step, for left step, and for sitting down (see Figure 16.9). Pulse-amplitude shaping is dynamic and varies per each of the six stimulation channels and per each menu, as does the distribution of output signal to each output channel [6, 27]. The time variation of the pulse amplitudes in each menu and per each channel, as in Figure 16.9, is therefore unique and is based on considerations of the executions of the given menu's function (for example, taking a right step), and of doing so safely, efficiently, and smoothly.

The stimulation signals are sequenced at the system's microcomputer chip as in Figure 16.9, where the envelope pattern of the stimulation impulses (and not the individual pulses) are illustrated. This pattern is automatically sequenced by the stimulator's computer to give a ramp increase of impulse levels for stand-up, to be followed by a lower constant level while the patient is standing, both being applied to the right and left quadriceps electrodes and to the right and left paraspinal/gluteal electrodes. When a left step menu is selected, either manually or through above-lesion (chest-level) EMG (as described below),

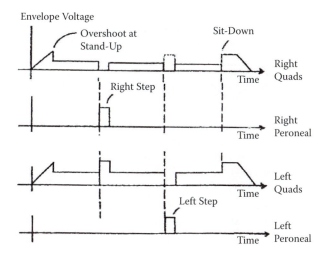

FIGURE 16.9 Synchronization scheme of envelopes of stimulation signals at various channels. (Note: Broken vertical lines divide between the four Parastep menus. Paraspinal signal envelopes correspond in shape to quadriceps channels except for peaks.)

then stimulation is stopped at the left quadriceps and paraspinals/gluteus while at the same time the common peroneal nerve is being stimulated to elicit a step. This lasts for a fixed duration of $T = 0.4$ to 1.0 s, as is preselected for the convenience of the patient. At the end of this period T, stimulation to the left common peroneal stops, and the left quadriceps and left paraspinals/gluteus are stimulated. However, during the step period, the level of stimulation at the right quadriceps is automatically increased by the sequencing program to compensate for the fact that full body weight is borne by the right leg over that period. If a right step is selected, then the same menu is employed, but with a reversal of roles of right and left. When a sit-down menu is selected, the sequencing program first triggers an audible and a visual warning to allow the patient to abort the sit-down if he is not ready to sit, and to allow for time to reach a chair and to comfortably sit down. Also, at that time, stimulation to the quadriceps is increased to compensate for possible weakening of the quadriceps that may have caused the patient to decide to sit.

The microchip also controls optical isolation chips to allow using a single power amplifier for all channels. This allows the Parastep to employ six rather than the usual four stimulation channels and to integrate them to reduce system weight, while facilitating full patient control of all channels. Furthermore, it facilitates considerable battery power savings. The additional two stimulation channels (at the paraspinals, for trunk stability) play a major role in enhancing standing time, ambulation distances, and speeds as compared with four-channel systems.

Control of the FES is performed by the stimulation signal's sequencing program of the Parastep's microcomputer, while selection of menus of that program is performed either manually (as in the Parastep commercial system) or via an above-lesion EMG control algorithm for menu selections [5, 7, 29].

16.2.2.4 Finger-Touch Menu Selection

Menu-selection finger-touch switches [27, 29] are located on the walker's handlebars for easy finger reach while normally holding the walker (or cane). They require only a light single and quick (short) finger-touch, without changing hand position on the bars. Adaptation and learning of balancing and of menu selection (only two menus during walking; of right and of left step, activated at right or left handle bar) is very easy and fast. Only finger-touch selection is available in the commercial Parastep system.

16.2.2.5 Above-Lesion EMG-Controlled Menu Selection

While the commercial Parastep system employs only manual touch buttons for menu selection, the laboratory Parastep system, that was tested on fourteen patients at Michael Reese Hospital, Chicago, Illinois, allowed also for above-lesion surface-EMG (electromyographic) menu selection with good results [7, 27, 29]. However, menu-selection EMG control was not incorporated in the commercial system (and is not covered by its FDA approval) because training is far lengthier and because it requires the donning of four more electrodes for EMG pick-up. This was felt to greatly limit the number of users. We comment that EMG electrode placement is much more critical than that of placing the stimulation electrodes themselves.

The above-lesion EMG-based menu selections employ surface-EMG signals from electrodes placed at above-lesion locations on the patient's chest. The thus-obtained EMG signal (see Figure 16.8) serves to map a pattern of upper-trunk posture that has been shown [5, 7, 27] to predict intended body function, corresponding to the four menus above (stand, left-step, right-step, and sit menus), with an accuracy of better than 99.8%. In this case, no finger controls are needed.

We emphasize that the relevant information from the above-lesion EMG signal is *not* based on the EMG level (power), but on the whole stochastic time-series pattern of that signal [27, 29]. Therefore, the patient does not have to produce a specific upper-body above-lesion (shoulder) movement to select a particular menu. The patient's natural walk and the natural changes in the above-lesion muscles, as are needed to move the walker and to otherwise balance when intending a particular step (or to stand-up or to sit down) that causes dynamic changes in the whole EMG stochastic pattern. These pattern changes are then recognized in the microchip's algorithm as a command to select a particular menu (from the menus above). In this way, the above-lesion EMG control differs from others that require unnatural pulling of shoulders or arms to produce an EMG-based command. Such intentional and unnatural movements divert the patient's concentration and yield an unnatural walk.

The mapping of the upper-trunk posture considers the above-lesion EMG signal, identified as $y(k)$, to satisfy a pure autoregressive (AR) time-series model [7, 29]:

$$y(k) = a(1)\ y(k{-}1) + a(2)\ y(k{-}2) + \ldots + a(n)\ y(k{-}n) + w(k) \tag{16.4}$$

where k is discrete time, such that $k = n{+}1, n{+}2, n{+}3$, etc., and where $w(k)$ denotes a discrete white-noise process.

We comment that both under finger-touch control (menu selection) and under EMG menu selection, the direction of step (and hence, of a walk) is determined by the shoulder movement of the walker as naturally performed when intuitively moving the walker to any desired direction.

16.2.2.6 Stimulation Signal Level Control

The stimulation level is adjusted at the Parastep's microcomputer chip, in accordance with either the patient's single finger-touch menu switching command or in response to an above-lesion EMG signal from the patent's chest that is interpreted in the same microchip.

16.2.2.7 Finger-Touch Force Level Control

The degree of recruitment of motor neurons determines the contraction force exerted by the muscle fibers that are associated with these neurons [27, 29]. The degree of recruitment depends, in turn, on the level of the stimulation signal when the motor neurons are triggered by FES stimuli. In the commercial Parastep, the stimulation level is controlled by the touch buttons at the left and right hand-side of the walker. Each finger touch raises the level by a single increment (out of ten possible level increments that are color marked). Most patients use one of the three lowest levels to start their walk, in order to minimize the rate of fatigue. During a half-hour walk, a patient will usually have to adjust the FES level only two or three times by a single discrete increment. The range of levels can be factory-set to suit special needs (patients).

16.2.2.8 Below-Lesion Response-EMG FES-Level Control

Alternatively, in the Parastep lab system tried at Michael Reese Hospital, a below-lesion surface-EMG stimulation-level control, denoted as Response-EMG Level Control, was successfully tested on some patients [27, 29].

Obviously, in complete upper-motor neuron paraplegics (thoracic level SCI patients), no EMG occurs below the level of the lesion because the lower-extremity neurons do not fire. However, under FES, action potentials arise at the stimulated motor neurons. These produce action potentials similar to those in nonimpaired situations. Furthermore, because any stimulation electrode-pair activates many hundreds of motor neurons simultaneously, all resultant action potentials are fully synchronized and appear as one very strong action potential due to this combined and synchronized firing (see Figure 16.1). This is in contrast to the surface-EMG above the lesion, which results from unsynchronized firing of hundreds of neurons. Furthermore, the resulting response-EMG increases with the degree of recruitment. It can thus serve to detect the progression of fatigue and to adjust stimulation levels accordingly. The commercial system uses only finger-touch level control because the reliability in calibration of relations between response-EMG level (and shape) against desired FES stimuli strength (to counter the fatigue) is still not sufficient. It also is not covered by Parastep's FDA approval. However, when using opto-isolators, the stimulation electrodes can simultaneously serve as response-EMG electrodes. This is due to the very short duration of the Parastep's stimulus in relation to the duration of a single action potential (which is due to the low stimulation pulse rate used by the Parastep system).

16.2.2.9 Peripheral Equipment

The Parastep FES system uses a walker (see Figures 16.3 and 16.4) or, in a few cases, a pair of elbow-support (Canadian) canes [27, 29]. Walkers are employed in all other FES ambulation systems, invasive or not. Walkers (or elbow-support canes) serve mainly for balance. Walkers carry (in the Parastep system) only 5% or less of body weight in trained FES users during standing and are crucial during the standing-up mode. Their balancing role is due to the fact that complete SCI paraplegics have no sensation (in addition to having no motor functions below their lesion). Hence, indirect sensation coming through their arms and hands, while holding the walker's handlebars, lets the users sense the ground to provide a certain psychological security. It thus allows the users to balance their bodies by slight shoulder and arm movements to better balance during standing and walking. The users are able to easily and rather naturally change direction of walking, at will, through shoulder positioning by which they turn their steps. One major function of the walker is during the stand-up phase from a seated position. The patient then gets up with the arms leaning on the walker. All theses reasons indicate the crucial role of walkers in achieving independent standing and walking.

The Parastep system is described in further detail by Graupe and Kohn [27, 28].

16.3 Patient Admissibility, Contraindications, and Training

16.3.1 Patient Admissibility Criteria

The cardiovascular status of the patients must be good. Hence, the criteria for a patient to be admitted to train and to use the Parastep standing/ambulation system are as follows [7, 27, 29]:

1. Must be in good general health and with a complete traumatic spinal cord lesion at levels no higher than C-7 and no lower than T-12
2. Intact lower motor units (lumbar level L-1 and below)
3. Must have a complete/near-complete SCI lesion that does not allow the patient to stretch his/her knees for standing up and where the patient has no substantial sensation (pain) of the stimuli
4. Surgery/wound following SCI must have healed, or as determined by the surgeon
5. Stable ortho-neuro-metabolic systems
6. No recent history of long bone stress fractures, osteoporosis, or severe hip or knee joint disease; a bone density test is advisable in cases of women over 40 years or patients who are many years (10 or more) beyond date of injury (the author had a patient 40 years postinjury who had no problem and was accepted to the FES ambulation program)
7. No history of cardiac or respiratory problems

8. Adequate trunk stability so that once quadriceps are stimulated, the patient can hold his upper trunk upright while supporting himself with a walker

9. Demonstrates appropriate muscle contractions in response to stimulation (absence of such response usually implies some lesions below T-12)

10. Standing tolerance: patient has adequate fatigue tolerance to practice and perform standing and walking functions after initial training

11. Balance and trunk control (at least when paraspinals are stimulated)

12. Must have adequate hand and finger control or voice control to manipulate the system controls (future systems may circumvent the need for finger control via speech recognition, to allow patients lacking hand/finger control to use the system)

13. Sufficient upper body and arm strength to lift oneself up to the walker for a second or two without stimulation and to grasp chair when stimulation is stopped for any reason

14. No severe scoliosis

15. No irreversible contractures

16. No morbid obesity

17. Patient is not pregnant

18. Motivation: the patient demonstrates and expresses appropriate desire and commitment to the training program

Also, interference of stimulation signal with an electronic cardiac pacemaker must be avoided.

16.3.2 Contraindications

Once the admission criteria above are met, no contraindications are known to the author from his personal observations or training experience with approximately 100 patients. Also, none are known to have been reported in the literature.

16.3.3 Patient Training

The experience of this author over twenty-three years of working with patients and of observing Parastep training programs outside his own training program, in the use of the Parastep FES system for standing indicates that, once the patient satisfies the criteria as above, the patient is able to stand and to ambulate if trained properly. Distances and speed vary widely; and even distances (and speeds) well below the averages cited in this chapter may be a major achievement for some patients, depending on their general health, level of lesion, age, and limitations they may have. The author had trained a 62-year-old T-3/T4 complete paraplegic patient (gunshot wounds) who was in a wheelchair for forty years and had never been stimulated. This patient stood up in his first session and took twelve steps in his third one-hour session. Motivation is definitely a key factor, and this also implies family/friends' (and physician's) encouragement and support. Family/friends' support is crucial. This should not just be verbal, but also in terms of helping the patient stand/walk at home after or between training sessions, by walking next to him/her to prevent the patient from a possible fall and moving (sharp) obstacles out of the way. The patient should have at least one strong armchair (possibly a metal chair) with armrests at an adequate height for the patient so that he/she can get up independently to the walker and then to sit down independently from the walker. Some patients do initially need help in placing the lower paraspinal or gluteus electrodes on the skin. It is very advisable that a family member or friend should observe at least part of one training session. The skin electrodes need replacement once every two weeks (or less if contact with the skin is inadequate). Bad electrodes or broken electrode connectors are the main reasons for stimulation failure.

Training programs vary widely and so do their respective results. This author is familiar with Parastep training programs that involve five to six hours a day of supervised training, over five to ten consecutive days; of Parastep programs of one hourly session every week or every two weeks over the course of one year; of Parastep programs of three one-hour sessions a week for eleven weeks ([31], the University of Miami program); and of Parastep programs of two hours per day, five days a week over four months

([32], the Vicenza program in Italy). Because all these programs use the same FES system (the Parastep system), the performance results shed light on their efficiency. However, they differ widely in cost and in the required time commitment by the patient. Therefore, the decision on which kind of program to attend is usually not a matter of choice.

Regardless of the training program, it is of utmost importance that the patient complements each supervised training session with after-hour home exercise of at least 15 min (many programs require much more).

In almost all training programs, training starts with reconditioning and strengthening of the muscles involved and also of arm muscles. First, the quadriceps muscles, which are those that are the most involved in stand-up and in standing, require strengthening. Treadmill exercises in walking are often used. In many programs, monitoring the heart rate and blood pressure is done during treadmill training. Parallel-bar standing and walking are sometimes used at the initial stages, during a muscle-strengthening phase. But parallel-bar exercise does not help in learning to rely on and to balance oneself with the walker and may therefore be counterproductive. Muscle strengthening while seated is a major part of the home-exercise routine throughout training, but takes place only in the first and/or second supervised sessions. It is psychologically extremely important to stand a patient up, even for 20 to 30 s (as long as is safe), already in the first session. This and the early taking of first few steps (even two or three) are great motivators. Hence, the first step should be taken after the patient can stand safely (with a walker) for about 3 min. Eventually, training and muscle strengthening should aim at standing for 10 min or more and at walking for as long as is possible. These sessions should start with treadmill standing and walking. At the last stages of training, patients should be taught to fall, by sudden power shutdown (they will learn to avoid an actual fall through proper use of a walker). They will also learn to lift themselves up from the ground with no help, to walk on rough ground, and on reasonable slopes. They will train in getting in and out of a car unaided. The most advanced T-9 to T-12 patients can then train on using an elbow-support cane.

Continuous walking after the end of training on a near-daily basis, for at least 45 min a day (not necessarily in one session per day), is essential for progress and for improved performance.

The first training session must involve obtaining sufficient quadriceps contraction to have each leg lift while the patient is in a seated position. Psychologically and motivationally, it is desirable for the patient to get up (to a walker, not to parallel bars) in the first or second session. However, this should not defer rigorous muscle strengthening in future sessions (actually, until end of the training program). For best results, daily (five days per week) training of one to two hours per day, followed by home exercises, yields far better outcomes at the end of training than a three-hours-per-week program, and even more if the whole training program is condensed over only one or two weeks. Still, whatever the training program, if after completion of training the patient continues to walk daily for 30 to 45 min, he will continue to improve and his performance will equal the best program (of course, with consideration for his individual status, lesion level, age, and general health).

Home exercise while undergoing training should be done when the patient stimulates while seated — except when, later in training (and with trainer's explicit permission) the patient is permitted to take a walker home and stands/takes steps *while* an able-bodied person is close at hand.

16.4 Walking Performance and Medical and Psychological Benefits Evaluation Results

16.4.1 Walking Performance Data

Walking distances covered by Parastep users vary with the individual user's level of injury, training, learned skills, and physical condition. Distance walked will vary and increase with practice and training. Individual goals are established for each user by the physical therapist. Studies conducted in different clinical settings reported distances walked by individual users ranging from a few feet to over a mile at a time, with the average distance being around 1450 ft (450 m per walk) for fully trained patients in certain training programs [32, 33].

TABLE 16.2   Ambulation performance results (Parastep users)

Ave. Speed	Ave. Distance	m/walk
Approx. 85 sessions daily over 4 months Vicenza (Cerrel-Bazo et al. [32])	444.3	14.5
32 sessions 3/week, 12 weeks Univ. of Miami (Klose et al. [31])	115	5.0

Performance is influenced by the training program, but more so by how rigorously the patient continues to actively stand and walk with the FES system after the end of training. Improvements in performance will be very noticeable one or two years after the end of formal training. Approximately 5% of Parastep users known to the author (from several U.S. training programs) can ambulate one mile per walk on occasions (usually one year or more after the end of formal training). The author expects this to be the case also for the Vicenza (Italy) training program.

The Miami Project to Cure Paralysis (University of Miami) reports average ambulation distances for Parastep users of 115 m per walk at a mean pace of 5 m/min, at the end of the training program of thirty-three sessions over eleven weeks [31]. For the Parastep training program of daily sessions over four months at the Centro di Rehabilitazione di Villa Margherita in Argugnano, Vicenza, Italy, an average distance of 444 meters per walk was reported, at a mean speed of 14.5 m/min and with a mean daily walk time of 90 min [32]. See also [33]. These performance differences are very significant.

Still, there is no reason to assume that persistent FES users in the eleven-week program cannot do as well as those in the four-month program at one year after the end of training. However, continuous use may be higher for patients whose performance at the end of training is considerably higher. This is the author's own experience (once weekly over one-year program). The Vicenza program reports zero dropout at fourteen to thirty-nine months after the end of training [32]. The author is not aware of other training programs with similar results.

We comment that the averages above are for patients whose SCI lesion levels are more or less evenly distributed between T-1 and T-12. Usually, performance is better if the SCI lesion is lower (toward T-12). However, motivation and persistence often make up for level of lesion. Still, patients who, for various medical or age reasons, cannot walk more than 10 m (per walk) at the end of training should still continue exercising, because the benefits of FES exercise are more than just a matter of distance or speed, as discussed below. Kralj et al. [16]) give general utilization statistics on their Ljubljana FES system by its developers.

However, these statistics do not include performance data or medical or psychological patient evaluation on that system. The data given below on the Parastep system are from independent centers (University of Miami Medical School, the Vicenza Rehabilitation Center, Italy), which are not connected with the system's manufacturers or its developers.

Table 16.2 gives further ambulation performance data.

A 14-minute video of complete thoracic-level paraplegic patients, while walking with the Parastep system, is available at www.ece.uic.edu/~graupe.

16.4.2   Evaluation Results on Medical Benefits for Walking with the Parastep System

The benefits of using the Parastep system go well beyond the benefits in the ability of walking, as discussed in the previous section. Medical and psychological evaluations published on Parastep users show several medical and psychological benefits to walking with the Parastep. These are discussed in this and in the next section. Most important medically is the major improvement in circulation at below the level of the SCI lesion. We discuss the medical and physiological evaluation results below, while psychological evaluation outcomes are summarized later.

16.4.2.1   Lower-Extremity Blood Flow

A study performed as a part of the Miami Project to Cure Paralysis of the Departments of Neurological Surgery and of Orthopedics and Rehabilitation of the University of Miami [34] and involving twelve

Parastep users reports an average increase in lower extremity blood inflow volume of 56% (from 417 mL/min to 650 mL/min) after twelve weeks (thirty-two sessions) of Parastep training. After paralysis due to thoracic-level SCI, blood flow to the lower extremities decreases considerably, with detrimental subsequent effects on kidney function and eventual cardiovascular effects. Cerrel-Bazo reported (verbally) to this author similar improvements at the Vicenza program in Italy. It is noted that within a few weeks after paralysis, thoracic-level paraplegics experience a severe loss in blood flow to below the lesion, with the related eventual cardiovascular and other consequences. Hence, such improvements are very significant.

16.4.2.2 Other Cardiovascular Effects

The above twelve-patient study at the Miami Project of the University of Miami [34] has shown that the average resting heartbeat of Parastep users decreased from 70.1 (prior to FES training) to 63.2 (posttraining). Also, the common femoral artery cross-sectional area increased by 50%, from 0.36 sq.cm (pretraining) to 0.48 sq.cm (post-FES training).

16.4.2.3 Physiological Responses to Peak Arm Ergometry

A study on physiological responses by fifteen Parastep users [31] to peak arm ergometry exercises have shown that average time to fatigue has improved from 15.3 min prestart of FES training to 19.2 min after thirty-three sessions of training. Also, the peak workload increased from 48.1 to 60 watts. Oxygen uptake at peak arm ergometry increased from 20.02 mL/kg/min pretraining to 23.01 mL/kg/min post-training, while the respiratory exchange ratio dropped from 1.26 pretraining to 1.18 posttraining, to indicate improvement in all these parameters. The patients (twelve men, three women) ranged in age from 21 to 45, in years from injury from 0.7 to 8.8, and in body weight from 53.6 kg to 83.5 kg.

16.4.2.4 Muscle Mass

A significant increase (10 to 22%) in thigh circumference was measured on Parastep users after three to six months of training at the University of Illinois/Michael Reese Hospital training program in Chicago [5].

16.4.2.5 Spasticity

Spasticity is common to all SCI patients with upper-motor lesions. In the author's experience in nineteen years of observing well over 100 patients training with or using the Parastep system, almost all patients who complained of spasticity commented on either considerable or some improvement in spasticity. This improvement was usually observed after the first two or three training sessions. Usually, the higher the degree of spasticity, the greater the improvement that was reported. This improvement was often reported as one of the reasons for participating in the FES program. The improvement in spasticity also is important due to the detrimental effect of medications (e.g., Baclofen, Valium, Lioresal), with respect to alertness and fatigue, and medication doses can then be reduced in many cases [5].

16.4.2.6 Bone Density

Practically all paraplegics suffer from reduced bone density. This happens right after injury and may be aggravated when the patient does not put weight on the legs. However, one Parastep patient in this author's program recorded a 50% bone density prior to training (but no bone injuries) with no improvement after one year; he continued to walk and reached one mile per walk. The only study published to date [36] does not show any improvement in bone density due to FES ambulation. However, this study refers to the end of eleven weeks of training. No study exists on patients who have consistently walked via FES for several years.

16.4.2.7 Pressure Ulcers (Decubitus Ulcers)

Almost all paraplegics suffer from decubitus ulcers. However, all but one patient at the author's FES program (at Michel Reese Hospital, Chicago) had no occurrence of a new ulcer while regularly using

TABLE 16.3 Medical and Physiological Evaluation Data (Parastep Users)

	Pre-FES-Training (Ave.)	Post-FES-Training (Ave.)	Ref.
Lower-extremity Blood Flow	417 mL/min	650 mL/min (improv.)	(Nash et al. [34]) 12 patient data/ U. of Miami
Heart Rate	70.1	63.2 (improv.)	(Nash et al. [34]) 12 patients/Miami
Time to Fatigue (at peak arm ergometry test)	15.3 min	19.2 min (impr.)	(Jacobs et al. [35]) 15 patients/Miami
Peak Workload Heart Rate (pk arm ergom. test)	188.5	183.1 (impr.)	(Jacobs et al. [35]) 15 patients/Miami
Oxyg. Uptake (pk arm ergom. test)	20mL/Kg/min	23mL/Kg/min (improv.)	(Jacobs et al. [35]) 15 patients/Miami
Spasticity		Usually improvement especially for very spastic pretraining	(Graupe and Kohn [6],[27], [29]) Michael Reese Hospital, Chicago
Bone Density		No follow-up data except for eleven weeks after start of training, where no significant change was reported	(Needham-Shropshire et al.[36])

TABLE 16.4 Psychological Evaluation Results for Parastep Users

	Pre-FES-Training (Ave.)	Post-FES-Training (Ave.)	Ref.
Physical self-concept (TSCS scores)	43.2 TSCS	52 TSCS (improv.)	[37], 15 patients/Miami
Depression scores (BDI scores)	8.8 BDI	5.4 BDI (improv.)	[37], 15 patients

FES. Improved blood circulation at below the lesion is most likely related to this [27]. The exception was due to a cut from a sharp object.

The medical and physiological evaluation data are summarized in Table 16.3.

16.4.3   Psychological Outcome Evaluation Results

16.4.3.1   Psychological Evaluation Results: Self-Concept Scores

A study of fourteen Parastep users after eleven weeks of training at the Miami program [37], concerning physical self-concept using the Tennessee Self-Concept Scale (TSCS), compares TSCS scores before the beginning of Parastep training against the score at the end of the eleven-week program. It shows that the average TSCS score improved in a statistically significant manner from 44.3 to 52.0. Furthermore, all patients with a score below 50 prior to FES training have improved, whereas no patient with an initial score above 50 dropped to below 50.

16.4.3.2   Psychological Observations: Depression Scores

The same study [37] as in the previous subsection reports on comparing Beck Depression Inventory (BDI) scores for measuring depression before and after eleven weeks of Parastep training. BDI scores of below 9 refer to no depression and scores between 9 and 18 indicate mild depression, and scores from 18 to 29 point to moderate depression. The results of the study show that all five patients who were initially at the mild or moderate depression score levels (one was initially even beyond the moderate range) did improve significantly. The patient who was initially beyond the moderate depression range (31 DBI score) improved to 24 (mild depression range). One of the two patients initially in the moderate range improved to the low–mild range, and the other to the no-depression range. All patients who were initially in the low depression range stayed in that range.

These psychological evaluation results are summarized in Table 16.4.

16.5   Regulatory Status

The Parastep I functional neuromuscular (electrical) system (referred to throughout this chapter as the Parastep system) for standing and for ambulation by thoracic-level paraplegics received FDA 21 approval on April 20, 1994 [8]. It was the first and is still the only noninvasive FES ambulation system to have received FDA approval. Furthermore, effective April 1, 2003, the Centers for Medicare and Medicaid Services (CMS) made a National Coverage Determination extending coverage to the Parastep I System for qualifying Medicare beneficiaries. Specific HCPCS codes have been assigned to cover costs associated with both the acquisition of Parastep I equipment [9] and for the physical therapy training services with Parastep I [10]. Medicare covers approximately 80% of equipment acquisition costs. Following the CMS example, most major medical insurers in the United States have already amended their policies to cover the Parastep system. The Parastep manufacturer, Sigmedics, Inc., of Fairborn, Ohio, has set up its own Patient Case Management Department [38] for the purpose of facilitating the insurance reimbursement process.

16.6   Conclusions

This chapter discussed the Parastep system, which is the first, and still the only, FDA-approved transcutaneous (noninvasive) FES system for ambulation by complete or near-complete thoracic-level paraplegics. It describes what this system can already do for the thoracic-level (complete) paraplegic patient in noninvasive FES. It gives concrete data from many studies on how that system performs. It also discusses its design, operation, admission criteria, contraindications, and training.

We thus conclude that a totally noninvasive FES for independent standing and mobility is already a reality for complete upper-motor-neuron, thoracic-level traumatic paraplegics. Furthermore, it is commercially available and it has received (2003) approval for reimbursement by the Center for Medicare and Medicaid Services (CMS), which regulates Medicare and Medicaid reimbursements policies in the United States and, subsequently, by practically all medical insurance companies in the United States. Training programs for the system exist in many hospitals and rehabilitation centers. As discussed above, upon completion of four months of daily training, ambulation distances for the Parastep system were reported to average 444 m per walk [32], or 115 m per walk in a thirty-three-session, eleven-week program [31]. Medical benefits have been documented in terms of greatly increased blood flow to the lower extremities [34], reduced spasticity [29], reduced incidence of decubiti [27], increased thigh circumference [27], and of psychological benefits (improved self-concept and depression scores) [37].

However, even ten years after FDA approval of such a noninvasive FES system, and two years after reimbursement was approved by Medicare, Medicaid, and most insurers, there is great ignorance in the paraplegic community about the availability of such a system and of its performance and benefits (see [38]). In [38], a statement by a patient is quoted (made in a recent symposium of prospective FES users, funded by the Whitaker Foundation): "in three to four different rehabilitation facilities and (having) talked to over 200 patients… none of them ever mentioned FES." This indicates ignorance regarding the role of FES in paraplegia, among physicians involved in caring for paraplegics and among the (physical and occupational) therapists and other related staff.

The consensus at the Symposium above (and which agrees with what this author repeatedly hears from patients) was that the desire to stand upright independently and to ambulate even short distances is the prime desire of paraplegics. Still, long-term compliance and long-term use of FES is also a problem. However, the circulatory benefits and the other medical and psychological benefits should play an important role, for patients, for physicians, and for insurance companies involved in the care of paraplegics.

All this does not detract in any way from the urgent need to repair the spinal cord through regeneration. Neither the Parastep nor any other FES approach can be a substitute for this because FES does *not* heal. It is an aid, just like eyeglasses or a hearing aid. It is hoped that regeneration will become a reality for human SCI patients. In the meantime, a realistic aid does exist that is already FDA approved and reimbursable. It can always be and will be improved, but its performance is usually pretty good.

References

1. Galvani, L. (1791). *Commentary on the Effect of Electricity on Muscular Motion,* translated by R.M. Green (1953), Elizabeth Licht Publishing Co., Cambridge, MA.

2. Lieberson, W.T., Holmquest, H.J., Scott, D., and Dow, H. (1961). Functional electrotherapy stimulation of the swing phase of the gait in hemiplegic patients. *Arch. Phys. Med. Rehab.,* p. 101.

3. Kantrowitz, A. (1960). *A Report of the Maimonides Hospital,* Brooklyn, NY, p. 45.

4. Kralj, A., Bajd, T., and Turk, R. (1980). Electrical Stimulation Providing Functional Use of Paraplegic Patients Muscles. *Med. Prog. Technol.,* 7: p. 3.

5. Graupe, D., Kralj, A., and Kohn, K.H. (1982). Computerized signature discrimination of above-lesion EMG for stimulating peripheral nerves of complete paraplegics. *Proc. IFAC Symp. Prosthetics Cont.,* Columbus, OH, March.

6. Graupe, D., Kohn, K.H., Basseas, S., and Naccarato, E. (1983). EMG-controlled electrical stimulation. *Proc. IEEE Frontiers of Eng. Comp. Health Care,* Columbus, OH.

7. Graupe, D., Kohn, K.H., Basseas, S., and Naccarato, E. (1984). Electromyographic control of functional electric stimulation in selected paraplegics. *Orthopedics,* 7:1134–1138.

8. FDA approval P900038 (1994). http://www.fda.gov/cdrh/pma94.html, April 20.

9. Centers for Medicare and Medicaid Services (CMS), Code K0600 (Parastep-I equipment acquisition), http://www.cms.hhs.gov/coverage, 2003.

10. Centers for Medicare and Medicaid Services (CMS), Code 97116 (physical training services with Parastep-I), http://www.cms.hhs.gov/coverage, 2003.

11. Marsolais, E.B. and Kobetic, R. (1983). Functional walking in paralyzed patients by means of electrical stimulation. *Clin. Orthop.,* p. 175:30–36.

12. Marsolais, E.B. and Kobetic, R. (1986). Implantation techniques and experience with percutaneous intramuscular electrodes in the lower extremities. *J. Rehab. Res. Dev.,* 23:

13. Holle, J., Frey, M., Gruber, H., Kern, H., Stoehr, H., Thoma, H. (1984). Functional electro-stimulation of paraplegics, experimental investigations and first clinical experience with an implantable stimulation device. *Orthopedics,* 7:1145–1155.

14. Davis, R., Kuzma, J., Patrick, J., Heller, J.W., McKendry, J., Eckhouse, R., and Emmons, E. (1992). Nucleus FES-22 stimulator for motor function in a paraplegic subject. *Proc. RESNA Int. Conf.,* June 6–11.

15. Kralj, A. and Bajd, T. (1989). *Functional Electrical Stimulation: Standing and Walking after Spinal Cord Injury,* CRC Press, Boca Raton, FL.

16. Kralj, A., Turk, R., Bajd, T., Stafancic, M., Sarvin, R., Benko, H., and Obreza, P. (1993). FES utilization statistics for 94 patients. *Ljubljana FES Conf.,* Ljubljana, Slovenia, p. 79–81.

17. Popovic, D. (1986). Control methodology for gait restoration. *Proc. 8th Annu. Conf. IEEE Eng. Med. Biol. Soc.,* Dallas-Ft. Worth, TX, p. 675–678.

18. Mayagoitia, R.E., Phillips, G.F., and Martinez, L.M. (1993). Mexican programmable eight channel surface stimulator. *Proc. Ljubljana FES Conf.,* Ljubljana, Slovenia, p. 169–170.

19. Phillips, G.F., Adler, J.R., and Taylor, S.J.G. (1993). A portable stimulator for surface FES. *Proc. Ljubljana FES Conf.,* Ljubljana, Slovenia, p. 166–168.

20. Jaeger, R. (1986). Design and simulation of closed-loop electrical stimulation orthoses for restoration of quiet standing in paraplegia. *J. Biomech.,* p. 825.

21. Kralj, A., Bajd, T., Turk, R., and Benko, H. (1989). Paraplegic patients standing by functional electrical stimulation. *Digest 12th Int. Conf. Med. Biol. Eng.,* Jerusalem, Israel, Paper 59.3.

22. Taylor, P.N., Ewins, D.J., and Swain, I.D. (1993). The Odstock closed-loop FES standing system — experience in clinical use. *Proc. Ljubljana FES Conf.,* Ljubljana, Slovenia, p. 97–100.

23. Tomovic, R., Vukobratovic, M., and Vodovnik, L. (1973). Hybrid actuators for orthotic systems — hybrid assistive system. *Proc. Int. Symp. External Cont. Hum. Extremities,* Dubrovnik, Yugoslavia, p. 73.

24. Andrews. B.J. and Bajd, T. (1984). Hybrid orthoses for paraplegics. *Proc. Int. Symp. External Cont. Hum. Extremities,* Dubrovnik, Yugoslavia, p. 55.

25. Solomonov, M., Best, R., Aguilar, E., Cetzee, T., D'Ambrosia, R., and Rarrata, R.V. (1997). Reciprocating gait orthosis powered with electrical muscle stimulation (RGO-2). *Orthopedics,* p. 315–324 (Part 1); p. 411–418 (Part 2).

26. Davis, R., MacFarland, W., and Emmons, S. (1994). Initial results of the nucleus FES-22 stimulator implanted system for limb movement in paraplegia. *Stereotat. Funct. Neurosurg.,* 63:192–197.

27. Graupe, D. and Kohn, K.H. (1994). *Functional Electrical Stimulation for Ambulation by Paraplegics,* Krieger Publishing Co., Malabar, FL.

28. Graupe, D. and Kohn, K.H. (1998). Functional neuromuscular stimulator for short-distance ambulation by certain thoracic-level spinal-cord-injured paraplegics. *Surg. Neurol.,* 36:202–207.

29. Graupe, D. and Kohn, K.H. (1997). Transcutaneous functional neuromuscular stimulation of certain traumatic complete thoracic paraplegics for independent short-distance ambulation. *Neurol. Res.,* 19:323–333.

30. Assoc. for Advancement of Med. Instrumentation/Amer. Natl. Standard Inst.: American National Standard for Transcutaneous Nerve Stimulators, *AINSllAAMI NS4* 0 1985, Arlington, VA. Approved May 20, 1986.

31. Klose, K.J., Jacobs, P.L., Broton, J.G., Guest, R.S., Needham-Shopshire, B.M., Lebwohl, N., Nash, M.S., and Green. B.A. (1997). Evaluation of a training program for persons with SCI paraplegia using the Parastep-I ambulation system. 1. Ambulation performance and anthropometric measures. *Arch. Phys. Med. Rehab.,* 78:789–793.

32. Cerrel-Bazo, H.A., Rizetto, A., Pauletto, D., Lucca, L., and Caldana, L. (1997). Assisting paraplegic individuals to walk by means of electrically induced muscle contraction: gait performance and patient compliance, Session 91, Paper 66, *Eighth World Congr. Int. Rehabil. Med. Assoc.,* Kyoto, Japan.

33. Chaplin, E. (1995). Functional neuromuscular stimulation for mobility in people with spinal cord injuries. The Parastep I system. *J. Spinal Cord Med.,* 19:99–105.

34. Nash, M.S., Jacobs, P.L., Montalvo, P.M., Klose, K.J., Guest, R.S., and Needham-Shropshire, B.M. (1997). Evaluation of a Training Program for Persons with SCI paraplegia using the Parastep-I ambulation system. 5. Lower extremity blood flow and hypermic responses to occlusion are augmented by ambulation training. *Arch. Phys. Med. Rehab.,* 78:808–814.

35. Jacobs, P.L., Nash, M.S., Klose, K.J., Guest, R.S., Needham-Shropshire, B.M., and Green, B.A. (1997). Evaluation of a training program for patients with SCI paraplegia using the Parastep-I ambulation system. 2. Effects on physiological responses of peak arm ergometry. *Arch. Phys. Med. Rehab.,* 78:794–798.

36. Needham-Shropshire, B.M., Broton, G.J., Klose, K.J., Lebwohl, N., Guest, R.S., and Jacobs, P.L. (1997). Evaluation of a training program for persons with SCI paraplegia using the Parastep-I ambulation system. 3. Lack of effect on bone mineral density. *Arch. Phys. Med. Rehab.,* 78:799–803.

37. Guest, R.L., Klose, K.J., Needham-Shropshire, B.M., and Jacobs, P.L. (1997). Evaluation of a training program for persons with SCI paraplegia using the Parastep-I ambulation system. Part 4. Effects on physical self-concept and depression. *Arch. Phys. Med. Rehab.,* 78: 804–807.

38. Kilgore, K.L., Scherer, M., Bobblit, R., Dettloff, J., Dombrowski, D.M., Goldbold, N., Jatich, J.W., Morris, R., Penko, J.S., Schremp, E.S., and Cash, L.A. (2001). Neuroprosthesis Consumers' Forum: consumer priorities for research directions. *Vet. Admin. J. Rehab. Res. Devel.,* p. 655–660.

17

Development of a Multifunctional 22-Channel Functional Electrical Stimulator for Paraplegia

Ross Davis, T. Johnston,
B. Smith, Randall R. Betz,
Thierry Houdayer, and
Andrew Barriskill

17.1 Introduction

The authors' aim has been to develop a generic FES implant for the restoration of functions in spinal cord injured (SCI) paraplegic individuals, the functions or modes of which can be matched to an individual's requirements: Upright Functional Mobility, Pressure Relief and Lower Extremity Exercise, Bladder and Bowel Control [1–6]. In addition, for bladder control, less invasive surgical procedures were proposed to avoid posterior conus rhizotomy, and sacral laminotomy in order to access the sacral nerve roots for stimulation [7, 8]. The hope was that this system would offer more functions and less surgery to patients with a cost–benefit ratio. This approach was termed "multifunctional."

Simple locomotor functions can complement the use of a wheelchair and be helpful in overcoming obstacles to wheelchair access, especially doorsteps and unadapted bathroom facilities. In addition, being able to stand up to reach objects and perform prolonged manual tasks would be convenient for many workplace and home situations [3–5]. Five paraplegic volunteers (two at the Neural Engineering Clinic (NEC) in Augusta, Maine; and three at the Shriners Hospital for Children (SHC) in Philadelphia, Pennsylvania) have participated in this device's evolution.

During 1983, Davis (NEC) became aware of the possibilities of modifying and using the twenty-two-channel cochlear implant technology (Cochlear Ltd., Lane Cove, N.S.W., Australia) as the basis for an implantable functional electrical stimulation (FES) system for the restoration of multiple functions in spinal cord injured (SCI) paraplegics.

The state of FES in paraplegia has been extensively reviewed [1–5]. These SCI individuals are unable to move their lower extremities or control bladder and bowel function. They must regularly self-catheterize (approximately three to six times per day). Secondary medical problems are prone to occur, such as pressure sores, osteoporosis, muscular atrophy in the lower limbs, muscle spasticity, deep vein thrombosis, cardiovascular disease, and depression. Although considerable FES achievements have been made, there has yet to be developed a safe, practical FES system for these multiple functions that is completely independent of the laboratory and is an energy-efficient mobility aid for prolonged use at home and in the workplace. The reason lies in the fact that FES is addressing complex problems requiring not only interdisciplinary knowledge from muscle and nerve physiology and electrical stimulation technology, but also implementation of biomechanical and control principles [6].

Other reasons that limit clinical application may also be significant, for example, cost–benefit considerations (especially for implanted systems). Although spinal injury results in loss of multiple physiological systems, neural implants to date have been developed to restore only specific functions. An approach was proposed to develop a generic FES implant the functions or modes of which can be matched to an individual patient's requirements. In addition, less invasive surgical procedures were proposed to avoid the posterior conus rhizotomies and sacral laminectomy associated with existing implanted bladder implants [7, 8].

Since 1984, three FES implant models have evolved from Cochlear's technology and its subsidiary: Neopraxis Pty. Ltd. The initial Nucleus FES-22 Stimulator was implanted in 1991 after animal and human studies, and with the U.S. Food and Drug Administration's approval (IDE# G87014) and Institutional Review Board (IRB) approval in a 21-year-old paraplegic subject (ASIA: T10).

17.2   Historical Aspect

In 1984, the Veterans Administration (VA) funded the initial animal studies at the Togus VA Medical Center (Augusta, Maine). These were aimed at determining what changes would be required to use a modified cochlear implant with a maximum pulse output of 4.3 mA and 0.4-ms pulse width, to be suitable for FES use in humans. An initial decision was taken to utilize epineurally placed electrodes (2.5 mm diameter platinum disks) in preference to epimysial or intramuscular electrodes because it was known that the stimulation currents would be lower and that there would be less movement of the electrodes. To determine exactly how low the stimulation currents would be and to determine the stimulation sites, initial anesthetized rabbits studies were conducted [9]. The threshold found for each branch of the split sciatic nerves of was 0.1 to 0.2 mA at 0.2 ms with 50 pps frequency. Maximal stimulation was achieved usually between 0.5 and 1.0 mA. Simultaneous dorsiflexion of both paws as well as co-contraction in the anterior and posterior muscle groups could be achieved.

At the Togus VA Medical Center, with the approval of their IRB and volunteer patients undergoing lower extremity amputation, stimulation studies were carried out at 0.2 ms pulse duration with 20 pps frequency, with a portable, battery-operated, calibrated constant-current unit (Cordis Corp., Miami, Florida, Model 910 A). The pulse amplitudes for producing maximal stimulation and contraction in the largest of the nerves (medial sciatic) ranged from 0.6 to 2.5 mA, which falls well within the range of the Cochlear receiver-stimulating unit to be used [9, 10]. Using the *Color Atlas of Human Anatomy, first edition* edited by McMinn and Hutchings (Yearbook Medical Publishers, Inc., Chicago, Illinois), whose dissections were reproduced as life-size photographs, allowed for measurements of the diameters to be made at different points along the nerves. These measurements were in relatively close agreement with the amputated nerve diameters of the nine volunteer patients [10].

17.3   Neural Engineering Clinic: Two Male Subjects with Nucleus FES-22 Stimulating System

As a first device, the FES-22 stimulator was only intended to provide its recipient with enhanced mobility functions. During 1985, Roger Avery (Custom Med Laboratories, Durham, New Hampshire) started

Nucleus FES-22

Praxis FES-24A

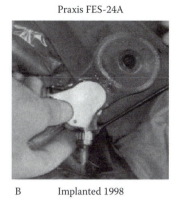

Implanted 1991   A   B   Implanted 1998

FIGURE 17.1 (A) Nucleus FES-22 system and (B) Praxis FES-24A system.

work on the design and manufacture for the implantable leads and electrodes. Because of the need for higher output currents, it was also necessary to design a new transmitter coil capable of delivering the higher power. To make each of the twenty-two output channels individually available, a circular epoxy housing was designed with twenty-two sockets around the perimeter (Figure 17.1A) with the diameter of the housing being determined by the diameter of the coil. During November/December 1991, the Nucleus FES-22 system was implanted in Subject A (21-year-old male paraplegic subject; ASIA: A T10) in three sessions at the Kennebec Valley Medical Center (now Maine General Medical Center), Augusta, Maine. The receiver/stimulator was placed subcutaneously at the lower right anterior intercostal margin with eleven connecting leads subcutaneously tunneled to the right and another eleven to the left hip areas. Following this, 2.5-mm diameter platinum disk electrodes were placed epineurally on the individual branches of the right and left femoral nerves by suturing the silicone elastomer ring around each electrode to the connective tissues on each side of the nerve branches. In the second and third procedures, electrodes were attached over gluteal, posterior tibial, peroneal, and sciatic nerves bilaterally [11]. A total of twenty electrodes were implanted epineurally, with one electrode placed subcutaneously in a Teflon bag in each of the femoral triangles, as a spare lead.

Six weeks following surgery (January 1992), the FES-22 system did produce threshold and maximal muscle contractions as tested in all twenty channels. At the second testing session in February 1992, the implanted system did not function properly, owing to a suspected electrostatic damage in the implant resulting in the loss of seven channels. Hardware and software changes were made, allowing the remaining fifteen channels to work. In December 1992, the fifteen channels were retested for threshold and maximal muscle contractions; the multivariate analysis did not show any change with time or body side, but a significant effect was seen with the electrode locations [12].

Subject A exercised his lower extremity muscles at home using a PC computer to control the implanted stimulator. In January 1997, he was provided with a battery-operated external Portable Conditioning System ($19 \times 11 \times 6$ cm), which he uses at home and at work sitting in his wheelchair. The exercise protocol stimulates the right and left knee extensors and ankle plantar/dorsi flexors alternately (4 s ON/4 s OFF), for a total of 20 min. After the muscles have been conditioned, dynamo-metric testing (isometric mode) has shown that implanted FES stimulation produces bilateral knee extension torque of 45 to 55 Nm at 30° and 65 Nm at 60° of knee flexion. Subject A exercises at least three days a week, and finds that if he does not do so, the spasticity in the lower extremities increases.

The laboratory PC-based FES-22 system implements a 10-ms duty-cycle state machine for open- and closed-loop control for use in prolonged standing mode. The controller is divided into three phases: (1) open-loop sit-to-stand; (2) closed-loop stand; (3) closed-loop stand-to-sit. To initiate standing up and sitting down, the subject uses a remote switch on a hand glove. The sensors used for closed-loop control are electrogoniometers across both knees, which respond to a 10° knee buckle, and accelerometers attached to the back at T6 level.

Controlled Nucleus FES-22 stimulation to the motor nerves of the quadriceps and gluteal muscles has resulted in uninterrupted standing for over 60 min [12]. This was achieved using the bilateral knee-angle goniometer sensors with Andrews' stabilizing Anterior Floor Reaction Orthosis (AFRO), which is an ankle–foot brace. With the knee goniometers sensing for a 10° buckle, the stimulator would come "ON" to correct the buckle; usually this occurred between 3 and 8% of the standing time. On recovery, the automatic switch "OFF" occurs when knee flexion has returned to less than 5°. Otherwise, lower extremity muscle activation is not required to maintain the upright posture [13].

17.4   Praxis FES-24A System

In 1998, Cochlear Ltd. formed a subsidiary company, Neopraxis Pty. Ltd., which decided to build on the knowledge gained from the FES-22 implant and to produce the Praxis FES-22A System. This system was designed to provide multiple functions — bladder and bowel control, enhanced mobility and seated pressure relief — in an effort to provide recipients with a cost-effective device that would address their most important needs.

17.4.1   Bladder Control

The traditional bladder stimulator, the Finetech–Brindley Stimulator, and now the Vocare (NeuroControl Corp., Cleveland, Ohio) operates by stimulating the sacral anterior roots [8]. This system has two primary drawbacks, which the Praxis system was designed to overcome: (1) posterior sacral rhizotomies are done, via a laminectomy, to achieve an areflexive bladder with increased capacity; (2) a sacral laminectomy is done to access the anterior sacral roots for fitting cuff-type electrodes. The rhizotomy procedure eliminates reflex erection in male recipients. Further, Creasey [8] states that "a patient who has the rhizotomies but does not use the implant (stimulator) would therefore be expected to become more constipated."

In August 1998, the Praxis FES-24A stimulator (Figure 17.1B) was developed by Neopraxis Pty. Ltd., and implanted in Subject B (35-year-old male paraplegic subject; ASIA: A, T10). Eighteen channels were used for stimulating individual nerves or branches for muscle contractions and limb movements, including exercise, pressure relief, standing, and stepping. The electrodes implanted for epineural stimulation were ten thin flexible platinum cuffs (Flexi-Cuff) that were sized, cut, and sutured closed with at least twice the diameter of encircled nerve. The other eight electrodes were 3-mm-diameter platinum buttons that were placed on the epineurium. Each button has an attached Dacron mesh surround that was sutured to the adjacent connective tissue on each side of the nerve.

Three channels for bilateral sacral root stimulation (S2-4) for bladder control (bowel control and erection, if possible) were provided. Sacral root stimulation was achieved by three pairs of LPR electrodes (10-mm long, solid platinum tubing of 1.0-mm diameter) inserted into the external sacral foramina in a lateral direction to follow and to stimulate the nerve roots epidurally. One further channel was connected to an epidural spinal cord stimulating electrode (Pisces Quad: Medtronic Inc., Minneapolis, Minnesota) for conus medullaris modulation of spastic bladder and bowel reflexes.

17.4.2   Praxis System Clinical Results

For the year prior to his implantation, Subject B was able to stand without knee bracing using a combination of the Andrews' Anterior Floor Reaction Orthosis and closed-loop skin surface FES applied directly over the femoral nerves, 2 to 3 cm below the inguinal ligament. With closed-loop control of stimulation, he would typically stand uninterrupted for 30 min, and up to 70 min. With training, Subject B did achieved the "C" posture and stood with the stimulation "OFF" for more than 50% of the standing time [14]. In December 1997, muscle strength tests done on the Biodex dynamometer (isometric mode) showed that surface stimulation of the right quadriceps (femoral nerve) was capable of eliciting 50 Nm of knee extension at 30° of knee flexion and 45 Nm at 45° [13–15].

Prolonged Standing (1 hour): Controlled FES + Andrews' AFO

FIGURE 17.2   Praxis FES-24A system: (A) prolonged standing (1 hr) and (B) controlled FES + Andrews' anterior floor orthosis.

After implantation of the Praxis FES 24-A system in August 1998, Subject B (FR) carried out an FES exercise routine that stimulated three separate sequences (quadriceps group, buttocks and posterior thigh group, and ankle group), each running initially for 5 min and extending to 15 min over a 2-week period. Each muscle in the sequence would be stimulated sequentially for 4 s on and off. Subject B (FR) found that daily stimulation decreased his muscle spasms and spasticity level.

When standing with the implanted system, he was able to perform a variety of one-handed tasks, including reaching for and holding a 2.2-kg object at arm's length. These tasks were achieved while in the "C" posture with closed-loop activation to the lower extremity muscles and balance maintained by the other upper extremity (15; Figure 17.2B).

17.4.2.1   Bladder Results

On September 4, 1998, in the Urodynamic Testing Laboratory, Subject 2 had his sacral roots (S3 and 4) bilaterally stimulated intermittently. This showed on three occasions that the bladder contracted with recorded pressures of between 45 and 50 cm of water. On December 14, 1998, urodynamic testing again showed consistent results from S3 and 4 sacral root stimulation, producing three sustained bladder contractions with pressures of 40 to 55 cm water and urination (Figure 17.3) with each stimulation pattern (5 s on / 5 s off, 20 Hz, 8 bursts). On April 2, 1999, urodynamic testing was repeated with two bladder reflex activations from each pattern of stimulation (5 s on / 5 s off, 20 Hz, 8 to 14 bursts). Pressures of 50 to 70 cm of water were recorded.

In April 1999, the internal FES-24-A unit's connecting wire between the internal antenna and the stimulator module broke as a result of Subject B (FR) repeated bending at the waist by Subject B (FR) [15]. The receiver/stimulator unit was removed in 1999, as Subject B (FR) complained of discomfort from the two connectors under his abdominal skin. The network of leads and electrodes were left for possible replacement of the newly designed system.

17.4.3   Praxis FES-24-B System

This third iteration system — the FES-24B System (Figure 17.4) — eliminates this internal wire breakage possibility and consists of:

- A body-worn controller "Navigator" capable of executing a wide variety of software control strategies
- A skin surface stimulator "ExoStim" to mimic an implant and to provide simple exercise functions prior to implantation

SCI: Bladder Voiding: Bilateral S3 + 4 Stimulation

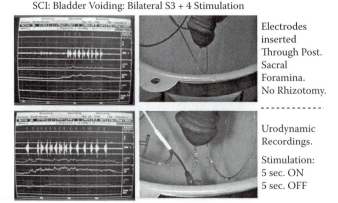

Electrodes
inserted
Through Post.
Sacral
Foramina.
No Rhizotomy.

Urodynamic
Recordings.

Stimulation:
5 sec. ON
5 sec. OFF

FIGURE 17.3 Bladder voiding with S3+4 sacral root stimulation.

TABLE 17.1 Muscles Implanted per Channel of Stimulation

Posterior adductor magnus
Biceps femoris — long head[a] or short head[b]
Gluteus maximus
Gluteus medius, minimus, and tensor fascia lata
Vastus lateralis and vastus intermedius
Vastus medialis and vastus lateralis
Tibialis anterior and extensor digitorum longus
Gastrocnemius, soleus, and flexor hallucis longus
Iliopsoas[c]

[a] Subjects 1 and 2.
[b] Subject 3.
[c] Subjects 2 and 3.

- Sensor packs incorporating accelerometers and a gyroscope to provide feedback information to control strategies
- A new implant receiver/stimulator was based on the latest cochlear implant control integrated circuit (IC), the "CIC3"
- A range of implantable electrode leads suitable for the system's multiple functions

The FES-24B System provides a maximum current output of 8 mA in a constant-current mode. Stimulation is achieved using biphasic (negative and positive phases, closely charge-matched) current pulses. Pulse widths can vary from 25 to 500 μs, and a per-channel pulse frequency of 0 to 400 Hz on each of the twenty-two channels can be obtained, which were designed as cathodes while the rear surface plate of the receiver/stimulator was connected to be the anode. The Stimulator provides real-time data telemetry functions, including the ability to measure the impedance of the current path through each electrode and the ability to transmit voltage measurements from each electrode [16].

17.4.4 Experience at Shriners Hospital for Children

Three males with paraplegia, ages 18, 21, and 21 years, underwent surgical implantation of the Muscle(s) Praxis FES-24B System between January 2002 Posterior and May 2003 at Shriners Hospital for Children, Philadelphia. Eighteen epineural electrodes (Table 17.1) were implanted for upright mobility in all three subjects, and three pairs of bifurcated linear pararadicular electrodes were placed extradurally on the bilateral S2, S3, and S4 mixed nerve roots for bladder and bowel function in the first two subjects.

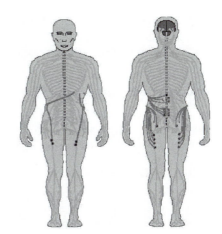

FIGURE 17.4 The FES-24B system.

A   B

FIGURE 17.5 Two subjects using the Praxis system for functional activities. (A) Subject 2 uses forearm crutches to descend stairs. (B) Subject 3 reaches for items on a shelf using a walker to support himself with one upper extremity.

17.4.4.1 Upright Mobility

Four weeks postimplantation, subjects participated in four weeks of strengthening and conditioning of the implanted muscles, followed by 17 to 22 weeks in which the focus was on programming of the upright mobility strategies and training for their functional use. Goals included achieving the transitions between sitting and standing, swing-through and/or reciprocal gait with a walker or crutches, and prolonged standing. For reciprocal gait, swing was achieved through stimulation to the iliopsoas, biceps femoris, and/or the tibialis anterior to create a flexor withdrawal response. Additional training goals included advanced activities, such as ascending and descending stairs (Figure 17.5A) and the achievement of subject-specific goals (Figure 17.5B). Bilateral ankle–foot orthoses were worn for all upright mobility activities.

Following training, data were collected for a variety of mobility activities, including transitions between sitting and standing, a short (6 m) and a long (6 min) walk, ascending and descending stairs, and maneuvering in an inaccessible bathroom stall. All subjects chose to use a swing-through gait pattern for the tested activities, except subject 2 who chose a reciprocal pattern for ascending stairs only. Subjects 1 and 3 each used a walker with wheels to perform the mobility activities, and subject 2 used forearm crutches. None of the subjects required physical assistance to complete the activities. Subjects 1 and 3 required supervision for all tested activities, and subject 2 was independent for all activities except stairs, with which he required supervision. Data for ascending and descending stairs were not collected with subject 1 as the activity was felt to be unsafe for him. Several activities could not be performed by subject 3 secondary to complaints of shoulder pain related to poor scapular muscle control.

17.4.4.2 Bladder and Bowel

Neuromodulation was attempted with subject 1 and acute suppression of reflexive bladder contractions during bladder filling was observed. When using stimulation to both S3 nerve roots throughout the day, this subject maintained a catheterized schedule (every 6 hours) comparable to that used when he took anticholenergic medication. This suggested that neurmodulation may have helped to suppress reflexive bladder activity on a daily basis as during the control period (without the neuromodulation or medication) he catheterized more frequently, on an average of every 4 hours. The ability to improve bowel evacuation was examined in subject 2, using two different stimulation paradigms: (1) low-frequency electrical stimulation (20 Hz, 350 μs, 8 mA) and (2) a combination of low-frequency and high-frequency stimulation (500 Hz, 350 μs, 8 mA). The daily use of electrical stimulation appeared to cause a reduction in the time to complete defecation by 40% with the first stimulation strategy and by 60% with the second strategy.

Despite numerous attempts with varying stimulation parameters to the sacral nerve roots, neither subject could obtain detrusor pressures sufficient to provide voiding with stimulation. Both subjects continued to catheterize for bladder emptying.

17.4.4.3 Electrode Stability

Three of the fifty-two electrodes placed for lower extremity stimulation experienced changes in the responses of the muscles. One of these was due to a disconnection at the connector site between the implant and the electrode lead. This was repaired and the electrode continued to function without further problems. The remaining two electrodes (biceps femoris and tibial nerve) were not replaced, as they did not impact function for the subjects involved.

17.4.4.4 Sensors

Closed-loop standing using sensor packs incorporating accelerometers and a gyroscope was attempted with Subject 1. Sensor packs were attached externally on the thigh and the calf (Figure 17.6) to detect the position of the knee while standing. Stimulation would decrease until a change in the knee joint angle was detected, at which time stimulation would again increase to prevent a knee buckle. Figure 17.7 demonstrates the use of the sensors for closed-loop feedback to the right quadriceps muscles during quiet standing. Using closed-loop control, the subject was able to stand with less stimulation to the quadriceps than what he had been using while standing with open-loop control. He was also able to stand for a longer period of time before the muscle fatigued, requiring him to sit. The algorithm for increasing and decreasing stimulation to the quadriceps did not create any balance disturbances for this subject.

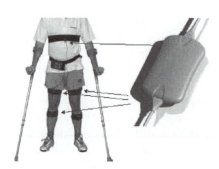

FIGURE 17.6 Sensor packs used for closed-loop standing.

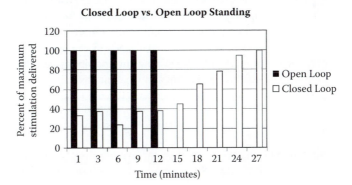

FIGURE 17.7 Using open-loop control, the stimulation remained at 100% for 12 minutes of standing after which the subject's muscles were too fatigued for him to remain upright. Using closed-loop control, stimulation could be maintained at a lower level, increasing over time as needed. With closed-loop control, standing time was more than doubled to 27 min.

17.4.5 Complications on Follow-up of the First Two Implanted at the NEC Site, Subjects (A and B)

In 2002, Subject A accidentally cut his left foot, which was treated superficially. In three to four days, his left lower extremity was swollen with an infection, which immediately was treated with intravenous antibiotics for two weeks. The swelling resolved but six weeks later the tissues around the Nucleus FES-22 system were swollen and inflamed. After three days of IV antibiotics, the implanted system was explanted, taking as much time as when it was implanted. The most difficult part was finding and dissecting the small 2.5-mm platinum electrodes and their silastic backing. He recovered well without further complications.

In 2001, Subject B was experiencing intermittent pain in the T7-8 vertebra at the postfractured site, after conservative treatment failed; the spine was fused in this area. By the time he was ready for implanting the Praxis FES-24B stimulator in 2003, the Neopraxis Company had been closed by Cochlear Ltd. However, he was offered the stimulator for implantation, but without further support from the company; he decided not to continue and elected to have the leads and electrodes removed.

At the SHC site, during the training period, Subject 2 sustained a stress fracture of the left proximal first metatarsal, which he believed happened when his left leg experienced greater impact at initial contact due to his poor control of swing for that step. The subject was immobilized for six weeks in a soft boot, after which he was able to return to training without further problems. At the end June of 2002, Subject 1 sustained an abrasion near his ankle and antibiotics were started once this was reported. Then at the beginning of August 2002, he began experiencing high fevers and complained of heat and inflammation around one of his surgical incisions. Despite treatment with IV antibiotics, this subject continued to experience problems with inflamed incisions, some of which resulted in open skin and fluid drainage. Antibiotic treatment appeared to temporarily suppress these reactions but problems continued. Due to this, the majority of the system has been removed, with future surgeries planned to remove the remainder.

17.5 Conclusion

In the developing field of FES and implantable neural prosthetic devices, there has been a need for reliable and safe, multichannel implantable stimulating systems to restore multiple functions in neurologically impaired patients. In paraplegic individuals, the stimulating systems' functions should be designed to modulate spasticity and precisely activate individual muscles for joint movement and control of bladder and bowel functions. The more channels available, the more nerves that can be activated and the more modes of functionality that can be restored. Our contribution to this goal has been continuous since 1983, and the two Praxis FES Systems [11–16] have provided the hope for a new rehabilitation aid for restoration of function in spinal cord injury paraplegia. Providing more functions with an FES system with a greater number of channels introduced new challenges to the subjects and research teams, including the need for multiple surgical procedures, new surgical approaches to placing electrodes, increased risk of infection, and greater hospitalization and rehabilitation times. Importantly, these challenges are being addressed through multiple research efforts [17] and Chapter 18 at various centers.

Acknowledgments

Our thanks to Cochlear Ltd. and Neopraxis Pty. Ltd. for making these studies possible. Our sincere thanks to our many collaborators at NEC: S.E. Emmons, J. McKendry, R. Eckhouse, A. Delehunty, W. MacFarland; and at SHC: M.J. Mulcahey, B. Benda, G. Creasey, and M. Pontari.

References

1. Kralj, A. and Bajd, T. *Functional Electrical Stimulation: Standing and Walking after Spinal Cord Injury.* CRC Press, Boca Raton, FL, 1989.
2. Davis, R. International functional electrical stimulation society: the development of controlled neural prostheses for functional restoration. *Neuromod.*, 3:1–5, 2000.
3. Agarwal, S., Triolo, R.J., Kobetic, R., Miller, M., Bieri, C., Kukke, S., Rohde, L., and Davis, J.A. Long-term user perceptions of an implanted neuroprosthesis for exercise, standing, and transfers after spinal cord injury. *J. Rehabil. Res. Dev.*, 40(3):241–252, 2003.
4. Bonaroti, D., Akers, J., Smith, B.T., Betz, R.R., and Mulcahey, M.J. Comparison of functional electrical stimulation to long leg braces for upright mobility for children with complete thoracic level spinal injuries. *Arch. Phys. Med. Rehab.*, 80:1047–1053, 1999.
5. Johnston, T.E., Betz, R.R., Smith, B.T., and Mulcahey, M.J. Implanted functional electrical stimulation: an alternative for standing and walking in pediatric spinal cord injury. *Spinal Cord*, 41(3):144–152, 2003.
6. Bajd, T. and Jaeger, R. FES for movement restoration. *BAM*, p. 228–229, 1994.
7. Brindley, G. The first 500 patients with sacral anterior root stimulator implants: general description. *Paraplegia*, 32:795–805, 1994.
8. Creasey, G. Managing bladder, bowel and sexual function after spinal cord injury. In *Handbook of Neuro-Urology*, Rushton, D., Ed. Marcel Dekker, New York, 1994, p. 233–251.
9. Davis, R., Eckhouse, J., Patrick, J., and Delehunty, A. Computerzed 22 channel stimulator for limb movement. *Appl. Neurophysiol.*, 50:444–448, 1987.
10. Davis, R., Eckhouse, J., Patrick, J., and Delehunty, A. Computer-controlled 22-channel stimulator for limb movement. *Acta Neurochir.*, 39:117–120, 1987.
11. Davis, R., Kuzma, J., Patrick, J., Heller, J., McKendry, J., Eckhouse, J., and Emmons, S. Nucleus FES-22 stimulator for motor function in a paraplegic subject. *RESNA Int.*, 1992, p. 228–229.
12. Davis, R., MacFarland, W., and Emmons, S. Initial results of the Nucleus FES-22 implanted stimulator for limb movement in paraplegia. *Stereotact. Funct. Neurosurg.*, 63:192–197, 1994.
13. Davis, R., Houdayer, T., Andrews, B., Emmons, S., and Patrick, J. Paraplegia: prolonged closed-loop standing with implanted Nucleus FES-22 stimulator and Andrews foot-ankle orthosis. *Stereotact. Funct. Neurosurg.*, 69:281–287, 1997.
14. Davis, R., Houdayer, T., Andrews, B., and Barriskill, A. Prolonged closed-loop functional electrical stimulation and Andrews ankle-foot orthosis. *Artif. Organs*, 23:418–420, 1999.
15. Davis, R., Houdayer, T., Andrews, B., Barriskill, A., and Parker, S. Paraplegia: implantable Praxis 24-FES System and external sensors for multi-functional restoration. *Proc. 5th Ann. Conf. Int. Funct. Electr. Stim. Soc.*, Aalborg, Denmark, June 18–21, 2000, p. 35–38.

16. Davis, R., Patrick, J., and Barriskill, A. Development of functional electrical stimulators utilizing cochlear implant technology. *Med. Electron. Phys.*, 23:61–68, 2001.
17. Schulman, J., Mobley, P., Wolfe, J., Voelkel, A., Davis, R., and Arcos, I. An implantable bionic network of injectable neural prosthetic devices: the future platform for functional electrical stimulation and sensing to restore movement and sensation. In *Engineering for Neural Enhancement and Replacement*, Walker, C.F. and DiLorenzo, D.J., Eds. CRC Press, Boca Raton, FL, 2005.

18

An Implantable Bionic Network of Injectable Neural Prosthetic Devices: The Future Platform for Functional Electrical Stimulation and Sensing to Restore Movement and Sensation

Joseph Schulman,
J. Phil Mobley,
James Wolfe, Ross Davis,
and Isabel Arcos

18.1 Introduction

Functional electrical stimulation (FES) is a rehabilitation technique for the restoration of lost neurological function, resulting from conditions such as stroke, spinal cord injury, cerebral palsy, head injuries, and multiple sclerosis. FES utilizes low-level electrical current applied in programmed patterns to different nerves or reflex centers in the central nervous system to produce functional movements. The stimulation may be triggered by a single switch (open-loop) or from sensor(s) or neuronal activity (closed-loop).

While FES has been used successfully to pace the heart[1] and to restore hearing[2] in the past, it has not been widely adopted as a means of reanimating paralyzed limbs that result from stroke and spinal cord injury (SCI). It is estimated by the U.S. National Institutes of Health (NIH) that there are more than 600,000 people who experience a stroke each year in the United States, with an associated comprehensive cost of $43 billion per year.[3] Of the more than 4 million stroke survivors alive today, many experience permanent impairments of their ability to move, think, understand, and use language, or speak — losses that compromise their independence and quality of life. Furthermore, stroke risk increases with age; and as the American population is growing older, the number of persons at risk for experiencing a stroke is increasing. There are also an estimated 250,000 Americans living with spinal cord injuries, with 10,000 to 12,000 new spinal cord injuries reported every year in the United States. The cost of managing the care of SCI patients approaches $4 billion each year.[4]

The potential of FES to restore function in these areas has been largely unfulfilled, mostly due to the limitations of the FES devices currently available. FES could also be used in limb loss applications to reduce phantom pain and to restore functional movement of prosthetic limbs. In 2000/2001, about 130,000 lower-limb amputations were performed annually in the United States.[5,6]

An optimal FES system should have the following fundamental characteristics. It should

1. Provide both stimulating and sensing capabilities
2. Be fully implantable
3. Be minimally invasive
4. Have real-time communication capability
5. Allow a practically unlimited number of stimulation and sensing channels
6. Function without external equipment or interconnected leads between components

This chapter describes a network of wireless implantable microstimulators/microsensors, also known as battery-powered BION®[1] (BIOnic Neuron) devices for functional electrical stimulation and sensing (FES-BPB system). This new platform was designed to overcome the limitations of the current FES technology by providing:

1. Microdevices that can be programmed to be either stimulators or sensors for use in closed-loop applications
2. Minimally invasive implantation procedures to reduce labor-intensive surgery and associated patient risks and to provide rapid recovery
3. Wireless bidirectional communications and telemetry to all stimulators and sensors, which eliminates the use of both transcutaneous leads (which are susceptible to infection) and surface-applied coils and stimulators
4. Real-time communication between the stimulators, sensors, and control unit to maintain continuous closed-loop control
5. Flexibility and functional expandability because there are no leads, and each implant has a full complement of programmable stimulators and sensors
6. A large number of channels, which allows the same system to be used for a variety of applications without interference in the same patient
7. Self-powered operation using rechargeable batteries to power the implantable devices; external equipment (e.g., power antennas) is only needed during battery recharging
8. Wireless sensors capable of measuring biopotentials, angle, position, pressure, temperature, and permanent magnet fields

18.2   Evolution of the Implantable BION Devices

In 1988, Loeb proposed and Heetderks mathematically showed that the concept of a wireless network of injectable microstimulators powered by an external antenna/coil was possible.[7,8] It was thought that these

---

[1] BION is a registered trademark of the Advanced Bionics Corporation, a Boston Scientific company.

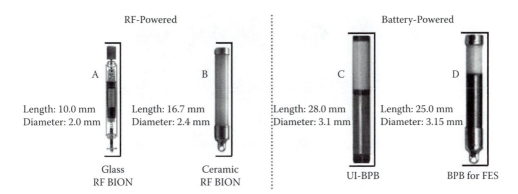

FIGURE 18.1 (**See color insert following page 15-4**). Evolution of BION devices.

microstimulators would eliminate many of the problems associated with the use of percutaneous electrodes, because they do not incorporate leads. Schulman, Loeb, and Troyk, under support contracts from the U.S. NIH (contract N01-NS-9-2327), the Alfred Mann Foundation (AMF, Santa Clarita, California), and the Canadian Network for Neural Regeneration and Functional Recovery, developed an injectable, glass-enclosed microstimulator (Figure 18.1A) that is powered and controlled by an external alternating magnetic field generated by a coil connected to a control unit. This first 255-channel stimulating system, later called the radio-frequency (RF) BION device, allowed instantaneous control of stimulation pulse amplitude, frequency, pulse width, pulse position timing, and pulse charge-recovery current.[8]

AMF continued to develop and improve the wireless RF BION device. As a result of these efforts, a second-generation RF BION device, which incorporates a ceramic case, an output capacitor, and Zener diodes to protect the device against electrostatic discharges was developed (Figure 18.1B).[9] These RF BION devices are currently being used or have been used in several clinical studies being conducted by AMF and its affiliated organizations, the Alfred Mann Institute (University of Southern California, Los Angeles) and Advanced Bionics Corporation (Sylmar, California). As of September 2004, thirty-three patients have been implanted with RF BION devices for treatment of urinary incontinence, obstructive sleep apnea, pain associated with shoulder subluxation, knee osteoarthritis, forearm contracture, and foot drop applications.[10–13]

While the wireless RF BION device eliminates the need for the leads associated with the percutaneous electrodes, it requires the patient to wear an external coil during use, to transmit power and data to the implanted device. Both to improve patient acceptance of this technology and to increase the reliability of the system, it was necessary to eliminate the need to constantly use an external coil to power and control the device. Thus, the idea for a battery-powered BION device (hereafter BPB) was conceived at the AMF.

AMF developed, with guidance from the Jet Propulsion Lab (JPL), a small cylindrical lithium ion rechargeable battery. Alfred Mann formed and financed a new company (Quallion, Inc.) to manufacture and improve these unique, highly reliable batteries. Today, these batteries can be safely recharged if discharged to 0 volts and are expected to operate for more than ten years. These batteries were designed specifically for the BPB (Figure 18.2).

AMF licensed the BPB technology to Advanced Bionics Corp., which designed and implemented the first BPB (stimulator only) for urinary incontinence (referred to as the UI-BPB). This was the first BION to have two-way telemetry (Figure 18.1C). The telemetry receiver in this UI-BPB turns on for a very short time interval every 1.5 s, to conserve battery power. Thus, rapid synchronization for limb control is not feasible with the UI-BPB. As of September 2004, the UI-BPB had been implanted in thirty-five patients for the treatment of urinary incontinence and migraine headaches.[14] Due to its lack of sensing capabilities, and slow communication response time, the UI-BPB is not well suited for FES applications.

AMF is currently developing the next-generation BPB. This BPB (Figure 18.1D) allows for the creation of a wireless FES network including both stimulation and sensing in each BPB for fully implantable

FIGURE 18.2 Quallion battery for the BPB.

closed-loop applications; data processing for sensed signals; high-speed bidirectional telemetry; wireless oscilloscope monitoring for fitting purposes, via back telemetry of voltage sensed signals; a rechargeable battery (enabling prolonged operation without external power); and the capability of communications with more than 850 BPBs simultaneously (effectively, 100 communications per second).[15]

The ceramic BION devices (Figures 18.1B, 18.1C, and 18.1D) use an extremely strong zirconia with 3% ytrium ceramic case. The FDA pointed out that long-term immersion in water significantly weakens this ceramic.[16] Over a three-year period, AMF came up with a process to improve the longevity of this ceramic. Today, accelerated life testing has shown that this ceramic will retain 80% of its strength after eighty years of soaking in saline solution.[17–19]

18.3 Battery-Powered BION System for Functional Electrical Stimulation and Sensing

The FES-BPB System is a wireless, multichannel network of separately implantable battery-powered BION devices that can be used for both stimulation and sensing. The system is comprised of a master control unit (MCU), a clinician's programmer, a recharging subsystem (charger and coil), and BPBs. Additional equipment, dissolvable suture material, and surgical insertion tools (used only during the implantation procedure) are also part of the system. The FES-BPB System can be set up for use in two configurations: (1) fitting mode and (2) stand-alone mode. A block diagram of the FES-BPB System is shown in Figure 18.3.

The MCU is the communication and control hub for the FES-BPB System. The MCU transmits commands to and receives data from all the BPBs and the recharging subsystem. There are two versions contemplated for the MCU packaging: (1) an external MCU, which will be outside the body and have a few controls accessible to the patient; and (2) an implantable MCU, which will be implanted in a convenient location in the body. The implantable MCU version will have a small patient control unit (PCU).

The clinician's programmer consists of software loaded on a computer that allows the clinician to configure and test the FES-BPB System for each patient.

The recharging subsystem (charger and coil) is used when the rechargeable battery of the BPB needs recharging. The charging process requires the placement of the coil close to the area on the patient's body where the BPB is implanted. Recharging is mandatory when the battery of the BPB is low. Depending on the frequency of use and stimulation levels delivered by the BPB, the battery could potentially run down in 1 to 8 days. Under normal stimulation conditions (nerve stimulation of 1 to 2 mA pulse amplitude, 15 to 100 μs pulse width, and 20 pulses per second (pps)), charging for about 5 to 20 min per day is required to charge the battery. The BPB maximum stimulation capability (20 mA pulse amplitude at 14 V compliance with 200 μs pulse width and 125 pps) can rapidly discharge the battery in a very short time and would result in the need to recharge the battery for longer periods of time and much more frequently.

The external charger coil transmits only power to the implanted devices. Charging and battery status are transmitted by each BPB to the MCU. Each BPB can accept a charging field twenty times the nominal

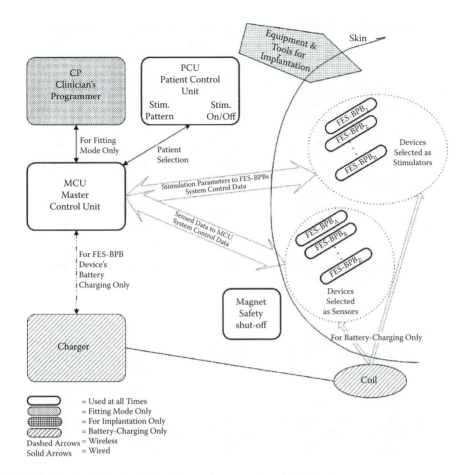

FIGURE 18.3 Functional electrical stimulation and sensor system (FES-BPB).

field to recharge without overheating. Upon completion of the charging process, the MCU then issues a stop-charging command to the charger.

The BPB has a sensor to detect the field from a permanent magnet. The function of the magnet is to hold off the stimulation. This is a safety feature in case the BPB is stimulating in an undesired manner and the patient does not have access to his control unit. When the magnet is positioned on the patient's body over the BPB implant area, a magnetic sensor inside the BPB detects the external magnetic field and holds off the stimulation. When the magnet is removed, the stimulation turns on again. This is the default mode of the magnetic detector. Other modes can be programmed during fitting.

The BPB implantation is performed in a minimally invasive procedure and is accomplished using a combination of specially designed insertion tools and commercially available items.

The functional description of the FES-BPB System components is presented below.

18.3.1 Master Control Unit (MCU)

The MCU is the communication and control hub for the FES-BPB System. The MCU transmits commands to and receives data from each one of up to a total of 850 BPBs in the system within one hundredth of a second. When the patient uses the FES-BPB System (in stand-alone mode), the MCU coordinates the activity of the BPBs by receiving data from implanted devices programmed as sensors, transmitting stimulation commands, and monitoring overall system status. It also serves as the basic user interface for the patient, providing system ON/OFF control and alarms, as well as program selection and limited parameter control. During fitting, the MCU acts as a conduit between the clinician's programmer and the rest of the system, enabling the transparent setup of each BPB and the coordination needed among BPBs to implement the desired functional movement.

The MCU also manages the recharging subsystem. The charger communicates with the system in the same manner as the BPBs, and it can be turned ON/OFF or checked for correct operation via the MCU.

The MCU contains the following safety mechanisms:

- An emergency STOP button on the external MCU, which when depressed, immediately issues a "stop stimulation" command to all BPBs.
- During recharging, if any BPB overheats or overcharges and cannot protect itself, it would communicate this information to the MCU, which would issue a "stop charging" command to the charger and alert the patient. If an external MCU is being used, it would produce a sound to alert the patient. In the case an implantable MCU is being used, the MCU would send a command to the BPBs to produce a specific stimulation pattern to alert the patient and would also communicate with the external PCU, if it is within communication range. The PCU would then generate an audible alert.

The MCU also stores patient usage data for the clinician. This data can be used to verify compliance and to analyze the stimulation and sensing parameters of each session. The approximate location of each BPB in the body is also maintained in this database.

18.3.2 Software and Firmware

The software for the FES-BPB System is divided into two components. One component is the clinician's programmer application, which runs on a laptop computer. The other component is the firmware running on the MCU. The personal computer (PC) with the clinician's programmer interfaces with the MCU via a serial communication link. During the fitting of the system to a particular patient, the two components work in concert to facilitate measurement and storage of the stimulation and sensor calibration parameters. Once these parameters have been gathered in a fitting session, the essential information can be stored in the MCU so that the MCU can operate in stand-alone mode to facilitate the desired functional movement. Ultimately, the MCU will modulate the stimulation output in response to information it receives from the BPBs programmed as sensors.

The clinician's programmer uses a graphical interface that contains screens to perform the following essential functions:

- Gather basic personal information for the patient, including information about the location in the body of his or her BPBs.
- Establish the stimulation range for each implanted BPB and allow selection of the stimulation parameters.
- Specify the details of the activity sequences that will be involved in the FES algorithm.
- Gather the trigger information that will be used to generate transitions between activity sequences in response to the sensor inputs.
- Compose the Finite State Machine functions that will drive a routine and download the complete program to the MCU for either immediate execution or later use.

18.3.3 Battery-Powered BION Device (BPB)

The BPB is a battery-powered microdevice capable both of delivering electrical stimulation and acting as general-purpose sensor for recording biopotential signals, pressure, distance, or angle between two BPBs and temperature. The following sections describe the functional building blocks of the BPB, battery, and packaging. The specifications of the BPB are provided in Table 18.1. Figure 18.4 shows the internal components of the BPB and Figure 18.5 shows a cross-section of an assembled BPB.

The BPB has the following subsystems.

TABLE 18.1   Battery-Powered BION Device Specifications

1. Physical	
Implant weight	0.6 g
Implant length with eyelet[++] and diameter	25 mm$_{max}$ length / 3.15 mm$_{max}$ diameter
Electrodes area	5.2 Sq. mm [0.008 sq. in.] stimulation electrode
	12.8 Sq. mm [0.019 sq. in.] return electrode
Case materials	Yttria-stabilized zirconia; titanium 6Al4V alloy
Electrodes material	Iridium
2. Stimulation Parameters	
Pulse amplitude	5 µA to 20 mA in 3.3% exponential steps (255 levels)
Pulse width	7.6 to 1953 µs in 7.6 or 15.2 µs steps
Pulse frequency	1 to 4096 pps
Stimulation control response time	10.6 ms maximum
Capacitor recharge current	10–500 µA
Compliance voltage	Up to 14 V automatically adjusted
Stimulation output capacitor	4 µF
Delay to start from a trigger	0 to 42.4 hr in 15.6 ms, 125 ms, 2s, 1 min, 10 min steps

	Min	Max	Step
Burst ON/OFF time			
Range 1	0.031s	0.9996 s	0.0156 s
Range 2	0.25 s	8.00 s	0.125 s
Range 3	4 s	128 s	2 s
Range 4	2 min	64 min	1 min

3. Sensors	
Temperature	16–50 °C with 0.3% accuracy
Magnetic field to trigger shut off	10.0 Gauss threshold
Goniometry (number of frequency channels)	8
Range	1–20 cm
Repeatable Accuracy Error	Less than 1% for 1–10 cm
Pressure (Range)	Readout = AC coupled. 300 to 900 mmHg absolute
Accuracy	±10 mmHg
Biopotential Sensing (Amplification)	10, 30, 100, 300, 1000
Low-frequency roll-off	1, 10, 30, 100, 300 Hz
High-frequency roll-off	300 Hz, 1 kHz, 3 kHz, 10 kHz
Notch filter	50 or 60 Hz
Input referred noise	5 µVrms
4. Communication	
Number of implants per patient	Up to 850 at 10 ms
ID, MCU/BPB	27/30 bits
Bandwidth	5 MHz
Sense-to-stimulate delay	10.6 ms maximum
Frequency band	100–500 MHz
MCU to BPB data rate 15 bits/6 µs	(15 bits data + 16 bits FEC[a]) / 6 µs
BPB to MCU data rate 8 bits/5 µs	(8 bits data + 8 bits FEC[a]) / 5 µs
Data streaming (oscilloscope mode)	39.8 K samples / sec × 3 channels (8 bit resolution)
5. Charging	
Frequency of charging field	127 kHz
Excessive magnetic field permissible	20 times nominal
6. Battery Type	
Lithium ion rechargeable, hermetically sealed	
Battery length and diameter	13 mm length, 2.5 mm diameter
Battery weight	0.21 g
Battery capacity	3.0 mA-hr, 10 mW-hr
Cell voltage range	3.0–4.0 V (3.6 V nom.)
Battery life	Nominally 10 years (usage dependent)

[a]   FEC = Forward error correction.

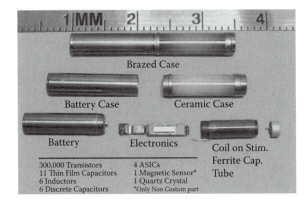

FIGURE 18.4 (See color insert following page 15-4). Battery-powered BION internal components.

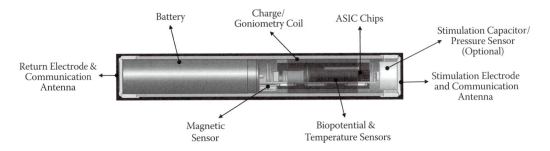

FIGURE 18.5 (See color insert following page 15-4). Battery-powered BION device cross-section.

FIGURE 18.6 MCU–BPB communication protocol.

18.3.3.1 Stimulation

The BPB is a single-channel, constant-current, charge-balanced stimulator. The stimulation output is capacitance-coupled, which also prevents direct connection between the battery or battery-generated DC voltages and the tissue. Stimulation pulse amplitude, width, and frequency can be independently adjusted. In addition, triggering events can cause the stimulation to be delivered continuously or in a pulse burst, which can be ramped up and/or down with a variety of start/stop times.

18.3.3.2 Communication

The bidirectional propagated wave RF communication between the MCU and the BPBs is established through a dipole antenna (Figure 18.5). This link operates at a frequency in the band 100 to 500 MHz, using quad-phase modulation with a 5-MHz bandwidth. The BPB communication module includes a crystal-controlled transmitter, receiver, and digital processing unit that synchronizes with and processes the MCU transmissions. The digital processing unit in the BPB also corrects small numbers of errors in the received data, decodes the MCU commands, and generates the responses to the MCU, including the reporting of higher numbers of communication errors that cannot be mathematically corrected. In this latter situation, the MCU would resend the message.

The communication protocol between the MCU and BPBs is shown in Figure 18.6. The timing of the frame is completely controlled by the MCU and every BPB with synchronization to its MCU's clock.

The header and trailer fields are used for frame synchronization and for carrying frame control data intended for all BPBs and/or for other MCUs. When an MCU detects another MCU, the one with the higher ID number shifts the time slots of all the BPBs it is controlling, to avoid communication interference. Once the MCU assigns the time slots for the downlink and uplink data packets to each one of the BPBs in the net, each BPB turns on its receiving or transmitting circuitry for only a few microseconds at the assigned times in each frame in order to save battery. The downlink data packets contain stimulation and/or sensing control data and forward error correction (FEC) bits to correct up to 4 or 5 bit errors. Bit errors beyond that number are reported to the MCU, which will then resend the message. If some messages are vital, the message would be sent twice or the value would be sent back to the MCU for the MCU to verify and authorize the command. Uplink data packets are transmitted by each of the BPBs and are used to carry information to the MCU (e.g., sensed data). The FEC in the uplink data packet only corrects 1 or 2 bit errors.

18.3.3.3   Power (Battery and Charging)

The main power source for the BPB is a 10-mW-hr rechargeable lithium-ion battery that allows the implanted device to operate as a stand-alone stimulator/sensor. Its special nonflammable lithium ion chemistry provides long life and permits the voltage to go to zero and be recovered safely without damage to the battery. The recharging process is achieved via a low-frequency (127 kHz) magnetic link with an external coil worn or placed nearby when charging. Assuming continuous stimulation pulses at 20 pps with 100-μs pulse width and 2-mA pulse amplitude into a 2kΩ load, the battery of a BPB selected as a stimulator will provide 100 hours of continuous operation. For a BPB selected as sensor, the battery will also provide 100 hours of continuous operation.

The lithium ion battery is specified to have a cycle life of 2000 cycles for a standard charge/discharge cycle, which is a fairly deep discharge of the battery before recharge occurs. The nominal stimulation/sensing requirements in many applications are such that the battery would not be discharged to the standard low level (if recharged daily). Thus, the 2000 cycles represent a lifetime of more than ten years if the battery is recharged daily.

18.3.3.4   Safety

The BPB includes the following safety features:

- A miniature magnetic sensor that detects the magnetic field from an external magnet and holds off the stimulation if, for some reason, it needs to be turned off.
- A temperature sensor that communicates with the charger, via the MCU, to terminate charging, if appropriate, and disconnects the battery when the temperature rises above a predetermined threshold.
- Battery safety circuitry that protects the battery from overvoltage, overdischarge, and overcharging.
- BPBs can protect themselves from magnetic fields in excess of twenty times the field necessary for maximum charging and, for a short time, for fields in excess of fifty times the field for maximum charging. This short time is more than sufficient for the BPB to send a message to the MCU to turn off the charger and to alert the patient of risk.

18.3.3.5   Biopotential Sensing, Data Display, and Data Analysis

The biopotential function is implemented to record neural or muscular electrical signals (EMG signals). Biopotential sensing is accomplished using a low-noise amplifier and bandpass filter circuit, followed by a digital post-processing circuit. The amplifier is adjustable from a gain of 10 to a gain of 1000. The low-frequency setting of the bandpass filter is adjustable from below 1 Hz to 300 Hz. The high-frequency setting is adjustable from 300 Hz to 10 kHz. Input referred noise is less than 5 μVrms (20 μVpeak).

18.3.3.5.1   Data Display: Oscilloscope Mode

During fitting, the analog signal from the amplifier/filter section can be digitized and transmitted from the BPB to the MCU to the clinician's programmer screen at rate of 40,000 samples per second. This "oscilloscope mode" (Figure 18.7) can be used when evaluating the placement of the BPB and during fitting, but it is not suitable for long-term use due to its high power demands.

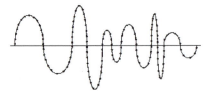

FIGURE 18.7 Biopotential sensing module: oscilloscope display mode.

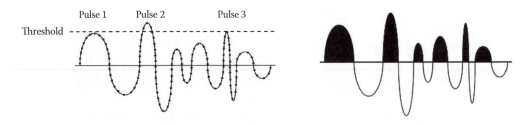

FIGURE 18.8 Data analysis with the BPB biopotential sensing module. Left: counting pulses above threshold line. Right: rectify and integrate neural signal.

18.3.3.5.2 Data Analysis

The analog output of the biopotential sensor also passes to a programmable window detection circuitry that can be set by the clinician to (1) count pulses that fall within (or above or below) the set thresholds (Figure 18.8, left); or (2) rectify and integrate the sensed signal (Figure 18.8, right). (1) Counting pulses: the neuronal pulses that occur are accumulated every 10 ms and relayed to the MCU. (2) Rectify and integrate: every 10 ms, if required, the circuit can rectify the amplitude of the biopotential sensor's analog output and sum up the average rectified signal. An output between 0 and 255 will be generated, indicating the average energy occurring every 10 ms.

18.3.3.6 Pressure Sensing

Some BPBs will be fabricated with a pressure transducer mounted at one end. The initial version of this sensor is about 3 mm in diameter and is sensitive to pressures along the axial dimension of the BPB. Future versions will be sensitive to lateral pressure, and may be mounted remotely from the BPB. The present full-scale, absolute pressure range is 400 to 900 mmHg. This signal can be read out either AC or DC coupled. When DC coupled, it reads the absolute pressure. Because ambient pressure varies with altitude changes, this offset can be accounted for by placing a reference sensor of the same type in the MCU, and then subtracting off this baseline.

18.3.3.7 Angle/Position Sensing (Goniometry)

The same internal coil that is used to receive the magnetic field to charge the BPB battery can also be programmed as a transmitter in any selected BPB or as a receiver in another selected BPB. The goniometry function is implemented using one BPB as a transmitter, and the other BPB as a receiver. The BPB programmed as a receiver detects and measures the signal strength of the received signal (Figure 18.9). The distance between two BPBs is derived from the intensity of the received magnetic field, which falls off approximately with the cube of the distance between the devices.

There are eight different programmable transmitter–receiver frequencies available for goniometry use. The eight frequencies are clustered around 127 kHz. This permits eight parallel goniometry systems consisting of one transmitter and any number of receivers. There is no limit to the number of BPB receivers that can process the signal strength to give distance measurements (from each of the transmitters). Each BPB receiver is able to send back a measurement 100 times per second. A goniometry pair (transmitter–receiver) can be used to measure distances between 1 and 20 cm.

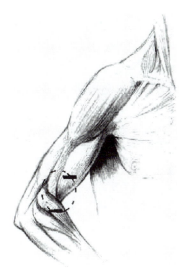

FIGURE 18.9 Use of BPBs for distance/angle measurements.

18.3.3.8 Temperature Sensing

An internal temperature sensor is incorporated as an additional safety mechanism to guard against overheating the BPB and to provide temperature data to patients, such as certain quadriplegics who do not sense temperature. The sensor is accurate to within 0.33°C and is operable over the range from 16 to 50°C. In the event a significant temperature rise is detected, the BPB can be shut down and/or communication with the MCU can be made to initiate appropriate external action (such as shutting down the charging field, if present). Readings are taken once per second and can be read by the MCU.

18.3.4 Recharging Subsystem (Charger and External Coil)

The charger produces a 12-kHz signal that generates a magnetic field in the charging coil. The MCU communicates with the charger to indicate when to turn on a charging field. The MCU interrogates each BPB to determine which BPB is going to be charged and when the BPB is fully charged. The MCU determines which BPBs are not being charged and indicates to the patient where the coil must be moved to charge those BPBs. If the charger is coupled to several coils but can only power one coil at a time, the MCU can then cause the charger to switch coils so uncharged BPBs can be charged. The MCU can also determine the state of the charge in each BPB and can initially select the most discharged devices to be charged first.

The recharging subsystem includes a temperature sensor that stops the recharging process if the external coil temperature adjacent to the patient skin rises higher than 41°C.

18.3.5 Magnet

The patient can stop stimulation by placing an external magnet near the location of the implanted device(s). A neodymium magnet is being used because it is small, lightweight, and it produces a very strong magnetic signal. The default mode of this magnet is to hold the stimulation off when the magnet is positioned on the patient's body, over the area where the BPB is implanted. When the magnet is removed, the stimulation turns back on. Other magnet control modes are available.

18.3.6 FES-BPB System Specifications

Table 18.1 provides the device specifications for the battery-powered BION device.

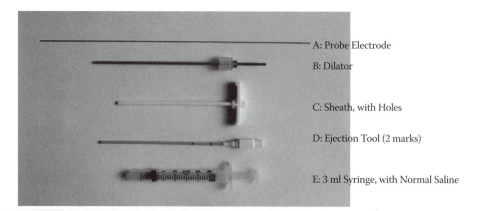

A: Probe Electrode

B: Dilator

C: Sheath, with Holes

D: Ejection Tool (2 marks)

E: 3 ml Syringe, with Normal Saline

FIGURE 18.10   (See color insert following page 15-4). BPB implantation tools.

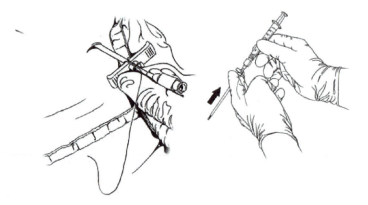

FIGURE 18.11   BPB implantation technique.

18.3.7   Minimally Invasive Procedure to Implant BPBs

To implant a BPB, a minimally invasive procedure is followed. The implantation procedure can be done in a clean Procedure Room, where the patient's implant sites can be surgically cleansed and draped with sterile towels and covered with adherent sterile plastic drapes. The implant physician scrubs his or her forearms and hands, is gowned and gloved, and wears a cap and mask.

The implantation (insertion) tools are shown in Figure 18.10. Under local anesthesia, a 5-mm skin incision is made. A sterile probe electrode (0.71 mm OD; insulated except at the tips, Figure 18.10) connected to an external stimulator is directed into the tissues to excite and find the target nerve/motor-point. With adjustments to the probe electrode, the optimal target muscle contraction is located. A customized introducer (dilator plus sheath) is then slid over the probe electrode. Stimulation with the probe electrode is repeated to ensure a similar optimal response and correct location. The probe electrode and dilator are then withdrawn, leaving the sheath in position.

The BPB has a dissolvable suture attached to the return electrode. The BPB's stimulation electrode end is inserted into the sheath and gently pushed by the ejection tool to the sheath tip, so that only the BPB stimulation electrode end protrudes. From the ejection tool tip, saline is infused into the sheath to allow the anodal end of the BPB to have electrical connection to the tissues through small holes in the distal sheath. The BPB is activated to test and confirm its optimal position relative to the target nerve/motor-point. By withdrawing the sheath over the ejection tool, the BPB is deposited into the tissues (Figure 18.11). The sheath and the ejection tool are then removed.

The BPB is retested to confirm that the optimal response is achieved. If this position is not satisfactory with regard to the responses to stimulation or recording, then the BPB can be retrieved by pulling on

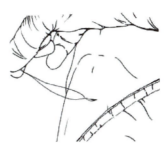

FIGURE 18.12 Retrieval of the BPB.

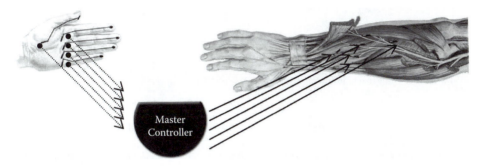

FIGURE 18.13 (**See color insert following page 15-4**). BPBs measuring pressure in fingers.

the suture attached to the BPB (Figure 18.12) and then reinserted. The emerging sutures are cut at the subcutaneous tissue level, and the wound is then closed.

The implanted BPB is tested one week after implantation to confirm that the responses are still adequate. If an inadequate response is observed, the wound could be reopened and the BPB retrieved by pulling on the sutures. A new BPB could then be reinserted to obtain proper response.

18.4 Applications

The different functions of the BPB (as a stimulator, biopotential signal sensor, goniometry sensor, pressure or temperature sensor) and the availability of multiple BPBs in one patient (up to 850 BPBs) gives the clinician many opportunities to restore neurological function, especially in poststroke syndrome, spinal cord injury, cerebral palsy, multiple sclerosis, traumatic brain injury, and for limb sensing in amputees to control fitted prostheses.

Take, for example, the case where a paralyzed upper extremity is implanted with multiple BPBs placed near motor-points or nerves of muscles in the arm, forearm, and hand. It will be possible to trigger sequential functional muscle actions to extend the arm and forearm, and open the hand to grasp an object. The limits of each functional action can be controlled from implanted BPBs working as goniometry sensors, measuring the angles of the elbow (see Figure 18.9) and wrist, and implanted BPBs working as pressure sensors, measuring the pressure at the finger tip (Figure 18.13).

The reverse of this extension can be similarly achieved using this stimulating and sensing system to bring the grasped object, for example, to the mouth. Similar closed-loop controls of stimulation could be used in the lower extremities for standing and ambulation. For partially paralyzed extremities, sensing of the muscle activities using BPBs would act as triggers to other BPBs to stimulate the motor-points of these muscles, thus augmenting the total action. Goniometry sensors would add the closed-loop controls to reduce or stop the actions. This approach could be used to augment swallowing, bladder control, and respiration.

Where pressure points need to be monitored — for example, at the heel (as a trigger for improving walking in stroke patients), the buttock (to avoid pressure sores), or hand (to detect the grasping of an

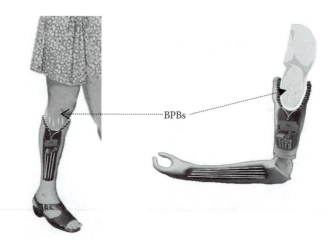

FIGURE 18.14   (See color insert following page 15-4). Use of BPBs in amputee patients.

object) — BPB devices placed in these sites can measure the pressure and either trigger motor-point functional stimulation to activate muscles, or stop a functional stimulation sequence.

18.4.1   FES-BPB System in Amputee Patients: Controlling Artificial Limbs

For amputee patients (Figure 18.14), BPBs working as biopotential sensors, can be inserted in the "stump" to pick up motor nerve signals, which can be used to control movement of the artificial limb's flexible components.

18.4.2   Cortical Interface Device: A Cortical Stimulator and Sensor Using the BPB System Technology

Individuals with spinal cord injury or disease that limits control over voluntary motion or sensing may be able to regain some of the ability of voluntary motion by monitoring the motor cortex and feeding back sensed response signals to the sensory cortex. Voluntary motion is expressed as neural activity in the motor cortex. The sensory cortex depends on muscle spindles and other sensors to help control the limb movement. By feeding back signals to the sensory cortex, the psychological use of the limb would be given back to the patient. The motor cortex signals can also be used to control wheelchairs and other helpful devices.

The miniaturized components developed for the BPB are used to create a cortical interface device (CID) with multiple stimulation and sensing electrodes within a single implantable package (Figure 18.15). The CID has the capability of monitoring up to several hundred electrodes that can be implanted or positioned in the motor cortex, sensory cortex, or a combination of both. The CID system consists of a base unit implanted in the skull, underneath the scalp, and one or several electrode arrays placed on the sensory or motor cortices.

The CID is equivalent to a group of sixty-four BPBs in its communication ability. It also includes an additional switching matrix that allows any amplifier to sense voltages either unipolar or bipolar from any two electrodes. The CID base unit dimensions and internal components are shown in Figure 18.16. The CID is constructed with the same technology developed for the BPB. It contains the same electronics as those used in the BPB as far as the communication, charging, power management, biopotential sensing, and stimulation modules are concerned. A CID base unit contains sixty-four biopotential sensing modules attached to one or more electrode arrays. The battery used in the CID provides 50 mA-h at 3.6 V.

The electrode arrays could be configured for sensing or stimulating purposes. The sensing electrode array includes signal processing capabilities using the same electronics as those in the biopotential sensing module in the BPB. The stimulating electrode array contains the same stimulation electronics as those

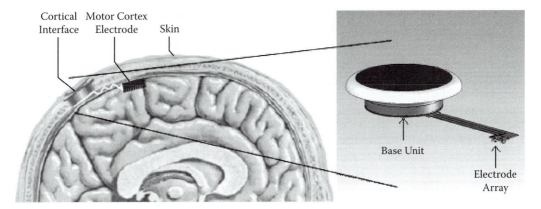

FIGURE 18.15   (See color insert following page 15-4). Cortical interface device (CID).

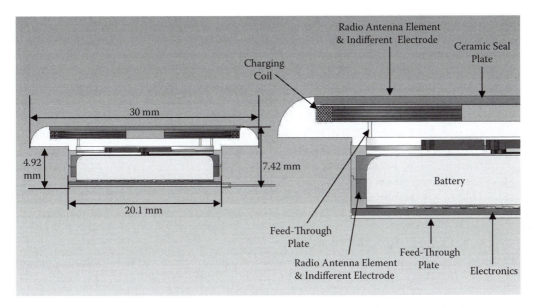

FIGURE 18.16   (See color insert following page 15-4). Cortical interface device: dimensions (left) and cross-section (right).

in the BPB stimulation module. The CID contains a powerful microprocessor to analyze the signals from the motor cortex and to reduce the data to sixty-four eight-bit messages that the MCU can use to control up to sixty-four muscles.

References

1. Heart Disease and Stroke Statistics – 2004 Update, American Heart Association.
2. http://www.bionicear.com/support/clinical_papers/supp_research_demo2.html
    http://www.bionicear.com/support/clinical_papers/supp_research_demo1.html
    http://www.bionicear.com/printables/Bilateral.pdf
    http://www.nidcd.nih.gov/health/hearing/coch_moreon.asp
    http://www.cochlear.com/896.asp
3. Stroke Testimony before the House Committee on Energy and Commerce Subcommittee on Health. NINDS opening statement to the House Committee on Energy and Commerce Subcommittee on Health, June 6, 2002. http://www.ninds.nih.gov/about_ninds/2002_stroke_testimony.htm#background

4. Facts and Figures at a Glance. National Spinal Cord Injury Statistical Center. Spinal Cord Injury: Hope through Research, May 2001. http://www.ninds.nih.gov/health_and_medical/pubs/sci.htm

5. Complications of Diabetes in the United States. National Diabetes Statistics. http://www.diabetes.niddk.nih.gov/dm/pubs/statistics/

6. Amputee Statistics (SAMPLE) from National Database http://rehabtech.eng.monash.edu/techguide/als/Stats.htm

7. Heetderks, W.J. RF powering of millimeter — and submillimeter-sized neural prosthetic implants. *IEEE Trans. Biomed. Eng.*, 35, 323–327, 1988.

8. Loeb, G.E., Zamin, C.J., Schulman, J.H., and Troyk, P.R. Injectable microstimulator for functional electrical stimulation. *Med. Biol. Eng. Comput.*, 29, NS13–NS19, 1991.

9. Arcos, I., Davis, R., Fey, K., Mishler, D., Sanderson, D., Tanacs, C., Vogel, M.J., Wolf, R., Zilberman, Y., and Schulman, J. Second-generation miscrostimulator. *Artif. Organs*, 26, 228–231, 2002.

10. Dupont, A.C., Bagg, S.D., Baker, L., Chun, S., Creasy, J.L., Romano, C., Romano, D., Waters, R.L., Wederich, C.L., Richmond, F.J.R., and Loeb, G.E. Therapeutic Electrical Stimulation with BIONS: Clinical Trial Report. *Proc. IEEE-EMBS Conference*, Houston, TX, 2002.

11. Richmond, F.J.R., Dupont, A.C., Bagg, S.D., Chun, S., Creasy, J.L., Romano, C., Romano, D., Waters, R.L., Wederich, C.L., and Loeb, G.E., Therapeutic electrical stimulation with BIONs to rehabilitate shoulder and knee dysfunction. *Proc. IFESS Conf.*, Ljubljana, Slovenia, 2002.

12. Buller, J.L., Cundiff, G.W., Noel, K.A., VanRooyen, J.A., Leffler, K.S., Ellerkman, R.M., and Bent, A.E. RF BION™: An Injectable Microstimulator for the Treatment of Overactive Bladder Disorders in Adult Females. European Association of Urology, February 2002.

13. Misawa, A., Shimada, Y., Matsunaga, T., Aizawa, T., Hatakeyama, K., Chida, S., Sato, M., Davis, R., Zilberman, Y., Cosendai, G., and Ripley, A.M. The use of the RF BION device to treat pain due to shoulder subluxation in chronic hemiplegic stroke patient — a case report. *Proc. IFESS Conference*, United Kingdom, 2004.

14. E-mail communication, Advanced Bionics Corporation, a Boston Scientific Company.

15. Schulman, J.H., Mobley, J.P., Wolfe, J., Regev, E., Perron, C.Y., Ananth, R., Matei, E., Glukhovsky, A., and Davis, R. Battery powered BION FES network. *Proc. IEEE-EMBS Conf.*, San Francisco, CA, 2004.

16. Personal communication, Joe Schulman.

17. Jiang, G., Fay, K., and Schulman, J. *In-vitro* and *in-vivo* aging tests of BION® micro-stimulator. Biomedical Engineering Department, University of Southern California, Los Angeles, CA, *7th Annu. Fred S. Grodins Graduate Res. Symp.*, March 2003.

18. Jiang, G., Purnell, K., and Schulman, J. Accelerated life tests and *in-vivo* tests of 3Y-TZP ceramics. *Mater. Processes for Medical Devices Conf.*, Proc. 2003.

19. Jiang, G., Mishler, D., Davis, R., Mobley, P., and Schulman, J. Ceramic to metal seal for implantable medical device. Biomed. Eng. Dept. USC, Los Angeles, CA. *Proc. 8th Annu. F. Grodins Graduate Res. Symp.*, p. 92–92, March 2004.

19

Tapping into the Spinal Cord for Restoring Function after Spinal Cord Injury

Lisa Guevremont and
Vivian K. Mushahwar

19.1 Introduction

Living organisms are intricate systems having thousands of ongoing processes interacting to maintain even the most basic life functions (i.e., energy production and protein synthesis). As the behavioral requirements of the organism increase, the complexity of the nervous system must also increase to initiate and coordinate a large variety of movements. For example, the *clione*, which has been studied extensively, has approximately 5000 neurons organized in five pairs of central ganglia. All the functional requirements of the *clione* can be regulated by this small nervous system, and electrical stimulation applied to specific groups of neurons can evoke stereotypical responses. Although humans possess a much larger repertoire of movements and therefore have significantly more complex nervous systems, levels of functional stereotypical organization remain within the central (brain and spinal cord) and peripheral nervous systems. Knowledge of the neural mechanisms underlying normal movements is an invaluable asset when designing a neuroprosthetic device because it enables one to tap into the neural circuitry residing at different levels to influence downstream events.

After spinal cord injury, the direct connection between the brain and periphery is severed, preventing voluntary control of the muscles below the level of the injury. However, the motor neurons below the lesion often remain viable, enabling muscle contractions to be evoked through functional electrical stimulation (FES) applied to the intact nerves. Functional electrical stimulation has achieved a broad range of applications in the field of neural prostheses, where it is used as an interface with the nervous system to restore function to muscles and organs after disease or injury.

This chapter provides a brief introduction to FES and existing systems for restoring standing and stepping after spinal cord injury. It then focuses on a technique called intraspinal microstimulation (ISMS) in which electrical stimulation is applied to the spinal cord below the level of the injury to activate the neural circuitry normally involved in the control of the lower limbs. By tapping into the nervous system at the level of the spinal cord, we are able to generate movements with some features that are consistent with those seen in intact individuals. An overview of the current work being done with ISMS is provided, along with specifications required for the future clinical implementation of the system.

19.2   General Introduction to FES

It has long been observed that electricity and physiological motor activity are correlated in some manner. As far back as 2000 BC, the Egyptians used electricity to treat paralysis and more recently electric fish were shown to cause muscle activation (Kellaway, 1946). By the time Hodgkin and Huxley formulated their model of action potential propagation down an axon (Hodgkin and Huxley, 1945), Hyman (1930) had already experimented with the use of electrical stimulation as a resuscitative technique after cardiac arrest. Further work on the cardiac system resulted in the development of the cardiac pacemaker (Zoll, 1952; Zoll and Linenthal, 1963), which has become one of the most well-known neural prostheses in clinical use today.

In 1961, Liberson et al. (1961) developed what is credited as the first clinically deployed device that uses electrical stimulation to activate muscles and restore function to a lower limb. Their "functional electrotherapy" technique was used to correct footdrop in individuals with hemiparesis. A foot switch consisting of a force sensing resistor (FSR) was used to detect periods when the foot was off the ground, during which time stimulation was applied through surface electrodes placed over the peroneal nerve. A year later, the term "functional electrical stimulation" (FES) was coined by Moe and Post (1962) and gained wider acceptance than Liberson's "functional electrotherapy."

19.2.1   Surface and Implantable FES Electrodes

Functional electrical stimulation can be applied using various techniques involving current delivered to the nervous tissue through electrodes situated either externally or internally. Perhaps the least invasive technique is surface stimulation in which electrodes are placed on the surface of the skin over the target nerve or muscle. To function appropriately, these electrodes require accurate placement that often translates into a long donning procedure every time the system is utilized. This procedure may be acceptable in systems requiring the activation of a single superficial muscle (e.g., footdrop [Dai et al., 1996]). Surface stimulation has other limitations, including the ability to stimulate only superficial muscles. Moreover, movement of the skin over the motor point of the muscle during motion can prevent consistent activation (Popovic, 2004). Poorly adhered electrodes or electrodes that are too small can cause burn injuries at high stimulation amplitudes (Popovic, 2004). Surface stimulation also has higher energy consumption than implanted systems due to the elevated current required when stimulating through the skin.

An alternative to surface stimulation is the use of percutaneous electrodes (Handa et al., 1989; Shimada et al., 1996) that are inserted into the muscle using a hypodermic needle. However, in this case, the connecting wires exit the skin at the implantation site, introducing a possible site of infection that requires continuous skin maintenance. The BION™ uses a similar insertion technique but consists of a fully implantable miniature stimulator enclosed completely in a small glass cylinder (Loeb et al., 2006). An external radio-frequency coil placed over the implanted device is used to power and control the BION™. Other fully implanted electrodes include intramuscular, epimysial, nerve-cuff (Naples et al., 1990; Grill and Mortimer, 1996), and flat interface nerve electrodes (FINEs) (Tyler and Durand, 2002), as well as longitudinal intrafascicular electrodes (LIFEs) (Yoshida and Horch, 1993b). Intramuscular and epimysial electrodes are used often in implanted neuroprostheses where muscle specificity is desired (Sharma et al., 1998). Intramuscular electrodes consist of wires penetrating into the muscle (i.e., percutaneous electrodes), whereas epimysial electrodes consist of flat meshes or disks that can be stitched to the muscle surface near the motor point (Grandjean and Mortimer, 1986). Cuff electrodes are designed to encircle the target nerve and can be constructed in mono-, bi-, or tri-polar configurations. Given that most major nerve trunks are composed of groups of axons (fascicles) innervating various muscles, it is often difficult to activate selective muscles with nerve cuff electrodes (Prochazka, 1993; Tai and Jiang, 1994; Grill and Mortimer, 1996). To activate a single muscle, the smallest branch of the nerve must be stimulated near the neuromuscular junction. While nerve cuffs can be implanted directly onto these nerve branches, they may result in nerve damage or mechanical failure if subjected to stresses and strains generated during muscle contractions (Naples et al., 1990). To achieve selective muscle activation with cuff electrodes implanted around proximal nerve trunks, cuffs with multiple stimulation sites distributed around the interior surface have been designed. By steering current between the multiple active sites, it is possible to activate selective fascicles within the nerve and thereby discriminate between the activation of various muscles (Veraart et al., 1993; Grill and Mortimer, 1996). Alternatively, FINEs can be used where the nerve is reshaped into a flat configuration to separate the fascicles (Tyler and Durand, 2002) and improve the selectivity of activation. The design of LIFEs provides muscle selectivity through the direct implant of fine microwires into specific nerve fascicles containing axons targeting the desired muscles (Yoshida and Horch, 1993b).

Other electrodes have been developed primarily for use in single-cell intracortical recordings. Electrode arrays such as the Utah and Michigan arrays have multiple electrodes mounted on a single silicon substrate. The Utah array is a 10×10 array of electrodes with either a single length or a slanted depth configuration (Nordhausen et al., 1996; Branner et al., 2001) and have been used for recording from the cortex and stimulating peripheral nerves, respectively. The Michigan array has a planar construction, with each probe having several active sites. Three-dimensional arrays can be constructed from various configurations of these subunits (Hoogerwerf and Wise, 1994).

19.3 Standing and Walking Systems Using FES

People with spinal cord injury have many needs, depending on the level and severity of the injury. The highest priority for people with quadriplegia is the restoration of arm and hand function. Improved trunk stability and restoration of bladder, bowel, and sexual functions are ranked among the top priorities of people with quadriplegia and paraplegia, followed closely by the desire for restored stepping (Anderson, 2004). Various approaches have been taken in addressing these needs, including the use of FES. This section reviews some of the main FES systems currently available for restoring standing and walking after spinal cord injury. The use of ISMS in achieving similar functions is discussed later.

To generate standing and stepping movements, a combination of muscles must be activated in a coordinated fashion to generate flexion and extension movements across the hip, knee, and ankle joints of the legs. Several FES approaches have been used to achieve this, including the use of stimulation applied through surface or implanted electrodes. These methods can be used alone or in conjunction with external braces across any or all of the joints of the leg. No system is complete without an appropriate means of controlling the stimulation. Therefore, a discussion of some of the control strategies currently in use or under development will be provided in the following section.

19.3.1 FES Systems for Restoring Standing

Standing systems must fulfill the functional requirements of generating consistent load-bearing force in both legs. One method of achieving this is through the use of knee, ankle, foot orthoses (KAFOs) to maintain the limb in an extended position. Alternatively, limb extension can be achieved through FES of the knee extensor muscles to prevent knee bucking. This stimulation can be provided through surface (Ewins et al., 1988; Bajd et al., 1999; Bijak et al., 2005) or implanted electrodes. An example of an implanted system for restoring standing is the CWRU/VA system (Davis et al., 2001). It consists of eight channels of stimulation applied through surgically implanted epimysial and intramuscular electrodes targeting the gluteal muscles, semimembranosus, vastus lateralis, and lumbar erector spinae. Standing is generated through open-loop application of continuous trains of FES to maintain extension of the lower limbs. Feedback-based control algorithms have also been developed that use sensor signals providing information about the joint angles (Wood et al., 1998) or the load taken by the hands (Donaldson Nde and Yu, 1996; Riener and Fuhr, 1998) to modify appropriately the amplitude of stimulation applied to the muscles during standing. The advantage of feedback control is that it delays the onset of muscle fatigue using the minimal level and duration of stimulation required to maintain functional standing (see Davis et al., Chapter 17).

19.3.2 FES Systems for Restoring Stepping

Stepping involves the coordinated activation of flexor and extensor muscles in the lower extremities, with full weight-bearing obtained during the stance (extensor) phase and ample foot clearance during swing (flexor) phase. Following spinal cord injury, external braces, such as KAFOs, and ankle–foot orthoses (AFOs) have been used alone (Merkel et al., 1984) or in conjunction with the application of FES (Andrews et al., 1988; Andrews et al., 1989; Solomonow et al., 1989; Davis et al., 1999) to restore stepping. A reciprocating gait orthosis (RGO) consisting of a set of hip, knee, ankle, and foot braces has also been used to provide mechanical coupling between the legs and maintain them in reciprocal phases during stepping in people with limited voluntary control. The use of RGOs in conjunction with FES has been shown to provide further improvements in stepping (Solomonow et al., 1989).

The Parastep (see Graupe, Chapter 16) and ETHZ-Paracare systems are examples of surface FES systems available clinically for restoring stepping (Graupe and Kohn, 1998; Popovic et al., 2001). Both systems use electrodes placed on the skin over the quadriceps, paraspinal (or gluteus) muscles, and common peroneal nerve to generate knee extension, hip extension, and full limb flexor reflex, respectively. The ETHZ-Paracare stepping system (Popovic et al., 2001) can be controlled using push-buttons for the initiation of swing, or alternatively through feedback signals obtained from foot-switches or gait detection sensors such as accelerometers or goniometers.

A sixteen-channel version of the CWRU/VA standing system (discussed above) can be implanted targeting extensor and flexor muscles to restore stepping and other leg exercise functions (Sharma et al., 1998). Another more extensive implant is the Praxis FES-22, which consists of one epidural, eight epineural, ten cuff, and three sacral root electrodes (see Davis et al., Chapter 17). Depending on which electrodes are used for stimulation, the implant can perform multiple functions, including the generation of leg movements for exercise, standing, or stepping. Distributed implant systems such as these have been relatively robust but require extensive surgery. Damage to the electrodes and lead breakage induced by the movements of the legs during simulation may also be encountered.

Stimulation applied in the central nervous system provides benefits over peripheral stimulation due to the ability to achieve a localized implant in a region of reduced movement. These benefits were the underlying drive for designing the lumbar anterior root stimulation implant (LARSI) for restoring standing and stepping after spinal cord injury. In this implant, stimulation is applied to the ventral roots exiting the spinal cord to evoke lower limb movements (Donaldson et al., 2003). However, this technique has been limited by the fact that each individual anterior root contains many populations of motoneuron axons innervating both agonist and antagonist muscles, at times generating inappropriate movements. For example, stimulation of a single ventral root evokes knee extension along with hip flexion, resulting in movements that are inappropriate for upright standing or stepping. Therefore, the implant has found the most success in generating leg movements for cycling.

Several strategies for controlling the amplitude and timing of the electrical stimuli to generate consistent stepping movements have been investigated. The majority of currently available FES systems employ the use of manual push-buttons built into walkers and crutches to initiate the swing phase of each step. Some systems use feedback control in which cyclic joint movements are tracked during the swing and stance phases of the stepping, and stimulation is continuously modulated using proportional/integral/derivative (PID) control (Veltink, 1991), or neural networks (Abbas and Triolo, 1997). State control of stepping has also been deployed in systems that use artificial learning to determine an appropriate set of sensor signals to generate appropriately timed transitions between the swing and stance phases (Andrews et al., 1988; Popovic et al., 2003). Alternatively, fuzzy logic is used to determine gait events during stepping (Skelly and Chizeck, 2001). An evaluation of control strategies for restoring stepping using FES was conducted by Popovic et al. (2003). Their results indicate that automatic control of ballistic walking has a lower metabolic cost than either automatic control of slow walking or push-button activated control. However, the volunteers preferred the latter forms due to the ability to coordinate upper and lower limb movements. These findings demonstrate the importance of developing an appropriate control scheme to be used in conjunction with FES systems.

19.4   Intraspinal Microstimulation

Intraspinal microstimulation is a technique that uses fine microwires to apply stimulation within the ventral horn of the spinal cord below the level of the injury to generate functional movements of the legs. The ISMS microwires are implanted in a compact and relatively motion-free region of the cord, and thus minimal strain is experienced by the implant during evoked limb movements. The target region for implantation is the lumbosacral enlargement, which is approximately 5 cm long in humans and contains all the motoneuron pools that innervate the muscles of the lower extremities as well as large proportions of stereotypical neuronal networks involved in the control of leg movements.

19.4.1   Historical Summary of the Use of ISMS

As early as the 1940s, Renshaw conducted experiments using penetration electrodes to stimulate motoneurons in the ventral gray matter of the spinal cord to demonstrate the time required for electrical excitation to cross a synapse (Renshaw, 1940). Intraspinal microstimulation has also been used for many years as an electrophysiological technique to demonstrate the reflex circuitry within the cord (Jankowska and Roberts, 1972).

In the late 1960s, ISMS was explored in animal experiments as a means of restoring micturition after spinal cord injury (Nashold et al., 1971). A pair of 0.4-mm diameter platinum–iridium wires with deinsulated conical tips was implanted into the spinal cord of cats and dogs, with and without spinal transections. The electrode tips were placed in the intermedio-lateral column to target the micturition center in the sacral (S1-S3) spinal cord. Starting in the early 1970s, twenty-seven patients in various centers (United States, France, and Sweden) were implanted with this spinal stimulation system. The system achieved a 60% success rate for effective bladder voiding but the stimulation was often associated with additional autonomic and motor responses (Nashold et al., 1981). In addition, a sphincterectomy was commonly needed in male patients to allow bladder voiding due to an inability to relax the external urethral sphincter using ISMS. Similar results have been more recently obtained during ISMS experiments in chronically and acutely implanted cats (Grill et al., 1999; McCreery et al., 2004; Tai et al., 2004) and rabbits (Pikov and McCreery, 2004), where high bladder pressures could be achieved, but often in conjunction with little relaxation of the bladder neck and sphincter resulting in incomplete voiding (Prochazka et al., 2003). In both cases, the technique was soon abandoned in favor of peripheral stimulation approaches providing similar outcomes. (See Gaunt and Prochazka (2006) for a full review of current neuroprostheses for urinary bladder function.)

Although ISMS has not been efficient in restoring micturition, it continues to show promising results as a means of restoring lower limb function after spinal cord injury (Pancrazio et al., 2006). In this case, we can tap into the residing circuitry within the lumbosacral enlargement of the spinal cord and activate neuronal networks and pools of motoneurons innervating the muscles of the legs.

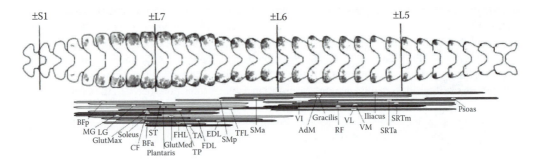

FIGURE 19.1 (See color insert following page 15-4). Distribution of motoneuron pools in the lumbosacral enlargement in the adult cat. The locations of the motoneuron pools are color coded on the cross-sections of the gray matter in the top panel. The extent of the pools along the lumbosacral enlargement is indicated by the bars below. Abbreviations (from left to right): BFp – biceps femoris posterior, MG – medial gastrocnemius, LG – lateral gastrocnemius, GlutMax – gluteus maximus, CF – caudofemoralis, BFa – biceps femoris anterior, ST – semitendinosus, GlutMed – gluteus medialis, FHL – flexor hallicus longus, TP – tibialis posterior, TA – tibialis anterior, FDL – flexor digitorum longus, EDL – extensor digitorum longus, SMp – semimembranosus posterior, TFL – tensor fascia latae, SMa – semimembranosus anterior, VI – vastus intermedius, AdM – adductor medialis, RF – rectus femoris, VL – vastus lateralis, VM – vastus medialis, SRTa – sartorius anterior, SRTm – sartorius medialis.

19.4.2 Use of ISMS for Inducing Functional Movements of the Limbs

Various forms of spinal cord stimulation have been used to investigate the organization and activate the locomotor networks within the spinal cord. These include the use of surface suction electrodes (Magnuson and Trinder, 1997; Vogelstein et al., 2006) in *in vitro* preparations and epidural electrodes *in vivo* in rats, cats, and humans (Dimitrijevic et al., 1998; Herman et al., 2002; Gerasimenko et al., 2003; Minassian et al., 2004).

Intraspinal microstimulation has been used in frogs, rats, and cats to investigate the presence of "movement primitives" in the spinal cord (Bizzi et al., 1991; Giszter et al., 1993; Tresch and Bizzi, 1999; Saltiel et al., 2001; Lemay and Grill, 2004). The concept of movement primitives refers to the hypothesized existence of a limited number of immutable neuronal networks that generate distinct movements of the limbs. These networks comprise the building blocks for all movement generation. However, a comparison of ISMS and peripheral FES responses indicate that these results may be attributable to the mechanical properties of the musculoskeletal system rather than a modular organization at the spinal cord level (Stein et al., 2002).

Over the past decade we have been developing the use of ISMS as a neuroprosthetic approach for restoring standing and stepping after spinal cord injury. Most of the development to date has been performed using animal models (primarily cats), with plans to conduct intraoperative investigations of ISMS in human volunteers within the next five years. Cats have been the animal model of choice due the size of their spinal cord relative to that of humans. The lumbosacral enlargement in humans is approximately 5 cm long and can be accessed by removing the T12 spinal process, whereas in cats this region spans 3 cm and can be accessed by removing the L5 spinal process.

19.4.2.1 The ISMS Implant

The ventral horn of the gray matter in the lumbosacral enlargement contains the cell bodies of all the motoneurons innervating the muscles of the legs. These cells are organized within the cord into pools of motoneurons with each pool innervating a single muscle. The relative locations of the motoneuron pools innervating specific muscles of the leg have been studied extensively in cats and have been shown to have a consistent spatial organization within the cord (Vanderhorst and Holstege, 1997; Mushahwar and Horch, 2000b) (Figure 19.1). This map is preserved across species, including humans (Sharrard, 1953). Stimulation applied through a single microwire (<300 µA) can directly or indirectly excite the motoneurons in these pools and generate contractions in the muscles they innervate. Electrical

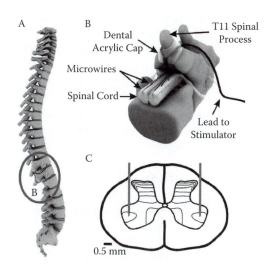

FIGURE 19.2 Cartoon illustration of the ISMS implant. A laminectomy performed at the T12 vertebral segment in humans will expose the L2-L5 spinal segments (A). The ISMS implant consists of an array of twenty-four microwires implanted individually through the dorsal surface of the cord (B). The entire array is secured to the T11 process using a dental acrylic cap and a lead travels to the stimulator interface. Motoneurons in the ventral horn of the gray matter are targeted by the microwire tips, as indicated in the cross-section of the spinal cord (C).

stimulation applied through individual electrodes targeting the ventral horn of the gray matter is capable of generating single joint movements as well as whole-limb flexion or extension synergies of the hindlimbs (Mushahwar et al., 2000; Mushahwar and Horch, 2000b). These coordinated multijoint responses that involve the activation of multiple motoneuron pools are seen even though the "direct" spread of current within the spinal cord has been shown to be less than 0.5 mm (Mushahwar and Horch, 1997; Lemay and Grill, 2004; Snow et al., 2006a). The possible mechanisms for this are addressed briefly below.

The ISMS implant and fixation techniques are based on a design originally developed for single-cell recording from dorsal root ganglia in awake and freely moving cats (Prochazka, 1984). The implant (Figure 19.2B) consists of eight to twenty-four individual microwires (30 μm diameter, 80%:20% platinum:iridium) arranged into an array and implanted bilaterally spanning a 3-cm-long region of the spinal cord (Figure 19.2A). Each platinum-iridium microwire is insulated except for a 30- to 60-μm tip that is sharpened and inserted through the dorsal surface of the spinal cord. The microwire tips target motoneuron pools in the ventral horn (Figure 19.2C) according to their previously mapped organization (Vanderhorst and Holstege, 1997).

19.4.2.2   ISMS and Fatigue Resistance

Results from multiple studies conducted in our laboratory demonstrate that ISMS is characterized by reduced fatigue (Mushahwar and Horch, 1997; Saigal et al., 2004) and a more graded force generation (Mushahwar and Horch, 2000a; Snow et al., 2006a) when compared to peripheral modes of stimulation. Electrical stimulation of peripheral nerves activates large fast, fatiguable fibers before smaller fatigue resistant fibers (Mortimer, 1981), leading to rapid muscle fatigue and a steep initial force production. This is a significant concern in situations where a muscle needs to maintain force for an extended period of time and may hinder the clinical use of FES. We recently demonstrated, in rats, that ISMS generates a mixed recruitment order of motor units within the activated muscle (Bamford et al., 2005). The fatigue resistance and graded force recruitment seen during ISMS is attributed to the initial recruitment of populations of slow, fatigue-resistant muscle fibers followed at higher stimulation amplitudes by the gradual activation of fast, fatiguable fibers. This sequence of activation resembles the physiological recruitment of motor units. Electrical stimulation (ISMS or peripheral FES) is applied extracellularly and activates axons in a reversed order due to the lower input resistance of the large axons. Peripheral FES activates motor axons directly, leading to the reversed recruitment of motor units. In contrast, ISMS primarily activates motoneurons trans-synaptically, which leads to a near-normal recruitment of slow to fast motor units. Computer simulations as well as recent experimental studies of ISMS demonstrated that the axons of afferents passing near the microwire tip are excited at a lower stimulus amplitude than the nearby motoneurons (McIntyre and Grill, 2000; Gaunt et al., 2006). This preferential activation of afferent networks by ISMS not only leads to the near-normal recruitment of motor units, but is also likely the main factor contributing to the spread of excitation to distributed but synergistic motoneuron pools. This could, in turn, explain the whole-limb coordinated movements elicited by stimulating through single ISMS wires.

Although ISMS provides improved fatigue resistance due to its mixed motor unit recruitment, increased force decay when high stimulation frequencies are used remains prevalent. During a normal contraction, motor units are activated in an asynchronous manner that allows fused muscle contraction (tetanus) to be achieved although the firing frequency of individual fibers (<20 Hz) is well below that required for fused muscle contraction (>40 Hz). However, electrical stimulation results in the simultaneous phase-locked activation of a large population of motor units (Mushahwar and Horch, 1997). Trains of stimulation at the fusion frequency are therefore required to generate a tetanic contraction of the muscle when the motor units are activated synchronously. This, in turn, causes an increased rate of fatigue and onset of muscle pain due to an increased production of metabolic byproducts and a reduction in the amount of time available to remove these waste materials between fiber contractions (Prochazka, 1993).

Multiple techniques for reducing the rate of fatigue during FES are used, including interleaving (Rack and Westbury, 1969; Yoshida and Horch, 1993a) or rotating stimulation sites, and performing postural switching. Interleaved stimulation more closely mimics the asynchronous firing of motor units seen in the natural activation of muscle in which low-frequency, nonoverlapping stimuli are applied to different subpopulations of fibers in a single muscle. This has been demonstrated using ISMS by implanting several microwires into a single target motoneuron pool. By alternating the stimuli through two or more electrodes, the frequency of stimulation applied through each electrode is only a fraction of the muscle fusion frequency. In this way, multiple populations of motor units can be activated asynchronously at subtetanic frequencies with the result being a summed smooth contraction (Mushahwar and Horch, 1997). This approach has also been demonstrated using LIFEs (Yoshida and Horch, 1993a) as well as the Utah slant electrode array (McDonnall et al., 2004).

Rotating stimulation uses a strategy similar to that of interleaving but instead of stimulating simultaneously through multiple electrodes, each as subfusion frequency, a train of 40-to 50-Hz stimulation is applied through a single electrode for a set duration. The stimulus through the first electrode is then turned off and delivered through another electrode for the same train duration. The stimulation continues to be rotated between electrodes in this fashion, providing the motor units activated by the first electrode adequate time to recover from fatigue. Rotating stimulation was demonstrated using a nerve cuff with multiple active sites, each recruiting different populations of motor units within the same muscle (Grill and Mortimer, 1996). This has also been applied in FES phrenic pacemakers for restoring breathing in people with high tetraplegia (Thoma et al., 1987).

Another solution to fatigue is the use of a technique called postural switching where different muscles can be activated to achieve the same overall functional outcome. This technique is generally used in standing, during which the weight of the individual can be shifted in such a way that the activation of postural muscles is periodically rotated, thus reducing the rate of fatigue (Krajl et al., 1986). Postural shifting can be applied to both ISMS and peripheral stimulation systems to further extend the duration of functional contraction.

19.4.2.3 ISMS Implant Stability

Special considerations have been taken to ensure that minimal damage is induced during the implantation of ISMS microwires and as a result of the continued presence of the implant. The material and diameter

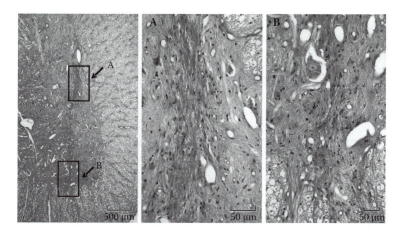

FIGURE 19.3 (See color insert following page 15-4). Histological evaluation of spinal cord tissue around a micro-wire implanted in a cat for six months. All panels show a 6-µm-thick cross-section of spinal cord tissue. The panel on the left shows the track generated by the implanted microwire. No lymphocytes or macrophages were found around the shaft (A) or tip (B) of the microwire. This indicates the absence of chronic inflammation over the six-month period of the implant.

of the microwires used in the ISMS implant developed in our laboratory were selected based on the requirements that the microwires (1) have sufficient stiffness to penetrate the cord yet cause minimal tissue displacement during implantation, (2) remain mechanically stable yet free to move ("float") with the tissue, (3) are electrochemically inert, and (4) can apply stimulation pulses that activate nervous tissue while remaining within the safety constraints of the interface. Histological techniques were used to evaluate the extent of the damage caused when ISMS microwires were chronically implanted in the spinal cord. Figure 19.3 shows a 6-mm-thick section of spinal tissue showing the track of a microwire that had been implanted for six months. Only mild gliosis was seen around the shaft (Figure 19.3A) and microwire tip (Figure 19.3B). Furthermore, the absence of lymphocytes and macrophages indicated no chronic inflammation occurred. These results demonstrate the advantage of designing a flexible implant in which each individual wire can "float" in the cord during the small translation, rotation, and compression of the spinal cord caused by postural movements. In contrast, in systems constructed from a single solid substrate, mechanical movement may cause increased tissue inflammation (Woodford et al., 1996; Biran et al., 2005; Polikov et al., 2005).

The maximum current that can safely be applied through an electrode is based on the amount of charge that can be injected into nervous tissue (McCreery et al., 1990), as well as the electrochemical processes that occur at the electrode tip (Mortimer, 1981). Duty cycle, charge per phase, charge density, and stimulation frequency are all factors that can influence the extent of tissue damage when stimulation is applied to the central nervous system (McCreery et al., 2004). It is particularly important in chronic implants to select stimulation limits and pulse shapes (mono- vs. bi-phasic) that prevent the initiation of electrochemical reactions that can compromise the integrity of the electrode (through corrosion or deposition) and produce byproducts that cause fluctuations in the pH that may damage the tissue surrounding the electrode tip (Mortimer, 1981). The maximum safe stimulation amplitude for the platinum–iridium wires used in our lab is 300 µA applied in biphasic charge balanced pulses with a 200-µs initial cathodic phase. These values were determined based on the surface area of the microwire tip and the acceptable charge density to maintain electrochemical stability (0.3 to 0.45 µC/mm^2 for platinum materials) (Mortimer, 1981). The values are also within the safe region of electrical stimulation applied to the central nervous system (Agnew et al., 1990).

The long-term efficacy of the ISMS implant was examined previously in intact (Mushahwar et al., 2000) and more recently in spinal cord injured cats. Three animals with complete spinal transections were implanted chronically with arrays of twenty-four platinum–iridium microwires targeting the

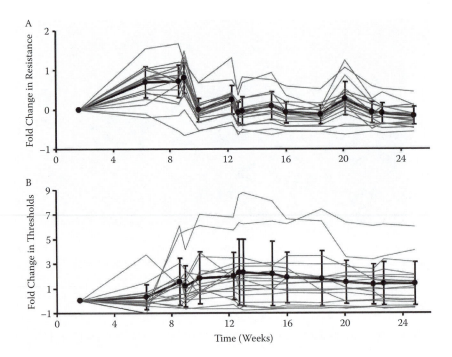

FIGURE 19.4 Microwire resistances and motor thresholds measured periodically throughout a six-month ISMS implant in a cat with spinal cord injury. The animal received a complete spinal transection at the T11 level and an ISMS implant during a single surgical session. Resistances and motor threshold values were normalized to the first recording taken ten days after surgery and are plotted with standard deviation bars. Microwire resistances varied only slightly over the duration of the implant, indicating an absence of significant microwire corrosion, material deposition, or insulation breakdown (A). The motor thresholds increased by twofold from start to termination of the experiment (B). This is consistent with the expected encapsulation around the microwire tip.

motoneuron pools in the lumbosacral enlargement (L5-S1). Two animals were implanted four to five weeks after receiving a spinal cord transection at T11. Both of these implants failed before the end of the six-month study due to mechanical failures at the spinal cord and connector levels. These failures were considered when designing future implants and will be discussed in greater detail in subsequent sections. The third animal received an ISMS implant and a complete transection of the spinal cord at the T11 level during a single surgical session. The implant remained functional for the duration of the six-month experiment. Figures 19.4 and 19.5 summarize results obtained from periodical measurements of the electrode impedances, motor thresholds, and motor responses conducted throughout the six-month period of implantation. We found that there were only small variations in the impedances of the microwires (Figure 19.4A), indicating that there was little physical change (corrosion or deposition) at the electrode tip. The small decrease in electrode impedance may suggest that the polyimide insulation around the microwire tips was degrading but there is no evidence to support this at this time and the effect is not substantial. Further evaluation should be conducted if these materials are to be used in future studies. The motor threshold of individual microwires (Figure 19.4B) increased by an average of twofold over the course of the six-month implant. However, the average motor threshold of all microwires increased from 76 to 83 μA, indicating that the absolute change in threshold (as opposed to relative to the initial threshold value recorded) was small. These increases are within the expected range and indicate a minor gliosis associated with the implant. No indication of chronic inflammation was observed.

The motor responses evoked through stimulation of single electrodes changed over the first ten weeks after implantation and simultaneous spinal transection. Early responses were more often characterized by flexion movements around the hip, knee, ankle, and toe joints. This time period corresponds to the expected duration of spinal shock during which flexor hyperreflexia is prominent. After this initial phase,

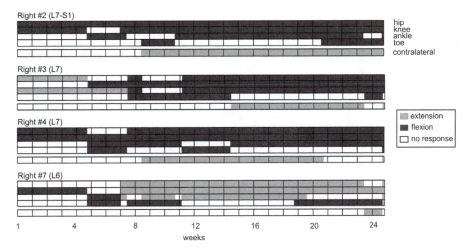

FIGURE 19.5 **(See color insert following page 15-4).** Motor responses evoked by stimulation applied through single microwires. Each band represents the time course of changes to the motor responses over the duration of a six-month implant in a cat that received a complete spinal cord injury on the same day as the implant. Each large band represents the responses of a single microwire. The horizontal subdivisions indicate the movement seen across each joint (flexion – purple and blue, extension – yellow and red, no response – green). Flexor responses were dominant for the first eight to ten weeks, after which time some responses switched to extension. After the ten-week point, the responses became increasingly stable. The flexor-dominated responses over the first eight weeks are presumably due to spinal shock and excessive hyper-reflexia soon after spinal cord injury.

a spectrum of responses similar to those seen in an intact animal was achieved, and the characteristics of the responses obtained from each microwire became increasingly consistent between recording sessions (i.e., stable responses). Figure 19.5 shows examples of the time course of motor responses evoked when stimulation was applied through four individual microwires. The data obtained from this animal provide further support for the stability of the ISMS implant in terms of both its physical integrity and functional responses. The changes in responses can be avoided in clinical application by excluding patients in the acute phase of spinal cord injury and selecting patients who have stable chronic injuries.

19.4.2.4 Restoring Standing and Stepping

The fatigue-resistant properties of ISMS make it particularly attractive for restoring standing after spinal cord injury, where prolonged stimulation is required to maintain a stable upright posture. Intraspinal microstimulation applied in an interleaved manner through eight microwires implanted bilaterally in the spinal cord resulted in prolonged standing in cats. The duration and quality of standing achieved using intramuscular stimulation and ISMS were compared and the results indicate that ISMS is capable of increasing the duration of standing achieved in comparison to peripheral FES (e.g., Figure 19.6). The addition of feedback rules that were used to modify the amplitude of stimulation based on the angles of the knee and ankle, and loading in the limbs was found to further improve the duration of standing using both stimulation techniques. Up to 28 min of stable standing was achieved using feedback-based control of ISMS (vs. 4.5 min for intramuscular stimulation), indicating that feedback control of ISMS could provide a substantial benefit in a neuroprosthetic device designed for restoring standing after spinal cord injury (Lau et al., 2006).

In addition to standing, the use of ISMS for restoring stepping has been explored. We define functional stepping as rhythmic alternations of the hindlimbs that provide full load-bearing support for the hind quarters of the animal during stance and ample foot clearance during swing. Functional overground locomotion has the additional requirement of the generation of substantial propulsive forces resulting in the forward progression of the animal.

The ISMS implant initially proposed for restoring lower limb function consisted of four rows of forty-eight electrodes (192 in total) implanted into each side of the spinal cord (Mushahwar and Horch, 1997).

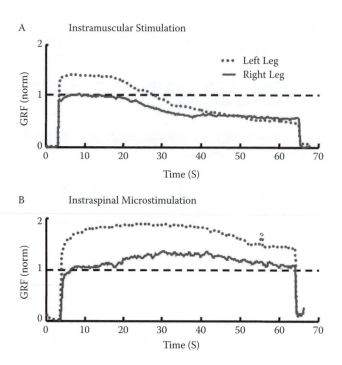

FIGURE 19.6 Comparison of standing generated using intramuscular stimulation and ISMS. Cats were made to stand by applying constant levels of stimulation for 1 min. Stimulation was applied either through intramuscular electrodes implanted in the quadriceps and gastrocnemius muscles (A), or intraspinal microwires targeting extensor motoneuron pools (B). The vertical ground reaction force (GRF) recorded using intramuscular stimulation decayed much more rapidly than that obtained using ISMS. The horizontal dashed line indicates the force levels necessary for full bearing (lifting) of the weight of the hindquarters.

The objective of this design was to ensure full selectivity and complete activation of all motoneurons within a pool with minimal current spread from the electrode tip. However, more recent studies have demonstrated that due to the activation of networks in the spinal cord, stimulation applied through a single electrode can generate full limb extensor or flexor synergies appropriate for the stance and swing phases of stepping, respectively. This means that locomotor-like stepping patterns can be generated by applying phasic stimulation through as few as four different electrodes eliciting flexion and extension responses in the left and right legs (Mushahwar et al., 2002; Saigal et al., 2004).

The fact that ISMS activates networks of neurons within the spinal cord has other interesting implications. We have found that 40- or 50-Hz trains of low-level (i.e., amplitude ≤ motor threshold) tonic stimulation can evoke in-place alternating stepping movements of the hindlimbs when applied through groups of microwires implanted in the intermediate and ventral areas of the spinal cord (Guevremont et al., 2006b). It is thought that the stimulation acts to excite intrinsic locomotor networks residing in the lumbosacral spinal cord (Brown, 1911; Cazalets et al., 1995; Kjaerulff and Kiehn, 1996). Figure 19.7 indicates the regions of the cord that generated various movements when stimulation was applied through single microwires. Microwires placed in the intermediate and medio-ventral areas (laminae VII and VIII) of the spinal cord were found most potent in eliciting alternating movements, and stimulation of the ventral horn (lamina IX) evoked single joint or whole-limb flexor or extensor synergies, depending on the rostro-caudal location of the microwire tip. In contrast, stimulation applied to the dorsal horn primarily evoked flexion movements and was found to cause a sensory response (i.e., paw shake or limb withdrawal) in intact awake animals (Mushahwar et al., 2002) due to activation of sensory fibers with ascending connections. The finding that ISMS in intermediate and medio-ventral areas was most potent in generating stepping movements is consistent with studies demonstrating that neurons in these regions

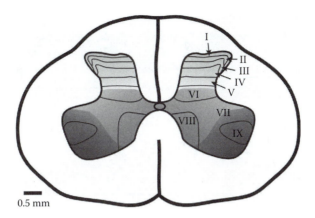

0.5 mm

FIGURE 19.7 (See color insert following page 15-4). Microwire tip locations within the lumbosacral spinal cord of cats. Stimulation applied through microwire tips in the dorsal horns of the gray matter (laminae I–VI) generated primarily flexion responses (yellow). Ventrally placed tips could evoke single joint movements or full limb flexion or extension synergies, depending on the specific target within the cord (purple). Stimulation of the intermediate and medio-ventral areas (red) often evoked unilateral or bilaterally alternating stepping movements of the hindlimbs.

play a role in rhythmogenesis during fictive locomotion (Noga et al., 1995; Kjaerulff and Kiehn, 1996; Tscherter et al., 2001).

While tonic stimulation was capable of producing rhythmic alternation of the hindlimbs, the evoked stepping-like movements were not fully load bearing and therefore did not meet the requirements of functional stepping stated above. To increase the level of load bearing, suprathreshold amplitudes of stimulation were phasically applied through alternating groups of microwires evoking flexion and extension movements (Saigal et al., 2004). Series of about 140 steps were achieved with predetermined patterns of stimulation corresponding to a travel distance of approximately 200 m in human locomotion. A high degree of fatigue resistance was seen over series of steps as indicated by low levels of force decay and kinematic stability (i.e., consistent joint movements) throughout the stepping sequences (Figure 19.8). Closed-loop control of stepping was also achieved using hip angle and limb loading (ground reaction force) to determine the appropriate timing for transitions from stance (extension) to swing (flexion). These control algorithms were further developed using intramuscular FES to achieve overground locomotion (Guevremont et al., 2006a), and will be applied to ISMS in future studies.

In summary, we have shown that ISMS in the ventral horn of the spinal cord produces single joint movements and coordinated multijoint synergies. It recruits motor units in a near-normal physiological order, which contributes to the increased fatigue resistance seen in ISMS-evoked responses. By tapping into the existing neuronal networks controlling leg movements, ISMS can produce prolonged durations of weight-bearing standing by stimulating through as few as four microwires within the lumbosacral enlargement. Similarly, by stimulating through as few as four microwires in each side of the spinal cord, bilateral stepping of the paralyzed hindlimbs can be produced. The responses evoked by ISMS are stable over time and the implantation procedure induces minimal tissue damage. Based on these findings, we believe that ISMS can be a viable neuroprosthetic approach for restoring leg function after spinal cord injury.

19.5 Requirements for Clinical Implementation of ISMS

19.5.1 Current Techniques for ISMS

The ISMS implant currently being used in our lab consists of twenty-four microwires arranged into a bilateral array. The microwires are implanted individually into the target motoneuron pools in the L5-S1 segments and the entire array is affixed to the L4 spinal process using cyanoacrylate and a dental

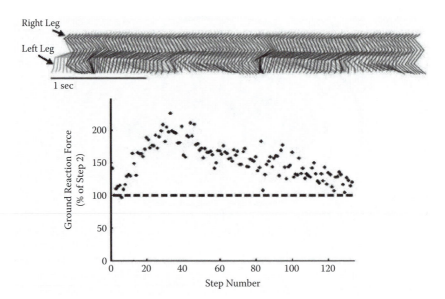

FIGURE 19.8 In-place stepping evoked using phasic ISMS. The stick figures were reconstructed from video taken of the left (gray) and right (black) legs. The graph plots the ground reaction force (vertical loading) generated during a series of more than 120 steps. The force has been normalized to the load bearing of the second step, which corresponded to 45 to 50% of the full weight of the animal (i.e., full weight bearing of the hindquarters — indicated by the horizontal dashed line). The number of steps taken by this animal would correspond to approximately 200 m traveled by an average North American man (*Source*: Modified from Saigal et al., 2004.)

acrylic cap. Figure 19.9A shows a cartoon illustration of the implanted end of the ISMS system. The array is constructed from twenty-four microwires, each deinsulated (30 to 60 μm) and sharpened at the tip (10 to 15°) before being bent to the appropriate depth to penetrate from the dorsal surface of the cord into the targeted motoneuron pool (dimension **a**). The microwires are then organized into two columns spanning the length of the lumbosacral enlargement (dimension **b**). The microwires are bundled and then drawn through a length of silastic tubing so that each electrode and its lead remain as one continuous wire. The ends of the tubing are sealed, with the array protruding in one direction and the connector, mounted on the animal's head, soldered to the leads at the other end. Finally, the array is bent (dimension **c**) so that the tube can be fixed to the spinal process with the wires running along the surface of the spinal cord. Figure 19.9B shows the implant after each of the individual wires has been implanted into the spinal cord. The wires lie along the dorsal surface of the cord and penetrate (direction into the page) to reach the motoneuron pools. Figure 19.9C shows the fixation technique used to secure the implant to the L4 spinal process. The spinal cord is covered with a layer of plastic thin film before the surgical wound is closed (Figure 19.9B).

There are advantages and disadvantages to using this implantation and fixation technique. One of the main advantages is that each microwire is implanted individually so the responses can be optimized by evaluating the responses during the implantation procedure. However, this comes at the cost of having a long and tedious surgery requiring extensive knowledge of spinal anatomy to adjust appropriately the microwire placement based on the response evoked. Although each wire is placed individually, the range of vertical and longitudinal targets along the cord is limited by the preset bend and single active site at each tip (dimension **a**), and also the longitudinal spacing (dimension **b**) of the array, which are determined during array construction. The wires must lie flat yet not taut and remain securely in place to minimize the tissue damage evoked due to the mismatch of mechanical properties of the microwires and the spinal cord tissue.

The array shown in Figure 19.9C was implanted into an intact cat and is an example of an implant that failed as a result of the protrusion of the silastic tubing from the dental acrylic fixation cap on L4.

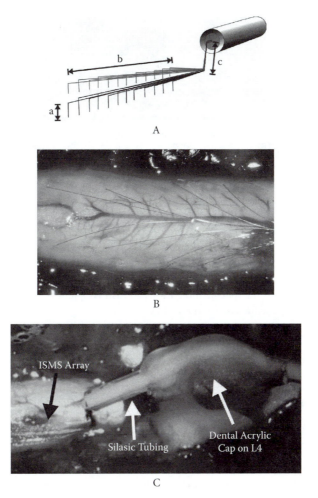

FIGURE 19.9 (**See color insert following page 15-4**). Implantation and fixation of the ISMS microwires. The implant consists of twenty-four individual microwires arranged into a bilateral array (A). Each microwire is cut and bent to target a motoneuron pool at a particular depth (dimension **a**) and longitudinal distance (dimension **b**) along the lumbosacral spinal cord. The array is bent at the point where it protrudes from the silastic tube so that the tube can be fixed to the L4 spinal process while allowing the array to lie flush with the surface of the cord (dimension **c**). Upon implantation, the wires lie along the dorsal surface of the cord and penetrate (into the page) to reach the ventral horn (B). An example of the fixation technique is shown in (C). This implant failed due to the long extension of the silastic tubing from the bone when bent and pulled the array during closure of the surgical wound.

Although the implant generated appropriate responses during surgery, no postoperative responses could be elicited and the animal suffered from motor deficits in its hindlegs. Postmortem analysis indicated that the region of silastic tubing extending beyond the edge of the remaining L4 vertebra was displaced during the closing of the incision. This caused pressure to be applied to the surface (circled region in Figure 19.9C) and the microwires to pull out of the cord. It has generally been found that the first twenty-four hours are critical in determining the success of the implant. During this time, initial processes of connective tissue formation take place around the implant that secure it in place, and only minimal movement of the animal can be tolerated safely during this period.

19.5.2   Design Specifications for a Clinical ISMS System

The key features for the development ISMS implant for clinical use include (1) movement toward a more standardized electrode array and implantation technique, and (2) use of electrode arrays tailored to the

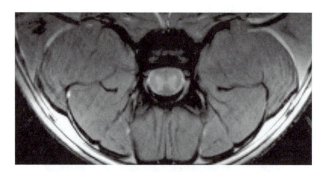

FIGURE 19.10 Magnetic resonance image (MRI) of the spinal cord of a cat. The white and gray matter can be distinguished using this technique (0.35 mm in-plane and 4 mm through-plane resolution).

specific dimensions of the lumbosacral spinal cord of each recipient. Our current implantation technique requires specialized skill in handling the microwires and extensive knowledge of spinal cord anatomy. Furthermore, to date, the microwire arrays have been constructed based on the average dimensions of the spinal cord, determined from large numbers of animals used during various procedures performed in our laboratory, rather than the specific dimensions of each recipient. This section discusses some of the developments required for the clinical use of ISMS.

19.5.2.1 Pre- and Postoperative Imaging

In the future, presurgical magnetic resonance imaging (MRI) of the lumbosacral enlargement will be utilized to obtain a more accurate representation of the dimensions of the spinal cord of each subject (e.g., Figure 19.10). Although this does not provide an indication of the functional organization of the particular motoneuron pools, imaging the cord prior to surgery guides the tailoring of the implant to the specific animal or human recipient.

Another consideration is the ability to image the ISMS implant *in vivo*, allowing us to determine changes in its position over time. Therefore, future ISMS designs should be composed of MRI-compatible, nonmagnetic materials such as platinum–iridium (e.g., Figure 19.11). The development of improved MRI techniques will result in increased image resolution and therefore more accurate measurements. This imaging technique could also significantly improve the surgical procedure if used intraoperatively for microwire implantation, similar to its use for the implantation of deep brain stimulation devices (Gibson et al., 2003; De Salles et al., 2004).

19.5.2.2 Electrode Array

The ultimate ISMS array would consist of at least two rows of electrodes implanted bilaterally into the spinal cord and spanning longitudinally the lumbosacral enlargement. In addition, each shaft would have more than one active site allowing the depth of the stimulated region to be selected once the array is implanted. These criteria can be met either using a custom tailored array design based on previous knowledge of the dimensions of the cord (from preoperative images) or, alternatively, using an implant consisting of several small, flexible, prefabricated arrays. The stiffness of the implanted device is critical due to the small, but present, motion of the cord during postural movements. Although this shift is minimal when compared to the displacement of the actual joints and muscle tissue, the cord does undergo translation, rotation, and compression.

The ISMS array currently being used in our lab is fabricated from individual microwires with exposed tips that have been constructed into a single array with predetermined dimensions (see above). This design allows the individual wires to be virtually mechanically independent of one another and to move freely in concert with the spinal cord. However, its construction limits the target locations that can be reached. Future versions of the implant will make use of multiple contact sites along the penetrating electrode shaft. This would improve motoneuronal recruitment by stimulating through a single active site or by interleaving the stimuli between multiple sites on the same shaft. We experimented with the

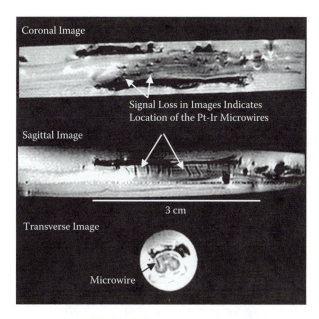

FIGURE 19.11 MRI of *ex-vivo* spinal tissue with an implanted ISMS array. The spinal cord of a cat implanted with an ISMS array was extracted postmortem and imaged using MRI. The microwires appear as dark regions (signal voids) in the images. The figure shows a coronal (from the dorsal surface), a sagittal, and a transverse view (0.39 mm in-plane and 1.6 mm through-plane resolution).

use of cylindrical lithography for fabricating stimulating electrodes with four active sites along each shaft (Figure 19.12A) (Snow et al., 2006a,b) and obtained a 100% yield in appropriate targeting of the motoneuron pools of interest. Further developments to reduce the electrode diameter (currently 85 µm) and stiffness and connection of lead wires are needed. Another technique involves the use of planar lithography where multiple stimulation sites can be placed on a flexible polyimide substrate which is inserted into the tissue using a silicon shuttle that is then removed, leaving the highly flexible device in place (Figure 19.12B). By controlling the surface area, and thus the impedance of each site, we could record or stimulate through different active sites using a single implanted device.

19.5.2.3 Insertion and Stabilization Techniques

To become a widely available FES technique, future designs of the ISMS system should include simple electrode insertion and fixation methods that can be used to provide highly reproducible results. For this reason we suggest the design of an array that provides a modular approach to the implantation. Groups of four to six electrodes on a flexible substrate (Figure 19.12B) could be fabricated in bilateral or unilateral strips to match the dimensions of the individual (based on spinal cord images). The surgeon could then implant several groups of these into the cord based on segmental landmarks and test the responses using intraoperative stimulation. Another possibility is the use of image-guided implantation to specifically target the substrate shafts into the ventral horn of the gray matter.

The use of cats as the animal model for the development of ISMS provides some additional challenges that may not be encountered in humans. For one, the human spinal cord is larger and is surrounded by a significantly stronger dura mater. This means that additional stability could be provided to the implant if modular sections of it are secured to the dura mater instead of relying solely on the attachment to the spinal process. A loop of wire near the implant would provide the required strain relief before being secured to the bone and traveling to the implanted stimulator.

19.5.2.4 Electronics Interface

The current state-of-the-art technology for FES systems makes use of implanted stimulators powered and programmed through inductive coupling, which eliminates the need for percutaneous connections.

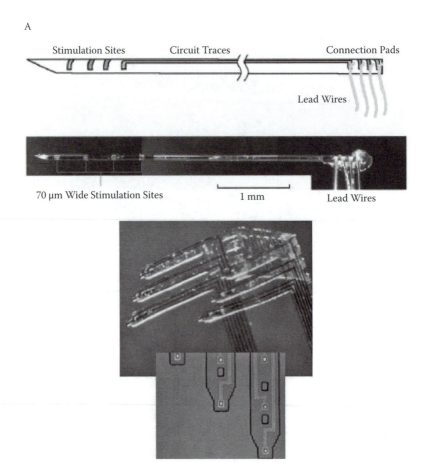

FIGURE 19.12 (See color insert following page 15-4). Alternative designs for ISMS electrodes. An electrode constructed using cylindrical lithography technology (A). The shaft is 85 μm and has insulated traces running to four distinct stimulation sites from lead wires bonded to connection pads at the nonimplanted end of the shaft. (*Source:* Modified from Figures 1 and 3 in Snow et al., 2006b.) First-generation polymer intracortical probe that can be modified to stimulate spinal tissue (B). (*Source:* Courtesy of Dr. Daryl Kipke, University of Michigan.) Multiple electrode sites are laid out on a thin-film polyimide planar substrate that can be folded into a three-dimensional array. Insertion is performed using a stiff silicone shuttle that is then removed, leaving only the highly flexible array in the spinal tissue.

A receiving (or transmitting) coil is implanted under the skin, typically below the clavicle (for deep brain stimulation) or below the last rib (for bladder and lower extremity FES systems). A transmitting (or receiving) coil can then be placed over the skin to communicate control signals to the implanted device. The clinical implementation of the ISMS technique could utilize similar technology.

The cabling from the implanted stimulator to the spinal array must be carefully designed so as to reduce bulk. The possibility of having twenty-four wires, each with three or four active sites, means that nearly 100 channels of communication are necessary. One possible design is to use 10-μm wires in a highly flexible ribbon cable. Alternatively, the signals picked up by the implanted receiver could be multiplexed and transmitted in series to a demultiplexing and stimulating stage closer to the implant that would translate the serial signals into meaningful stimulation pulses to each of the active electrode sites.

19.5.2.5 Stimulator

The studies conducted for restoring standing and stepping using ISMS suggest that a minimum of four channels are required (Mushahwar et al., 2002; Saigal et al., 2004). Increasing the number of available channels improves the possible repertoire of movements that can be generated and allows for the use of interleaving techniques to provide additional fatigue resistance. It also allows for necessary redundancy in the system. For this reason the stimulator should have the capability of generating at least sixteen independent channels of stimulation. Currently available implantable stimulators (e.g., Davis et al., 1987; Kljajic et al., 1992; Bijak et al., 2001) will be adopted.

The control system should be user friendly while remaining highly flexible. Independently programmable signals are required for each of the stimulation channels, which can be set by a therapist based on the motor thresholds and range of movements evoked by stimulation applied through each electrode. In addition, the control unit should have input for a number of feedback signals that can be used to determine the appropriate stimulation parameters during standing and stepping sequences.

19.6 Future Directions

Several future developments of the ISMS system can be envisioned. The use of a biomimetic interface instead of three-dimensional electrode structures or microwires would provide a seamless interaction with the spinal cord. Integration of neuromorphic electronic chips such as the silicon central pattern generator (Lewis et al., 2003) to control the timing and amplitude of stimulation would further enhance the quality and metabolic efficiency of functional leg movements produced by ISMS.

Perception of movements evoked by ISMS would be a valuable addition to the system. Continuous awareness of the position of the limbs, foot placement, and loading would provide a higher level of interaction between the implant and the user. It would also increase the users' confidence in utilizing the ISMS system and improve their adaptability to it. Finally, the incorporation of cortical control of ISMS (Mushahwar et al., 2006) would grant the user a volitionally driven system that mimics, to a large extent, the natural control of movements.

Intraspinal microstimulation can also be used in combination with other interventions such as regeneration, neuropharmacology, and body weight supported treadmill training for restoring function after spinal cord injury. The effect of ISMS before and after the application of neuromodulatory drugs (e.g., the α_2-noradrenergic receptor agonist, clonidine) in cats with spinal cord injury has been investigated to evaluate the functional benefits borne by the combination of these two approaches (Barthelemy et al., 2006; Guevremont et al., 2006b). The evoked movements were enhanced, and sites that produced only flexor-dominated responses prior to the delivery of the drug were capable of eliciting coordinated weight-bearing stepping after its administration.

The use of ISMS can also be deployed for the restoration of arm movements in individuals with cervical-level injuries. This exciting extension of the applicability of ISMS is currently in its early stages of investigation (Moritz et al., 2006).

19.7 Conclusion

This chapter discussed the strengths and current weaknesses involved in utilizing the ISMS technique for restoring lower limb function after spinal cord injury. The main advantages of the technique include not only the localization of the implant in a 5-cm region of the lumbosacral spinal cord, but also its ability to activate neural circuitry in a more physiological manner than peripheral systems. This leads to the generation of synergistic movements and the recruitment of muscle fibers in a near-normal order. We have investigated the use of ISMS for restoring standing and stepping and, although there are technical improvements that need to be made to the design of the implant, the benefits seen provide evidence of the potential efficacy of ISMS. Our results indicate that with the future developments suggested here, ISMS will be able to translate into a successful, clinically implemented FES technique.

Acknowledgments

We thank Daniel Hallihan, Jan Kowalczewski, Bernice Lau, Jonathan Norton, Enid Pehowich, Costantino Renzi, Rajiv Saigal, and Sean Snow for their contributions to the work presented in this chapter. Funding was provided by the Alberta Heritage Foundation for Medical Research (AHFMR), the Canadian Fund for Innovation (CFI), the Canadian Institutes for Health Research (CIHR), the International Spinal Research Fund (ISRT), and the National Institutes of Health (NIH). L. Guevremont has been supported by Graduate Research Scholarships from the Faculty of Medicine and Dentistry at the University of Alberta and V.K. Mushahwar is an AHFMR Scholar.

References

Abbas, J.J. and Triolo, R.J. (1997). Experimental evaluation of an adaptive feedforward controller for use in functional neuromuscular stimulation systems. *IEEE Trans. Rehab. Eng.*, 5, 12–22.

Agnew, W., McCreery, D., Yuen, T.G., and Bullara, L. (1990). Effects of prolonged electrical stimulation of the central nervous system. In *Neural Prostheses: Fundamental Studies.* McCreery, D., Ed. Prentice Hall, Englewood Cliffs, NJ, p. 226–251.

Anderson, K.D. (2004). Targeting recovery: priorities of the spinal cord-injured population. *J. Neurotrauma*, 21, 1371–1383.

Andrews, B.J., Barnett, R.W., Phillips, G.F., and Kirkwood, C.A. (1989). Rule-based control of a hybrid FES orthosis for assisting paraplegic locomotion. *Automedica*, 11, 175–199.

Andrews, B.J., Baxendale, R.H., Barnett, R., Phillips, G.F., Yamazaki, T., Paul, J.P., and Freeman, P.A. (1988). Hybrid FES orthosis incorporating closed loop control and sensory feedback. *J. Biomed. Eng.*, 10, 189–195.

Bajd, T., Munih, M., and Kralj, A. (1999). Problems associated with FES-standing in paraplegia. *Technol. Health Care*, 7, 301–308.

Bamford, J.A., Putman, C.T., and Mushahwar, V.K. (2005). Intraspinal microstimulation preferentially recruits fatigue-resistant muscle fibres and generates gradual force in rat. *J. Physiol. (London)*, 569, 873–884.

Barthelemy, D., Leblond, H., Provencher, J., and Rossignol, S. (2006). Non-locomotor and locomotor hindlimb responses evoked by electrical microstimulation of the lumbar cord in spinalized cats. *J. Neurophysiol.*, 96, 3273–3292.

Bijak, M., Mayr, W., Girsch, W., Lanmuller, H., Unger, E., Stohr, H., Thoma, H., and Plenk, H., Jr. (2001). Functional and biological test of a 20 channel implantable stimulator in sheep in view of functional electrical stimulation walking for spinal cord injured persons. *Artif. Organs*, 25, 467–474.

Bijak, M., Rakos, M., Hofer, C., Mayr, W., Strohhofer, M., Raschka, D., and Kern, H. (2005). Stimulation parameter optimization for FES supported standing up and walking in SCI patients. *Artif. Organs*, 29, 220–223.

Biran, R., Martin, D.C. and Tresco, P.A. (2005). Neuronal cell loss accompanies the brain tissue response to chronically implanted silicon microelectrode arrays. *Exp. Neurol.*, 195, 115–126.

Bizzi, E., Mussa-Ivaldi, F.A. and Giszter, S. (1991). Computations underlying the execution of movement: a biological perspective. *Science*, 253, 287–291.

Branner, A., Stein, R.B., and Normann, R. A. (2001). Selective stimulation of cat sciatic nerve using an array of varying-length microelectrodes. *J. Neurophysiol.*, 85, 1585–1594.

Brown, T.G. (1911). The intrinsic factors in the act of progression in the mammal. *Proc. Roy. Soc. London*, Ser. B 84, 309–318.

Cazalets, J.R., Borde, M., and Clarac, F. (1995). Localization and organization of the central pattern generator for hindlimb locomotion in newborn rat. *J. Neurosci.*, 15, 4943–4951.

Dai, R., Stein, R.B., Andrews, B.J., James, K.B., and Wieler, M. (1996). Application of tilt sensors in functional electrical stimulation. *IEEE Trans. Rehab. Eng.*, 4, 63–72.

Davis, J.A., Triolo, R.A., Ulhir, J., Bieri, C., Rohde, L., Lissy, D., and Kukke, S. (2001). Preliminary performance of a surgically implanted neuroprosthesis for standing and transfers Where do we stand? *J. Rehab. Res. Devel.*, 38, 609–617.

Davis, R., Eckhouse, R., Patrick, J.F., and Delehanty, A. (1987). Computerized 22-channel stimulator for limb movement. *Appl. Neurophysiol.*, 50, 444–448.

Davis, R., Houdayer, T., Andrews, B., and Barriskill, A. (1999). Paraplegia: prolonged standing using closed-loop functional electrical stimulation and Andrews ankle-foot orthosis. *Artif. Organs*, 23, 418–420.

De Salles, A.A., Frighetto, L., Behnke, E., Sinha, S., Tseng, L., Torres, R., Lee, M., Cabatan-Awang, C., and Frysinger, R. (2004). Functional neurosurgery in the MRI environment. *Minim. Invasive. Neurosurg.*, 47, 284–289.

Dimitrijevic, M.R., Gerasimenko, Y., and Pinter, M.M. (1998). Evidence for a spinal central pattern generator in humans. *Ann. NY Acad. Sci.*, 860, 360–376.

Donaldson, N., Rushton, D.N., Perkins, T.A., Wood, D.E., Norton, J., and Krabbendam, A.J. (2003). Recruitment by Motor nerve root stimulators: significance for implant design. *Med. Eng. Phys.*, 25, 527–537.

Donaldson Nde, N. and Yu, C.-N. (1996). FES standing: control by handle reactions of leg muscle stimulation (CHRELMS). *IEEE Trans. Rehab. Eng.*, 4, 280–284.

Ewins, D.J., Taylor, P.N., Crook, S.E., Lipczynski, R.T., and Swain, I.D. (1988). Practical low cost stand/sit system for mid-thoracic paraplegics. *J. Biomed. Eng.*, 10, 184–188.

Gaunt, R.A. and Prochazka, A. (2006). Control of urinary bladder function with devices: successes and failures. *Prog. Brain Res.*, 152, 163–194.

Gaunt, R.A., Prochazka, A., Mushahwar, V.K., Guevremont, L., and Ellaway, P.H. (2006). Intraspinal microstimulation excites multisegmental sensory afferents at lower stimulus levels than local {alpha}-motoneurons. *J. Neurophysiol.*, 96, 2995–3005.

Gerasimenko, Y.P., Avelev, V.D., Nikitin, O.A., and Lavrov, I.A. (2003). Initiation of locomotor activity in spinal cats by epidural stimulation of the spinal cord. *Neurosci. Behav. Physiol.*, 33, 247–254.

Gibson, V., Peifer, J., Gandy, M., Robertson, S., and Mewes, K. (2003). 3D visualization methods to guide surgery for Parkinson's disease. *Stud. Health Technol. Inform.*, 94, 86–92.

Giszter, S.F., Mussa-Ivaldi, F.A., and Bizzi, E. (1993). Convergent force fields organized in the frog's spinal cord. *J. Neurosci.*, 13, 467–491.

Grandjean, P.A. and Mortimer, J.T. (1986). Recruitment properties of monopolar and bipolar epimysial electrodes. *Ann. Biomed. Eng.*, 14, 53–66.

Graupe, D. and Kohn, K.H. (1998). Functional neuromuscular stimulator for short-distance ambulation by certain thoracic-level spinal-cord-injured paraplegics. *Surg. Neurol.*, 50, 202–207.

Grill, W.M., Bhadra, N., and Wang, B. (1999). Bladder and urethral pressures evoked by microstimulation of the sacral spinal cord in cats. *Brain Res.*, 836, 19–30.

Grill, W.M., Jr. and Mortimer, J.T. (1996). Quantification of recruitment properties of multiple contact cuff electrodes. *IEEE Trans. Rehab. Eng.*, 4, 49–62.

Guevremont, L., Norton, J., and Mushahwar, V. (2006a). Open- and closed-loop control strategies for restoring overground locomotion using FES. In *11th Annual Conference of the International FES Society*, Zao, Japan, p. 23–25.

Guevremont, L., Renzi, C.G., Norton, J.A., Kowalczewski, J., Saigal, R., and Mushahwar, V.K. (2006b). Locomotor-related networks in the lumbosacral enlargement of the adult spinal cat: activation through intraspinal microstimulation. *IEEE Trans. Neural Syst. Rehab. Eng.*, 14, 266–272.

Handa, Y., Hoshimiya, N., Iguchi, Y., and Oda, T. (1989). Development of percutaneous intramuscular electrode for multichannel FES system. *IEEE Trans. Biomed. Eng.*, 36, 705–710.

Herman, R., He, J., D'Luzansky, S., Willis, W., and Dilli, S. (2002). Spinal cord stimulation facilitates functional walking in a chronic, incomplete spinal cord injured. *Spinal Cord*, 40, 65–68.

Hodgkin, A.L. and Huxley, A.F. (1945). Resting and action potentials in single nerve fibres. *J. Physiol.*, 104, 176–195.

Hoogerwerf, A.C. and Wise, K.D. (1994). A three-dimensional microelectrode array for chronic neural recording. *IEEE Trans. Biomed. Eng.*, 41, 1136–1146.

Hyman, A.S. (1930). Resuscitation of the stopped heart by intracardiac therapy. *Arch. Intern. Med.*, 46, 553–568.

Jankowska, E. and Roberts, W.J. (1972). An electrophysiological demonstration of the axonal projections of single spinal interneurones in the cat. *J. Physiol.*, 222, 597–622.

Kellaway, P. (1946). The part played by electric fish in the early history of bioelectricity and electrotherapy. *Bull. Hist. Med.,* 20, 112–137.

Kjaerulff, O. and Kiehn, O. (1996). Distribution of networks generating and coordinating locomotor activity in the neonatal rat spinal cord *in vitro*: a lesion study. *J. Neurosci.,* 16, 5777–5794.

Kljajic, M., Malezic, M., Acimovic, R., Vavken, E., Stanic, U., Pangrsic, B., and Rozman, J. (1992). Gait evaluation in hemiparetic patients using subcutaneous peroneal electrical stimulation. *Scand. J. Rehabil. Med.,* 24, 121–126.

Krajl, A., Bajd, T., Turk, R., and Benko, H. (1986). Posture switching for prolonging functional electrical stimulation standing in paraplegic patients. *Paraplegia,* 24, 221–230.

Lau, B., Guevremont, L., and Mushahwar, V.K. (2007). Open- and closed-loop control strategies for restoring standing using intramuscular and intraspinal stimulation, *IEEE Trans. Neural Syst. Rehabil. Eng.,* 15, 273–285.

Lemay, M.A. and Grill, W.M. (2004). Modularity of motor output evoked by intraspinal microstimulation in cats. *J. Neurophysiol.,* 91, 502–514.

Lewis, M.A., Etienne-Cummings, R., Hartmann, M.J., Xu, Z.R., and Cohen, A.H. (2003). An in silico central pattern generator: silicon oscillator, coupling, entrainment, and physical computation. *Biolog. Cybernetics,* 88, 137–151.

Liberson, W.T., Holmquest, H.J., Scot, D., and Dow, H. (1961). Functional electrotherapy: stimulation of the peroneal nerve synchronized with the swing phase of the gait of hemiplegic patients. *Arch. Phys. Med. Rehabil,* 42, 101–105.

Loeb, G.E., Richmond, F.J., and Baker, L.L. (2006). The BION devices: injectable interfaces with peripheral nerves and muscles. *Neurosurg. Focus,* 20, E2.

Magnuson, D.S.K. and Trinder, T.C. (1997). Locomotor rhythm evoked by ventrolateral funiculus stimulation in the neonatal rat spinal cord *in vitro*. *J. Neurophysiol.,* 77, 200–206.

McCreery, D., Pikov, V., Lossinsky, A., Bullara, L., and Agnew, W. (2004). Arrays for chronic functional microstimulation of the lumbosacral spinal cord. *IEEE Trans. Neural Syst. Rehabil. Eng.,* 12, 195–207.

McCreery, D.B., Agnew, W.F., Yuen, T.G., and Bullara, L. (1990). Charge density and charge per phase as cofactors in neural injury induced by electrical stimulation. *IEEE Trans. Biomed. Eng.,* 37, 996–1001.

McDonnall, D., Clark, G.A., and Normann, R.A. (2004). Selective motor unit recruitment via intrafascicular multielectrode stimulation. *Can. J. Physiol. Pharmacol.,* 82, 599–609.

McIntyre, C.C. and Grill, W.M. (2000). Selective microstimulation of central nervous system neurons. *Ann. Biomed. Eng.,* 28, 219–233.

Merkel, K.D., Miller, N.E., Westbrook, P.R., and Merritt, J.L. (1984). Energy expenditure of paraplegic patients standing and walking with two knee-ankle-foot orthoses. *Arch. Phys. Med. Rehabil.,* 65, 121–124.

Minassian, K., Jilge, B., Rattay, F., Pinter, M.M., Binder, H., Gerstenbrand, F., and Dimitrijevic, M.R. (2004). Stepping-like movements in humans with complete spinal cord injury induced by epidural stimulation of the lumbar cord: electromyographic study of compound muscle action potentials. *Spinal Cord,* 42, 401–416.

Moe, J.H. and Post, H.W. (1962). Functional electrical stimulation for ambulation in hemiplegia. *J. Lancet,* 82, 285–288.

Moritz, C.T., Lucas, T.H., Perlmutter, S.I., and Fetz, E.E. (2006). Forelimb movements and muscle responses evoked by microstimulation of cervical spinal cord in sedated monkeys. *J Neurophysiol.,* 97, 110–120.

Mortimer, J.T. (1981). Motor prostheses. In *Handbook of Physiology II, Section 1 Motor Control.* Brooks, V.B., Ed., Williams and Williams, Baltimore, p. 155–187.

Mushahwar, V.K., Collins, D.F., and Prochazka, A. (2000). Spinal cord microstimulation generates functional limb movements in chronically implanted cats. *Exp. Neurol.,* 163, 422–429.

Mushahwar, V.K., Gillard, D.M., Gauthier, M.J., and Prochazka, A. (2002). Intraspinal micro stimulation generates locomotor-like and feedback-controlled movements. *IEEE Trans. Neural Syst. Rehabil. Eng.,* 10, 68–81.

Mushahwar, V.K., Guevremont, L., and Saigal, R. (2006). Could cortical signals control intraspinal stimulators? A theoretical evaluation. *IEEE Trans. Neural Syst. Rehabil. Eng.*, 14, 198–201.

Mushahwar, V.K. and Horch, K.W. (1997). Proposed specifications for a lumbar spinal cord electrode array for control of lower extremities in paraplegia. *IEEE Trans. Rehabil. Eng.*, 5, 237–243.

Mushahwar, V.K. and Horch, K.W. (2000a). Muscle recruitment through electrical stimulation of the lumbo-sacral spinal cord. *IEEE Trans. Rehabil. Eng.*, 8, 22–29.

Mushahwar, V.K. and Horch, K.W. (2000b). Selective activation of muscle groups in the feline hindlimb through electrical microstimulation of the ventral lumbo-sacral spinal cord. *IEEE Trans. Rehabil. Eng.*, 8, 11–21.

Naples, G.G., Mortimer, J.T., and Yuen, T.G. (1990). Overview of peripheral nerve electrode design and implantation. In *Neural Prostheses: Fundamental Studies*. Agnew, W. and McCreery, D., Eds., Prentice Hall, Englewood Cliffs, NJ, p. 108–145.

Nashold, B.S., Jr., Friedman, H., and Boyarsky, S. (1971). Electrical activation of micturition by spinal cord stimulation. *J. Surg. Res.*, 11, 144–147.

Nashold, B.S., Jr., Friedman, H., and Grimes, J. (1981). Electrical stimulation of the conus medullaris to control the bladder in the paraplegic patient. A 10-year review. *Appl. Neurophysiol.*, 44, 225–232.

Noga, B., Fortier, P., Kriellaars, D., Dai, X., Detillieux, G., and Jordan, L. (1995). Field potential mapping of neurons in the lumbar spinal cord activated following stimulation of the mesencephalic locomotor region. *J. Neurosci.*, 15, 2203–2217.

Nordhausen, C.T., Maynard, E.M., and Normann, R.A. (1996). Single unit recording capabilities of a 100 microelectrode array. *Brain Res.*, 726, 129–140.

Pancrazio, J.J., Chen, D., Fertig, B.S., Miller, R.L., Oliver, E., Peng, G.C.Y., Shinowara, N.L., Weinrich, M., and Kleitman, N. (2006). Toward neurotechnology innovation: report from the 2005 Neural Interfaces Workshop. An NIH-Sponsored Event. *Neuromodulation*, 9, 1–7.

Pikov, V. and McCreery, D.B. (2004). Mapping of spinal cord circuits controlling the bladder and external urethral sphincter functions in the rabbit. *Neurourol. Urodyn.*, 23, 172–179.

Polikov, V.S., Tresco, P.A., and Reichert, W.M. (2005). Response of brain tissue to chronically implanted neural electrodes. *J. Neurosci. Methods*, 148, 1–18.

Popovic, D. (2004). Neural prostheses for movement restoration. In *Biomedical Technology and Devices Handbook*. Moore, J. and Zouridakis, G., Eds., CRC Press, London, p. 1–47.

Popovic, D., Radulovic, M., Schwirtlich, L., and Jaukovic, N. (2003). Automatic vs hand-controlled walking of paraplegics. *Med. Eng. Phys.*, 25, 63–73.

Popovic, M.R., Keller, T., Pappas, I.P., Dietz, V., and Morari, M. (2001). Surface-stimulation technology for grasping and walking neuroprosthesis. *IEEE Eng. Med. Biol. Mag.*, 20, 82–93.

Prochazka, A. (1984). Chronic techniques for studying neurophysiology of movement in cats. In *Methods in the Neurosciences*, Vol. 4. Lemon, R., Ed., Wiley, New York, p. 113–128.

Prochazka, A. (1993). Comparison of natural and artificial control of movement. *IEEE Trans. Rehabil. Eng.*, 1, 7–17.

Prochazka, A., Mushahwar, V., Downie, J.W., Shefchyk, S.J., and Gaunt, R.A. (2003). Functional Microstimulation of the Lumbosacral Spinal Cord: Quarterly Report #5. NIH-NINDS contract N01-NS-2-2342.

Rack, P.M. and Westbury, D.R. (1969). The effects of length and stimulus rate on tension in the isometric cat soleus muscle. *J. Physiol.*, 204, 443–460.

Renshaw, B. (1940). Activity in the simplest spinal reflex pathway. *J. Neuophysiol.*, 3, 373–387.

Riener, R. and Fuhr, T. (1998). Patient-driven control of FES-supported standing up: a simulation study. *IEEE Trans. Rehabil. Eng.*, 6, 113–124.

Saigal, R., Renzi, C., and Mushahwar, V.K. (2004). Intraspinal microstimulation generates functional movements after spinal-cord injury. *IEEE Trans. Neural Syst. Rehabil. Eng.*, 12, 430–440.

Saltiel, P., Wyler-Duda, K., D'Avella, A., Tresch, M.C., and Bizzi, E. (2001). Muscle synergies encoded within the spinal cord: evidence from focal intraspinal NMDA iontophoresis in the frog. *J. Neurophysiol.*, 85, 605–619.

Sharma, M., Marsolais, E.B., Polando, G., Triolo, R.J., Davis, J.A., Jr., Bhadra, N., and Uhlir, J.P. (1998). Implantation of a 16-channel functional electrical stimulation walking system. *Clin. Orthopaed. Related Res.*, 347, 236–242.

Sharrard, W.J. (1953). Correlation between changes in the spinal cord and muscle paralysis in poliomyelitis; a preliminary report. *Proc. Roy. Soc. Med.*, 46, 346–349.

Shimada, Y., Sato, K., Kagaya, H., Konishi, N., Miyamoto, S., and Matsunaga, T. (1996). Clinical use of percutaneous intramuscular electrodes for functional electrical stimulation. *Arch. Phys. Med. Rehabil.*, 77, 1014–1018.

Skelly, M.M. and Chizeck, H.J. (2001). Real-time gait event detection for paraplegic FES walking. *IEEE Trans. Neural Syst. Rehabil. Eng.*, 9, 59–68.

Snow, S., Horch, K.W. and Mushahwar, V.K. (2006a). Intraspinal microstimulation using cylindrical multielectrodes. *IEEE Trans. Biomed. Eng.*, 53, 311–319.

Snow, S., Jacobsen, S.C., Wells, D.L., and Horch, K.W. (2006b). Microfabricated cylindrical multi-electrodes for neural stimulation. *IEEE Trans. Biomed. Eng.*, 53, 320–326.

Solomonow, M., Baratta, R., Hirokawa, S., Rightor, N., Walker, W., Beaudette, P., Shoji, H., and D'Ambrosia, R. (1989). The RGO Generation II: muscle stimulation powered orthosis as a practical walking system for thoracic paraplegics.[erratum appears in *Orthopedics*, (1989), 12, 1522]. *Orthopedics*, 12, 1309–1315.

Stein, R.B., Aoyagi, Y., Mushahwar, V.K., and Prochazka, A. (2002). Limb movements generated by stimulating muscle, nerve and spinal cord. *Arch. Italiennes de Biologie*, 140, 273–281.

Tai, C., Booth, A.M., de Groat, W.C., and Roppolo, J.R. (2004). Bladder and urethral sphincter responses evoked by microstimulation of S2 sacral spinal cord in spinal cord intact and chronic spinal cord injured cats. *Exp. Neurol.*, 190, 171–183.

Tai, C. and Jiang, D. (1994). Selective stimulation of smaller fibers in a compound nerve trunk with single cathode by rectangular current pulses. *IEEE Trans. Biomed. Eng.*, 41, 286–291.

Thoma, H., Gerner, H., Holle, J., Kluger, P., Mayr, W., Meister, B., Schwanda, G., and Stohr, H. (1987). The phrenic pacemaker. Substitution of paralyzed functions in tetraplegia. *ASAIO Trans.*, 33, 472–479.

Tresch, M.C. and Bizzi, E. (1999). Responses to spinal microstimulation in the chronically spinalized rat and their relationship to spinal systems activated by low threshold cutaneous stimulation. *Exp. Brain Res.*, 129, 401–416.

Tscherter, A., Heuschkel, M.O., Renaud, P., and Streit, J. (2001). Spatiotemporal characterization of rhythmic activity in rat spinal cord slice cultures. *Eur. J. Neurosci.*, 14, 179–190.

Tyler, D.J. and Durand, D.M. (2002). Functionally selective peripheral nerve stimulation with a flat interface nerve electrode. *IEEE Trans. Neural Syst. Rehabil. Eng.*, 10, 294–303.

Vanderhorst, V.G. and Holstege, G. (1997). Organization of lumbosacral motoneuronal cell groups innervating hindlimb, pelvic floor, and axial muscles in the cat. *J. Compar. Neurol.*, 382, 46–76.

Veltink, P. (1991). Control of FES-induced cyclical movements of the lower leg. *Med. Biolog. Eng. Comput.*, 29, NS8–NS12.

Veraart, C., Grill, W.M., and Mortimer, J.T. (1993). Selective control of muscle activation with a multipolar nerve cuff electrode. *IEEE Trans. Biomed. Eng.* 40, 640–653.

Vogelstein, R.J., Etienne-Cummings, R., Thakor, N.V., and Cohen, A.H. (2006). Phase-dependent effects of spinal cord stimulation on locomotor activity. *IEEE Trans. Neural Syst. Rehabil. Eng.*, 14, 257–265.

Wood, D.E., Harper, V.J., Barr, F.M.D., Taylor, P.N., Phillips, G.F., and Ewins, D.J. (1998). Experience in using knee angles as part of a closed-loop algorithm to control FES-assisted paraplegic standing. In *6th Vienna International Workshop on Functional Electrostimulation*, p. 137–140.

Woodford, B.J., Carter, R.R., McCreery, D., Bullara, L.A., and Agnew, W.F. (1996). Histopathologic and physiologic effects of chronic implantation of microelectrodes in sacral spinal cord of the cat. *J. Neuropathol. Exp. Neurol.*, 55, 982–991.

Yoshida, K. and Horch, K. (1993a). Reduced fatigue in electrically stimulated muscle using dual channel intrafascicular electrodes with interleaved stimulation. *Ann. Biomed. Eng.,* 21, 709–714.

Yoshida, K. and Horch, K. (1993b). Selective stimulation of peripheral nerve fibers using dual intrafascicular electrodes. *IEEE Trans. Biomed. Eng.,* 40, 492–494.

Zoll, P.M. (1952). Resuscitation of the heart in ventricular standstill by external electric stimulation. *N. Engl. J. Med.,* 247, 768–771.

Zoll, P.M. and Linenthal, A. J. (1963). External and internal electric cardiac pacemakers. *Circulation,* 28, 455–466.

Fundamental Science and Promising Technologies

20

Theory and Physiology of Electrical Stimulation of the Central Nervous System

Warren M. Grill

20.1 Introduction

Electrical stimulation is a widespread method to study the form and function of the nervous system and a technique to restore function following disease or injury. The central nervous system (CNS) includes the brain and spinal cord (Figure 20.1). Both the spinal cord and brain include regions primarily populated by cell bodies (somas) of neurons, and termed gray matter for its color, and regions primarily populated by axons of neurons, and termed white matter. The diversity of neuronal elements and the complexity of the volume conductor make understanding the effects of stimulation more challenging in the case of CNS stimulation than in the case of peripheral stimulation. Specifically, it is unclear, in many cases, what neuronal elements (axons, cell bodies, presynaptic terminals; Figure 20.1) are activated by stimulation [Ranck, 1975]. Further, it is unclear how targeted neural elements can be stimulated selectively without coactivation of other surrounding elements. This chapter presents a review of the properties of CNS stimulation as required for rational design and interpretation of therapies employing electrical stimulation.

Electrical stimulation has been used to determine the structure of axonal branching [Jankowska and Roberts, 1972], examine the strength of connections between neurons, and determine the projection

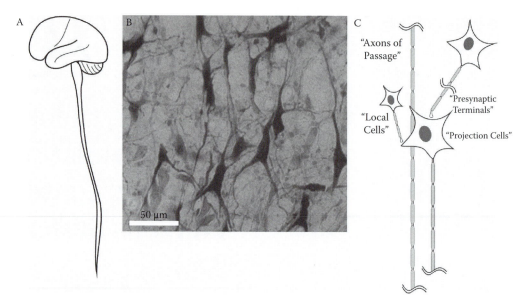

FIGURE 20.1 Structure of the central nervous system. (A) The central nervous system (CNS) includes the brain and spinal cord. (B) The gray matter of the CNS contains the cell bodies of neurons as well as dendritic and axonal processes. (C) When an electrode is placed within the heterogeneous cellular environment of the CNS, it is unclear which neuronal elements are affected by stimulation.

patterns of neurons [Lipski, 1981; Tehovnik, 1996]. Examples of the application of CNS stimulation in treatment of neurological disorders include the treatment of pain by stimulation of the brain [Coffey, 2001] and spinal cord [Cameron, 2004], treatment of tremor and the motor symptoms of Parkinson's disease [Gross and Lozano, 2000], as an experimental treatment for epilepsy [Velasco et al., 2001; Hodaie et al., 2002], as well as a host of other neurological disorders [Gross, 2004]. In addition, CNS stimulation is being developed for restoration of hearing by electrical stimulation of the cochlear nucleus [Otto et al., 2002] and for restoration of vision [Brindley and Lewin, 1968; Schmidt et al., 1996; Troyk et al., 2003].

A nerve cell or a nerve fiber can be artificially stimulated by depolarization of the cell's membrane. The resulting action potential propagates to the terminal of the neuron, leading to release of neurotransmitters that can impact the postsynaptic cell. Passage of current through extracellular electrodes positioned near neurons creates extracellular potentials in the tissue. The resulting potential distribution can result in transmembrane current and depolarization. Alternately, extracellular potentials may modulate or block ongoing neuronal firing, depending on the magnitude, distribution, and polarity of the potentials.

The objective of this chapter is to present the biophysical basis for electrical stimulation of neurons in the CNS. The focus is on using a fundamental understanding of both the electric field and its effects on neurons to determine the site of neuronal excitation or modulation in the CNS where electrodes are placed among heterogeneous populations of neuronal elements, including cells, axons, and dendrites.

20.2   Generation of Potentials in CNS Tissues

Passage of current through tissue generates potentials in the tissue (recall Ohm's law: $V = IR$). The potentials depend on the electrode geometry, the stimulus parameters (current magnitude), and the electrical properties of the tissue. For example, the potential generated by a monopolar point source can be determined analytically using the relationship $V_e(r) = \dfrac{I}{4\pi\sigma r}$, where I is the stimulating current, σ is the conductivity of the tissue medium (Table 20.1), and r is the distance between the electrode and the

TABLE 20.1 Electrical Conductivity of CNS Tissues

Tissue Type	Electrical Conductivity (S/m)	Ref.
Dura	0.030	Holsheimer et al., 1995
Cerebrospinal fluid	1.5; 1.8	Crile et al., 1922; Baumann et al., 1997
Gray matter	0.20	Ranck, 1963; Li et al., 1968; Sances and Larson, 1975
White matter	Anisotropic	
Transverse		
	0.6	Ranck and BeMent, 1965 (cat dorsal columns)
	1.1	Nicholson, 1965 (cat internal capsule)
Longitudinal		
	0.083	Ranck and BeMent, 1965
	0.13	Nicholson, 1965
Encapsulation tissue	0.16	Grill and Mortimer, 1994

measurement point. The point source model is a valid approximation for sharp electrodes with small tips [McIntyre and Grill, 2001]. Larger electrodes are typically used for chronic stimulation of the CNS, and the spatial distribution of the potentials in the tissue differs from those produced by a point source electrode (Figure 20.2). Examples of the spectrum of electrode types used for CNS stimulation (and recording) are shown in Figure 20.3.

The extracellular potentials generated by the passage of current depend on the electrical properties of the tissue. The electrical properties of the CNS tissue are both inhomogeneous and anisotropic (Table 20.1), and the distribution of potentials within the CNS tissue will depend strongly on the CNS tissue and electrode geometries. In general, biological conductivities have a small reactive component [Ackman and Seitz, 1984; Eisenberg and Mathias, 1980], and thus a relatively small increase in conductivity at higher frequencies [Ranck, 1963; Nicholson, 1965; Ranck and BeMent, 1965].

Spatial variations in the electrical properties of the tissue can cause changes in the patterns of activation (Grill, 1999). In most cases, to calculate accurately the extracellular potentials generated by extracellular stimulation requires a numerical solution using a discretized model, for example with the finite element method (e.g., Veltink et al., 1989; McIntyre and Grill, 2002).

20.3   Response of Neurons to Imposed Extracellular Potentials

As described in the previous section, the distribution of extracellular potentials depends on the electrode geometry, the electrical properties of the extracellular tissue, and the stimulation amplitude. The effect of the potentials on neurons depends on the nerve cell type, its size and geometry, as well as the temporal characteristics of the stimulus. During stimulation of peripheral nerves, it is clear that it is the axons in the vicinity of the electrodes that are activated. However, the CNS contains a heterogeneous population of neuronal elements, including local cells projecting locally around the electrode, as well as those projecting away from the region of stimulation, axons passing by the electrode, and presynaptic terminals projecting onto neurons in the region of the electrode (Figure 20.1C). The effects of stimulation can be mediated by activation of any or all of these elements and include both the direct effects of stimulation of postsynaptic elements, as well as the indirect effects mediated by electrical stimulation of presynaptic terminals that mediate the effects of stimulation via synaptic transmission.

From this complexity arise two principal questions during stimulation of the CNS [Grill and McIntyre, 2001]: (1) what neuronal elements are activated by extracellular stimulation?, and (2) how can targeted elements be stimulated selectively? Computational modeling provides a powerful tool to study extracellular excitation of CNS neurons. The volume of tissue stimulated, both for fibers and cells, and how this changes with electrode geometry, stimulus parameters, and the geometry of the neuronal elements are quite challenging to determine experimentally. Using a computer model enables examination of these parameters under controlled conditions, and enables simultaneous determination of the effects of stimulation on all the different neural elements around the electrode. Computational modeling of the effects

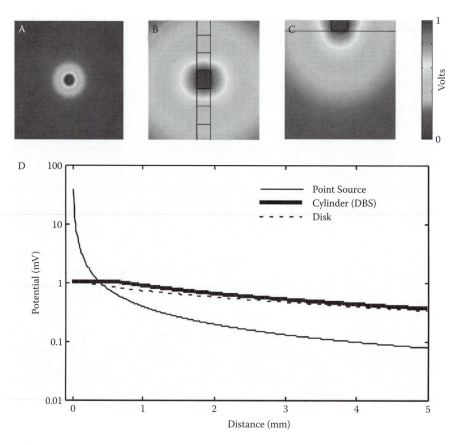

FIGURE 20.2 (See color insert following page 15-4). Electric fields generated by the passage of current in CNS tissue. The first step in determining the response of CNS neurons to extracellular stimulation is to calculate the electric potentials generated in the tissue by passing current through the electrode. Potentials produced by passing current into a homogenous region of the CNS ($\sigma = 0.2$ S/m) using a point-source electrode (A), a cylindrical electrode (B) as used for deep brain stimulation, and a disk electrode (C) as used for epidural or cortical surface stimulation. (D) Although a simple analytical solution exists for the potentials generated by a point-source electrode, they differ substantially from the potentials generated by larger cylindrical or disk electrodes.

of extracellular stimulation on neurons involves a two-step approach. The first step is to calculate the electric potentials generated in the tissue by passage of current through the electrode. The second step is to determine the effect (or effects) of those potentials on the surrounding neurons.

20.4   From Cell to Circuit: Construction of Models of CNS Neurons

Electrical circuits are used to model the electrical behavior of neurons. These electrical-equivalent circuits, often referred to as cable models, represent the neuron as a series of cylindrical elements. Each cylinder is, in turn, replaced by a "compartment," representing the neuronal membrane, and a resistor representing the intracellular space. Thus, the model becomes a series of membrane compartments, connected by resistors. Each compartment is itself an electrical circuit that includes a capacitor representing the membrane capacitance of the lipid bilayer, resistors representing the ionic conductances of the trans-membrane proteins (ion channels), and batteries representing the differences in potential (Nernst potential) arising from ionic concentration differences across the membrane. The process of constructing an equivalent electric circuit model of a CNS neuron is illustrated in Figure 20.4.

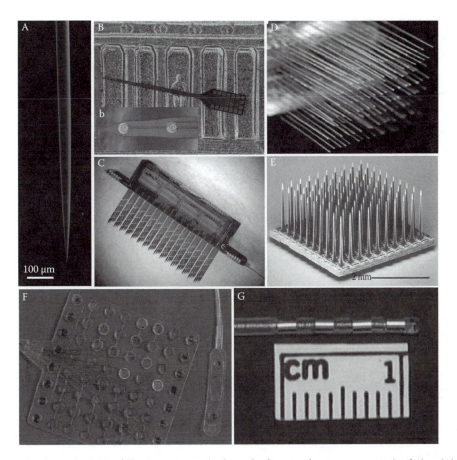

FIGURE 20.3 (See color insert following page 15-4). Electrodes for central nervous system stimulation. (A) Single iridium microwire electrode developed at Huntington Medical Research Institutes that can be used for extracellular recording from single units or extracellular microstimulation of small populations of neurons [McCreery et al., 1997] (B) Multisite silicon microprobe developed at the University of Michigan and higher magnification view (b) of two electrode sites near the tip. (Images courtesy of J.F. Hetke, University of Michigan.) (C) Three-dimensional assembly of multisite silicon microprobes [Bai et al., 2000]. The array is four probes, 256 sites on 400-μm centers in three dimensions. There are 16 parallel stimulating channels (16 sites active at any time) with off-chip current generation. The array is fed by a 7-lead ribbon cable at a data rate of up to 10 Mbps. It operates from ±5V supplies. (Image courtesy of K.D. Wise, University of Michigan.) (D) Arrays of up to 128 microwires enable simultaneous extracellular recording from multiple single neurons. Each wire is 50-μm diameter stainless steel, insulated with Teflon [Nicolelis et al., 2003]. (E) Multielectrode silicon array developed at the University of Utah [Normann et al., 1999]. (F) Subdural grid and strip electrode arrays used for cortical stimulation and recording (PMT Corporation, Chanhassen, Minnesota). (G) Quadrapolar electrode used for deep brain stimulation (Medtronic Inc., Minneapolis, Minnesota).

One can readily calculate the values of circuit elements from the geometry of the neurons and the specific values of neuron electrical properties. Consider a cylindrical representation of a segment of neuronal element, with diameter d and length l (Figure 20.4E). If one cuts and "unrolls" the cylinder, then the membrane resistance, R_m, is calculated as:

$$R_m = \text{Specific membrane resistance / Area of segment}$$

$$= r_m / \pi * l * d$$

where typical values for the specific membrane resistance range from 1000 to 5000 Ω-cm^2. The membrane resistance is nonlinear, its value depending on the voltage across the membrane (transmembrane potential).

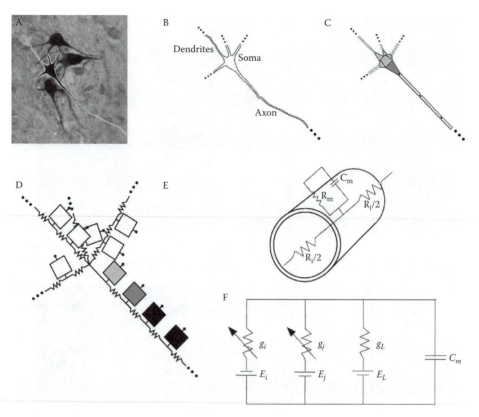

FIGURE 20.4 Construction of models of central nervous system (CNS) neurons. (A) Examples of stained neurons in the CNS. (B) The morphology of stained neurons can be reconstructed in three dimensions. (C) The morphology is then converted into a series of equivalent cylindrical elements. (D) The cylindrical elements are subsequently replaced by electrical equivalent circuits with resistive elements representing the intracellular space, and compartmental models representing the membrane. (E) Each cylindrical segment includes a representation of the membrane and the intracellular space, and the values of the equivalent circuit elements can be calculated from the geometry of the cylinder and specific parameter values. (F) Each compartment model of the membrane may contain several nonlinear ionic conductances (g_i, g_j) and a linear ionic conductance (g_L) representing various ionic channels in the membrane, batteries representing the Nernst potential arising from the difference in concentration of ions on the inside and outside of the membrane (E_i, E_j, E_L), and a capacitor (C_m) representing the capacitance arising from the lipid bilayer of the cell.

Further, separate elements (typically calculated as conductances) are used to represent the transmembrane paths for different ionic species, and the model of a patch of membrane includes several of these in parallel (Figure 20.4E). Similarly, the membrane capacitance, C_m, can be calculated as:

$$C_m = \text{Specific membrane capacitance} * \text{Area of segment}$$

$$= c_m * \pi * 1 * d$$

where typical values of the specific membrane capacitance range between 1 and 2 $\mu F/cm^2$. The intracellular resistance, R_i, can be calculated as:

$$R_i = \text{Intracellular resistivity} * \text{Segment length} / \text{Cross-sectional area of segment}$$

$$= \rho_i * 1 / (\pi * (d/2)^2)$$

where typical values of the intracellular resistivity range from 50 to 400 Ω-cm.

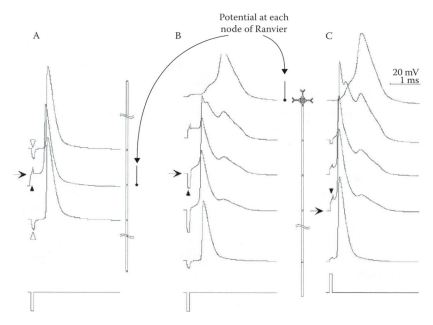

FIGURE 20.5 Action potential initiation by extracellular stimulation in CNS neurons by cathodic and anodic stimuli. Each trace shows transmembrane voltage as a function of time for different sections of the neuron. (A) Stimulation with a monophasic cathodic stimulus pulse from an electrode positioned 1 mm over a node of Ranvier of the axon. Depolarization occurs in the node directly beneath the electrode (solid arrowhead), and hyperpolarization occurs in the adjacent nodes of Ranvier (open arrowhead). Action potential initiation occurs in the node of Ranvier directly under the electrode (arrow), and the action potential propagates in both directions. (B) During threshold stimulation with an electrode positioned 1 mm over the cell body, action potential initiation occurs at a node of Ranvier of the axon. With cathodic stimuli (duration 0.1 ms), action potential initiation occurred at the second node of Ranvier from the cell body (arrow). (C) With anodic stimuli (duration 0.1 ms), action potential initiation occurred in the third node of Ranvier from the cell body (arrow).

20.5 Sites of Action Potential Initiation in CNS Neurons

The response of a cable model, representing a CNS neuron, to extracellular electrical stimulation is shown in Figure 20.5. The transmembrane potential as a function of time, in different segments of the neuron, is shown for a cathodic electrode positioned over the axon (Figure 20.5A), for a cathodic electrode positioned over the cell body (Figure 20.5B), and for an anodic electrode positioned over the cell body (Figure 20.5C).

During stimulation over the axon with a cathodic current, the axon is depolarized immediately beneath the electrode and hyperpolarized in regions lateral to the electrode (arrowheads in Figure 20.5A). Action potential initiation occurs in the most depolarized node of Ranvier, immediately beneath the electrode (arrow) and then propagates in both directions.

The response of a CNS neuron is more complex. With both cathodic and anodic stimuli delivered through an electrode placed 1 mm above the cell body, action potential initiation occurred in the axon, although the electrode is positioned directly over the soma. With 0.1-ms duration cathodic stimuli, action potential initiation occurred at the second node of Ranvier from the cell body (arrow); and with 0.1-ms duration anodic stimuli, action potential initiation occurred in the third node of Ranvier from the cell body (arrow). During the cathodic stimulus pulse, the node of Ranvier where action potential initiation occurred was hyperpolarized by the stimulus (arrowhead). Following termination of the stimulus, the cell body and dendritic tree discharged through the axon, leading to action potential initiation [McIntyre and Grill, 1999]. This finding in a computational model is consistent with contemporary *in vitro* results from cortex [Nowak and Bullier, 1998a,b]. Thus, with cathodic stimuli, action potential initiation occurred in a part of the neuron that was hyperpolarized by the stimulus, and this indirect mode of activation increases the threshold for activation of local cells with cathodic stimuli. Conversely, with anodic stimuli, the site of action potential initiation was at the node that was most depolarized by the stimulus (arrowhead). These position-dependent thresholds also are reflected in exciting populations of neurons (see below).

20.6 Excitation Properties of CNS Stimulation

The finding that action potential initiation occurs in the axon has several important implications for CNS stimulation. First, because excitation occurs in the axon, there is little difference in the extracellular chronaxie times for excitation of local cells and excitation of passing axons (see "Strength–Duration Relationship" below). Therefore, chronaxie time is not a sensitive indicator of the neuronal element that is activated by extracellular stimulation [Miocinovic and Grill, 2004]. Second, because action potential initiation occurs at some distance from the site of integration of synaptic inputs, the effects of co-activation of presynaptic fibers may be less than expected, and the axon may still fire even when the cell body is hyperpolarized (e.g., by inhibitory synaptic inputs) (see "Indirect Effects" below). Therefore, extracellular unit recordings of cell body firing may not accurately reflect the output of the neuron [Grill and McIntyre, 2001; McIntyre et al., 2004]. Finally, the difference in the mode of activation of local cells by cathodic stimuli and anodic stimuli is the basis for the difference in threshold between cathodic and anodic stimuli (see "Effect of Stimulus Polarity" below).

20.6.1 Strength–Duration Relationship

The stimulus amplitude necessary for excitation, I_{th}, increases as the duration of the stimulus decreases. The strength–duration relationship describes this phenomena and is given by:

$$I_{th} = I_{rh}[1 + T_{ch}/PW]$$

where the parameter I_{rh} is the rheobase current and is defined as the current amplitude necessary to excite the neuron with a pulse of infinite duration, and the parameter T_{ch} is the chronaxie and is defined as the pulse duration necessary to excite the neuron with a pulse amplitude equal to twice the rheobase current.

Measurements with intracellular stimulation have demonstrated that the temporal excitation characteristics, including chronaxie (T_{ch}) and refractory period, of cells and axons differ (Figure 20.6A). However, during extracellular stimulation of neurons, action potential initiation occurs in the axon, even with the stimulating electrode positioned over the cell body or dendrites (see above). Although with intracellular activation the chronaxies of many cell bodies exceed 1 ms, with extracellular activation they are below 1 ms [Stoney et al., 1968; Ranck, 1975; Asanuma et al., 1976; Swadlow, 1992] and lie within the ranges determined for extracellular activation of axons [Ranck, 1975; Li and Bak, 1976; West and Wolstencroft, 1983]. For stimulation of cortical gray matter, although the mean T_{ch} for *intracellular* stimulation of cells (15 ms) was substantially longer than T_{ch} for *extracellular* stimulation of axons (0.27 ms), the mean T_{ch} for *extracellular* stimulation of local cells (0.38 ms) was comparable to that for extracellular stimulation of axons [Nowak and Bullier, 1998a]. Further, during extracellular stimulation, the chronaxies measured with extracellular stimulation depended on a number of factors other than the neuronal element that was stimulated [Miocinovic and Grill, 2004]. The chronaxies of different neuronal elements determined with extracellular stimulation overlap and do not enable unique determination of the neuronal element stimulated.

20.6.2 Current–Distance Relationship

The current required for extracellular stimulation of neurons (threshold, I_{th}) increases as the distance between the electrode and the neuron (r) increases. This is described by the current–distance relationship [Stoney et al., 1968]:

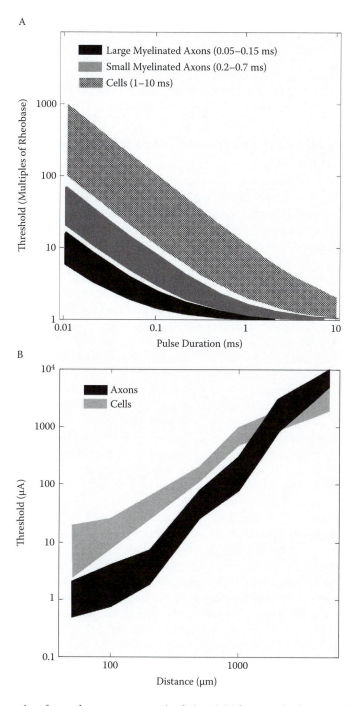

FIGURE 20.6 Properties of central nervous system stimulation. (A) The strength–duration relationship describes the amplitude required for stimulation as a function of the stimulation pulse duration. Strength–duration curves for intracellular stimulation of different neural elements were constructed from data summarized by Ranck (1976). (B) The current–distance relationship describes the threshold intensity required for stimulation as a function of the distance between the electrode and the neuron. Current distance curves for axons and cells were constructed from data summarized by Ranck (1976).

$$I_{th} = I_R + k \cdot r^2$$

where the offset (I_R) determines the absolute threshold and the slope (k) determines the threshold difference between neurons at different distances from the electrode. The current–distance relationships for excitation of axons and cells in the CNS have been measured in a large number of preparations, and current distance curves for these two populations are summarized in Figure 20.6B.

20.6.3 Effect of Stimulus Polarity and Stimulus Waveform on CNS Stimulation

During excitation of axons in the peripheral nervous system, different stimulus polarities produce changes in the threshold as well as changes in the site of action potential initiation, and similar but more pronounced effects occur during CNS stimulation. Figure 20.7 shows the results of a computational study to determine which neuronal elements are activated by extracellular stimulation in the CNS. A model including populations of local cells and axons of passage, randomly positioned around a point source stimulating electrode, was used to compare the activation of local cells to the activation of passing fibers with different stimulation waveforms [McIntyre and Grill, 2000]. Using cathodic pulses, the threshold for activation of passing axons was less than the threshold for activation of local neurons; and when 70% of the axons were activated, approximately 10% of the local cells were also activated. When using anodic pulses, the threshold for activation of local cells is less than the threshold for activation of passing axons, and the stimulus amplitude that activated 70% of the local cells also activated 25% of the passing axons. The basis for this effect can be understood by comparing action potential initiation in local cells using cathodic and anodic stimuli described above.

To prevent the possible degradation of the stimulating electrode(s) or damage to the tissue, chronic stimulation is conducted with biphasic stimulus pulses [Lilly et al., 1955; Robblee and Rose, 1990]. The response of passing axons and local cells to symmetric biphasic pulses is shown in Figures 20.7C and 20.7D. Using either anodic-phase first or cathodic-phase first pulses, the threshold for activation of passing axons was less than the threshold for activation of local neurons, and the relative selectivity for axons was lower with either pulse than with monophasic cathodic pulses.

These results and previous experimental evidence demonstrates that different neuronal elements have similar thresholds for extracellular stimulation [Roberts and Smith, 1973; Gustafsson and Jankowska, 1976] and illustrates the need for the design of methods that enable selective stimulation. Stimulus waveforms can be designed explicitly to take advantage of the nonlinear conductance properties of neurons and thereby increase the selectivity between activation of different neuronal elements. Biphasic asymmetrical stimulus waveforms capable of selectively activating either local cells or axonal elements consist of a long-duration, low-amplitude prepulse followed by a short-duration, high-amplitude stimulation phase. The long-duration prepulse phase of the stimulus is designed to create a subthreshold depolarizing prepulse in the nontarget neurons and a hyperpolarizing prepulse in the target neurons [Grill and Mortimer, 1995; McIntyre and Grill, 2000]. Recall that during cathodic stimulation, the site of excitation in axons is the depolarized node of Ranvier, while the site of excitation in local cells is a node of Ranvier that is hyperpolarized by the stimulus (Figure 20.5). Conversely, with anodic stimuli, the site of excitation in local cells is a depolarized node of Ranvier, and the most polarized node of passing axons is hyperpolarized by the stimulus. Thus, the same polarity prepulse will produce opposite polarization at the sites of excitation in local cells and passing axons. The effect of this subthreshold polarization is to decrease the excitability of the nontarget population and increase the excitability of the target population via alterations in the degree of sodium channel inactivation [Grill and Mortimer, 1995]. Therefore, when the stimulating phase of the waveform is applied, neuronal population targeted for stimulation will be activated with greater selectivity [McIntyre and Grill, 2000]. Asymmetrical charge-balanced, biphasic, cathodic phase first stimulus waveforms result in selective activation of local cells, while asymmetrical charge-balanced, biphasic, anodic phase first stimulus waveforms result in selective activation of fibers of passage. Further, charge balancing is achieved as required to reduce the probability

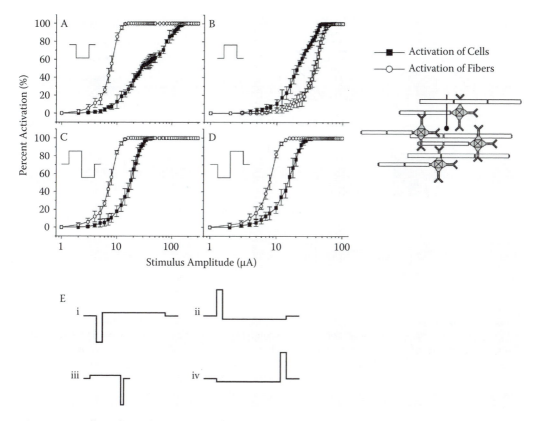

FIGURE 20.7 Effect of stimulus polarity and waveform on excitation of populations of local cells and passing axons. (A to D) Input–output curves from a population model containing 50 passing axons and 50 local cells randomly positioned around a point-source stimulating electrode. The curves are the percent of neurons (passing axons, local cells) activated as a function of the stimulation amplitude for excitation with (A) 0.2-ms duration monophasic cathodic pulses, (B) 0.2-ms duration monophasic anodic pulses, (C) anodic phase first biphasic symmetric pulses (0.2 ms per phase), and (D) cathodic phase first biphasic symmetric pulses (0.2 ms per phase). (Modified from McIntyre and Grill, 2000.) (E) Examples of asymmetric charge-balanced biphasic pulses. Cathodic-phase first (i) and anodic-phase first pseudomonophasic pulses have low-amplitude second phases and exhibit excitation properties similar to monophasic cathodic and anodic stimuli, respectively. Novel asymmetric pulses that manipulate neuronal excitability via a subthreshold first phase also provides charge balancing. The anodic pre-pulse (0.2 ms) followed by a cathodic stimulus phase (0.02 ms) enables preferential excitation of passing axons, while a cathodic prepulse (1.0 ms) followed by an anodic stimulus pulse (0.1 ms) enables preferential excitation of local cells [McIntyre and Grill, 2000].

of tissue damage and electrode corrosion. Note that these prepulse waveforms differ from the pseudo-monophasic waveforms used in some stimulators in that the low-amplitude, long-duration phase of the waveform precedes rather than follows the high-amplitude, short-duration phase of the waveform (Figure 20.7E).

20.6.4 Indirect Effects of Extracellular Stimulation

The thresholds for excitation of presynaptic terminals and subsequent indirect effects on local neurons (mediated by synaptic transmission) are similar to thresholds for direct effects (mediated by stimulus current) during extracellular stimulation (Figure 20.8A) of spinal cord motoneurons [Gustafson and Jankowska, 1976]; rubrospinal neurons [Baldissera et al., 1972]; and corticospinal neurons [Jankowska et al., 1975]. Further, the chronaxie of presynaptic terminals (~0.14 ms in frog spinal cord, Tkacs and Wurster, 1991; 0.06 to 0.54 ms in rat subthalamic nucleus, Hutchison et al., 2002) is comparable to that

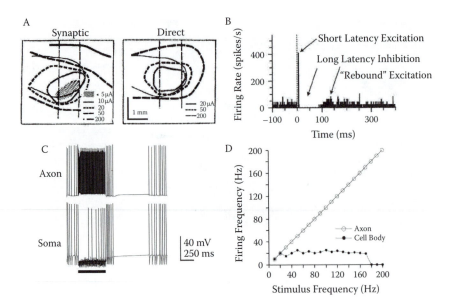

FIGURE 20.8 Central nervous system (CNS) stimulation results in direct effects and indirect effects on CNS neurons. (A) Two-dimensional maps of thresholds for indirect (synaptic) and direct activation of neurons in the red nucleus. (From Baldissera et al., 1972.) (B) Complex polyphasic changes in the firing rate of a cortical neuron in response to extracellular stimulation. (From Butovas and Schwarz, 2003.) (C) Transmembrane potential in the axon (top trace) and cell body (bottom trace) of a model thalamocortical neuron before, during (black bar at bottom), and after extracellular stimulation. (From McIntyre et al., 2004.) Extracellular stimulation results in simultaneous inhibition of the cell body, as a result of activation of presynpatic terminals and subsequent indirect effects, and excitation of the axon, as a result of direct action potential initiation in a node of Ranvier. (D) Firing rate in the cell body and axon during extracellular stimulation of a model thalamocortical neuron (modified from McIntyre et al., 2004). The firing rate in the cell body is lower than that in the axon, as a result of simultaneous indirect synaptic effects on the soma and direct excitation of the axon.

of passing axons, and thus effects may be attributed to activation of passing axons when in fact they arise from activation of presynaptic elements. These "indirect" effects of stimulation must be considered when electrodes are placed within the heterogeneous environment of the CNS. During extracellular stimulation, release of inhibitory and/or excitatory neurotransmitters from presynaptic terminals can result in complex polyphasic changes in the firing rate of postsynpatic neurons (Figure 20.8B) [Butovas and Schwarz, 2003] and modulate the threshold for excitation of the postsynaptic neuron [Swadlow, 1992; McIntyre and Grill, 2002]. Thus, indirect effects mediated by synaptic transmission may alter the direct effects of stimulation on the postsynaptic cell. Furthermore, antidromic propagation of action potentials originating from activation of axon terminals can lead to widespread activation or inhibition of targets distant from the site of stimulation through axon collaterals. However, recall that action potential initiation occurs at some distance from the soma, where integration of synaptic inputs occurs, and thus the axon may be excited even when the cell body is hyperpolarized (Figure 20.8C). Therefore, extracellular unit recordings of firing in the soma may not accurately reflect the output of the neuron (Figure 20.8D) [Grill and McIntyre, 2001; McIntyre et al., 2004].

20.7 Summary

This chapter described electrical activation of neurons within the central nervous system (CNS). Electrical stimulation is used to study the form and function of the nervous system and as a technique to restore function following disease or injury. Successful application of electrical stimulation to treat nervous system disorders as well as interpretation of the results of stimulation require an understanding of the cellular-level effects of stimulation. Quantitative models provide a means to understand the response of neurons to extracellular stimulation. Further, accurate quantitative models provide powerful design tools that can be used to engineer stimuli that produce a desired response.

The fundamental properties of the excitation of CNS neurons were presented with a focus on what neural elements around the electrode are activated under different conditions. During CNS stimulation, action potentials are initiated in the axons of local cells, even for electrodes positioned over the cell body. The threshold difference between cathodic and anodic stimuli arises due to differences in the mode of activation. Anodic stimuli cause depolarization of the axon and excitation via a "virtual cathode," while cathodic stimuli cause hyperpolarization at the site of excitation and the action potential is initiated during repolarization. The threshold for activation of presynaptic terminals projecting into the region of stimulation is often less than or equal to the threshold for direct excitation of local cells, and indirect effects mediated by synaptic transmission may alter the direct effects of stimulation on the postsynaptic cell. The fundamental understanding provided by this analysis enables the rational design and interpretation of studies and devices employing electrical stimulation of the brain or spinal cord.

Acknowledgments

Research in Dr. Grill's laboratory and preparation of this chapter were supported by NIH Grant R01 NS-40894.

References

Ackman, J.J. and Seitz, M.A. (1984). Methods of complex impedance measurement in biologic tissue. *CRC Crit. Rev. Biomed. Eng.*, 11:281–311.

Asanuma, H., Arnold, A., and Zarezecki, P. (1976). Further study on the excitation of pyramidal tract cells by intracortical microstimulation. *Exp. Brain Res.*, 26:443–461.

Bai, Q., Wise, K.D., and Anderson, D.J. (2000). A high-yield microassembly structure for three-dimensional microelectrode arrays. *IEEE Trans. Biomed. Eng.*, 47(3):281–289.

Baldissera, F., Lundberg, A., and Udo, M. (1972). Stimulation of pre- and postsynaptic elements in the red nucleus. *Exp. Brain Res.*, 15:151–167.

Baumann, S.B., Wozny, D.R., Kelly, S.K., and Meno, F.M. (1997). The electrical conductivity of human cerebrospinal fluid at body temperature. *IEEE Trans. Biomed. Eng.*, 44(3):220–223.

Brindley, G.S. and Lewin, W.S. (1968). The sensations produced by electrical stimulation of the visual cortex. *J. Physiol.*, 196:479–493.

Butovas, S. and Schwarz, C. (2003). Spatiotemporal effects of microstimulation in rat neocortex: a parametric study using multielectrode recordings. *J. Neurophysiol.*, 90(5):3024–3039.

Cameron, T. (2004). Safety and efficacy of spinal cord stimulation for the treatment of chronic pain: a 20-year literature review. *J. Neurosurg.*, 100(3 Suppl.):254–267.

Coffey, R.J. (2001). Deep brain stimulation for chronic pain: results of two multicenter trials and a structured review. *Pain Med.*, 2:183–192.

Crile, G.W., Hosmer, H.R., and Rowland, A.F. (1922). The electrical conductivity of animal tissues under normal and pathological conditions. *Am. J. Physiol.*, 60:59–106.

Eisenberg, R.S., and Mathias, R.T. (1980). Structural analysis of electrical properties of cells and tissues. *CRC Crit. Rev. Biomed. Eng.*, 4:203–232.

Grill, W.M. (1999). Modeling the effects of electric fields on nerve fibers: influence of tissue electrical properties. *IEEE Trans. Biomed. Eng.*, 46:918–928.

Grill, W.M. and McIntyre, C.C. (2001). Extracellular excitation of central neurons: implications for the mechanisms of deep brain stimulation. *Thalamus & Relat. Syst.*, 1:269–277.

Grill, W.M. and Mortimer, J.T. (1994). Electrical properties of implant encapsulation tissue. *Ann. Biomed. Eng.*, 22:23–33.

Grill, W.M. and Mortimer, J.T. (1995). Stimulus waveforms for selective neural stimulation. *IEEE Eng. Med. Biol.*, 14:375–385.

Gross, R.E. and Lozano, A.M. (2000). Advances in neurostimulation for movement disorders. *Neurol. Res.*, 22, 247–258.

Gross, R.E. (2004). Deep brain stimulation in the treatment of neurological and psychiatric disease. *Expert Rev. Neurother.*, 4, 465–78.

Gustafsson, B. and Jankowska, E. (1976). Direct and indirect activation of nerve cells by electrical pulses applied extracellularly. *J. Physiol.*, 258:33–61.

Hodaie, M., Wennberg, R.A., Dostrovsky, J.O., and Lozano, A.M. (2002). Chronic anterior thalamus stimulation for intractable epilepsy. *Epilepsia*, 43, 603–608 (2002).

Holsheimer, J., Struijk, J.J., and Tas, N.R. (1995). Effects of electrode geometry and combination on nerve fibre selectivity in spinal cord stimulation. *Med. Biol. Eng. Comput.*, 33:676–682.

Hutchison, W.D., Chung, A.G., and Goldshmidt, A. (2002). Chronaxie and refractory period of neuronal inhibition by extracellular stimulation in the region of rat STN. Program No. 416.3. Abstract Viewer/Itinerary Planner CD-ROM, Society for Neuroscience, Washington, D.C.

Jankowska, E., Padel, Y., and Tanaka, R. (1975). The mode of activation of pyramidal tract cells by intracortical stimuli. *J. Physiol.*, 249:617–636.

Jankowska, E., and Roberts, W.J. (1972). An electrophysiological demonstration of the axonal projections of single spinal interneurones in the cat. *J. Physiol.*, 222:597–622.

Li, C.-H., Bak, A.F., and Parker, L.O. (1968). Specific resistivity of the cerebral cortex and white matter. *Exp. Neurol.*, 20:544–557.

Li, C.L. and Bak, A. (1976). Excitability characteristics of the A- and C-fibers in a peripheral nerve. *Exp. Neurol.*, 50:67–79.

Lilly, J.C., Hughes, J.R., Alvord, E.C., Jr., and Galkin, T.A. (1955). Brief noninjurious electric waveform for stimulation of the brain. *Science*, 121:468–469.

Lipski, J. (1981). Antidromic activation of neurones as an analytic tool in the study of the central nervous system. *J. Neurosci. Methods*, 4:1–32.

McCreery, D.B., Yuen, T.G., Agnew, W.F., and Bullara, L.A. (1997). A characterization of the effects on neuronal excitability due to prolonged microstimulation with chronically implanted microelectrodes. *IEEE Trans. Biomed. Eng.*, 44(10):931–939.

McIntyre, C.C. and Grill, W.M. (1999). Excitation of central nervous system neurons by nonuniform electric fields. *Biophys. J.*, 76:878–888.

McIntyre, C.C. and Grill, W.M. (2000). Selective microstimulation of central nervous system neurons. *Ann. Biomed. Eng.*, 28:219–233.

McIntyre, C.C. and Grill, W.M. (2001). Finite element analysis of the current-density and electric field generated by metal microelectrodes. *Ann. Biomed. Eng.*, 29(3):227–235.

McIntyre, C.C. and Grill, W.M. (2002). Extracellular stimulation of central neurons: influence of stimulus waveform and frequency on neuronal output. *J. Neurophysiol.*, 88:1592–1604.

McIntyre, C.C., Grill, W.M., Sherman, D.L., and Thakor, N.V. (2004). Cellular effects of deep brain stimulation: model-based analysis of activation and inhibition. *J. Neurophysiol.*, 91:1457–1469.

Miocinovic, S. and Grill, W.M. (2004). Sensitivity of temporal excitation properties to the neuronal element activated by extracellular stimulation. *J. Neurosci. Methods*, 132:91–99.

Nicholson, P.W. (1965). Specific impedance of cerebral white matter. *Exp. Neurol.*, 13:386–401.

Nicolelis, M.A.L., Dimitrov, D., Carmena, J.M., Crist, R., Lehew, G., Kralik, J.D., and Wise, S.P. (2003). Chronic, multisite, multielecrode recording in macaque monkeys. *Proc. Natl. Acad. Sci., U.S.A.*, 100:11041–11046.

Normann, R.A., Maynard, E.M., Rousche, P.J., and Warren, D.J. (1999). A neural interface for a cortical vision prosthesis. *Vision Res.*, 39(15):2577–2587.

Nowak, L.G. and Bullier, J. (1998a). Axons, but not cell bodies, are activated by electrical stimulation in cortical gray matter. I. Evidence from chronaxie measurements. *Exp. Brain Res.*, 118:477–488.

Nowak, L.G. and Bullier, J. (1998b). Axons, but not cell bodies, are activated by electrical stimulation in cortical gray matter. II. Evidence from selective inactivation of cell bodies and axon initial segments. *Exp. Brain Res.*, 118:489–500.

Otto, S.R., Brackmann, D.E., Hitselberger, W.E., Shannon, R.V., and Kuchta, J. (2002). Multichannel auditory brainstem implant: update on performance in 61 patients. *J. Neurosurg.*, 96:1063–1071.

Ranck, J.B., Jr. (1963). Analysis of specific impedance of rabbit cerebral cortex. *Exp. Neurol.*, 7:153–174.

Ranck, J.B., Jr. (1975). Which elements are excited in electrical stimulation of mammalian central nervous system: a review. *Brain Res.*, 98:417–440.

Ranck, J.B., Jr. and BeMent, S.L. (1965). The specific impedance of the dorsal columns of the cat: an anisotropic medium. *Exp. Neurol.*, 11:451–463. 440.

Robblee, L.S. and Rose, T.L. (1990). Electrochemical guidelines for selection of protocols and electrode materials for neural stimulation. In *Neural Prostheses: Fundamental Studies*, Agnew, W.F. and McCreery, D.B., Eds., Prentice-Hall, Englewood Cliffs, NJ, p. 25–66.

Roberts, W.J. and Smith, D.O. (1973). Analysis of threshold currents during microstimulation of fibers in the spinal cord. *Acta Physiol. Scand.*, 89:384–394.

Sances, A., Jr. and Larson, S.J. (1975). Impedance and current density studies. In *Electroanesthesia: Biomedical and Biophysical Studies*, Sances, A. and Larson, S.J., Eds., Academic Press, New York, 1975, p. 114–24.

Schmidt, E.M., Bak, M.J., Hambrecht, F.T., Kufta, C.V., O'Rourke, D.K., and Vallabhanath, P. (1996). Feasibility of a visual prosthesis for the blind based on intracortical microstimulation of the visual cortex. *Brain*, 119:507–522.

Stoney, S.D., Jr., Thompson, W.D., and Asanuma, H. (1968). Excitation of pyramidal tract cells by intracortical microstimulation: effective extent of stimulating current. *J. Neurophys.*, 31:659–669.

Swadlow, H.A. (1992). Monitoring the excitability of neocortical efferent neurons to direct activation by extracellular current pulses. *J. Neurophysiol.*, 68:605–619.

Tehovnik, E.J. (1996). Electrical stimulation of neural tissue to evoke behavioral responses. *J. Neursci. Methods*, 65:1–17.

Tkacs, N.C. and Wurster, R.D. (1991). Strength-duration and activity-dependent excitability properties of frog afferent axons and their intraspinal projections. *J. Neurophysiol.*, 65:468–476.

Troyk, P., Bak, M., Berg, J., Bradley, D., Cogan, S., Erickson, R., Kufta, C., McCreery, D., Schmidt, E., and Towle, V. (2003). A model for intracortical visual prosthesis research. *Artif. Organs*, 27:1005–1015.

Velasco, F., Velasco, M., Jimenez, F., Velasco, A.L., and Marquez, I. (2001). Stimulation of the central median thalamic nucleus for epilepsy. *Stereotact. Funct. Neurosurg.*, 77, 228–232.

West, D.C. and Wolstencroft, J.H. (1983). Strength-duration characteristics of myelinated and non-myelinated bulbospinal axons in the cat spinal cord. *J. Physiol.*, 337:37–50.

21

Transient Optical Nerve Stimulation: Concepts and Methodology of Pulsed Infrared Laser Stimulation of the Peripheral Nerve *In Vivo*

Jonathon D. Wells,
Anita Mahadevan-Jansen,
C. Chris Kao,
Peter E. Konrad, and
E. Duco Jansen

21.1 Introduction

Transient optical stimulation of neural tissue is a fundamentally new approach to stimulate the peripheral nerve that has distinct advantages when compared to other peripheral nerve excitation modalities, including standard electrical nerve stimulation. This novel methodology for nerve excitation relies on irradiation of the nerve surface, in a noncontact fashion, with a pulsed (infrared) laser operating at an optimized radiant exposure and wavelength for the generation of compound action potentials and associated physiological effect (i.e., muscle contraction or sensory response). The response is extremely spatially precise, providing the possibility for selective targeting of individual nerve fascicles, and does not result in tissue damage at laser radiant exposures (energy/area) at least two times that required for a visible muscle contraction. Thus, optical stimulation presents an innovative contact-free approach to neural activation that may benefit clinical nerve stimulation as well as fundamental neurophysiology and neuroscience. This chapter defines and characterizes optical nerve stimulation as well as details the methodology of this technique. The fundamental principles and experiments that help define the optimal parameters needed for safe and effective stimulation of the peripheral nerve are described. The biophysical mechanistic considerations and working hypotheses are briefly discussed. Finally, the potential impact of optical stimulation is noted in terms of applications and clinical utility for this technology.

21.1.1 Limitations of Standard Electrical Nerve Stimulation

Throughout modern history, electricity has served as the standard method for excitation of neural tissue, both clinically, in diagnostics and therapeutics, and for basic scientific research, in the pursuit of clinical utility as well as conceptual understanding of action potential propagation/signaling and even nerve regeneration (Sisken et al., 1993; Lu and Waite, 1999; Roehm and Hansen, 2005). Although this method sees widespread use, it has several fundamental limitations. To stimulate the nerve electrically requires physical contact with a metal electrode, often pierced into the tissue, which can give rise to tissue damage. Spatial precision of stimulation is inadequate due to the size of electrodes and, more importantly, the inherent induction of an electric field spanning a spatial area much greater than the size of the electrode. This will ultimately initiate a population response due to the recruitment of multiple axons. This spread of electrical current in a graded fashion beyond the electrode causes poor spatial specificity with this technique (Palanker et al., 2005). Many applications of neural stimulation require precision of the stimulus in a small target tissue. Electrodes designed to deliver precise stimulation have inherently high impedance characteristics, which, in turn, impose higher voltage requirements to deliver the same charge as dictated by Ohm's law. In addition, in applications where electrophysiological recordings are performed to observe the response to stimulation, an inescapable "stimulation artifact" exists owing to the fact that the stimulation technique occurs in the same domain as the recording technique. Hence, recordings of these extremely small electrical signals (action potentials) in the vicinity of the point of electrical stimulation are inherently contaminated by the electric field involved in the stimulation. The result is a recorded signal that contains electrical artifact from the stimulus, which is usually much larger than the recorded action potential, and precise interpretation is often not possible, or at the very least requires significant processing to obtain meaningful results (McGill et al., 1982; Miller et al., 2000). These limitations in the current technique have driven researchers to pursue other means for neural stimulation. The literature thoroughly documents the notion that action potentials can be triggered in neurons using many different stimuli; these include electrical, magnetic, mechanical, thermal, chemical, and optical means. Here we focus attention on optical methods for the stimulation of neural tissue.

21.1.2 Definition of Optical Stimulation

We define *optical stimulation* as the direct induction of an evoked potential (EP/AP) in response to a transient targeted deposition of optical energy. This implies that only a pulsed source can be used for stimulation of neural tissue, and that continuous wave (CW) irradiation will not lead to compound action potential generation.

Now that we have a strict definition for optical stimulation, let us briefly review some other laser application to help illustrate what is and is not considered optical stimulation. Typically, the use of lasers in biomedicine relies on high energy effects such as tissue ablation and photoacoustic wave generation (Welch et al., 1991; Wietholt et al., 1992; Jansen et al., 1996; Vogel and Venugopalan, 2003; Kanjani et al., 2004). In contrast, a number of low-power laser applications have been described. For example, in an application known as biostimulation or low-level light therapy (LLLT), low fluence levels at laser wavelengths that are weakly absorbed in tissue are applied continuously for several minutes. Due to largely unknown mechanisms, the laser radiation modulates biological processes such as inflammation, cell proliferation, and others. This modality has been shown to improve wound healing, stimulate hair growth, alter pain perception, and promote regeneration of neural tissue (Walsh, 1997; Chen et al., 2005). Although the words "laser," "optical," and "stimulation" may appear in these texts relating to neural tissue, we do not consider this "optical nerve stimulation" as it explicitly does not fit with our stated definition of transient deposition of energy leading to activation of a potential as a direct result of incident light.

This technique uses radiant exposures at wavelengths that are more strongly absorbed than in LLLT to directly stimulate neural tissue. As discussed later in this chapter, we have preliminary evidence that the induction of a temperature gradient (dT/dz or dT/dt) is, in fact, required to induce an action potential. Similarly, the term "optical stimulation" in neural tissue can be used to describe the use of light to activate caged compounds or phototransduction in visual cortex mapping; using the above definition, we do not consider these applications a form of optical stimulation. Nevertheless, we have demonstrated that the radiant exposure needed to induce neural stimulation is well below the threshold for inducing permanent damage to the tissue. We refer to the radiant exposure needed for optical stimulation of neural tissue as "low level" relative to the conventional therapeutic laser applications that lead to tissue coagulation and ablation. A final distinction arises in the literature for modulation of the excitability of nerves using light (Wu et al., 1987; Balaban et al., 1992; Bragard et al., 1996). Here, laser stimulation means applied *light acting to modulate that signal* or potential, which is *produced spontaneously or by some other means* (electrical stimulation), rather than light stimulation being the primary source of that signal. In contrast, our definition of laser "stimulation" involves the direct incidence of light on the neural tissue *resulting in an evoked potential* from the neural tissue. In this case, the laser light is not modulating an existing potential; rather, it is the *means by which a signal is produced*. This distinction clearly separates these two uses for a laser incident on neural tissue.

21.1.3 Previous Work in Optical Stimulation

Although no reports of low-level, direct laser stimulation of neural tissue exist, it is instructive to review literature pertaining to high-energy, transient laser irradiation of the nervous system. Optical stimulation was first reported (Fork, 1971) as action potentials generated in *Aplysia* neurons (pigmented) through a reversible mechanism. This was the first indication that optical irradiance of nerve cells could perhaps induce neural stimulation in the form of an elicited action potential. In a different study, a bundle of rat CNS fibers in the medial lemniscus and cuneate bundle in the spinal cord (recording from the thalamic VPN) was reported as a side effect to ablation using a short pulse, ultraviolet excimer laser (Allegre et al., 1994). The stimulation radiant exposures (1.0 J/cm^2) were greater than the tissue damage threshold (0.9 J/cm^2); nonetheless, animal movements were observed in response to pulsed laser energy. Hirase et al. (2002) reported that a high-intensity, mode-locked infrared femtosecond laser induced depolarization and subsequent action potential firing in transiently irradiated pyramidal neurons. However, prior to our work described in the subsequent sections of this chapter, there had been no systematic studies published on the application of optical energy for neural activation. In particular, there is no evidence in the literature on the concept of using low levels of pulsed infrared light to chronically stimulate neural potentials *in vivo* for future clinical as well as research applications.

21.2 Optical Stimulation

The basis of this work is that delivery of pulsed laser light can be used for contact-free, damage-free, artifact-free stimulation of discrete populations of neural fibers. We have previously shown that a pulsed, low-energy laser beam elicits compound nerve and muscle action potentials, with resultant muscle contraction, which is indistinguishable from responses obtained with conventional bipolar, electrical stimulation of the rat sciatic nerve *in vivo* (Wells et al., 2005a). The stimulation threshold (0.3 to 0.4 J/cm^2) at optimal wavelengths in the infrared (1.87, 2.1, 4.0 μm) is at least two times less than the threshold at which any histological tissue damage occurs (0.8 to 1.0 J/cm^2). Optical nerve stimulation has three fundamental advantages over electrical stimulation (Wells et al., 2005b) that make it ideal for a number of procedures that currently employ electrical means as the standard of care:

1. The precision of optically delivered energy is far superior to electrical stimulation techniques and can easily be confined to individual nerve fascicles without requiring separation between the area of stimulation and other areas.

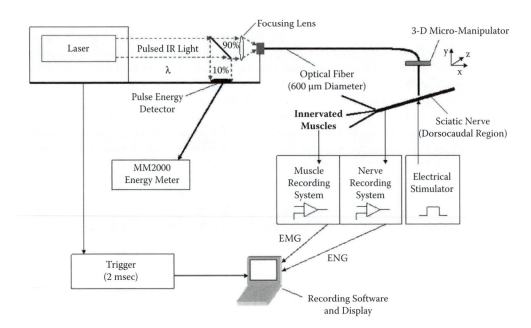

FIGURE 21.1 Typical experimental setup for optical stimulation and recording in the rat sciatic nerve.

2. Optical stimulation does not produce a stimulation artifact, whereas electrical stimulation inherently results in noise in the recorded signal.
3. Optical stimulation is achieved in a noncontact fashion, a technical advantage that can minimize the risk of nerve trauma or metal–tissue interface concerns.

The following section describes the methodology and fundamental considerations that one must understand to benefit from these advantages without causing tissue damage. It should be noted that the work described here primarily deals with the peripheral nervous system. To date we have focused on inducing motor responses. In other studies in collaboration with Richter and Walsh at Northwestern University, this has been extended to the sensory nervous system (spiral ganglion cells in the cochlea) (Izzo et al., 2005; Richter, 2005a,b; Izzo, 2006a,b).

21.2.1 Introduction to the Feasibility, Methodology, and Physiological Validity

Initially, to demonstrate the ability to stimulate peripheral nerves with a pulsed laser, a proof of concept study was performed *in vivo* on the sciatic nerve of a frog. Shortly thereafter, we demonstrated feasibility within our current mammalian peripheral nerve model, the rat sciatic nerve. The typical experimental setup to perform optical stimulation with electrical recording of the nerve and muscle potentials is depicted in Figure 21.1.

In general, an infrared pulsed laser source is optically manipulated to a small focal spot utilized for optical stimulation of the peripheral nerve. For these experiments, the holmium:YAG laser operating at a wavelength of 2.12 μm and pulse duration of 350 μs was used. This wavelength has been shown to be optimal for peripheral nerve stimulation. The importance of this parameter is discussed in detail in Section 21.2.4. Delivery to the tissue is accomplished with an optical fiber, waveguide, or simply a free-beam incident on the nerve surface. Wavelengths that transmit through optical fibers (<2.5 μm) are considered ideal because the tip of the fiber can be easily manipulated in three dimensions for precise delivery to the nerve. Stimulation experiments in the rat sciatic nerve reveal that a 400- to 600-μm fiber diameter can most efficiently result in excitation while maintaining precision in stimulation, although the optimal fiber diameter will vary according to the thickness of the given peripheral nerve bundle. While not discussed here, the theoretical limits for both delivery methods are on the order of a few micrometers. Radiant exposures required to stimulate vary, depending on the wavelength of the laser source used (see wavelength dependence section). Electrical stimulation and recording of the compound nerve and muscle potentials can be employed to verify the validity of the evoked response from laser stimulation and compare this to the standard electrical stimulation methods.

Several experiments were performed *in vivo*, initially on the frog sciatic nerve, and subsequently in mammals using a rat model, to verify the physiologic validity of optical stimulation. To confirm the direct stimulatory effect of low-level optical energy, the nerve was optically isolated from its surrounding tissues using an opaque material and stimulated. A consistent evoked response was recorded, indicating that the incident light is directly responsible for the compound nerve (CNAP) and muscle action potentials (CMAP) observed. Both signals were lost when the delivery of optical energy was blocked with a shutter, indicating that stimulation was not due to artifacts associated with the trigger pulse or other electrical interference synchronous with acquisition. Application of a depolarizing neuromuscular blocker (succinylcholine) resulted in a measurable CNAP and loss of CMAP, confirming the involvement of normal propagation of impulses from nerve to muscle upon optical stimulation.

In a proof of principle study, CNAPs and CMAPs were consistently observed and recorded using conventional electrical recordings (Figure 21.2) from both electrical and optical peripheral nerve excitation methods. CNAP responses were amplified 5000X and filtered using a high-pass filter (>20 Hz) and a low-pass filter (<3 kHz). CMAP responses were amplified 1000X and filtered using a high-pass filter (>0.05 Hz) and low-pass filter (<5 kHz). The similarity in the shape and timing of the signals from optical and electrical stimulus in Figure 21.2 show that conduction velocities, represented by the time between the CNAP and CMAP, are equal. These traces imply that the motor fiber types recruited and seen in the recorded compound action potentials are identical, regardless of excitation mechanism. That is, based solely on observation of the physiologic portions of recorded signals (nerve and muscle), one cannot discern between the two stimulation techniques. However, two important signal characteristics manifest in Figure 21.2 that allow one to differentiate between optically and electrically evoked potentials. One is the inherent electrical stimulation artifact that is only seen in the electrically stimulated peripheral nerve recordings. The other is the superior spatial selectivity, or precise and localized number of axons recruited with optical stimulation when compared to electrical stimulation. This phenomenon is realized by the order of magnitude difference in amplitude (proportional to the number of axons recruited) between electrical and optical recordings. In the following sections, each of these unique advantages associated with optical stimulation is explored in more detail.

21.2.2 Generation of an Artifact-Free Nerve Potential Recording

The standard method for peripheral nerve stimulation requires that the stimulation technique occurs in the same domain as the recording technique, through electrical means. Therefore, an inescapable artifact, the amplitude of which is much greater than the physiological signal, is inherent to any electrically stimulated nerve recording for the first 1 to 2 ms. Considering the speed at which action potentials are propagated, it is clear that this artifact may obscure measurement of this signal. The lack of stimulation artifact intrinsic to traditional electrical methodology for nerve stimulation is a unique advantage with the optical stimulation methods. The artifact associated with electrical stimulation prevents scientists from recording neural potentials near the site of stimulation. The electrical noise magnitude increases proportionally to the stimulus intensity. Consequently, it is not possible to make interpretations or observations on excitability characteristics of tissue with recording electrodes near the stimuli. This fundamental limitation of adjacent electrical stimulation and recording processes is demonstrated in Figure 21.2b. This plot contains the CNAP response recorded from the rat sciatic nerve following electrical stimulus. Recording occurs 22 mm away from the site of stimulation. A large electrical artifact completely conceals the nerve response for over 1 ms following stimulation. Thus, the onset time — and in some

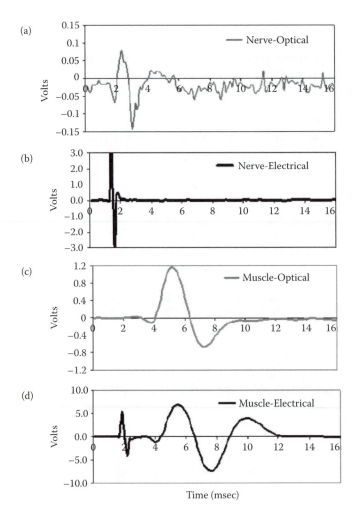

FIGURE 21.2 Compound nerve and muscle action potentials recorded from sciatic nerve in rat. (a) CNAP recorded using optical stimulation at 2.12 μm; (b) CNAP from electrical stimulation; (c) biceps femoris CMAP recorded using optical stimulation at 2.12 μm; and (d) biceps femoris CMAP using electrical stimulation. The stimulation time for all recordings occurred at $t = 1.8$ ms.

cases peak amplitude of the response — is very difficult to distinguish from background, and therefore no relevant response characteristics or signal processing can be inferred.

In contrast, Figure 21.2a depicts the nerve response to optical stimulation (same stimulation and recording site as electrical) using laser radiant exposures above stimulation threshold intensities, which do not contain a noise artifact. Now the nerve conduction velocities from the fast and slower conducting motor fibers within the sciatic nerve can be quantified in terms of timing and amplitude. The distance from stimulation to recording in the nerve was 22 mm, and two peaks are seen at 0.6 and 2.5 ms following the laser stimulus ($t = 1.8$ ms) yielding conduction velocities measured to be 36.7 m/s with fast conducting axons and slower conduction fiber velocity of 8.8 m/s. Peak amplitudes of the CNAP response from all three fiber types are manifest. It is worth noting that the velocity of conduction within the nerve subsequent to laser stimulation falls within the normal range for the rat sciatic nerve fast-conducting Aα motor neurons and slower-conducting Aγ motor neurons. Thus, this new modality for nerve excitation enables simultaneous stimulation and recording from adjacent portions of a nerve, a phenomenon that is infeasible using electrical means for activation. These results also imply that all motor fiber types are excitable with pulsed laser irradiation using optimal laser parameters.

21.2.3 Spatial Selectivity in Optical Stimulation

It is well known in electrophysiology that electrical stimulation has an unconfined spread of charge radiating far from the electrode. In the case of peripheral nerve stimulation, as the injected current required for stimulation increases, the volume of tissue affected by the electric field increases proportionally. Therefore, modulating the electrical stimulation intensity will lead to a graded response when stimulating excitable tissue (for a review, see Palanker et al., 2005). From data obtained with electrical stimulation, the greater the energy applied, the more fibers recruited, resulting in larger amplitude compound potentials. Thus, the CNAP and CMAP represent a population response to stimulation, made up of individual all-or-none responses from constituent axons, where a linear relationship exists between stimulation intensity and strength of the CNAP response (Geddes and Bourland, 1985a). The electrical current density necessary to evoke potentials in this tissue is significant, and the associated extent of the electric field affects tissue a considerable distance from the electrode. Thus, a minimum value for spatial selectivity in activation exists, and appreciably limits the precision of electrical stimulation (Geddes and Bourland, 1985b). In contrast, lasers excel in applications necessitating a precisely controlled and quantifiable volume of action in biological tissue (van Hillegersberg, 1997; Vogel and Venugopalan, 2003). Laser distribution in tissue (i.e., volume of excited axons) depends on penetration depth, spot size, and laser radiant exposure. Each of these constitutes a variable parameter. The wavelength of light determines the penetration depth of photons from a laser; thus, the depth of axons recruited in optical stimulation can be controlled very precisely over a large range of depth by modifying the laser wavelength. The next section discusses this phenomenon in detail. The laser spot size incident on the nerve can be decreased to an extremely small area (several micrometers). As a result of a small spot and lack of radial diffusion in tissue, optical stimulation allows for more selective excitation of fascicles, resulting in isolated, specific muscle contraction. Thus, theoretically, these parameters (wavelength, spot size, and radiant exposure) can be optimized for efficient stimulation of any tissue geometry by changing the wavelength, optical fiber diameter, or laser intensity used.

As a demonstration of the spatial discrimination innate to optical stimulation, CMAP recordings from electrical and optical stimulation were compared within the rat sciatic nerve using threshold energies for each modality. Figure 21.3 depicts the difference in selective activation for electrical vs. optical stimulation. CMAP recording electrodes were placed within the gastrocenemius and biceps femoris approximately 40 and 55 mm from the site of stimulation, respectively. Electrical stimulation with threshold energy (1.02 A/cm^2) was delivered proximal to the first nerve branch point on the fascicle leading to the gastrocenemius and the muscular responses within gastrocenemius and biceps femoris were simultaneously recorded. Note that using the minimum energy required to stimulate contraction of the gastrocenemius still results in stimulation of the neighboring biceps femoris fascicle (causing biceps femoris contraction). The change in voltage for these CMAPs was 1.495 and 0.492 V, respectively, seen in Figure 21.3a. Laser stimulation at threshold (0.4 J/cm^2) is shown for comparison with a voltage change 0.102 V recorded in the gastrocenemius and no response observed in the biceps femoris (Figure 21.3b). Grossly, the electrical stimulation results in excitation of the entire nerve and a subsequent twitch response from all innervated muscles. In contrast, the optical stimulation results in a muscle twitch of the muscle innervated by the targeted nerve fascicle. By moving the laser spot across the nerve, different individual muscle groups can indeed be stimulated. The precision and spatial specificity with optical activation demonstrates selective recruitment of nerve fibers, as indicated by comparing the relative magnitudes of nerve and muscle potentials (Figure 21.3) elicited from optical and electrical stimulation. These results collected with optical nerve stimulation in mammals unequivocally confirm that optical nerve activation exhibits significant spatial specificity, or lack of spread of stimulus to axons not directly irradiated by the optical source.

Another noteworthy observation from our studies is that optical stimulation with less than 1 J/cm^2 can produce extremely precise stimulation of individual fascicles in a volume of axons considerably smaller than that attainable with threshold electrical stimulation. As in electrical stimulation, increasing optical energy results in a linear increase in recruitment of axons. The linear relationship suggests that

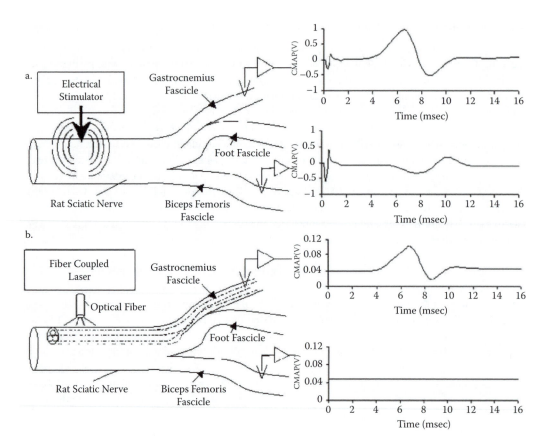

FIGURE 21.3 Selective recruitment of isolated nerve fascicles within a large peripheral nerve using electrical vs. optical stimulation techniques. (a) Electrical stimulation with threshold energy (1.02 A/cm²) delivered to the fascicle leading to the gastrocnemius. Muscular responses within gastrocnemius and biceps femoris were simultaneously recorded. (b) Laser stimulation at threshold (0.4 J/cm²) recorded in the gastrocnemius and no response observed in the biceps femoris.

the energy is confined to a tissue volume immediately beneath the laser spot and has limited diffusion to surrounding tissue, unlike electrical stimulation. A limit to laser excitation does exist at about 2 J/cm² stimulation radiant exposures, where a decrease in the physiologic response occurs. This is attributed to axon damage within the nerve as stimulation energies approach the laser thermal damage and ablation threshold, affecting the tissue's ability to generate and propagate action potentials.

21.2.4   Threshold for Stimulation Dependence on Wavelength

When applying laser light to biological tissue, a variety of complex interactions can occur. Although a comprehensive review of all aspects of laser–tissue interaction is clearly beyond the scope this chapter, some important concepts must be discussed to understand the light distribution in neural tissue. Both tissue characteristics and laser parameters contribute to this diversity. Tissue optical properties, refractive index and the wavelength-dependent coefficients of absorption and scattering, govern how light will interact with and propagate within the irradiated tissue. Alternatively, the following parameters are given by the laser radiation itself: wavelength, exposure time, laser power, applied energy, spot size, radiant exposure (energy/unit area), and irradiance (power/unit area).

In describing the optical properties and light propagation in tissues, light is treated as photons. The primary reason for this approach is that biological tissue is an inhomogeneous mix of compounds, many with unknown properties. Hence, analytical solutions to Maxwell's equations (basic electromagnetic [EM] theory that treats light as an EM wave induced by an oscillating dipole moment) in this medium poses an intractable mathematical problem. The representation of light as photons presents the opportunity to apply probabilistic approaches that lend themselves particularly well to numerical solutions that are manageable in computer simulations. Photons in a turbid medium such as tissue can move randomly in all directions and may be scattered (described by its scattering coefficient μ_s [m^{-1}]) or absorbed (described by its absorption coefficient μ_a [m^{-1}]). These coefficients, along with anisotropy (i.e., the direction in which a photon is scattered) and index of refraction, are referred to as the optical properties of a material. If photons impinge on tissue, several things can happen; some photons will reflect off the surface of the material (Fresnel reflection) and the majority of the photons will enter the tissue. In the latter case, the photon is absorbed (and can be converted to heat, trigger a chemical reaction, or cause fluorescence emission), or the photon is scattered (bumps into a particle and changes direction but continues to exist and has the same energy). Although light scattering does occur in soft biological tissues, such as the peripheral nerve, in the infrared (IR). For the purposes of this discussion we assume that scattering is negligible relative to absorption. Thus, as a first-order approximation, light penetration in peripheral nerve tissue can be described by the wavelength-dependent property of tissue absorption. Because of this, we can also assume that the light propagation into the tissue will be confined to regions directly under the irradiated spot on the nerve surface.

In tissue optics, absorption of photons is a crucial event because it allows a laser to cause a potentially therapeutic (or damaging) effect on a tissue. Without absorption, there is no energy transfer to the tissue and the tissue is left unaffected by the light. Molecules that absorb light are called *chromophores*. In the IR tissue absorption is dominated by water absorption, so the major chromophore in the peripheral nerve is water. The absorption of light can be characterized using Beer's law, which predicts that the light intensity in a material decays exponentially with depth (z):

$$E(z) = E_0\, e^{-\mu_a(\lambda)z}$$

where E_0 is the incident irradiance [W/m^2], $E(z)$ is the irradiance through some distance z of the medium, and $\mu_a(\lambda)$ is the wavelength-dependent absorption coefficient.

For a photon traveling over an infinitesimal distance Δz, the probability of absorption is given by $\mu_a * \Delta z$, where μ_a is defined as the absorption coefficient (m^{-1}) (i.e., $1/\mu_a$ is the mean free path a photon travels before an absorption event takes place) (Welch and Gemert, 1995). A related and useful parameter is the penetration depth, defined as the depth in the medium at which the energy or irradiance is reduced to $1/e$ times (~37%) the incident irradiance at the surface. By definition, the penetration depth equals $1/\mu_a$ in cases where there is no scattering.

The irradiance (power per unit area [W/m^2]) gives us information about how much light made it to a certain point in the tissue, but it does not tell us how much of that light is absorbed at that point. We define a new term called the heat source term or "rate of heat generation" (S) as the number of photons absorbed per unit volume [W/m^3]. Note that number of photons absorbed can be related to amount of heat generated, that is, heat source. Mathematically, heat source can be written as the product of the irradiance at some point in the tissue, $E(z)$, and the probability of absorption of that light at that point, μ_a:

$$S(z) = \mu_a\, E_0\, e^{-\mu_a z} = \mu_a\, E(z)$$

Once the power density $S(z)$ [W/m^3] is known, the energy density $Q(z)$ [J/m^3] is easily calculated by multiplying the power density by the exposure duration, Δt:

$$Q(z) = S(z)\, \Delta t$$

Then the laser induced temperature rise is given by:

$$\Delta T(z) = \frac{Q(z)}{\rho c}$$

where ρ is the density [kg/m^3] and c is the specific heat [$J/kg \bullet K$] of the irradiated material.

With this as background, theoretically the most appropriate wavelengths for stimulation will depend on the tissue geometry of the target tissue (i.e., here the peripheral nerve). A typical rat sciatic nerve section stimulated in this study was approximately 1.5 mm in diameter, with a 100- to– 200-μm epineural and perineural sheath between the actual axons and the nerve surface. Despite the fact that the number of fascicles per nerve varies greatly across all mammalian species, the typical fascicle thickness is constant and tends to be between 200 and 400 μm (Paxinos, 2004). Thus, to theoretically achieve selective stimulation of individual fascicles within the main nerve the penetration depth of the laser must be greater than the thickness of the outer protective tissue (200 μm) and in between the thickness of the underlying fascicle (penetration depth of 300 to 500 μm). In general, ultraviolet wavelengths ($\lambda = 100$ to 400 nm) are strongly absorbed by tissue constituents such as amino acids, fats, proteins, and nucleic acids, while in the visible part of the spectrum ($\lambda = 400$ to 700 nm), absorption is dominated by (oxy)hemoglobin and melanin. The near-infrared part of the spectrum (700 to 1300 nm) represents an area where light is relatively poorly absorbed (this is referred to as the tissue absorption window, allowing deep penetration) while in the mid- to far-infrared (> 1400 nm), absorption by tissue water dominates and results in shallow penetration (Vogel and Venugopalan, 2003). By irradiating the nerve surface overlying the target fascicle for stimulation within the main branch, *infrared laser light* may provide profound selectivity (in terms of spot size and optical penetration depth) in excitation of individual fascicles, resulting in isolated muscle contraction without thermal damage to tissue if the appropriate wavelength and spot size are utilized.

To test this hypothesis, a continuously tunable, pulsed infrared laser source in the form of a free electron laser (FEL) was employed (Edwards and Hutson, 2003). The FEL is a tunable laser that operates in the 2- to 10-μm IR region, and emits a pulse with a duration of 5 μs. Wavelengths at or near relative peaks and valleys of the IR tissue absorption spectrum ($\lambda = 2.1, 3.0, 4.0, 4.5, 5.0,$ and 6.1 μm) (Hale and Querry, 1973) were chosen for this study to facilitate recognition of general trends in stimulation thresholds compared to tissue absorption. While the FEL is an excellent source for gathering experimental data and exploring the wavelength dependence of the interaction owing to its tunability, it is neither easy to use nor clinically viable. Nevertheless, experimental data gathered with this tunable light source can provide guidance for the design of an appropriate and optimized turnkey benchtop laser system for optical nerve stimulation.

The *stimulation threshold* is defined as the minimum radiant exposure required for a visible muscle contraction occurring with each laser pulse. The *ablation threshold* is defined as the minimum radiant exposure required for visible cavitation or ejection of material from the nerve, observed using an operating microscope, with ten laser pulses delivered at 2 Hz. The stimulation threshold exhibits a wavelength dependence that mirrors the inverse of the soft tissue absorption curve. This trend is clearly illustrated in Figure 21.4a, which shows the stimulation and ablation threshold radiant exposures for five trials with each of the six wavelengths used in this study. The water absorption spectrum is included to discern general trends. Suitable wavelengths for optimal stimulation, those with maximum efficacy and minimum damage, can be inferred. The wavelength dependence of the optical stimulation thresholds yields pertinent wavelengths for the most favorable stimulation values based on the optical properties of the target neural tissue. Because absorption dominates scattering in the IR, the hypothesis was that at wavelengths where absorption is least, light penetration depth (i.e., 1/absorption) is maximized; thus, the nerve is more efficiently stimulated with less damage because photons are distributed over a greater tissue volume to minimize thermal injury. As one would expect based on a photothermal mechanism, the ablation threshold for neural tissue is inversely proportional to the water absorption curve, or directly proportional to the depth of laser penetration in the tissue. We see that the stimulation threshold is lower

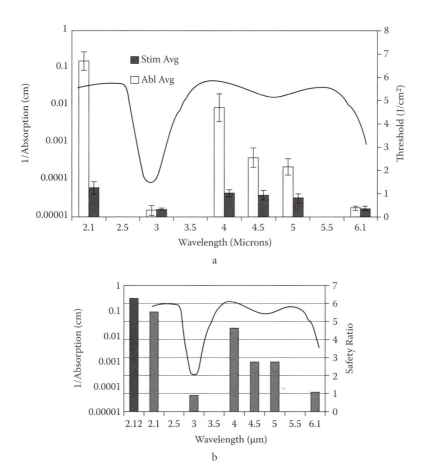

FIGURE 21.4 Wavelength dependence of the (a) stimulation vs. the ablation thresholds (b) the safety ratio = ablation threshold/stimulation threshold. The solid line in both figures indicates the optical penetration depth (left y-axis). In figure (b), the safety ratio obtained for the Ho:YAG laser is shown in stripes.

at wavelengths with high absorption, but it is also easier to ablate tissue (less radiant exposure required) at these wavelengths. Thus, a more useful indicator of optimal wavelengths is the *safety ratio*, defined as the ratio of threshold radiant exposure for ablation to that for stimulation.

This ratio (Figure 21.4b) identifies spectral regions with a large margin between radiant exposures required for excitation and damage, and thus of safety. Results indicate that the highest safety ratios (>6) are obtained at 2.1 and 4.0 μm, which correspond to valleys in tissue absorption and have nearly equivalent absorption coefficients. We can conclude that clinically relevant wavelengths for optimal stimulation, at least in the peripheral nerves and their anatomy/geometry, will not occur at peaks in tissue absorption because the energy required to produce action potentials within the nerve is roughly equal to the energy at which tissue damage occurs. For example, the penetration depth at $\lambda = 3$ μm is roughly 1 μm in soft tissue. In this case, the axons can only be stimulated by heat that has diffused from the point of absorption in the outer layers of connective tissue surrounding the nerve or from the propagation of a laser-induced pressure wave. We can also predict that absolute valleys in the absorption curve (i.e., visible and NIR region, 400 to 1400 nm) will not yield optimal wavelengths because the low absorption, owing to lack of endogenous chromophores for these wavelengths in neural tissue, will distribute the light over a large volume, leading to insufficient energy being delivered to the nerve fibers for an elicited response. Results show that the most appropriate wavelengths for stimulation of the sciatic nerve occur at relative valleys in IR soft tissue absorption, which produce an optical penetration depth of 300 to 500 μm (corresponding to the optical penetration depth at $\lambda = 2.12$ μm). In this scenario, the optical penetration depth matches up with the target geometry to stimulate one fascicle within the nerve. Note that the laser spot size can be adjusted to give precision of stimulation in all three dimensions of tissue volume.

By matching the absorption values of the wavelengths yielding the highest safety ratio with commercially available pulsed lasers, a clinically useful benchtop laser becomes a possibility. There are few lasers that emit light at 4.0 μm in wavelength, and fiber-optic delivery at this wavelength is problematic as regular glass fibers do not transmit beyond 2.5 μm. However, the holmium:YAG (Ho:YAG) laser at 2.12 μm is commercially available and is currently used for a variety of clinical applications (Razvi et al., 1995; Topaz et al., 1995; Kabalin et al., 1998; Fong et al., 1999; Jones et al., 1999). Although the inherent pulse duration and pulse structure of this laser differs from the FEL, light at this wavelength can be delivered via optical fibers, thus facilitating the clinical utility of this laser. The Ho:YAG laser was successfully used for neural stimulation, with an average stimulation threshold radiant exposure of 0.32 J/cm^2 and an associated ablation threshold of 2.0 J/cm^2 ($n = 10$), yielding a safety ratio of greater than 6.

21.2.5 Nerve Histological Analysis

Information obtained from the wavelength dependence study clearly suggests that the penetration depth in nerve tissue using the Ho:YAG can provide the desired stimulatory effect with the lowest radiant exposure compared to that required for tissue ablation. While tissue ablation served as a good indicator for safe wavelengths by allowing calculation of a safety ratio for stimulation, this phenomenon is not a synonym for thermal damage resulting in altered tissue morphology and function. It is essential to define an exact range of "safe" laser radiant exposures, or the values between threshold and the upper end of radiant exposures, which do not result in permanent tissue damage to strictly define what is appropriate for clinical use. To this end, nerves were prepared for histological analysis by a neuropathologist specializing in assessment of thermal changes in tissue resulting from laser irradiation.

To quantify (thermal) damage induced by optical stimulation in peripheral nerve tissue, histological analysis was performed on excised rat sciatic nerves, extracted acutely (less than one hour after stimulation) or three to five days following stimulation. In acute studies, the radiant exposure was varied but always larger than the stimulation threshold, and ten laser pulses at this radiant exposure were delivered to each site. For a positive control, a damaging lesion was induced using radiant exposures well over the ablation threshold in a location adjacent to the stimulation site. In survival studies, muscle and skin were sutured following stimulation and the animal was allowed to survive for a period of three to five days before nerves were harvested to assess any delayed neuronal damage and Wallerian degeneration. A sham procedure with no stimulation was performed in the contra-lateral leg as negative control. None of the shams showed any signs of damage, verifying a sound surgical technique and minimal tissue dehydration due to the surgical procedure alone. Indications of damage include, but are not limited to, collagen hyalinization, collagen swelling, coagulated collagen, decrease or loss of birefringence image intensity, spindling of cells in perineurium and in nerves (thermal coagulation of cytoskeleton), disruption and vacuolization of myelin sheaths of nerves, disruption of axons, and ablation crater formation. These criteria help define a four-point grading scheme assigned by a pathologist blinded to the treatment of a given sample to each acute specimen indicating extent of damage at the site of optical stimulation: 0 – no visible thermal changes, 1 – thermal changes in perineurium, no nerve damage, 2 – thermal damage in perineurium extending to the interface of the perineurium and the nerve, 3 – thermal damage in perineurium and in nerve. Survival scoring was reported as damage or no damage to the nerve.

Figure 21.5 shows sample histological images (H&E stain) of the rat sciatic nerve from the acute experiments following Ho:YAG laser stimulation. Results indicate that none of the ten nerves studied showed any signs of acute thermal tissue damage at the site of stimulation with radiant exposures up to two times stimulation threshold (Wells et al., 2005a,b). Histological examination of nerves from the survival study do not reveal damage to the nerve or surrounding perineurium in eight of the ten specimens, with damage occurring at radiant exposures above two times threshold. These histological findings suggest that nerves can be consistently stimulated using optical means at or near threshold

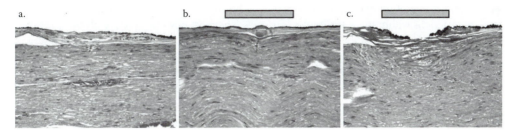

FIGURE 21.5 **(See color insert following page 15-4).** Histological images (H&E stain or 5 μm tissue section) of the rat sciatic nerve from the acute experiments following Ho:YAG laser stimulation. (a) Normal nerve tissue sample with no laser irradiation. (b) Laser irradiation with ten pulses at 2 Hz using radiant exposures slightly above stimulation threshold (0.5 J/cm²). (c) Laser irradiation with ten pulses at 2 Hz using radiant exposures above damage threshold producing a lesion in neural tissue (2.5 J/cm). Shaded boxes represent relative size of laser spot (<1 mm diameter).

without causing any neural tissue damage. These findings are further corroborated by a functional analysis of toe spreading in the survival animals. No functional neurological deficits were seen in any of the animals stimulated at less than two times the stimulation threshold.

21.3 Mechanism

While our studies have shown that optical stimulation is an effective and advantageous method for stimulation of neural tissue, the obvious and intriguing question of the underlying mechanism is largely unanswered. Exactly what biophysical stimulus is induced in the tissue by the absorbed laser light that ultimately results in an action potential and given this biophysical stimulus, what is the biological mechanism responsible for the transduction into action potentials? To a large extent, unraveling these mechanisms is still in its infancy. To get a grasp on this question, it is important to build a conceptual understanding of the laser tissue interactions that occur during optical stimulation to refine the optimal parameter set for this technique, as well as identify both the possible clinical applications and limitations for this nerve stimulation modality. The best strategy for determining the biophysical mechanism responsible for optical stimulation is to take a process of elimination approach to prove or disprove the possibility of the various types of photobiological interactions that may occur. Before we discuss our hypothesis for the underlying photobiological effect resulting in laser excitation of the peripheral nerve, it is appropriate to review some basic concepts regarding light–tissue interactions.

   In general, many studies have been conducted investigating potential interaction effects using all types of laser systems and tissue targets. Although the number of possible combinations for the experimental parameters is unlimited, three main interaction mechanisms are classified today: (1) photochemical, (2) photothermal, and (3) photomechanical. It is worth noting here that chemical, thermal, and mechanical means have all been previously shown to produce action potentials in neurons. Before going into detail, an interesting observation deserves to be stated. All these seemingly different interaction types share a common property: the characteristic radiant exposure [J/cm²] ranges from approximately 1 to 1000 J/cm². This is surprising because the irradiance itself [W/cm²] varies over more than fifteen orders of magnitude. Thus, a single parameter distinguishes and primarily controls these processes: the duration of the laser exposure, which is largely similar to the interaction time itself (Niemz, 2004). According to a graph of the laser radiant exposure vs. the duration of pulse width the time scale can roughly be divided in three major sections; (1) continuous wave or exposure times greater than 1 s for photochemical interactions, (2) 100 s down to 1 μs for photothermal interactions, and (3) 1 μs and shorter for photomechanical interactions. It should be clear, however, that these boundaries are not strict, and adjacent interaction types cannot always be separated. Thus, overlap in these main regions does exist. For example, in the range of 1 to several hundreds of microseconds, the interaction mechanisms typically have photothermal as well as photomechanical components to them, while many photochemical interactions also exhibit photothermal components.

In brief, the group of photochemical interactions is based on the fact that light can induce chemical effects and reactions within macromolecules or tissues. The most obvious example of this is created by nature itself: photosynthesis. In the field of medical laser applications, photochemical interaction mechanisms play a role during photodynamic therapy (PDT) (Takahashi et al., 2002; Ionita et al., 2003; Yamamoto et al., 2003). Frequently, biostimulation is also attributed to photochemical interactions, although this is not scientifically ascertained. Photochemical interactions take place at very low irradiances (typically 1 W/cm²) and long exposure times ranging from seconds to tens of minutes. Recent and exciting developments that rely on photochemical interactions include experimental applications in the field of photostimulation of neurons where light may be used to activate "caged compounds (McCray and Trentham, 1989; Eder et al., 2002, 2004). In this scenario, stimulatory neurotransmitters are linked to an inactivating group (a "caged" compound). Upon UV light exposure, cleavage of the neurotransmitter from its "cage" is achieved, rendering the active form of the stimulatory neurotransmitter only there where light exposure is activated. This technique takes advantage of the high spatial resolution of uncaging molecules with light.

We examined whether the mechanism for optical nerve stimulation is a result of photochemical effects from laser–tissue interaction. In essence, the stimulation thresholds in the infrared part of the spectrum follow the water absorption curve (Wells, 2005a,b), suggesting that no "magical wavelength" has been identified, effectively excluding a single tissue chromophore responsible for any direct photochemical effects. This also provides some evidence that the effect is directly thermally mediated or a secondary effect to photothermal interactions (i.e., photomechanical effects) as tissue absorption from laser irradiation can be directly related to the heat load experienced by the tissue. Theoretically, one can predict that a photochemical phenomenon is not responsible because infrared photon energy (<0.1 eV) is too low for a direct photochemical effect of laser–tissue interactions and the laser radiant exposures used are insufficient for any multiphoton effects (Thomsen, 1991).

Maxwell's EM theory suggests an inherent electric field exists within laser light, which is associated with the propagation of light itself and driven by a time and space varying electric and magnetic field (Waldman, 1983). We questioned whether the electric field within the light beam used to irradiate and stimulate the peripheral nerve is large enough to directly initiate action potentials, considering the standard method of stimulation is through electrical means. To test this proposition, we used an alexandrite laser operating at 750 nm (near-infrared light) to attempt stimulation of the peripheral nerve. This wavelength, unlike the Ho:YAG wavelength, has minimal absorption in soft tissue; however, the electric field of intensity is similar regardless of wavelength. Thus, any stimulation reported with a low absorption wavelength would indicate that the electric field of the laser light mediates stimulation. Results explicitly prove a direct electrical field effect due to laser radiation traversing the tissue is highly unlikely as a means for optical stimulation because light from the alexandrite laser did not stimulate even at radiant exposures fifty times higher than those used for the Ho:YAG laser. Experimental calculations further illustrate this point. Consider the equation: $S_{threshold} = \frac{1}{2} c\varepsilon_o E_{max}^2$, where the threshold laser radiant exposure ($S_{threshold}$) = 0.32 J/cm², the speed of light (c) = 3.10⁸ m/s, and the permittivity of neural tissue (ε_o) in units of A-s/V-m ($c\varepsilon_o$ = 0.002634). The calculated value for the maximum instantaneous intensity of the electric field (E_{max}) at the tissue surface is 0.155 V/mm², or 0.05 mA/mm². This theoretical prediction is well below the electrical stimulation threshold of the peripheral nerve found in our previous studies, where 0.95 ± 0.58 A/cm² was required for surface stimulation. Moreover, it is important to realize that the electric field owing to light oscillates at 10¹⁴ to 10¹⁵ Hz, which is an order of magnitude higher than the typical electrical stimulation field oscillator frequency.

Photomechanical effects are secondary to rapid heating with short laser pulses (<1 μs) that produce forces, such as explosive events and laser-induced pressure waves, able to disrupt cells and tissue. Because we are operating well below the ablation threshold, ablative recoil can be excluded as a source of mechanical effects. In contrast, tissue heating will always result in thermoelastic expansion. Nerve stimulation using pressure waves (rapid mechanical displacement, ultrasound) is well documented in the literature (Shusterman et al., 2002; Norton, 2003). We sought to prove or disprove photomechanical effects (thermoelastic expansion or pressure wave generation) leading to optical stimulation.

Contributions from pressure waves to optically stimulate the peripheral nerve were studied by examining the effect of pulse duration on stimulation threshold. It is clear from our results that the stimulation threshold radiant exposure required for stimulation at this wavelength does not change with pulse width through almost three orders of magnitude (5 μs to 5 ms). Moreover, all pulse durations lie well outside the stress confinement zone. Given that, there is strong evidence that laser-induced pressure waves are not implicated in the optical stimulation mechanism. Because pressure wave generation has been discarded as a plausible means, tissue displacement during the laser pulse was measured next using a phase-sensitive OCT setup (Rylander et al., 2004) to test the actual magnitude of thermoelastic expansion of the tissue resulting from optical stimulation. The change in surface displacement of the rat sciatic nerve (*ex vivo*) upon irradiation with Ho:YAG radiant exposures slightly above threshold (0.4 J/cm^2) were measured to be 300 nm. Displacement of 300 nm in a 350-μsec pulse width is small, but not negligible. Nevertheless, while at this point we cannot exclude contributions of the thermoelastic expansion, this effect is thermal in origin.

Through this process of elimination we have systematically shown that the electric field, and photochemical and photomechanical effects from laser tissue interactions do not result in excitation of neural tissue. Thus, we have arrived at the hypothesis that laser stimulation of neural tissue is mediated by some photothermal process resulting from transient irradiation of peripheral nerves using infrared light.

Photothermal interactions include a large group of interaction types resulting from the transformation of absorbed light energy to heat, leading to a local temperature increase and thus a temperature gradient both in time and space. While photochemical processes are often governed by a specific reaction pathway, photothermal effects generally tend to be nonspecific and are mediated primarily by absorption of optical energy and secondly governed by fundamental principles of heat transport. Depending on the duration and peak value of the temperature achieved, different effects such as coagulation, vaporization, melting, or carbonization may be distinguished. An excellent overview of these interaction regimes can be found in Jacques (1992). It is essential to emphasize that thermal interactions in tissue are typically governed by rate processes; that is, it is not just the temperature that plays a role, but also the duration for which the tissue is exposed to a particular temperature is a parameter of major importance. Once deposited in tissue and given sufficient time, the traditional mechanism of heat transfer applies to laser-irradiated biological tissues. Heat flows in biological tissue whenever a temperature difference exists. The transfer of thermal energy is governed by the laws of thermodynamics: (1) energy is conserved, and (2) heat flows from areas of high temperature to areas of low temperature. The primary mechanisms of heat transfer to consider include conduction, convection, and radiation (Incropera, 2002).

Two-dimensional radiometry of the irradiated tissue surface was performed to gain a better understanding of the thermal processes and actual tissue temperature values required for optical nerve stimulation. Using this technique, the temperature profile in space and time was observed. We measured a peak temperature rise at the center of the spot of 8.95°C, yielding an average temperature rise of 3.66°C across the Gaussian laser spot. The peripheral nerve temperature profile in time was also observed using the infrared camera from laser stimulation. The thermal relaxation time is defined as the time required for the temperature of the tissue to return to 1/e (37%) of the maximum tissue temperature change. In the case of the rat peripheral nerve, we measured the thermal relaxation time to be about 90 ms, which corresponds well with the theoretical value of about 100 ms. We can infer that the pulse width of light delivered to the tissue must be less than 90 ms in duration to result in the desired stimulation effect. We can also infer that temperature superposition will begin to occur at higher repetition rates (>5 Hz) as the tissue requires slightly greater than 200 ms to return to baseline temperature. At repetition rates greater than 5 Hz, tissue temperatures will become additive with each ensuing laser pulse and resulting tissue damage may begin to occur with long-term stimulation.

Based on results from measurements of tissue temperature as a function of radiant exposure, nerve temperature clearly increases linearly with laser radiant exposure. Recent literature suggests that slight thermal changes to mitochondria begin to occur as low as 43°C (protein denaturation begins at tissue temperature close to 57°C). This temperature corresponds to the onset of thermal damage radiant exposure found from histological analysis of short-term laser nerve stimulation (0.8 to 1.0 J/cm^2). These results imply that optical stimulation of peripheral nerves is mediated through a thermal gradient as a result of laser tissue interaction and that this phenomenon is safe at radiant exposures of at least two times the threshold required for action potential generation. In the case of nonhydrated tissue, the temperature as a function of radiant exposure shifts upward 6°C. Here, the mitochondrial damage will theoretically begin to occur between 0.5 and 0.6 J/cm^2, thus illustrating the importance of tissue hydration for safe and efficient nerve excitation.

21.4 Impact

21.4.1 Applications

Optical neural activation has three fundamental advantages over electrical stimulation that make it ideal for a number of procedures that currently employ electrical stimulation as the standard of care. First, the precision of optically delivered energy is far superior to electrical stimulation techniques. Examples of limitations in electrical stimulation techniques arise when precision stimulation of neural structures are required for peripheral nerve surgery, during which small clusters of nerve fibers are stimulated to determine their viability in peripheral nerve repair (Weiner, 2003). In peripheral nerve surgery, electrical stimulation is utilized to identify the connectivity and functionality of specific nerve roots to selectively avoid or resect. This usually requires dissecting apart the nerve bundles to determine which ones conduct through a damaged area and which bundles do not. The optical method could confine the stimulation easily to segments of a nerve without requiring separation between the intended area to be stimulated and other areas. Similarly, surgeries involving cranial nerves would benefit from precise functional testing, such as differentiating nerve tissue from tumor in small areas such as the central pontine angle through which the vestibular and facial nerve traverse. Auditory nerve stimulation could be significantly enhanced with a larger number of distinct stimulation sites along the cochlea than is currently possible using electrical means. This suggests that it may be possible to develop a better cochlear implant.

Second, optical stimulation does not produce an electrical artifact during stimulation, whereas electrical stimulation inherently results in an artifact in the recorded signal. To record small nerve potentials in response to the electrical stimulation, usually the recording electrodes are located a sufficient distance away from the stimulation source. Furthermore, signal averaging techniques are frequently used for discerning electrical responses contained within large electric field artifacts (Fiore et al., 1996; Wagenaar and Potter, 2002; Andreasen and Struijk, 2003). Optical stimulation, on the other hand, produces no stimulation artifact in the recorded response, and therefore the recorded response can be very close to the stimulation source. Clinically, this results in neural potentials that can be more easily recorded near the source of stimulation. Also, fewer stimuli need to be applied due a decreased need for signal averaging (which requires usually hundreds of stimuli), which in turn facilitates higher throughput of mapping.

Third, electrical stimulation requires contact between the electrode and the tissue being stimulated. It is susceptible to all the properties of impedance, current shunting, and field distortion around the area of contact between the electrode and the tissue in the acute setting (e.g., in current cochlear implants, a maximum of six to nine channels/electrodes are used, owing to the fact that along the basilar membrane each electrode affects an area of approximately 4 to 8 mm) (Palanker et al., 2005). In the chronic setting, issues of half-cell potential differences, metal toxicity, and tissue reaction to various implanted electrodes significantly limit the materials and sizes of materials used for chronic electrode implants. Optical stimulation, on the other hand, does not require direct contact with the tissue being stimulated, thereby minimizing tissue disturbance. Furthermore, plating and deplating of metal in an ionic medium (interstitial fluid) is not an issue with optical stimulation. It would seem more likely that chronically implanted optical stimulating probes would be more precise due to the lack of any stimulating current spread, and also more tolerable as a chronic implant due to longer tissue stability (no unstable impedance characteristics) and safer (i.e., inert) interface materials (glass/fiber-optic cable vs. metal) (Agnew et al., 1989).

In summary, the capability of optical energy to yield a contact-free, spatially selective, artifact-free method of stimulation has significant advantages over electrical methods for a variety of diagnostic and therapeutic clinical applications.

21.4.2 Future Directions

Optical stimulation presents a paradigm shift in neural activation that has major implications for clinical neural stimulation as well as fundamental neurophysiology and neuroscience. To date, this concept has been demonstrated using large, cumbersome, and expensive laboratory laser sources (FEL, Ho:YAG). For optical stimulation to find its way to practical utility and clinical use, a simple, user-friendly, portable, reliable, and low-cost device must be developed. Aculight, a company that specializes in the design and manufacturing of innovative solid-state lasers for the defense and medical markets, in collaboration with Vanderbilt University, has demonstrated the utility and unique capability of this concept by developing a portable optical stimulator for routine use in laboratory settings, with the ultimate goal of delivering a clinically usable product. The prototype of the optical stimulator is a pulsed 1.85 to 1.87 μm (resulting in similar absorption as Ho:YAG laser in soft tissue) diode laser with a fiber-coupled output, representing a >95% reduction in size compared to the device used for initial testing at Vanderbilt University. The laser-based stimulator, which appears similar to telecom products offered today, has the advantages of compactness and portability, high reliability, and low life-cycle cost. Furthermore, the unit plugs into a regular power outlet (110 V) and has no special power or cooling requirements Experiments described above have been repeated with this laser system and, as expected, show very similar results.

In summary, we have shown a novel alternative to electrical stimulation to interface with the neural system using light. This method provides several unique advantages over traditional methods. However, transient optical stimulation of neural tissue is in its infancy and many questions remain open with regard to the underlying mechanism, the limitations of its utility, and applications that have not even been thought of at this time. Moving this field forward will require multidisciplinary approaches and intense research efforts on all fronts.

References

Agnew, W. F., McCreery, D.B., et al. (1989). Histologic and physiologic evaluation of electrically stimulated peripheral nerve: considerations for the selection of parameters. *Ann. Biomed. Eng.*, 17(1):39–60.

Allegre, G., Avrillier, S., et al. (1994). Stimulation in the rat of a nerve fiber bundle by a short UV pulse from an excimer laser. *Neurosci. Lett.*, 180(2):261–264.

Andreasen, L.N. and Struijk, J.J. (2003). Artefact reduction with alternative cuff configurations. *IEEE Trans. Biomed. Eng.*, 50(10):1160–1166.

Balaban, P., Esenaliev, R., et al. (1992). He-Ne laser irradiation of single identified neurons. *Lasers Surg. Med.*, 12(3):329–337.

Bragard, D., Chen, A.C., et al. (1996). Direct isolation of ultra-late (C-fibre) evoked brain potentials by CO_2 laser stimulation of tiny cutaneous surface areas in man. *Neurosci. Lett.*, 209(2):81–84.

Chen, Y.S., Hsu, S.F., et al. (2005). Effect of low-power pulsed laser on peripheral nerve regeneration in rats. *Microsurgery*, 25(1):83–89.

Eder, M., Zieglgansberger, W., et al. (2002). Neocortical long-term potentiation and long-term depression: site of expression investigated by infrared-guided laser stimulation. *J. Neurosci.*, 22(17):7558–7568.

Eder, M., Zieglgansberger, W., et al. (2004). Shining light on neurons-elucidation of neuronal functions by photostimulation. *Rev. Neurosci.*, 15(3):167–183.

Edwards, G. S. and Hutson, M.S. (2003). Advantage of the Mark-III FEL for biophysical research and biomedical applications. *J. Synchrotron. Radiat.*, 10(Pt. 5):354–357.

Fiore, L., Corsini, G., et al. (1996). Application of non-linear filters based on the median filter to experimental and simulated multiunit neural recordings. *J. Neurosci. Methods*, 70(2):177–184.

Fong, M., Clarke, K., et al. (1999). Clinical applications of the holmium:YAG laser in disorders of the paediatric airway. *J. Otolaryngol.*, 28(6):337–343.

Fork, R.L. (1971). Laser stimulation of nerve cells in Aplysia. *Science*, 171(974):907–908.

Geddes, L.A. and Bourland, J.D. (1985a). The strength-duration curve. *IEEE Trans. Biomed. Eng.*, 32(6):458–459.

Geddes, L.A. and Bourland, J.D. (1985b). Tissue stimulation: theoretical considerations and practical applications. *Med. Biol. Eng. Comput.*, 23(2):131–137.

Hale, G.M. and Querry, M.R. (1973). Optical-constants of water in 200-nm to 0.2-mm wavelength region. *Appl. Optics*, 12(3):555–563.

Hirase, H., Nikolenko, V., et al. (2002). Multiphoton stimulation of neurons. *J. Neurobiol.*, 51(3):237–247.

Incropera, F.P. (2002). *Fundamentals of Heat and Mass Transfer*. New York: John Wiley & Sons.

Ionita, M.A., Ion, R.M., et al. (2003). Photochemical and photodynamic properties of vitamin B2 — riboflavin and liposomes. *Oftalmologia*, 58(3):29–34.

Izzo, A.D., Jansen, E.D., and Walsh, J.T. (2006b). Laser stimulation of the auditory nerve. *Lasers Surg. Med.*, (submitted).

Izzo, A.D., Suh, E., Walsh, J.T., et al. (2006a). Selectivity of optical stimulation in the auditory system. *SPIE*, San Jose, CA.

Izzo, A.D., Walsh, J.T., and Jansen, E.D. (2005). Safe ranges for optical cochlear neuron stimulation. *Mid-Winter Meeting for the Association for Research in Otolaryngology (ARO)*, New Orleans, LA.

Jacques, S.L. (1992). Laser-tissue interactions. Photochemical, photothermal, and photomechanical. *Surg. Clin. N. Am.*, 72(3):531–558.

Jansen, E.D., Asshauer, T., et al. (1996). Effect of pulse duration on bubble formation and laser-induced pressure waves during holmium laser ablation. *Lasers Surg. Med.*, 18(3):278–293.

Jones, J.W., Schmidt, S.E., et al. (1999). Holmium:YAG laser transmyocardial revascularization relieves angina and improves functional status. *Ann. Thorac. Surg.*, 67(6):1596–1601; discussion 1601-2.

Kabalin, J.N., Gilling, P.J., et al. (1998). Application of the holmium:YAG laser for prostatectomy. *J. Clin. Laser Med. Surg.*, 16(1):21–22.

Kanjani, N., Jacob, S., et al. (2004). Wavefront- and topography-guided ablation in myopic eyes using Zyoptix. *J. Cataract Refract. Surg.*, 30(2):398–402.

Lu, J. and Waite, P. (1999). Advances in spinal cord regeneration. *Spine*, 24(9):926–930.

McCray, J.A. and Trentham, D.R. (1989). Properties and uses of photoreactive caged compounds. *Annu. Rev. Biophys. Biophys. Chem.*, 18:239–270.

McGill, K., Cummins, K.L., et al. (1982). On the nature and elimination of stimulus artifact in nerve signals evoked and recorded using surface electrodes. *IEEE Trans. Biomed. Eng.*, 29(2):129–137.

Miller, C.A., Abbas, P.J., et al. (2000). An improved method of reducing stimulus artifact in the electrically evoked whole-nerve potential. *Ear Hear.*, 21(4):280–290.

Niemz, M.H. (2004). *Laser-Tissue Interactions*, Berlin: Springer.

Norton, S.J. (2003). Can ultrasound be used to stimulate nerve tissue? *Biomed. Eng. Online*, 2: 6.

Palanker, D., Vankov, A., et al. (2005). Design of a high-resolution optoelectronic retinal prosthesis. *J. Neural Eng.*, 2(1):S105–S120.

Paxinos, G. (2004). *The Rat Nervous System*. Sydney, Australia: Elsevier.

Razvi, H.A., Chun, S.S., et al. (1995). Soft-tissue applications of the holmium:YAG laser in urology. *J. Endourol.*, 9(5):387–390.

Richter C.P., Walsh, J.T., and Jansen, E.D. (2005a). Optical Stimulation of the Auditory System. *NIH Symposium on Neural Interfaces*, Bethesda, MD.

Richter, C.P., Walsh, J.T., and Jansen, E.D. (2005b). Optically-evoked acoustic nerve activity. *Mid-Winter Meeting for the Association for Research in Otolaryngology (ARO)*, New Orleans, LA.

Roehm, P.C. and Hansen, M.R. (2005). Strategies to preserve or regenerate spiral ganglion neurons. *Curr. Opin. Otolaryngol. Head Neck Surg.*, 13(5):294–300.

Rylander, C.G., Dave, D.P., et al. (2004). Quantitative phase-contrast imaging of cells with phase-sensitive optical coherence microscopy. *Opt. Lett.*, 29(13):1509–1511.

Shusterman, V., Jannetta, P.J., et al. (2002). Direct mechanical stimulation of brainstem modulates cardiac rhythm and repolarization in humans. *J. Electrocardiol.*, 35 Suppl:247–256.

Sisken, B.F., Walker, J., et al. (1993). Prospects on clinical applications of electrical stimulation for nerve regeneration. *J. Cell. Biochem.*, 51(4):404–409.

Takahashi, M., Nagao, T., et al. (2002). Roles of reactive oxygen species in monocyte activation induced by photochemical reactions during photodynamic therapy. *Front Med. Biol. Eng.*, 11(4):279–294.

Thomsen, S. (1991). Pathologic analysis of photothermal and photomechanical effects of laser-tissue interactions. *Photochem. Photobiol.*, 53(6):825–835.

Topaz, O., Rozenbaum, E.A., et al. (1995). Laser-assisted coronary angioplasty in patients with severely depressed left ventricular function: quantitative coronary angiography and clinical results. *J. Interv. Cardiol.*, 8(6):661–669.

van Hillegersberg, R. (1997). Fundamentals of laser surgery. *Eur. J. Surg.*, 163(1):3–12.

Vogel, A. and Venugopalan, V. (2003). Mechanisms of pulsed laser ablation of biological tissues. *Chem. Rev.*, 103(2):577–644.

Wagenaar, D.A. and Potter, S.M. (2002). Real-time multi-channel stimulus artifact suppression by local curve fitting. *J. Neurosci. Methods*, 120(2):113–120.

Waldman, G. (1983). *Introduction to Light: The Physics of Light, Vision, and Color.* Englewood Cliffs, NJ: Prentice Hall.

Walsh, L.J. (1997). The current status of low level laser therapy in dentistry. 1. Soft tissue applications. *Aust. Dent. J.*, 42(4):247–254.

Weiner, R.L. (2003). Peripheral nerve neurostimulation. *Neurosurg. Clin. N. Am.*, 14(3):401–408.

Welch, A.J. and v. Gemert, M.J.C. (1995). *Optical-Thermal Response of Laser-Irradiated Tissue.* New York: Plenum Press.

Welch, A.J., Motamedi, M., et al. (1991). Laser thermal ablation. *Photochem. Photobiol.*, 53(6):815–823.

Wells, J.D., Kao, C., et al. (2005). Application of Infrared Light for in vivo Neural Stimulation. *J. Biomed. Optics*, 10: 064003.

Wells, J.D., Kao, C., et al. (2005). Optical Stimulation of Neural Tissue in vivo. *Optics Letters*, 30(5): 504–507.

Wietholt, D., Alberty, J., et al. (1992). Nd-YAG laser-photocoagulation — acute electrophysiological, hemodynamic, and morphological effects in large irradiated areas. *Pace-Pacing and Clin. Electrophysiol.*, 15(1):52–59.

Wu, W.H., Ponnudurai, R., et al. (1987). Failure to confirm report of light-evoked response of peripheral nerve to low power helium-neon laser light stimulus. *Brain Res.*, 401(2):407–408.

Yamamoto, M., Nagano, T., et al. (2003). Production of singlet oxygen on irradiation of a photodynamic therapy agent, zinc-coproporphyrin III, with low host toxicity. *Biometals*, 16(4):591–597.

Transcranial Magnetic Stimulation of Deep Brain Regions

Yiftach Roth and
Abraham Zangen

22.1 Introduction

Transcranial magnetic stimulation (TMS) is a noninvasive technique used to apply brief magnetic pulses to the brain. The pulses are administered by passing high currents through an electromagnetic coil placed upon the scalp that can induce electrical currents in the underlying cortical tissue, thereby producing a localized axonal depolarization. Neuronal stimulation by TMS was first demonstrated in 1985 (Barker et al., 1985), when a circular coil was placed over a normal subject vertex and evoked action potentials from the abductor digiti minimi. Since then this technique has been applied to studying nerve conduction, excitability, and conductivity in the brain and peripheral nerves, and to studying and treating various neurobehavioral disorders, primarily mood disorders (Kircaldie et al., 1997; Wassermann and Lisanby, 2001).

The ability of the TMS technique to elicit neuronal response has until recently been limited to brain cortex. The coils used for TMS (such as round or a figure-of-eight coil) induce stimulation in cortical regions mainly just superficially under the windings of the coil. The intensity of the electric field drops dramatically deeper in the brain as a function of the distance from the coil (Maccabee et al., 1990; Tofts, 1990; Tofts and Branston, 1991; Eaton, 1992). Therefore, to stimulate deep brain regions, a very high intensity would be necessary. Such intensity cannot be reached by standard magnetic stimulators, using the regular figure-of-eight or circular coils. Stimulation of regions at depths of 3 to 4 cm, such as the

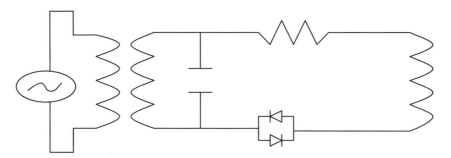

FIGURE 22.1 Typical magnetic stimulation circuit, including high-voltage transformer, capacitor, resistor, thyristor trigger, and stimulating coil.

leg motor area, can be achieved using coils such as the double-cone coil (Stokic et al., 1997; Terao et al., 1994, 2000), which is a larger figure of eight with an angle of about 95° between the two wings. However, the intensity needed to stimulate deeper brain regions effectively would stimulate cortical regions and facial nerves over the level that might lead to facial pain, facial and cervical muscle contractions, and may cause epileptic seizures and other undesirable side effects.

This chapter describes the principles and design of TMS coils for deep brain stimulation. The construction of such coils should meet several goals simultaneously:

1. High enough electric field intensity in the desired deep brain region that will surpass the threshold for neuronal activation
2. High percentage of electric field in the desired deep brain region relative to the maximal intensity in the cortex
3. Minimal aversive side effects during stimulation, such as pain and activation of facial muscles

22.2 Basic Principles of TMS

The TMS stimulation circuit consists of a high-voltage power supply that charges a bank of capacitors, which are then rapidly discharged via an electronic switch into the TMS coil to create the briefly changing magnetic field pulse. A typical circuit is shown in Figure 22.1, where low-voltage AC is transformed into high-voltage DC, which charges the capacitors. A crucial component is the thyristor switch, which must traverse very high current at very short times of 50 to 250 μs. The cycle time depends on the capacitance (typically 10 to 250 μF) and on the coil inductance (typically 10 to 30 μH). Typical peak currents and voltages are 5000 A and 1500 V, respectively.

Most TMS stimulators produce a biphasic pulse of electric current. During the discharge cycle, the TMS circuit behaves like an RCL circuit, and the current I is given by:

$$I(t) = \frac{V}{wL}\exp(-\alpha t)\sin(wt) \tag{22.1}$$

where $\alpha = R/2L$, $w = \sqrt{(LC)^{-1} - \alpha^2}$, and R, C, and L are the total values of the resistance, capacitance, and inductance, respectively, in the circuit. The inductance is mainly the coil inductance but there is an additional contribution from the cables, and the resistance includes contributions from the thyristor and the coil.

Biologically, the most relevant parameter for neuronal activation is the induced electric field, which is proportional to the rate of change of the current (dI/dt). The brief strong current generates a time-varying magnetic field B. An electric field E is generated in every point in space with direction perpendicular to the magnetic field, with amplitude proportional to the time-rate of change of the vector potential $A(r)$.

The vector potential $A(r)$ in position r is related to the current in the coil I by the expression:

$$A(r) = \frac{u_0 I}{4\pi} \int \frac{dl'}{|r - r'|} \tag{22.2}$$

where $\mu_0 = 4\pi * 10^{-7}$ *Tm/A* is the permeability of free space, the integral of dl' is over the wire path, and r' is a vector indicating the position of the wire element. The magnetic and electric fields are related to the vector potential through the expressions:

$$B_A = \nabla \times A \tag{22.3}$$

$$E_A = \frac{-\partial A}{\partial t} \tag{22.4}$$

The only quantity that is changing with time is the current I. Hence, the electric field E_A can be written as:

$$E_A = \frac{-\mu_0 \partial I}{4\pi \partial t} \int \frac{dl'}{|r - r'|} \tag{22.5}$$

Because brain tissue has conducting properties, while the air and skull are almost complete insulators, the vector potential will induce accumulation of electric charge at the brain surface. This charge is another source for electric field, which can be expressed as:

$$E_\Phi = -\nabla \Phi \tag{22.6}$$

where Φ is the scalar potential produced by the surface electrostatic charge.

The total field in the brain tissue E is the vectorial sum of these two fields:

$$E = E_A + E_\Phi \tag{22.7}$$

The influence of the electrostatic field E_Φ is, in general, to oppose the induced field E_A and consequently to reduce the total field E. The amount of surface charge produced and hence the magnitude of E_Φ depend strongly on coil configuration and orientation. This issue will be elaborated in the following sections.

Figure 22.2 demonstrates the electric field pulse produced by a figure-of-eight coil, as measured by a two-wire probe in a brain phantom filled with saline solution at physiologic concentration. In repetitive TMS (rTMS), several such pulses are administered in a train of between 1 and 20 Hz.

This electric field produces action potential in excitable neuronal cells, which might result in activation of neuronal circuits when applied above a certain threshold. The neuronal response depends not only on the electric field strength, but also on the pulse duration, through a strength-duration curve of the form:

$$E_{th} = b(1 + c/\tau) \tag{22.8}$$

where E_{th} is the threshold electric field required to induce neuronal response, and τ is the duration the field was above this threshold. The biological parameters determining neural response are the threshold at infinite duration, called the rheobase (b, measured in V/m), and the duration at which the threshold is twice the rheobase, called the chronaxie (c, in μs). Motor and sensory curves as reported by Bourland

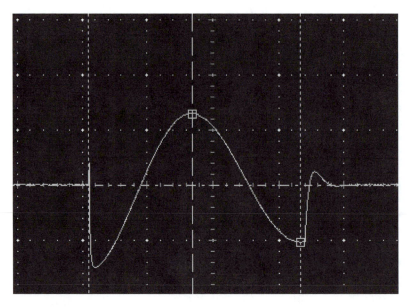

FIGURE 22.2 The induced electric field of a figure-of-eight (figure-8) coil vs. time over a TMS pulse cycle. The time scale is 100 μs.

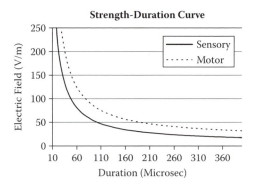

FIGURE 22.3 Neural strength-duration curve depicting stimulation threshold vs. duration.

et al. (1996) are shown in Figure 22.3. These curves should be treated as illustrative only, because the chronaxie and rheobase depend on many biological and experimental factors, such as whether or not the nerves are myelinated (hence peripheral and cortical parameters should be different), or train frequency in rTMS, which in general reduces the threshold for stimulation.

As shown by Heller and Van Hulstein (1992), the three-dimensional maximum of the electric field intensity will always be located at the brain surface, for any configuration or superposition of TMS coils. It is possible, however, to increase considerably the depth penetration and the percentage of electric field intensity in deep brain regions, relative to the maximal field at the cortex. The next section outlines the construction principles for efficient deep brain stimulation, and subsequent sections demonstrate several examples of TMS coils designed to accomplish this goal.

22.3 Deep TMS Coils: Design Principles

While the activation of peripheral nerves depends mainly on the derivative of the electric field along the nerve fiber (Maccabee et al., 1993), the most relevant parameter for activation of brain structures seems to be the electric field intensity (Thielscher and Kammer, 2002; Amassian et al., 1992). In both cases, however, physiological studies indicate that optimal activation occurs when the field is oriented in the same direction as the nerve fiber (Durand et al., 1989; Roth and Basser, 1990; Basser and Roth, 1991; Brasil-Neto et al., 1992; Mills et al., 1992; Pascual-Leone et al., 1994; Niehaus et al., 2000; Kammer et al., 2001). Hence, to stimulate deep brain regions, it is necessary to use coils in such an orientation that they will produce a significant field in the preferable direction to activate the neuronal structures or axons under consideration.

In light of these findings, the geometrical features of each specific design depend primarily on:

- The location and size of the deep brain region or regions intended for activation
- The preferred direction or directions one wants to stimulate

The design of a specific coil is dictated by these goals. Nevertheless, all deep TMS coils have to share the following important features:

1. *Base complementary to the human head.* The part of the coil close to the head (the base) must be optimally complementary to the human skull at the desired region. In some coils, the base may be flexible and able to receive the shape of an individual patient, and in other coils it may be more robust, namely arcuated in a shape that fits the average human skull at the desired region. In the latter case, there may be a few similar models designed to fit smaller and larger heads.
2. *Proper orientation of stimulating coil elements.* Coils must be oriented such that they will produce a considerable field in a direction tangential to the surface, which should also be the preferable direction to activate the neurons under consideration. That is, the wires of the coils are directed in one or more directions, which results in a preferred activation of neuronal structures orientated in these particular directions. In some cases, there is one preferred direction along the length or width axis; and in other cases, there are two preferred directions along both the length and width axes.
3. *Summation of electric impulses.* The induced electric field in the desired deep brain regions is obtained by optimal summation of electric fields, induced by several coil elements with common direction, located in different locations around the skull. The principle of summation can be applied either in time or in space, or in a combination of both. The main kinds of summation include:
   a. *One-point spatial summation.* In this kind of summation, coil elements carrying current in the desired direction are placed in various locations around the head, in such a configuration to create a high electric field intensity in a specific deep brain region, which is simultaneously a high percent of the maximal electric field at the brain cortex.
   b. *Morphological line spatial summation.* The goal of this summation is to induce an electric field at several points along a certain neuronal structure. This line should not be straight and may have a complex bent path. The application of diffusion tensor imaging (DTI) in MRI for fiber tracking is an evolving field that may improve significantly the efficacy of TMS treatment. If, for example, we know the path of a certain axonal bundle, a coil can be designed in a configuration that will produce a significant electric field at several points along the bundle. This configuration can enable induction of an action potential in this bundle, while minimizing the activation of other brain regions. For example, the TMS coil can be activated in an intensity that will induce a subthreshold electric field at most brain regions, which will not induce an action potential; while the induction of a subthreshold field along the desired path can induce an action potential in this bundle, thus increasing the specificity of the TMS treatment.
   c. *Temporal summation.* The various coil elements can be stimulated consecutively and not simultaneously. As shown in Figure 22.3, the neuronal activation threshold depends on both electric field intensity and stimulation duration. The TMS coil can be designed in such a configuration that the various elements are scattered around the desired region or path, so that passing a current in each element will produce a significant field at the desired deep brain region. In such a case, the coil can be stimulated consecutively so that at each time period only a certain element or a group of elements is activated. In this way, a significant electric field will be induced in the desired deep brain region at all time periods; while in more cortical

regions, a significant field will be induced mainly at certain periods, when proximate coil elements are activated. This will enable stimulation of the deep brain structure while minimizing stimulation of other brain regions, and specifically of cortical regions. A detailed study of the neural response to trains with interpulse intervals of milliseconds (instead of hundreds of milliseconds as in rTMS) will aid in refining this technique.

4. *Minimization of radial components.* Coil construction is meant to minimize wire elements carrying current components that are nontangential to the skull. The electric field intensity in the tissue to be stimulated and the rate of decrease of the electrical field as a function of distance from the coil depend on the orientation of the coil elements relative to the tissue surface. It has been shown that coil elements that are nontangential to the surface induce accumulation of surface charge, which leads to the cancellation of the perpendicular component of the induced field at all points within the tissue, and reduction of the electrical field in all other directions. At each specific point, the produced electric field is affected by the lengths of the nontangential components, and their distances from this point. Thus, the length of coil elements that are not tangential to the brain tissue surface should be minimized. Furthermore, the nontangential coil elements should be as small as possible and placed as far as possible from the deep region to be activated.

5. *Remote location of return paths.* The wires leading the current in a direction opposite to the preferred direction (the return paths) should be located far from the base and the desired brain region. This enables higher absolute electric field in the desired brain region. In some cases, the return paths may be in the air, namely far from the head. In other cases, part of the return paths may be adjacent to a different region in the head that is distant from the desired brain region.

6. *Shielding.* Feature 5 enables the possibility of screening. Because the return paths are far from the main base, it is possible to screen all or part of their field by inserting a shield around them or between them and the base. The shield consists of a material with high magnetic permeability, capable of inhibiting or diverting a magnetic field, such as mu metal, iron, or steel core. Alternatively, the shield is comprised of a metal with high conductivity, which can cause electric currents or charge accumulation that can oppose the effect produced by the return portions.

Specific deep TMS coils for stimulating different deep brain regions are described in the next sections.

22.4   A Coil for Stimulation of Deep Brain Regions Related to Mood Disorders: Simulations and Phantom Measurements

Accumulating evidence suggests that the nucleus accumbens plays a major role in mediating reward and motivation (Self and Nestler, 1995; Schultz et al., 1997; Breiter and Rosen, 1999; Ikemoto and Panksepp, 1999; Kalivas and Nakamura, 1999). Functional MRI (fMRI) and positron emission tomographic (PET) studies showed that the nucleus accumbens is activated in cocaine addicts in response to cocaine administration (Lyons et al., 1996; Breiter et al., 1997). Other brain regions are also associated with reward circuits, such as the ventral tegmental area, amygdala, and medial prefrontal, cingulate, and orbitofrontal cortices (Breiter and Rosen, 1999; Kalivas and Nakamura, 1999). Moreover, neuronal fibers connecting the medial prefrontal, cingulate, or orbitofrontal cortex with the nucleus accumbens may have an important role in reward and motivation (Jentsch and Taylor, 1999; Volkow and Fowler, 2000). The nucleus accumbens is also connected to the amygdala and the ventral tegmental area. Therefore, activation of these brain regions may affect neuronal circuits, mediating reward and motivation. In rats and monkeys and even in humans, electrical stimulation of the median forebrain bundle is rewarding; and when a stimulating electrode is inserted into various parts of that bundle (including the ventral tegmental area, the median prefrontal cortex, and the nucleus accumbens septi), compulsive self-stimulation can be obtained (Milner, 1991; Jacques, 1999). The new coil (called the Hesed coil, H-coil) is designed to stimulate effectively deeper brain regions without increasing the electrical field intensity in the superficial cortical regions. Numeric simulations and phantom measurements of the total electrical field produced by the Hesed coil inside a homogeneous spherical volume conductor are presented and compared with

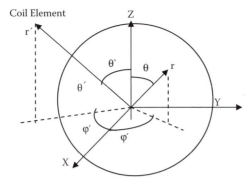

FIGURE 22.4 The relation between the spherical coordinate system and the Cartesian coordinate system in which the field components in every point were calculated. R is the radius vector to the point inside the sphere where the field is computed, and r' is the vector to the differential coil element on which the integration is performed.

results from a circular coil in different orientations and from the double-cone coil. The decrease in the electrical field in the brain as a function of the distance from the new coil is much slower compared with previous coils. It is hoped that such a coil can stimulate deeper regions, such as the nucleus accumbens and the fibers connecting the medial prefrontal or cingulate cortex with the nucleus accumbens. Activation of these fibers may induce reward, and chronic treatment may have antidepressant properties or serve as a new strategy against drug addiction.

22.4.1 Methods

22.4.1.1 Numerical Simulations

The simulations were conducted using a Mathematica program (Wolfram, 1999). The head was modeled as a spherical homogeneous volume conductor with a radius of 7 cm. The induced and electrostatic field at a specific point inside the spherical volume were computed for several coil configurations, using the method presented by Eaton (1992), and the total electric fields in the x, y, and z directions were calculated.

The vector potential A and scalar potential Φ can be expanded in terms of spherical harmonic functions up to N order. After enforcing the boundary conditions at the sphere boundary, the final expressions for the total electric field in the three Cartesian directions are:

$$E_j = E_{Aj} + E_{\Phi j} \quad j = x, y, z \tag{22.9}$$

where the induced field in each direction is given by:

$$E_{Aj} = \frac{-\mu_0 \partial I}{\partial t} \sum_{l=0}^{N} \sum_{m=-1}^{l} Y_{lm}(\theta, \varphi) C_{lm}^{j} \quad j = x, y, z \tag{22.10}$$

where $Y_{lm}(\theta, \varphi)$ are spherical harmonic functions; r, θ, and φ are spherical coordinates of the point inside the conductive sphere where the electric field is calculated (see Figure 22.4); and C_{lm}^{j} are j-components of the integration over the coil path:

$$C_{lm}^{j} = \int_{coil} \frac{Y_{lm}^{*}(\theta', \varphi') dl_j}{(2l+1) r'^{l+1}} \quad j = x, y, z \tag{22.11}$$

where * means complex conjugate; r', θ', and φ' are spherical coordinates of the coil element (Figure 22.4); and dl_j is the j-component of the differential element of the coil.

The electrostatic fields in the *x*-, *y*-, and *z*-directions are given by:

$$
E_{\Phi x} = -\sin(\theta)\cos(\varphi)\sum_{l=1}^{N+1}\sum_{m=-1}^{l} V_{lm} h r^{l-1} Y_{lm}(\theta,\varphi) +
$$

$$
\cos(\theta)\cos(\varphi)\sum_{l=1}^{N+1}\sum_{m=-1}^{l} V_{lm} h r^{l-1} \frac{1}{2}\Big[\exp(i\varphi)\sqrt{((l-m+1)(l+m))}Y_{l,m-1}(\theta,\varphi) -
$$

$$
\exp(-i\varphi)\sqrt{((l+m+1)(l-m))}Y_{l,m+1}(\theta,\varphi)\Big] +
$$

$$
\frac{\sin(\varphi)}{\sin(\theta)}\sum_{l=1}^{N+1}\sum_{m=-1}^{l} V_{lm} h r^{l-1} im Y_{lm}(\theta,\varphi)
\tag{22.12}
$$

$$
E_{\Phi y} = -\sin(\theta)\sin(\varphi)\sum_{l=1}^{N+1}\sum_{m=-1}^{l} V_{lm} h r^{l-1} Y_{lm}(\theta,\varphi) +
$$

$$
\cos(\theta)\sin(\varphi)\sum_{l=1}^{N+1}\sum_{m=-1}^{l} V_{lm} h r^{l-1} \frac{1}{2}\Big[\exp(i\varphi)\sqrt{((l-m+1)(l+m))}Y_{l,m-1}(\theta,\varphi) -
$$

$$
\exp(-i\varphi)\sqrt{((l+m+1)(l-m))}Y_{l,m+1}(\theta,\varphi)\Big] -
$$

$$
\frac{\cos(\varphi)}{\sin(\theta)}\sum_{l=1}^{N+1}\sum_{m=-1}^{l} V_{lm} h r^{l-1} im Y_{lm}(\theta,\varphi)
\tag{22.13}
$$

$$
E_{\Phi z} = -\cos(\theta)\sum_{l=1}^{N+1}\sum_{m=-1}^{l} V_{lm} h r^{l-1} Y_{lm}(\theta,\varphi) -
$$

$$
\sin(\theta)\sum_{l=1}^{N+1}\sum_{m=-1}^{l} V_{lm} h r^{l-1} \frac{1}{2}\Big[\exp(i\varphi)\sqrt{((l-m+1)(l+m))}Y_{l,m-1}(\theta,\varphi) -
$$

$$
\exp(-i\varphi)\sqrt{((l+m+1)(l-m))}Y_{l,m+1}(\theta,\varphi)\Big]
\tag{22.14}
$$

where $i=\sqrt{-1}$, and V_{lm} is a complex function of the integrals over coil path C_{lm}^{j} :

$$
V_{lm} = \frac{-\mu_0}{l}\frac{\partial I}{\partial t}\Big(\sqrt{[(l+m-1)(l+m)/(2l+1)(2l-1)]}\,0.5\big(C_{l-1,m-1}^{y}i - C_{l-1,m-1}^{x}\big) +
$$

$$
\sqrt{[(l-m-1)(l-m)/(2l+1)(2l-1)]}\,0.5\big(C_{l-1,m+1}^{y}i + C_{l-1,m+1}^{x}\big) +
$$

$$
\sqrt{[(l-m)(l+m)/(2l+1)(2l-1)]}\,C_{l-1,m}^{z}\Big)
\tag{22.15}
$$

The simulations were performed using 10th-order approximation. The summations in Equations 22.10 to 22.15 were computed up to $N = 10$. The convergence rate depends on the distance from the coil elements and on coil configuration, and, in general, is faster for more remote points. For the new coil design, the convergence rate was faster than for the circular coil. For points close to the coils (up to 1.5 cm), the induced field was corrected by the exact formula (Equation 22.5). For more remote points, the error was less than 1%. In all the calculations, the rate of current change was taken as 10,000 A/100 μs. The field is given in volts per meter (V/m).

22.4.2 Measurements of the Electrical Field Induced in a Phantom Brain

The electrical field induced by the new coil and the double-cone coil (Magstim; Whitland, United Kingdom) was measured in a saline solution placed in a hollow glass model of the human head

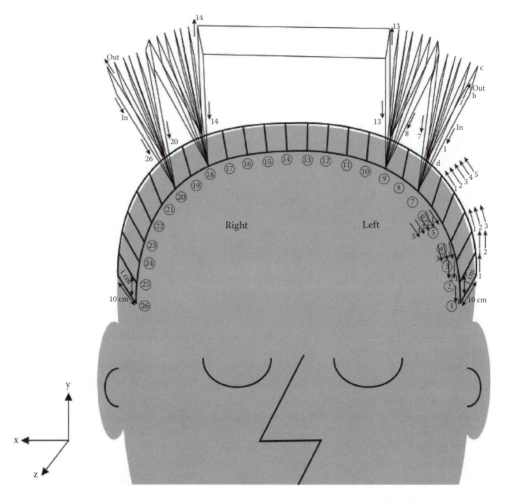

FIGURE 22.5 The Hesed coil shape when applied over the human head. The same coil can be placed around the forehead to stimulate nerve fibers in the supero-inferior direction. The only elements that produce an electrical field in the z-direction are the 26 strips attached to the head (numbered 1 through 26), where the current is in the $+z$ direction, and the 26 return paths at the edges of the fans where the current is in the $-z$ direction.

(15 × 17 × 20 cm; Cardinal Industries, Inc., Milwaukee, Wisconsin), using a two-wire probe. The distance between the noninsulated edges of the two wires of the probe was 14 mm. Voltage measured divided by the distance between the wire edges gives the induced electrical field figure. Stimulation was delivered using the Magstim Model 200 stimulator at 100% power level. The coils were placed on the glass surface and the electrical field was measured at numerous points within the saline solution.

22.4.3 Results

The simulations revealed that, in general, the presence of accumulating surface charge induced by coil configurations having a radial current component changes the total field in a nontrivial way. The presence of an electrostatic field not only reduces the total field at any point, but also leads to a significant reduction in the percentage of the total field in depth, relative to total field at the surface. Moreover, both the total field and the percentage relative to the surface at any specific point depend on its distance from the nontangential coil elements.

The basic concept of the new coil design is to generate a summation of the electrical field in depth by inducing electrical fields at different locations around the surface of the head, all of which have a

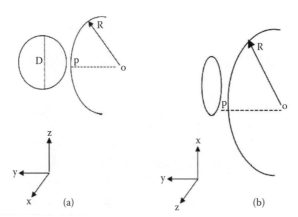

FIGURE 22.6 (A) A circular coil with diameter D placed perpendicular to the head surface. The head is modeled as a sphere with a radius $R = 7$ cm. The coil has a current component in the y-direction, which is perpendicular to the head, and a component in the z-direction, which is completely parallel to the head surface only at the attachment point p. (B) A circular coil with diameter D placed tangential to the head surface. The coil has current components in the x- and z-directions, which are completely parallel to the head surface only at the attachment point p.

common direction. Such an approach increased the percentage of electrical field induced in depth, relative to the field in the surface regions. In addition, because a radial component had a dramatic effect on the percentage of the electrical field in depth, an effort was made to minimize the overall length of nontangential coil elements, and to locate them as distantly as possible from the deep region to be activated. This region simulated the location of the nucleus accumbens. Calculations for several coil configurations were made and the optimal configuration (called the *Hesed coil*) was compared with standard circular coils and with the double-cone coil. We compared simulation results of field distribution of the Hesed coil design (Figure 22.5), of a double-cone coil, and of a circular coil oriented both perpendicular (Figure 22.6A) and parallel (Figure 22.6B) to the head. Figure 22.5 shows the coil design when applied on the human head. The coil contains several strips (26 in the example in Figure 22.5) attached to the head, all connected serially, and having wires that induce stimulation in the desired direction. This desired direction is the anteroposterior direction in the example shown in Figure 22.5 (z-direction). For each strip there is a return path wire having a current component at the opposite direction (z-direction), located 5 cm above the head. These return paths are located at the top edges of four fans to remove the currents flowing through them away from the deep regions of the head. The specific design of the fans is meant to reduce the inductance of the coil. The fans are connected to the frame near strips 7, 9, 18, and 20 (see Figure 22.5). These loci were chosen to remove the return paths as much as possible from the deep brain region to be activated most effectively. The only wires with currents that have radial components are those connecting the strips that are attached to the head with their return paths along the sides of the fans. An optimized coil would have a flexible frame allowing all elements of the coil that are touching the head to be tangential to the head surface (see Figure 22.5).

In the calculations of the field produced by the Hesed coil design, we assumed that the only coil elements carrying current components that are not tangential to the surface are the wires connecting the return paths with the strips that are attached to the head (along the fans). This is a plausible assumption in the realistic case where the coil is attached to the skull. In the human head, the cerebral spinal fluid is approximately parallel to the skull everywhere, and one can assume that the conductive properties of the cerebral spinal fluid are similar to those of the brain. The electrostatic field resulting from the contribution of the nontangential elements was calculated for each point and subtracted from the induced field of the coil. To obtain maximal efficacy, given the limitations of the stimulator and the need for a specific range of the coil inductance (15 to 25 µH), the average lengths of the strips were taken as 8 cm. The simulations were made for strip lengths of 9 cm over one hemisphere and 7 cm over the other hemisphere (to obtain a slight preference for one hemisphere stimulation and to have the opportunity to reach stimulation threshold in one hemisphere only). The wires connecting the head strips to their return paths (the nontangential elements) were taken as 5 cm long. The locations of the strips were determined to fit the human head, as in Figure 22.5. Hence, the distances of strips 13 and 14 from the sphere center were taken as approximately 6 cm; strips 3 and 24 were located approximately 7 cm from the sphere center. For the orientation shown in Figure 22.5, the maximal total field in the anteroposterior direction (z-direction) was produced at the cortex near the center of strips 1 and 26 at the sides. The field at the top of the head was reduced considerably because of the influence of the return paths and of the nontangential wires along the sides of the fans.

Figure 22.7 shows the induced (Figure 22.7A) and total (Figure 22.7B) field in the z-direction (Ez, defined in Figure 22.6) of a one-turn, 5.5-cm-diameter circular coil placed perpendicular and parallel to a 7-cm-radius spherical volume conductor, as a function of distance from the coil edge. The fields were calculated along the line connecting the sphere center to the coil point closest to the surface (line o-p in Figure 22.6). It is clear that the reduction in total field resulting from charge accumulation is much larger when the coil is oriented perpendicular to the surface (Figure 22.7). In addition, a comparison was made with the induced and total fields of one winding from the new Hesed coil, including strip 1 with its connection to the return path and its return path itself (and taking strip lengths of 5.5 cm). The simulations show that although the induced field of the strip is slightly larger than that of a circular coil with similar dimensions (see Figure 22.7), the difference in the total field is much larger (see Figure 22.7). This results from the fact that the field reduction due to electrostatic charge accumulation in the case of the winding of strip 1 is very small because the only elements carrying radial current components are the wires along paths a-b and c-d (see Figure 22.5), which are a relatively small fraction of the winding length, and are distant from the points under consideration. The induced and total field in the z-direction (Ez) resulting from the entire Hesed coil compared with the double-cone coil is shown in Figure 22.8. The field of the Hesed coil is computed along the line from strip 26 (where it is maximal) to the sphere center. The field of the double-cone coil is computed along the line from the junction at the coil center (where it is maximal) to the sphere center. Although the double-cone coil produces a much larger induced field than the Hesed coil (see Figure 22.8A), the rate of decay of the effective total field with distance is much smaller for the Hesed coil (see Figure 22.8B). Hence, at a depth of 6 cm, the total electrical field of the Hesed coil is already a little larger than that of the double-cone coil (see Figure 22.8B).

Figure 22.9 shows the z-component of the electrical field as a function of distance, relative to the field at a distance of 1 cm, for the Hesed coil, a double-cone coil with 14-cm diameter for each wing, and the 5.5-cm-diameter circular coil oriented tangential and perpendicular to the head surface. The field produced by the Hesed coil at a depth of 6 cm is approximately 35% of the field at a depth of 1 cm near the middle of strip 26 (where the field induced by the Hesed coil is highest throughout the brain). The field produced by the double-cone coil at a depth of 6 cm is only about 8% of the field 1 cm from the coil. The field produced by the 5.5-cm-diameter circular coil at this depth is less than 2% of the field 1 cm from the coil. For a larger circular coil, the percentage of field in depth is somewhat higher, but still smaller than that of the double-cone coil (data not shown).

Actual measurements of the electrical fields in a phantom brain using the first manufactured version of the Hesed coil and a double-cone coil basically confirmed our theoretical calculations. Both coils produced slightly lower fields at any point in the phantom brain compared with the theoretical calculations. However, this was more evident in the case of the Hesed coil, and the percentage of field in depth relative to the surface was slightly lower compared with our calculations. The results for the total field and the percentage in depth are presented in Figure 22.10.

It is clear that the total field induced by the double-cone coil, using the maximal output of the stimulator (10,000 amps/100 µs) produces a markedly greater electrical field up to 6 cm depth, compared with the Hesed coil (see Figure 22.10A), but the percentage in depth is markedly greater when the Hesed coil is used (see Figure 22.10B).

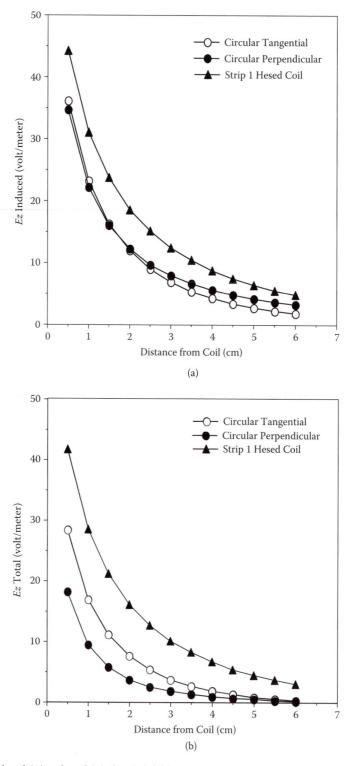

FIGURE 22.7 Induced (A) and total (B) electrical field in the *z*-direction plotted as a function of distance from a one-turn circular coil of 5.5-cm diameter placed tangential or perpendicular to the head surface. In addition, the induced and total fields of the winding of the Hesed coil connected to strip 1 are shown for the case of strip length of 5.5 cm.

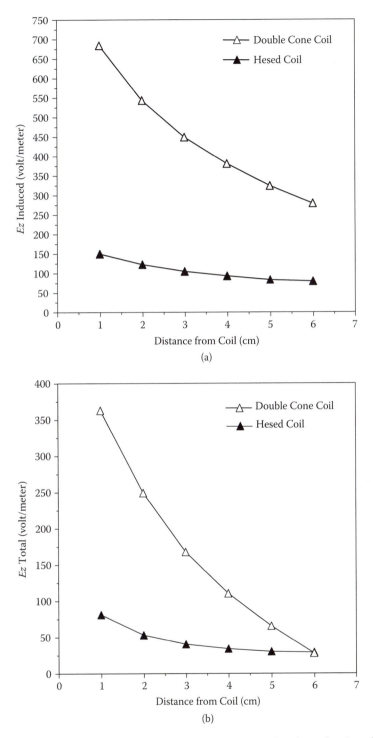

FIGURE 22.8 Induced (A) and total (B) electrical field in the *z*-direction plotted as a function of distance for the double-cone coil and the Hesed coil. The electrical fields were calculated for a six-turn double-cone coil with a diameter of 14 cm for each wing, an opening angle of 95°, and a central linear section of 3 cm, and for the Hesed coil with strip lengths of 9 cm over the right hemisphere and 7 cm over the left hemisphere. The field of the Hesed coil is computed along the line from strip 26 (where *Ez* is maximal) to the sphere center. The field of the double-cone coil is computed along the line from the central linear section (where *Ez* is maximal) to the sphere center.

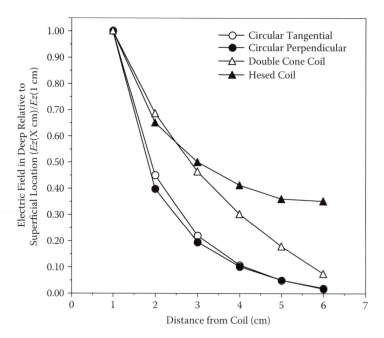

FIGURE 22.9 Electrical field in the *z*-direction relative to the field 1 cm from the coil as a function of distance. Data are presented for the Hesed coil, the double-cone coil, and the 5.5-cm-diameter circular coil oriented tangential and perpendicular to the head surface. The total electrical field at each point along the line from the point of maximal *Ez* to the sphere center was divided by the *Ez* value calculated at a 1-cm distance.

22.5 Transcranial Magnetic Stimulation of Deep Brain Regions: Evidence for Efficacy of the H-Coil

The biological efficacy of the H-coil was tested (Zangen et al., 2005) using the motor threshold as a measure of biological effect. The rate of decrease of the electric field as a function of the distance from the coil was measured by gradually increasing the distance of the coil from the skull and measuring the motor threshold at each distance. A comparison was made to the figure-8 coil.

22.5.1 Methods

22.5.1.1 Subject

Six healthy, right-handed volunteers (four men and two women, mean age 36 years, range 25 to 45 years) gave written informed consent for the study, which was approved by the National Institute of Neurological Disorders and Stroke Institutional Review Board. Subjects were interviewed and examined by a neurologist and found to be free of any significant medical illness or medications known to affect the CNS.

22.5.1.2 TMS Coils

The TMS coils used in this study were a specific version of the H-coil and a figure-of-eight (figure-8) coil. The H-coil version used in this study allows a comfortable placement above the hand motor cortex. The theoretical considerations and design principles of H-coils are explained in a previous study (Roth et al., 2002). In short, the coil is designed to generate a summation of the electric field in a specific brain region by locating coil elements at different locations around this region, all of which have a common current component that induces an electric field in the desired direction (termed the +*z*-direction). In addition, because a radial component has a dramatic effect on the electric field magnitude and on the rate of decay of the electric field with distance, the overall length of coil elements that are nontangential to the skull should be minimized, and these elements as well as coil elements having a current component

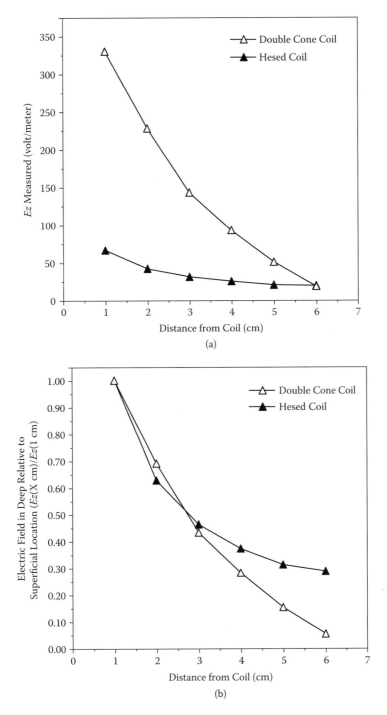

FIGURE 22.10 Measurements of the electrical field induced by the Hesed coil and the double-cone coil in a phantom brain. The electrical field induced in the *z*-direction (A) and the electrical field in the *z*-direction *relative* to the field 1 cm from the coil (B) are plotted as a function of distance from the coil. For both coils, the data show the measurements along the line from the point where the maximal *Ez* value is obtained (as described previously) to the sphere center.

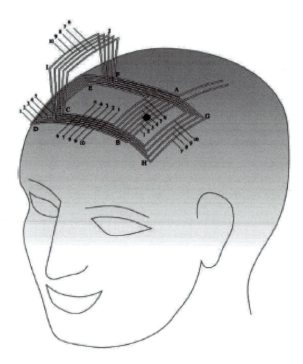

FIGURE 22.11 Sketch of the H-coil version used in this study placed on a human head. The coil orientation shown in the figure is designated for optimal stimulation of the left APB (indicated by a black spot).

in the opposite direction (the $-z$-direction) should be located as distant as possible from the brain region to be activated.

The H-coil version used in the present study is shown in Figure 22.11. The coil has 10 strips carrying a current in a common direction ($+z$-direction) and located around the desired motor cortex site (segments A–B and G–H in Figure 22.11). The average length of the strips is 11 cm. The only coil elements having radial current components are those connected to the return paths of five of the strips (segments C–I and J–F in Figure 22.11). The length of these wires is 8 cm. The return paths of the other five strips are placed on the head at the contralateral hemisphere (segment D–E in Figure 22.11). The wires connecting between the strips and the return paths (segments B–C and F–A in Figure 22.11) are, on average, 9 cm long. The H-coil was compared to a standard commercial Magstim figure-8 coil with internal loop diameters of 7 cm.

22.5.1.3 Experimental Setup

Subjects were seated with the right forearm and hand supported. Motor evoked potentials (MEPs) of the right abductor pollicis brevis (APB) muscle were recorded using silver/silver chloride surface electrodes. Subjects were instructed to maintain muscle relaxation throughout the study. The EMG amplitude was amplified using a conventional EMG machine (Counterpoint, Dantec Electronics, Skovlunde, Denmark) with a bandpass between 10 and 2000 Hz. The signal was digitized at a frequency of 5 kHz and fed into a laboratory computer.

A Magstim Super Rapid stimulator (The Magstim Company, New York City), which produces a biphasic pulse, coupled with either the figure-8 coil or the H-coil, was used. Preliminary studies showed the H-coil to have a loudness of 122 dB when activated, similar to other coils used in our laboratory. As standard laboratory practice, subjects were fitted with foam earplugs to attenuate the sound.

The coil was placed on the scalp over the left motor cortex. The intersection of the figure-8 coil was placed tangential to the scalp with the handle pointing backward and laterally at a 45° angle away from the midline. Thus, the current induced in the neural tissue was directed approximately perpendicular to the line of the central sulcus and therefore optimal for activating the corticospinal pathways

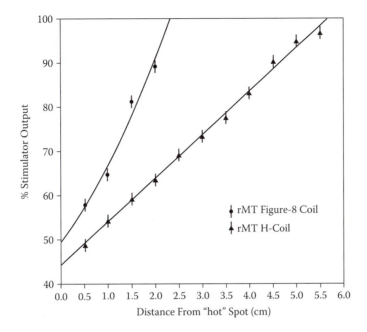

FIGURE 22.12 Intensity needed for APB stimulation at different heights above the scalp. Resting motor threshold of the APB was measured at different distances above the "hot spot" when using either the H-coil or the figure-8 coil. The percentage of stimulator power needed to reach the resting motor threshold vs. the distance of the coil from the "hot spot" on the skull is plotted. The points represent the means and SDs (standard deviations) of six healthy volunteers.

transsynaptically (Brasil-Neto et al., 1992; Kaneko et al., 1996). Similarly, the H-coil was placed on the scalp with the handle pointing backward in such a way that the center of the strips covered the motor cortex and in such a direction that the current induced in the neural tissue would be perpendicular to the line of the central sulcus. With a slightly suprathreshold stimulus intensity, the stimulating coil was moved over the left hemisphere to determine the optimal position for eliciting MEPs of maximal amplitude (the "hot spot"). The optimal position of the coil was then marked on the scalp to ensure coil placement throughout the experiment. The resting motor threshold was determined to the nearest 1% of the maximum stimulator output and was defined as the minimal stimulus intensity required to produce MEPs greater than 50 μV in more than 5 of 10 consecutive trials at least 5 s apart.

The coils were held in a stable coil holder that could be adjusted to different heights above the "hot spot" on the scalp. The resting motor threshold was determined at different distances above the scalp, using increments of 0.5 cm.

22.5.1.4 Safety Measurements

Because the H-coil was not used in previous clinical TMS studies, we asked the subjects to report any side effects, including pain, anxiety, or dizziness, and we performed cognitive and hearing tests before and after the TMS session. For the cognitive testing, we used the CalCap computer program to test immediate and delayed memory, as described previously (Wassermann et al., 1996).

22.5.2 Results

None of the six subjects who participated in the study reported any significant side effects after the TMS session. We did not find any change in cognitive or hearing abilities in these six subjects. A slight and short-lasting headache was reported by one of the six subjects. In a different experiment done subsequently, and not reported in this chapter, we used the H-coil to deliver single or paired pulses at 1 Hz during 20 s over five different locations on the scalp in three additional subjects. The third of these subjects experienced some hearing loss in his left ear, a 30-dB loss at 4000 Hz, which has been stable for 10 months and appears permanent. The ear protection had fallen out transiently during the study.

The percentage of stimulator output required for APB activation by each coil is plotted in Figure 22.12 as a function of distance from the "hot spot" on the scalp. It can be seen that the efficacy of the H-coil at large distances from the scalp was significantly greater as compared to the figure-8 coil. When using the maximal stimulation power output, the figure-8 coil can be effective (reach stimulation threshold) up to 2 cm away from the coil, while the H-coil can be effective 5.5 cm away from the coil. Moreover, the rate of decay of effectiveness as a function of the distance from the coil is much slower in the H-coil relative to the figure-8 coil (Figure 22.12).

22.5.3   Discussion

The findings confirm our theoretical calculations and phantom brain measurements (Roth et al., 2002) indicating the ability of the H-coil to stimulate brain structures at a large distance from the coil. The comparison between the TMS coils demonstrated a significantly improved depth penetration, and a much slower rate of decay of effectiveness as a function of the distance from the coil, when using the H-coil relative to the regular figure-8 coil. This indicates that when stimulating deep brain regions using the H-coil, the cortical stimulation is not much higher for the same activation in depth.

The H-coil produces a summation of the electric field from several coil elements carrying current in the same direction. In contrast, the electric field of the figure-8 coil is produced by a concentrated region in the center of the coil. In addition, the relative fraction of the figure-8 field that is produced by coil elements that are nontangential to the skull surface is much larger than in the H-coil. These two reasons lead to the fact that although the figure-8 field is more focal, it has more significant reduction both in absolute field magnitude at any point and in the percentage of the deep region field relative to field at the surface.

According to our calculations and phantom brain measurements (Roth et al., 2002), the field induced in cortical regions by the H-coil is much lower than that of the double-cone coil. Therefore, it is likely that the excitation threshold can be reached at 4 to 6 cm using the H-coil without inducing pain and other side effects.

It should be emphasized that although the structure stimulated in this study was in the motor cortex, and the medium between the coils and the "hot spot" was mainly air; the rate of decay within the brain itself should be very similar. This is a delicate point that should be elaborated. The electric conductivity of the brain is much greater than that of air. In conductive materials such as the brain, radial current components of the coil would lead to charge accumulation on the surface of the brain, which would cause a decrease of the field at any point inside the brain. Hence, the rate of decay of the electric field with distance would be faster in the brain than that measured in air. Nevertheless, the field distribution inside the brain is *independent* of the location of the interface between the conductive and insulating media (Branston and Tofts, 1990; Tofts, 1990; Eaton, 1992). As a result, the rate of decay within the brain when attaching the coil to the skull would be similar to that measured in this study, where the coil was raised above the skull. Small changes are expected due to the fact that coil configuration relative to the skull is somewhat different when it is raised. The H-coil is designed to minimize radial current components when attached to the scalp; hence, the amount of radial components may be slightly different and probably larger when the H-coil is raised above the scalp. Therefore, the advantage of the H-coil as compared to the figure-8 coil, in terms of the rate of decay of the field as a function of distance, may be even greater when the coils are attached to the scalp.

Although the H-coil has a remarkable ability to penetrate into deeper brain regions, due to the slower decay of the electric field as a function of distance, none of the subjects in the present study reported any side effects and cognitive or hearing abilities were not affected. Nevertheless, it should be emphasized that subjects in the present study experienced only twenty to thirty single pulses at intensities greater than those needed for minimal APB activation (when looking for the hot spot for APB activation) and that the rest of the pulses were given just at the minimal level for APB activation when the coil was placed either on the scalp or at different heights above the scalp. Future studies will address the safety and efficacy of the H-coil when used in higher doses. As reported, one subject has experienced some hearing loss in a subsequent experiment. Our past safety studies have not demonstrated that hearing loss is expected (Pascual-Leone et al., 1992). As the loudness of the H-coil does not appear different from other coils, this result may be due, in part, to particularly sensitive hearing in this subject and some lapse in hearing protection. However, as we cautioned before, the event does emphasize the need to take the necessary measures to protect hearing in all TMS studies.

22.6   Transcranial Magnetic Stimulation of Deep Prefrontal Regions

Medial prefrontal and orbitofrontal regions are known to be associated with reward circuits. The H1 and H2 coils are designed to stimulate deep prefrontal structures, with minimal undesired side effects such as pain, motor stimulation, and facial muscles activation. The coils are wound with a double 14 AWG (American Wire Gauge) insulated copper wire wound into several windings, and connected in series. Detailed illustrations of the wiring patterns of H1 and H2 are shown in Figure 22.13 and Figure 22.14, respectively; and coil specifications for H1 and H2 are presented in Table 22.1 and Table 22.2, respectively.

The effective part of the H1 coil in contact with the patient scalp has the shape of half a donut, with 14 strips of 7 to 12 cm in length (Figure 22.13). These strips produce the most effective field of the coil, and are oriented on an anterior–posterior axis. The 14 strips are distributed above the prefrontal cortex of the left hemisphere, with a separation of 1 cm between them. Three strips (8, 9, and 10 in Figure 22.13) are elongated toward the forehead, and their continuations pass in the left–right direction along the orbitofrontal cortex (segments I–J 8 to 10 in Figure 22.13), with a separation of 1 cm between them. The return paths of strips 1 to 7 are attached to the head in the right hemisphere (segments D–E 1 to 7 in Figure 22.13), with a separation of 0.8 cm between them. The return paths of strips 8 to 14 are remote 7 cm from the head (segments M–G 8 to 14 in Figure 22.13), with a separation of 0.3 cm between them.

The frame of the inner rim of the half donut is flexible in order to fit the variability in human scull shape.

The paths of the 14 windings of the H1 coil are shown in Figure 22.13. The windings of strips 1 to 7 traverse the path A-B-C-D-E-F-G-H-A. The windings of strips 8 to 10 traverse the path A-I-J-K-L-M-G-H-A. The windings of strips 11 to 14 traverse the path A-N-L-M-G-H-A.

Figure 22.14 shows a diagram of the H2 coil, which is designed to stimulate deep brain regions and fit the human head. The effective part of the coil, in contact with the patient scalp has a shape of half a donut, with 10 strips of 14 to 22 cm in length (Figure 22.14). These strips produce the most effective field of the coil and are oriented in a right–left direction (lateral–medial axis). Three strips pass in front of the forehead along the orbitofrontal cortex (1 to 3 in Figure 22.14), with a separation of 1 cm between them. Seven strips pass above the forehead along the prefrontal cortex (4 to 10 in Figure 22.14), with a 0.8-cm separation.

The paths of the 10 windings are shown in Figure 22.14. The windings of strips 1 to 3 traverse the path A-B-C-D-E-F-G-H-I-J-Q-R-S-T-K-L-A. The winding of strip 4 traverses the path A-B-G-H-I-J-Q-R-S-T-K-L-A. The winding of strip 5 traverses the path A-M-N-B-G-O-P-H-I-J-Q-R-S-T-K-L-A. The windings of strips 6 and 7 traverse the path A-M-N-B-G-O-P-H-I-J-K-L-A. The windings of strips 8 and 9 traverse the path A-B-G-H-I-J-K-L-A. The winding of strip 10 traverses the path A-H-I-J-K-L-A.

22.6.1   Comparison of Electric Field Distributions

The electric field distributions produced by H1 and H2 coils were measured in a phantom brain with the general dimensions $23 \times 19 \times 15$ cm, which was filled with 0.9% weight/volume saline. To determine the output of the coil at some depth, the induced electric field was measured using a two-wire probe. The distance between the ends of the two-wire probe was measured as 12.7 ± 0.2 mm. Voltage measured divided by the distance between the wires gives the induced electric field value. The two coils were compared to a standard Magstim figure-8 coil with an internal loop diameter of 7 cm, and a Magstim

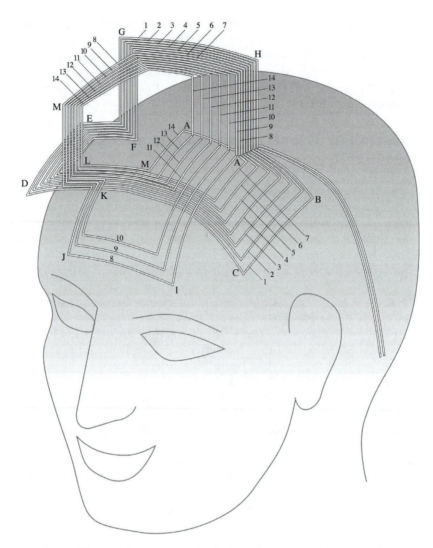

FIGURE 22.13 Sketch of the H1 coil near a human head. The coil orientation shown in the figure is designated for activation of structures in the prefrontal cortex, in the anterior–posterior direction.

TABLE 22.1 H1 Coil Specifications

Number of windings:	14 wire loops, including strips and return paths
Strips length:	Strips 1–7: 11–14: 7 cm
	Strips 8–10: 10–12 cm
Main induced field direction:	Anterior–posterior axis
Strips separation:	1 cm (typical)
Connecting cable:	2m ± 0.5 m
Coil inductance (including cable):	30 μH ± 1 μH
Max. magnetic field strength:	3.2 T
Max. electric field strength 0.5 cm from coil:	200 V/m
Wire size (circular section copper):	Two 14 AWG insulated wires in parallel
Wire length:	750 cm

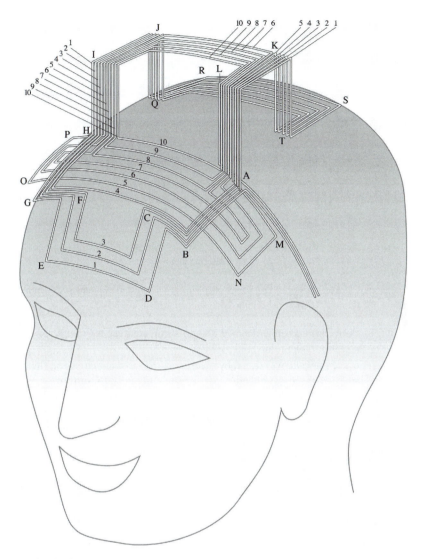

FIGURE 22.14 Sketch of the H2-coil near a human head. The coil orientation shown in the figure is designated for activation of structures in the prefrontal cortex, in the lateral–medial direction.

TABLE 22.2 H2 Coil Specifications

Number of windings:	10 wire loops, including strips and return paths
Strips length:	14–22 cm
Main induced field direction:	Lateral–medial axis
Strips separation:	0.8 cm (typical)
Connecting cable:	2m ± 0.5 m
Coil inductance (including cable):	25 μH ± 1 μH
Max. magnetic field strength:	3.0 T
Max. electric field strength 0.5 cm from coil:	190 V/m
Wire size (circular section copper):	Two 14 AWG insulated wires in parallel
Wire length:	800 cm

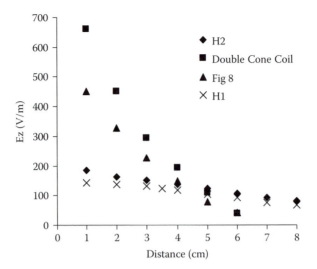

FIGURE 22.15 Phantom measurements of the electric field in the *z*-direction (up–down direction), plotted as a function of distance, for H1 and H2 coils, the double-cone coil, and the figure-8 coil.

double-cone coil. The double-cone coil is deemed able to stimulate deeper brain regions compared to other coils (Terao et al., 1994, 2000; Stokic et al., 1997).

The depth penetration of the coils was tested by measuring the electric field along the up–down line (*z* axis) beneath the center of the most effective part of the coil, at 100% output of Magstim Rapid stimulator. In H1, the most effective part was under strip 8 (under third of A–I 8 segment in Figure 22.13), where the probe is oriented in an anterior–posterior direction (*y* axis). In H2, the most effective part was the center of strip 5 (center of C–F 5 segment in Figure 22.14), where the probe is oriented in a lateral–medial direction (*x* axis). In the double-cone coil and the figure-8 coil, the most effective part was the junction at coil center, where the probe is oriented in an anterior–posterior direction (*y* axis). Plots of total electric field as a function of distance are shown in Figure 22.15.

Figure 22.16 shows the electric field as a function of distance, relative to the field at a distance of 1 cm, and the figure-8 coils.

It can be seen that the total electric field induced by the double-cone coil, and by the figure-8 coil, using the maximal output of the stimulator, is markedly greater than the field produced by the H1 and H2 at short distances of 1 to 2 cm. Yet, at distances of above 5 cm, the fields of the H1 and H2 coils become greater, due to their much slower rate of decay. Figure 22.16 reveals that the percentage in depth for the H-coils is greater than the two other coils already at the 2-cm distance, and this advantage of the H-coils becomes more prominent with increasing distance. In comparing between the two H-coils, it can be seen that H1 produces a slightly smaller absolute field magnitude, but a larger percentage in depth, relative to H2. The fields produced by the H1 and H2 at the 6-cm depth are about 63 and 57% of the field 1 cm from the coil, respectively, while the fields of the double-cone coil and the figure-8 coil attenuate to 8 to 10% at this distance.

Stimulation of brain structures at 3 to 4 cm depth with the double-cone coil is painful because a much higher field is induced at superficial cortical areas and at the facial muscles. To reach the stimulation threshold at 5 to 8 cm depth, a much higher intensity would be needed, which would increase pain and the risk for other side effects (such as convulsions). The total field induced by the H1 and H2 coils, even at maximal power output, will be 3 to 4 times lower than the double-cone coil in cortical regions. The H1 and H2 fields at 5 to 6 cm depth are not much smaller than their fields in the cortex, and greater than the double-cone coil field at that depth. Therefore, it is likely that the excitation threshold can be reached at 6 to 7 cm using the H-coils without the induction of pain or other side effects. The percentage of the electric field in depth produced by the standard figure-8 coils is similar to the double-cone coil,

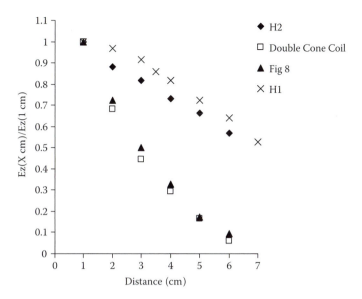

FIGURE 22.16 Electric field relative to the field 1 cm from coil, as a function of distance, for the H1 and H2 coils, the double-cone coil, and the figure-8 coil, according to the phantom brain measurements.

but the absolute field magnitude is much smaller. Therefore, the figure-8 coil would not only cause greater side effects, but could not reach stimulation threshold in depth, even at maximal power output.

References

Amassian, V.E., Eberle, L., Maccabee, P.J., and Cracco, R.Q. (1992). Modeling magnetic coil excitation of human cerebral cortex with a peripheral nerve immersed in a brain-shaped volume conductor: the significance of fiber bending in excitation. *Electroenceph. Clin. Neurophysiol.*, 185:291–301.

Barker, A.T., Jalinous, R., and Freeston, I.L. (1985). Non-invasive magnetic stimulation of the human motor cortex. *Lancet*, 1:1106–1107.

Barker, A.T., Garnham, C.W., and Freeston, I.L. (1991). Magnetic nerve stimulation-the effect of waveform on efficiency, determination of neural membrane time constants and the measurement of stimulator output, in magnetic motor stimulation: Basic principles and clinical experience. *Electroenceph. Clin. Neurophysiol.*, 43(Suppl.):227–237.

Basser, P.J. and Roth, B.J. (1991). Stimulation of a myelinated nerve axon by electromagnetic induction. *Med. Biol. Eng. Comput.*, 29:261–268.

Bohning, D.H. (2000). Introduction and overview of TMS physics. In *Transcranial Magnetic Stimulation in Neuropsychiatry*, Washington, D.C.: American Psychiatric Press, p. 13.

Bourland, J.D., Nyenhuis, J.A., Noe, W.A., Schaefer, J.D., Foster, K.S., and Geddes, L.A. (1996). Motor and sensory strength-duration curves for MRI gradient fields. In *Proc. Int. Soc. Magn. Reson. Med. 4th Scientific Meeting and Exhibit*, New York, p 1724.

Branston, N.M. and Tofts, P.S. (1990). Magnetic stimulation of a volume conductor produces a negligible component of induced current perpendicular to the surface. *J. Physiol. (London)*, 423:67p.

Brasil-Neto, J.P., Cohen, L.G., Panizza, M., Nilsson, J., Roth, B.J., and Hallett, M. (1992). Optimal focal transcranial magnetic activation of the human motor cortex: effects of coil orientation, shape of the induced current pulse, and stimulus intensity. *J. Clin. Neurophysiol.*, 9:132–136.

Breiter, H.C., Gollub, R.L., Weisskoff, R.M., et al. (1997). Acute effects of cocaine on human brain activity and emotion. *Neuron*, 19:591–611.

Breiter, H.C. and Rosen, B.R. (1999). Functional magnetic resonance imaging of brain reward circuitry in the human. *Ann. N.Y. Acad. Sci.*, 877:523–547.

Cohen, D. and Cuffin, B.N. (1991). Developing a more focal magnetic stimulator. Part I: some basic principles. *J. Clin. Neurophysiol.*, 8:102–111.

Cohen, L.G., Roth, B.J., Nilsson, J., et al. (1990). Effects of coil design on delivery of focal magnetic stimulation. Technical considerations. *Electroencephal. Clin. Neurophysiol.*, 75:350–357.

Durand, D., Ferguson, A.S., and Dalbasti, T. (1989). Induced electric fields by magnetic stimulation in non-homogeneous conducting media. *IEEE Eng. Med. Biol. Soc. 11th Annu. Int. Conf.*, Seattle, WA. 6:1252–1253.

Eaton, H. (1992). Electric field induced in a spherical volume conductor from arbitrary coils: application to magnetic stimulation and MEG. *Med. Biol. Eng. Comput.*, 30:433–440.

Heller, L. and Van Hulstein, D.B. (1992). Brain stimulation using electromagnetic sources: theoretical aspects. *Biophys. J.*, 63:129–138.

Ikemoto, S. and Panksepp, J. (1999). The role of nucleus accumbens dopamine in motivated behavior: a unifying interpretation with special reference to reward-seeking. *Brain Res. Rev.*, 31:6–41.

Jacques, S. (1999). Brain stimulation reward: "pleasure centers" after twenty five years. *Neurosurgery*, 5:277–283.

Jentsch, J.D. and Taylor, J.R. (1999). Impulsivity resulting from frontostriatal dysfunction in drug abuse: implications for the control of behavior by reward-related behaviors. *Psychopharmacology*, 146: 373–390.

Kalivas, P.W. and Nakamura, M. (1999). Neural systems for behavioral activation and reward. *Curr. Opin. Neurobiol.*, 9:223–227.

Kammer, T., Beck, S., Thielscher, A., Laubis-Herrmann, U., and Topka, H. (2001). Motor thresholds in humans. A transcranial magnetic stimulation study comparing different pulseforms, current directions and stimulator types. *Clin. Neurophysiol.*, 112:250–258.

Kaneko, K., Kawai, S., Fuchigami, Y., Morita, H., and Ofuji, A. (1996). The effect of current direction induced by transcranial magnetic stimulation on the corticospinal excitability in human brain. *Electroencephalogr. Clin. Neurophysiol.*, 101:478–482.

Kirkcaldie, M.T., Pridmore, S.A., and Pascual-Leone, A. (1997). Transcranial magnetic stimulation as therapy for depression and other disorders. *Aust. N. Z. J. Psychiatry*, 31:264–272.

Lyons, D., Friedman, D.P., Nader, M.A., and Porrino, L.J. (1996). Cocaine alters cerebral metabolism within the ventral striatum and limbic cortex of monkeys. *J. Neurosci.*, 16:1230–1238.

Maccabee, P.J., Eberle, L., Amassian, V.E., Cracco, R.Q., Rudell, A., and Jayachandra, M. (1990). Spatial distribution of the electric field induced in volume by round and figure '8' magnetic coils: relevance to activation of sensory nerve fibers. *Electroenceph. Clin. Neurophysiol.*, 76:131–141.

Maccabee, P.J., Amassian, V.E., Eberle, V.E., and Cracco, R.Q. (1993). Magnetic coil stimulation of straight and bent amphibian and mammalian peripheral nerve *in vitro*: locus of excitation. *J. Physiol.*, 460: 201–219.

Mills, K.R., Boniface, S.J., and Schubert, M. (1992). Magnetic brain stimulation with a double coil: the importance of coil orientation. *Electroenceph. Clin. Neurophysiol.*, 85:17–21.

Milner, P.M. (1991). Brain-stimulation reward: a review. *Can. J. Psychol.*, 45:1–36.

Niehaus, L., Meyer, B.U., and Weyh, T. (2000). Influence of pulse configuration and direction of coil current on excitatory effects of magnetic motor cortex and nerve stimulation. *Clin. Neurophysiol.*, 111:75–80.

Pascual-Leone, A., Cohen, L.G., Shotland, L.I., Dang, N., Pikus, A., Wassermann, E.M., Brasil-Neto, J.P., Valls-Sole, J., and Hallett, M. (1992). No evidence of hearing loss in humans due to transcranial magnetic stimulation. *Neurology*, 42:647–651.

Pascual-Leone, A., Cohen, L.G., Brasil-Neto, J.P., and Hallett, M. (1994). Non-invasive differentiation of motor cortical representation of hand muscles by mapping of optimal current directions. *Electroenceph. Clin. Neurophysiol.*, 93:42–48.

Ren, C., Tarjan, P.P., and Popovic, D.B. (1995). A novel electric design for electromagnetic stimulation: the slinky coil. *IEEE Trans. Biomed. Eng.*, 42:918–925.

Roth, B.J. and Basser, P.J. (1990). A model of the stimulation of a nerve fiber by electromagnetic radiation. *IEEE Trans. Biomed. Eng.*, 37:588–597.

Roth, B.J., Cohen, L.G., Hallet, M. Friauf, W., and Basser, P.J. (1990). A theoretical calculation of the electric field induced by magnetic stimulation of a peripheral nerve. *Muscle Nerve*, 13:734–741.

Roth, Y., Zangen, A., and Hallett, M. (2002). A coil design for transcranial magnetic stimulation of deep brain regions. *J. Clin. Neurophysiol.*, 19:361–370.

Ruhonen, J. and Ilmoniemi, R.J. (1998). Focusing and targeting of magnetic brain stimulation using multiple coils. *Med. Biol. Eng. Comput.*, 38:297–301.

Schultz, W., Dayan, P., and Montague, P.R. (1997). A neural substrate of prediction and reward. *Science*, 275:1593–1599.

Self, D.W. and Nestler, E.J. (1995). Molecular mechanisms of drug reinforcement and addiction. *Annu. Rev. Neurosci.*, 18:463–495.

Stokic, D.S., McKay, W.B., Scott, L., Sherwood, A.M., and Dimitrijevic, M.R. (1997). Intracortical inhibition of lower limb motor-evoked potentials after paired transcranial magnetic stimulation. *Exp. Brain Res.*, 117:437–443.

Terao, Y., Ugawa, Y., Sakai, K., Uesaka, Y., and Kanazawa, I. (1994). Transcranial magnetic stimulation of the leg area of motor cortex in humans. *Acta Neurol. Scand.*, 89:378–383.

Terao, Y., Ugawa, Y., Hanajima, R., et al. (2000). Predominant activation of I1-waves from the leg motor area by transcranial magnetic stimulation. *Brain Res.*, 859:137–146.

Thielscher, A. and Kammer, T. (2002). Linking physics with physiology in TMS: a spherical field model to determine the cortical stimulation site in TMS. *Neuroimage*, 17:1117–1130.

Tofts, P.S. (1990). The distribution of induced currents in magnetic stimulation of the brain. *Phys. Med. Biol.*, 35:1119–1128.

Tofts, P.S. and Branston, N.M. (1991). The measurement of electric field, and the influence of surface charge, in magnetic stimulation. *Electroenceph. Clin. Neurophysiol.*, 81:238–239.

Volkow, N.D. and Fowler, J.S. (2000). Addiction, a disease of compulsion and drive: involvement of the orbitofrontal cortex. *Cereb. Cortex*, 10:318–325.

Wassermann, E.M., Grafman, J., Berry, C., Hollnagel, C., Wild, K., Clark, K., and Hallett, M. (1996). Use and safety of a new repetitive transcranial magnetic stimulator. *Electroencephalogr. Clin. Neurophysiol.*, 10:412–417.

Wassermann, E.M. and Lisanby, S.H. (2001). Therapeutic application of repetitive transcranial magnetic stimulation: a review. *Clin. Neurophysiol.*, 112:1367–1377.

Watson, D., Clark, L.A., and Tellegen, A. (1988). Development and validation of brief measures of positive and negative affect: the PANAS scales. *J. Pers. Soc. Psychol.*, 54:1063–1070.

Zangen, A., Roth, Y., Voller, B., and Hallett, M. (2005). Transcranial magnetic stimulation of deep brain regions: evidence for efficacy of the H-coil. *Clin. Neurophysiol.*, 116(4):775–779.

Zimmermann, K.P. and Simpson, R.K. (1996). "Slinky" coils for neuromagnetic stimulation. *Electroencephalogr. Clin. Neurophysiol.*, 101:145–152.

23

Application of Neural Plasticity for Vision Restoration after Brain Damage

Imelda Pasley,
Sandra Jobke,
Julia Gudlin, and
Bernhard A. Sabel

23.1 Introduction

23.1.1 Vision Restoration: A New Paradigm in Neuroscience

Our ability to see the world depends not only on the function of the eyes, but also to a large extent on the processing of neuronal information in the brain. Loss of vision may not only be the consequence of damage of the eye or the retina, but is a frequent consequence of brain damage due to stroke or trauma, which afflicts the visual system in about 20% of the cases. Vision loss due to damage of the central nervous system (CNS) is generally assumed to be irreversible. Once blind, so goes the generally held notion, regions of the visual field can no longer recover and the patient is left with no hope. However, over the past two decades, neuropsychological research has shown that the "blind" regions of the visual field have a hitherto little-recognized ability to process residual vision; and in more recent years, efforts have been made to "reactivate" such residual visual potential through vision training methods. This chapter describes how regular activation of such residual vision with appropriate training paradigms gradually improves some (although not all) visual capacities. This fundamental discovery of visual field restoration now sets a new stage for the development of effective therapies for the treatment of partial blindness in patients with damage of the brain's visual system.

23.1.2 Structure and Function of the Visual System

Most information about the characteristic features of the visual world, such as color, shape, distance, size, brightness, and texture can be gathered by our sense of vision. Perception of visual stimuli may seem like such an effortless task but a detailed layout of the structure of the visual system reveals a highly meticulous and intricate visual processing network. Unlike in reptiles and amphibians where most visual processing occurs in the retina, much of the visual processing in mammals takes place in the brain. It has been suggested that the brain might have expanded to accommodate a highly detailed visual processing system that is adaptive to pressures from natural selection (Thompson, 2000). Evolution has bestowed upon us a legacy: the ability to visually process the relationships of stimuli within a context that provides us with an up-to-date schema that optimally guides goal-directed behavior. This may explain why about half the human cortex is involved in some aspect of visual processing.

At the peripheral level, the visual system starts with the eye, where light stimuli enter the cornea, pass through the lens, and get reflected as a projected image onto the retina. The retina, although peripherally located, is actually part of the central nervous system (CNS) as embryologically it is formed from an out-pouching neural tube (Reid, 1999). The retina is composed of three layers consisting of the photo-receptors, the interneurons, and the retinal neurons (retinal ganglion cells). When light reaches the retina, it falls on the photoreceptors with its rods and cones. These distinguish two functional systems: scotopic and photopic, respectively (Rosenzweig, Leiman, and Breedlove, 1999). Both systems have properties that are different yet complementary to each other. For example, the scotopic system has approximately 100 million rods (per eye) that are located outside the fovea. They are sensitive to low lighting conditions and have large receptive fields making for lower acuity. On the other hand, the photopic system contains approximately four million cones that contain three classes of opsin that provide the basis of color vision. These cones concentrate in the immediate vicinity of the fovea, the central focal point of the eye, and have low sensitivity to light, thus requiring a relatively strong stimulation.

From the photoreceptors, information is relayed to the bipolar cells located within the layer of interneuron, which pass the signals to the retinal ganglion cells. Information leaves the retina and is sent to the brain via the axons of the M and P retinal ganglion cells (RGCs). These axons eventually converge at the optic disc (the blind spot) where they form the optic nerve. The axons from the temporal halves of each retina continue into the optic tract, while the axons from the nasal halves cross to the optic tracts to the opposite hemisphere. Axons in the optic tract either terminate in the superior colliculus (SC) or in the lateral geniculate nucleus (LGN). From the magnocellular and parvocellular layers of the LGN, axons travel to the primary visual cortex through the optic radiation. Cells from these two types of layers in the six-layered LGN terminate mainly in layer 4 of the primary visual cortex but within different sublayers. Basically, the M and P cells of the RGCs that respectively lead to the magnocellular and parvocellular layers of the LGN and terminate at the primary visual cortex, represent the M and P pathways. These pathways convey different information about the visual stimulus, with the former providing luminance contrast and temporal frequency while the latter provides color contrast and spatial frequency.

The primary visual cortex (also known as the striate cortex, Brodmann's area 17, or visual area 1 (V1)), is located at the occipital lobe and has a point-to-point representation of the retinal area (Thompson, 2000). Input to V1 comes from the M and P pathways of the LGN. Each pathway projects to either one of two different visual subsystems: the M pathway projects strongly to the dorsal stream while the P pathway projects strongly to the ventral stream. The dorsal visual pathway (the "where" pathway) from V1 to the posterior parietal cortex is involved with localizations of objects in space (Ungerleider and Mishkin, 1982). In contrast, the ventral visual pathway (the "what" pathway) from V1 to the infero-temporal cortex is involved in processing form and color information or object identification (Ungerleider and Mishkin, 1982). A later interpretation of the roles of these two streams indicated that both streams process localization and object identification but that each stream uses the visual information differently (Goodale and Milner, 1992). The dorsal stream with its visuo-motor modules uses the information to program and control particular movements that action entails while the ventral stream forms long-term perceptual representation that allows an organism to choose a course of action appropriate to objects in the world (Goodale, 2000).

To fully appreciate how the striate cortex becomes the primary player in visual processing, a basic knowledge of its functional architecture is necessary. V1 contains two basic classes of cells: (1) nonpyramidal and (2) pyramidal cells (Kandel et al., 2000). Nonpyramidal cells are local inhibitory interneurons that are confined to V1 and pass information from the LGN to the striate cortex. Pyramidal cells, which are excitatory, project not only to local interconnecting neurons, but also to other brain regions. These cells integrate activity within the six layers of V1 through their axons that extend upward and downward within the cortex. Moreover, these cells have horizontal connections that synapse on other immediate as well as remote pyramidal cells. These horizontal connections link columnar units that represent vertically oriented systems across the layers of V1. At least three such systems have been delineated: (1) orientation columns, (2) blobs, and (3) ocular dominance columns (Kandel et al., 2000). The columnar organization that is found in the primary visual cortex, as well as in the secondary visual areas, allows for unfolding of several dimensions of the visual information.

From area V1, signals are sent to other visual cortical areas (that project back to V1) such as area V2, V4, and the inferior temporal area. In monkeys, there are at least thirty-two visual areas in the visual cortex and it is believed that there are more in humans. Different visual areas are specialized in processing certain features of the visual stimuli. For example, cells in area V2 are involved in the perception of contours and send their axons to area V4. Area V4 responds strongly to sinusoidal frequency gratings as well as to concentric and radial stimuli and to wavelength differences. It has been suggested that V4 cells may play an intermediate role between spatial-frequency processing in V1 and V2 cells and the recognition of pattern and form in cells of the inferior temporal area. Area V5 is associated with motion perception while area V6 is involved in shape perception. Another secondary visual area that incidentally appears not to have the usual retinal map is the visual–temporal (TE) area. The TE, as the name suggests, is located in the temporal lobe and receives input from area V6 as well as other secondary visual areas. Monkey studies have shown that this area has a role in object recognition (Phillip et al., 1988) and that cells in this area respond to complex shapes (Nakamura et al., 1994).

23.1.3 Epidemiology of Visual System Dysfunction

Vision loss is perhaps one of the more, if not the most, debilitating sensory deficit for humans given that we rely heavily on our sense of sight in gathering information from the external environment. This even becomes evident when considering the amount of cortical tissue allocated to our visual system, which imposes a high risk for visual loss whenever brain damage occurs. There is a 20 to 30% chance of losing some amount of visual capacity leading to visual field disorder (VFD) after stroke, brain trauma, or brain surgery (Zihl and Cramon, 1986; Hollwich, 1988; Koelmel, 1988; Rossi et al., 1990; Huber, 1991). Prosiegel (1988) found a 24% incidence of VFD in a sample of approximately 400 patients who either had cerebrovascular disease, closed head trauma, cerebral hypoxia, tumors, or encephalitis. Children may experience VFDs as well such as those who have cerebral hypoxia (Lambert et al., 1987; Werth and Moehrenschlager, 1999).

A substantial amount of cell loss after brain damage has been attributed to secondary neuronal death more than the primary injury itself. Immediately after mechanical injury, there is disruption of the blood–brain barrier that sometimes can lead to extracellular edema (Povlishock et al., 1978). Edema can spread outside the injury site and extend as far as the whole hemisphere or the whole brain, instigating secondary brain damage. Secondary brain damage has been defined as cell injury that is not immediately evident until after a delay of hours or days (Siesjoe and Siesjoe, 1996); this phenomenon is reflected in stroke patients who do not spontaneously report visual problems when examined within twenty-four hours after the incident (Celesta et al., 1997) yet they do so some weeks after (Kerkhoff et al., 1990). This secondary response to injury can include delayed axon injury and axon swelling. These swollen and ballooned axons are found not only around the contusion area, but also at great distances from the injured area (Cervos-Navarro and Lafuente, 1991); hence, they are called diffuse axonal injury (DAI). In DAI, two types of lesions have been identified: (1) disruptive or axotomy, and (2) nondisruptive, which refers to internal axon damage. Axons that have nondisruptive damage are more likely to respond to treatment. These spared fibers of cells damaged within the visual system provide the basis for residual vision (Sabel, 1999).

The location and/or severity of the injury would determine the kind of visual field defect. For example, a complete unilateral optic nerve injury would lead to monocular blindness, while an incomplete unilateral optic nerve injury would lead to monocular diffuse VFD, and so on. Kerkhoff (1999) listed three categories of visuo-behavioral deficits that patients with VFD have to deal with: (1) hemianopic (referring to blindness in one half of the visual field) reading deficits (Zihl, 1995b); (2) visual–spatial exploration (Kerkhoff, 1994; Zangemeister et al., 1995; Zihl, 1995a); and (3) visual–spatial judgment (Kerkhoff, 1993; Barton, 1998). Some 50 to 90% of patients with VFD complain about reading difficulties, and 17 to 70% reported visual exploration difficulties.

23.2 Natural Plasticity of the Visual System

23.2.1 Plasticity in Early Development

One of the most influential forefathers of neuroscience, whose pioneering work perhaps provided the greatest impetus for an exponential explosion of subsequent studies, put forth the idea of plasticity sometime during the early part of the 20th century. Santiago Ramon y Cajal stated, "The morphology of nerve cells does not correspond to an unchanging and inevitable pattern laid down by heredity. It depends entirely on environmental, physical and chemical circumstances." (See overview by Whitaker-Azmitia, 2002.) Konorski (1948) went on to promote the plasticity hypothesis by indicating that "…certain permanent functional transformations arise in particular systems of neurons as a result of appropriate stimuli or their combination, we shall call *plasticity* and the corresponding changes *plastic changes*."

Although plasticity can occur throughout one's lifetime, it has been widely accepted that the developing cortex is more plastic than the mature brain. Animals with early damage to the visual cortex demonstrate less serious deficits than animals with adult lesions (Spear, 1995; Payne et al., 1996). Spear (1996) pointed out that neonatal brain possesses ample pathways that would disappear during development but would otherwise remain following neonatal damage. Moreover, the young brain can develop new projections or physiological properties that the adult brain cannot. The author further indicated that neurotrophins are readily upregulated after early brain trauma, which facilitates growth of new projections to a target or prevents the retraction of current ones. Why recovery from brain damage seems to be more promising in young animals than in adults is the enigma that led to the microscopic scrutiny of changes that occur during early vs. late development.

Previous studies of axonal growth in the retina have revealed two distinct stages that would transpire during development (Jhaveri et al., 1991) as represented by distinct growth characteristics (Schneider et al., 1985, 1987; Jhaveri et al., 1990): (1) the elongation mode and (2) the arborization mode. These two modes are defined by differences in rates of extension, in fasciculation or defasciculation of the growing axons, and by dramatic alterations in collateral formation. Regeneration of axons or death of cells would depend on which stage the cell would be at the time of injury. In hamsters, for example, RGCs are still able to regenerate damaged axons if transected before P3, prior to focalization of terminal arbors (So et al., 1981; Wikler et al., 1986; Carman et al., 1988; Carman, 1989). Transection at later times would result in cell death (Perry and Cowey, 1982; Linden et al., 1983) when the elaboration of arbors has proceeded and trophic dependence of ganglion cells upon their targets have escalated (Armson et al., 1987). In adult axons, the elongation growth can be readily prompted, while the arborization growth mode demands intricate interactions with target structures to be reinduced (Jhaveri et al., 1991). Arborization potential has been suggested to depend on ability of cells to synthesize proteins necessary for growth (Moya et al., 1988). Growth associated proteins (GAPs), implicated in axonal growth, are differentially expressed time-wise such that some show changes in expression in animals with early lesions (Moya et al., 1986) although one, GAP-43, is expressed throughout adulthood (Moya et al., 1990). Still another influential factor is the glial environment that, for example, guides developing retinal axons towards their targets (Silver and Robb, 1979; Silver and Sidman, 1980). Glial responses of mature oligodendrocytes to axonal injury are nonpermissive for axonal growth (Ard et al., 1988; Schwab and Caroni, 1988). Nevertheless, the idea of CNS regeneration in the adult brain should not be completely dismissed. Studies have demonstrated that elongation can be obtained even in an adult brain; however, arborization is an unrelenting problem that requires further investigation (Jhaveri et al., 1991). It is interesting to note that the extent of arborization, at least in the developing brain, has some upper limit but also that neurons attempt to maintain a specific number of connections (i.e., when pruned, axons sprout in corresponding amounts elsewhere) (Sabel and Schneider, 1988).

As discussed, the visual system undergoes remarkable changes during early development, but this process of "plasticity" is not over after some critical period or after a certain age. Plasticity, that is, the ability of the brain to adapt to change, is the lifelong ability of the brain and perceptual learning, which can take place throughout adulthood, is an example of how brain plasticity is part of the normal brain's repertoire to adapt to change throughout life.

23.2.2 Perceptual Learning after Development Is Complete

Goldstone (1998) has defined perceptual learning as involving relatively long-lasting changes to an organism's perceptual system that improve its ability to respond to its environment and are caused by this environment (Goldstone, 1998). The author further clarified that changes not due to environmental inputs derive from maturation rather than learning. Perceptual learning entails the strengthening and weakening of associations (Hall, 1991) that may also sometimes be disadvantageous in fulfilling certain perceptual tasks (Samuel, 1981).

During development, the brain goes through remodeling of neural circuitry marked by growth and loss of axons, dendrites, synapses, and neurons (Neville and Bavelier, 2000). Environmental influences have the greatest impact on cortical organization during this remodeling period as sensory inputs determine which axons, dendrites, synapses, and neurons are retained that shape purposeful neural circuits (Rakic, 1976; Hubel and Wiesel, 1977; Sur, Pallas, and Roe, 1990). For example, closing an eyelid throughout a critical period will result in the deterioration of ocular dominance columns that correspond to the visually deprived eye. Otherwise, visual deprivation after the critical period is over has hardly any effect on the pattern of ocular dominance (Hubel and Wiesel, 1977; Blakemore, Garey, and Vital-Durand, 1978; Horton and Hocking, 1997). The time period during which experience can influence specific neural systems subserving respective behavioral capabilities varies depending on the rate of maturation of these systems. For example, the retinal-based visual processing has a short sensitive phase while the relatively late development of cortical areas allows more time for it to be modified by experience. Sensory experience provides input that increases neuronal activity, resulting in augmentation of dendritic branching and spine density. While there seems to be a critical period wherein deprivation of experience can affect cortical organization, it has been demonstrated, however, that experience-induced plasticity can be observed even in older animals (Altman and Das, 1964; Juraska et al., 1980; Green et al., 1983; Katz and Davies, 1984; Greenough et al., 1986; Jones et al., 1997). Experience may provide not only the basis of learning, but also a basis for recovery observed postlesion, regardless of age.

23.2.3 Spontaneous Recovery after Lesions

Traditionally, it has been held that the cortical organization of the adult brain is fixed and that the effects of lesion damage are irreparable. This implied that any injury the brain sustains during the course of a lifetime would translate to a given functional deficit that is irreversible. Damage to the visual area, for example, would mean permanent visual loss. In relatively more recent years, however, this view has been challenged by the demonstration of plasticity in the adult cortex (Chino, 1997; Gilbert, 1998; Kaas, 1994).

Approximately a few weeks to three months after visual cortical damage, spontaneous functional recovery may occur, although to a limited degree (Zihl and von Cramon, 1985; Gray et al., 1989; Tiel-Wilck and Koelmel, 1991; Zhang et al., 2006). Poppelreuter (1917) had reported spontaneous recovery of visual functions found in soldiers with gunshot lesion. Subsequent studies have found similar spontaneous visual improvement (Trobe and Miekle, 1973; Bogousslavsky et al., 1983; Hier et al., 1983; Koelmel, 1984, 1988; Zihl and von Cramon, 1985, 1986; Messing and Gaenshirt, 1987; Tiel-Wilck, 1991). This improvement has been explained as a result of recovery from the swelling around the area of injury, or of functional revival of partially disrupted neural circuitry. The amount of improvement can range from 7% to as much as 85%. Such inconsistency may be reflected in differences in testing method and criteria of what constituted visual recovery. Interestingly, a common finding was that patients with relatively large transition zones would reliably show a more significant amount of spontaneous recovery.

The question arises as to the neurobiological mechanisms of recovery of vision following lesions. In this context, the reorganization of receptive fields is relevant.

23.2.4 Plasticity of Receptive Field Reorganization

Hartline (1938) defined the receptive field as the retinal region that must be illuminated to obtain a response in any given fiber. Although he was referring to the axon of a retinal neuron, it should be noted that all the cells in the visual system — photoreceptors, bipolar cells, the ganglion cells, geniculate and cortical neurons — have receptive fields. The definition of receptive field was modified later on to also consider the specific properties of the stimulus that visual neurons preferentially respond to, such as color or the direction of motion. Much of the credit for this work belongs to Hubel and Wiesel, who were awarded the Nobel Prize for their outstanding discoveries of how the visual system processes information in the brain (see Hubel and Wiesel, 2005, for a collection of twenty-five years of their scientific papers, but also Zigmond et al., 1999). It should be further noted that visual images falling in the surround region of cortical neurons exert a modulatory influence on the response of cortical neurons to stimuli that fall in their receptive field, despite their lack of response to stimulation of the surround region alone (Blakemore and Tobin, 1972; Maffei and Fiorentini, 1976; Nelson and Frost, 1978; Allman et al., 1985; Gilbert and Wiesel, 1990; DeAngelis et al., 1994; Li and Li, 1994; Sillito et al., 1995; Levitt and Lund, 1997; Walker et al., 1999). When considering the response of a given V1 neuron, it should be kept in mind that the size of its receptive field and modulatory surround field is determined by thalamic, lateral intraareal, and feedback interareal inputs (Angelucci et al., 2002).

Lesions anywhere in the visual system, be it retinal or cortical, lead to receptive field reorganization. Eysel et al. (1999) indicated that these two types of lesions generate different reorganization tasks and thus lead to functionally distinct outcomes. For example in retinal lesions, cortical cells are deprived of afferent input from the respective retinal area. Reorganization occurs in such a way that there is redundant representation of information. Information from intact afferents is represented on their corresponding cortical cells as well as on the adjacent deafferented cells. In cortical lesions, the afferent cells are deprived of cortical targets and thus the input is relayed to the neighboring intact cells of the damaged target. Nevertheless, both lesion types share a common early lesion-induced change that seems adaptive for functional reorganization. The authors have explained that the initial effect after damage is diminished activity, either close to the border of the cortical lesion or in the deafferented cortical region. An increase in glutamate levels (Arckens et al., 1997) as well as a decrease in GABA (Rosier et al., 1995) facilitates synaptic reorganization of these cells. The increased excitation and reduced inhibition catalyzes of the transformation of the preexistent subthreshold synapses into suprathreshold synapses. Finally, the

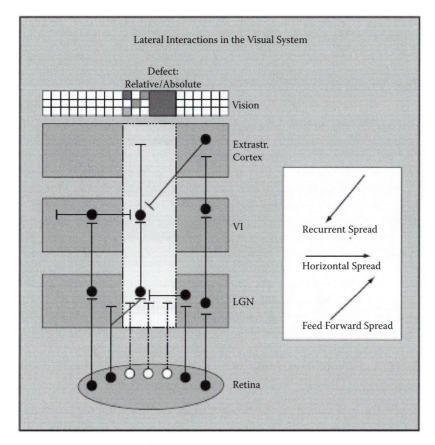

FIGURE 23.1 At various places in the visual system (retina, lateral geniculate nucleus of the thalamus [LGN], and visual cortex, V1), visual information can spread laterally. The lateral interactions are part of normal brain function, but following lesions, they contribute to the known receptive field plasticity found in animals. The lateral spread of information can be a feed-forward, horizontal, or recurrent (feedback) spread. Receptive field plasticity emphasizes that the visual system is not organized in a strict point-to-point (or retinotopic) fashion, but rather that there are numerous paths along which information can travel from the retina to visual cortex. If parts of visual cortex are deafferented, these silenced regions (gray areas) may become activated again through lateral interactions via the neuroplasticity process [11]. LGN = lateral geniculate nucleus; V1 = visual cortex (From Sabel and Kasten, 2000).

modified synaptic connections can be stabilized by axonal sprouting that accompanies functional reorganization (Darian-Smith and Gilbert, 1994; Obata et al., 1999).

In addition to axonal sprouting, receptive field reorganization has been attributed to lateral influences such as long-range intracortical horizontal connections (Pettet and Gilbert, 1992; Darian-Smith and Gilbert, 1995; Chen et al., 2005; Kennerley et al., 2005). These connections induce activation in neurons within their target zone in the deafferented area. These horizontal connections may extend up to 3 or 4 mm, allowing them to connect to distant cells that respond to the same specific stimulus characteristic, such as vertical orientation. Additionally, other lateral influences include feed-forward spread and recurrent spread. (See Figure 23.1.)

23.3 Technological Applications to Induce Vision Restoration

23.3.1 Definition

This section focuses on a discussion of vision restoration therapy (VRT), a rehabilitative treatment that aims to improve a patients' vision field deficit by stimulating areas of residual vision typically located between areas of the seeing visual field and the damaged (blind) field. Its development was based on prior work by many other laboratories, which led to the foundation upon which vision restoration could be developed as a concept.

23.3.2 How Vision Restoration Therapy (VRT) Works

VRT runs on personal computers so that visual field training can conveniently be performed at home without much assistance from medical staff and thus a large number of training sessions can be achieved. Briefly, the subject's head is positioned in front of a computer monitor with a chin-rest. To be sure that the subject does not make excessive eye movements, the subject is asked to fixate on a fixation stimulus located in the middle of the monitor, which occasionally changes its color for a short time in irregular time intervals. The subject must respond to each color change by pressing the space bar. The number of such color changes is recorded as a measure of fixation quality. Furthermore, the subject is asked to also respond to additional white "target" stimuli presented at random locations on a dark gray background anywhere in the visual field. The subject is instructed to respond to these and to the color changes of the fixation point as fast as possible, but also as exactly as possible. The VRT program records both the correct detections of the target stimuli and the fixation spot color changes, as well as the reaction time. If the subject presses the space bar even if there is no stimulus or outside a permissible time window, a false hit is registered. For a time period of six months or more, the subject performs two training sessions, half an hour each, on a daily schedule. Treatment results are stored so that changes in visual field size can be recorded. The training area is adapted monthly based on the subject's progress. (See Figure 23.2.)

23.3.3 Historical Considerations

23.3.3.1 Early Studies

There have been controversies about the viability of successfully rehabilitating visual field loss. In brain-damaged persons, the treatment of visual deficits from cerebral injury is traditionally not regarded as possible. In contrast to this view, several studies indicate that restitution may indeed be possible. Poeppel et al. (1973, 1987), for example, pointed out that the human visual system has some capacities for plasticity after damage. The first to show that training may improve visual functions were Zihl and von Cramon (1985), who found a significant increase of visual field size in patients with postchiasmatic damage of the visual system when luminance thresholds were measured systematically at the same position of visual field border. Their results were criticized, however, because alternative explanations of the effect, such as methodological artifacts, could not be ruled out at that time (e.g., spontaneous recovery or a change in the patient's fixation) (Bach-y-Rita, 1990; Balliett, Blood, and Bach-y-Rita, 1985). But also, Kerkhoff et al. (1994) found not only an improvement in the visual gaze by 30°, but also an expansion of the visual field by 6.7°.

23.3.3.2 Theoretical Basis

Before discussing vision restoration in more detail, first consider which mechanisms of action might be involved when vision restoration takes place.

Receptive fields in the primary visual cortex, the fundamental building blocks of visual processing, are the most fundamental physiological signs of plasticity. Especially after retinal lesions, the receptive fields in the visual center rearrange to establish new functional areas that have not previously received information from the damaged site (Kaas et al., 1990; Gilbert and Wiesel, 1992; Daas, 1997). This mechanism of receptive field plasticity might be involved in vision restoration, particularly with regard to the therapy success in patients with lesions of the optic nerve (Wuest, 1997). Plasticity may thus result from central "rewiring" of neuronal information processing, but this is probably not the only mechanism of vision restoration.

Attention also plays an important role in the activation of the surviving neurons. It is well known that cells in the primary visual pathway change their spatial and temporal behavior, depending on the EEG

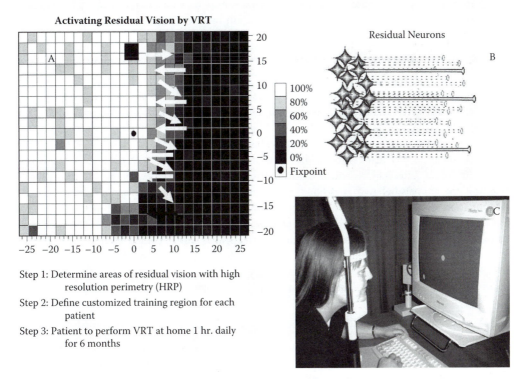

FIGURE 23.2 Before therapy starts, visual field charts (A) are created that reveal blind regions (black) and seeing regions (white). At the visual field border, areas can be found (gray) in which patients sometimes respond to stimuli but sometimes do not respond. These areas of partial function are also called "areas of residual vision." Their neurobiological substrate is presumably surviving neurons in primary visual cortex that have managed to survive the injury (B). Vision restoration is achieved by having the patient train with visual stimuli as displayed according to a three-step process. (C) Patients carry out daily training using a computer monitor that presents visual stimuli primarily in these areas of residual vision (shown by arrows in A).

synchronization state, which is a function of alertness (Wörgötter et al., 1998). Receptive field size is larger during a more "drowsy" EEG state but smaller (more defined) under conditions of general brain activation (desynchronization). Indeed, attention increases the activation of specific neuronal areas as shown electrophysiologically or by magnetic resonance imaging (Heinze, 1994). It is possible that the activation of cortical neurons of the visual system can be raised when attention is increased, particularly under conditions where attention is also directed to areas of residual vision. We presume that the activating effects of orientating the attention and the visual stimulation add up so that the perception is increased.

Restoration of vision may be mediated by activating all elements of the damaged visual system, including those visual pathways that survive the injury. Zihl and Cramon (1986), Ungerleider and Haxby (1994), and Sincich and Horton (2003) report the existence of extrastriate pathways bypassing V1 that presumably remain undamaged after cortical injury. In this manner, visual information can be processed without involving the primary visual cortex. Experiments with monkeys have provided evidence for the existence of two pathways in the visual extrastriate cortex. The extrastriate pathways run over tectum and pulvinar, bypassing V1, and transfer information directly to higher extrastriate cortex areas. Both anatomical (Cowey and Stoerig, 1989; Standage and Benevento, 1983; Benevento and Yoshida, 1981) and physiological (Girard and Bullier, 1989; Rodman et al., 1989; Girard et al., 1991) studies have shown these projections in animals, and they directly innervate higher cortical areas (e.g., area MT that is responsible for perception of motion; Zeki et al, 1991; Tootell et al., 1995; Schoenfeld et al., 2002) without activating V1. To summarize, three fundamental mechanisms may be involved in visual system plasticity: (1) receptive field reorganization, (2) neuronal activation by attention, and (3) activation of residual visual pathways.

23.3.3.3 Animal Studies

The question arises as to how the brain is able to handle visual functions after a lesion or after deprivation. Already in the 1960s, Hubel and Wiesel (1962, 1963, 1965, 1970; Wiesel and Hubel, 1963, 1965) had shown the impact of visual deprivation on the normal development of the visual system. But one of the most important experiments was carried out by Chow and Stewart (1972), who investigated if it was possible to improve visual functions in animals with early deprivation. In five newborn cats, one eyelid was closed for up to two years. After reopening this eye, the animals did not seem to use it and only relied on the other (intact) eye. Then, Chow and Stewart closed the intact eye in three of the cats. In the beginning, the animals behaved as if blind; but after a little while, they were able to solve a pattern discrimination task again; that is, the eye deprived of input recovered some of its functions.

In several perimetric studies of visual field defects in monkeys following removal of various portions of striate cortex, it was reported that although the defects were almost complete, the animals were still able to detect visual stimuli within the "blind" areas of visual field (Cowey and Weiskrantz, 1963; Cowey, 1967; Anderson and Symmes, 1969; Wurtz and Mohler, 1976; Mohler and Wurtz, 1977). This points toward the existence of alternate pathways that are able to sustain some visual functions.

23.3.4 The Role of Partially Damaged Brain Regions (Areas of Residual Vision)

Areas of residual vision are sectors of the visual field that do not function properly but in which some visual capacities have survived the injury. In these regions of the visual field, patients are able to detect visual stimuli albeit unreliably. For example, if five stimuli are presented in a given region of the visual field, the patient only can respond one time or two times. These areas are not absolutely blind and are long known as areas of "relative defect." However, just as a glass of water can be viewed as half-full or half-empty, the term "relative defect" emphasizes the deficit, whereas the term "residual vision" stresses the residual capacities that provide a structural substrate for restoration (Kasten et al., 1998a,b, 1999). The most important determinant of how much residual vision remains depends on the relative number of fibers surviving the damage. If only very few fibers survive, blindsight may occur, whereas a larger degree of neuronal sparing may allow conscious, although unreliable, responses (Sabel and Kasten, 2000). Furthermore, ARVs serve as a major predictor of outcome after VRT. The size of ARV prior to training is correlated with the increase in visual field borders after VRT training (see below).

23.4 Mechanisms of Action

23.4.1 VRT and the Role of Eye Movements in VRT

23.4.1.1 The Role of Eye Movements in VRT

One might suspect that any vision improvement (visual field border shift) is an artifact of eye movements. By moving the eyes every time stimuli are presented, the patient might be able to "cheat" and thus detect stimuli by moving the seeing field toward the stimulus. To study this possibility, it is necessary to track the eyes of the subjects during perimetric assessments. The simplest method to track eye movements is direct observation of the subject by an experimenter. However, this method does not allow eye movements of less than 1° to be noticed and, of course, it is not a quantitative method.

A better method to register eye movements is to use a video camera or a so-called "eye tracker" system (see below). In our studies on VRT, we use both — the direct observation method and the eye tracker system — to control for eye movements. Kasten et al. (2006) recorded eye movements of patients with hemianopia before and after VRT. They found that saccades were directed equally to the right or the left side; that is, with no preference toward the blind hemifield. Furthermore, many patients showed a smaller variability of horizontal eye movements after VRT. These results argue against the theory that the visual field enlargements are artifacts induced by eye movements. Furthermore, the patients were asked to fixate on a point in the middle of the monitor and to press a button if this point changed colors. We reasoned that detecting color changes would be an indirect measure of fixation performance. In our clinical studies, only patients who are able to perceive at least 90% of all fixation point color changes are included — to avoid possible confounding by eye movements — and after VRT, the performance in detecting such color changes of the fixation spot was unchanged (e.g., Sabel et al., 2004). Thus, the evidence indicates that VRT-induced visual field improvements are not caused by altered eye movements of fixation behavior.

23.4.1.2 Blind Spot Analysis

Trauzettel-Klosinski and Reinhard (1998) stated that the lack of a shift in blind spot position is a good indicator that fixation is not eccentric and remains stable after the lesion. Sabel et al. (2004) analyzed the blind spot position in sixteen patients and found that the position of the blind spot remained identical in twelve of the sixteen patients. Only four patients showed a small shift of the blind spot. In two patients the blind spot was located more temporally, and the other two patients showed a slight blind spot shift nasally. In fact, these two latter patients did not belong to the category of patients who improved their visual field defect. In summary, the blind spot analysis confirms once again that visual field enlargements are not explainable by altered fixation behavior, that is, by cheating on the part of the patients.

23.4.1.3 Eye Tracker Recordings

Kasten et al. (2006) were the first to examine the size of eye movements before and after a visual training. The aim was to study if eye movements increase after training and if eye movements can explain the visual field enlargements. For this purpose, a two-dimensional eye tracking system (Chronos Vision, Germany) with a measurement range between –40° and +40° and a resolution of <0.05° was used. Fifteen patients were examined before and after a three-month VRT session. Before the training, Kasten found eye position variability (SD) of 0.93° horizontally and 1.16° vertically. After the training, an SD of 0.71° horizontally and 1.39° vertically was measured. Most of the patients even showed smaller eye movements after training. Therefore, eye movements cannot explain the visual field enlargements after training.

23.4.2   Cognitive Mechanisms

23.4.2.1 Effects of Acute Attention

Single cell recordings (Gilbert, 1998; Ito & Gilbert, 1999), electrophysiological (Mangun and Hillyard, 1987) and brain imaging studies (Somers et al., 1999; Martinez et al., 1999) have found that attention increases neuronal activation which then facilitates performance in visual processing tasks. Faster reaction times are observed as stimuli become more easily detected and discriminated more efficiently with focused attention than when attention is diffused across the entire visual field (Eriksen and Rohrbaugh, 1970; Posner, 1980; Treisman and Gelade, 1980; Nakayama and Makeben, 1989). Attention enhances (Corbetta et al., 1990; Hillyard and Anllo-Vento, 1998; Treue and Martinez Trujillo, 1999; Kastner et al., 1999) and suppresses (Moran and Desimone, 1985; Chelazzi et al., 1993; Luck et al., 1997; Smith et al., 2000; Vanduffel et al., 2000; Serences et al., 2004) neural activity so that spatially attending to a location in the visual field would facilitate preferential processing of selected information and filtering out unwanted information (Pinsk et al., 2004). This creates an attentional spotlight (Brefczynski and DeYoe, 1999) that is neurally represented by a receptive field that shrinks around the attended stimulus, reducing the impact of unwanted stimuli (Connor et al., 1996; Moran and Desimone, 1985; Reynolds and Desimone, 1999).

23.4.2.2 Training with Attention Cues

Poggel et al. (2001) used visuospatial cues to direct patients to attend at the visual border. This led not only to an acute enhancement of perceptual performance in areas of residual vision, but long-term training with such a cueing paradigm also enhanced vision restoration beyond what standard VRT achieved (Poggel et al., 2004). This was accomplished by the following experimental setup. A large, dim gray cue frame enclosing a predetermined area (that included an intact, ARV and blind sections) was presented before the training stimulus appeared. The attentional cue was presented only in the upper visual field so that later within subject comparison of possible visual border improvement could be made using the patient's lower visual field. A prominent expansion of the visual field border was observed in the area where the cue was presented, which was significantly greater than the visual field expansion in the noncued lower visual field. An additional interesting observation is that the improvement in performance of patients given attentional cueing did not depend on the size of the areas of residual vision (ARV) assessed at the start of the study. On the other hand, the improvement in performance of the group that did not receive attentional cueing depended on the initial size of the ARV prior to training.

23.4.3 Factors Affecting Efficacy

As repeatedly shown, training outcome does not depend on the age of the patient (Mueller, 2002, 2003). Older patients benefit from VRT as much as younger patients do. Furthermore, gender does not play a role either in visual field enlargements, and nor does the type or the age of the lesion. Only response times were significantly correlated with recovery of vision and might indicate a possible plasticity potential in patients using VRT (Mueller, 2002). Patients with longer response times in the visual field test experience a more pronounced increase in intact visual field than patients with faster reaction to suprathreshold stimulation. Increased reaction times at pretraining baseline could be a sign of increased visual thresholds, which are generally found in areas of residual vision; that is, in patients with large transition zones, VRT might be especially helpful (Poggel et al., 2001). Additionally, pre-post response time improvements in computer perimetry were significantly correlated with training success.

23.5 Efficacy Studies

As discussed above, evidence for training-induced recovery from visual field impairments in brain-damaged patients has been reported in several studies (e.g., Zihl et al., 1985; Kerkhoff et al., 1994; Kasten and Sabel, 1995; Kasten et al., 1998; Julkunen et al., 2003). Visual-field enlargement due to VRT on a computer monitor has been shown for the first time in the laboratory of the Institute of Medical Psychology in Magdeburg. Meanwhile, VRT has been used for many years in brain-damaged patients achieving both visual field enlargements (Kasten et al., 1995; Kasten and Sabel, 1998; Sabel et al., 2004) and reaction time gains (Mueller et al., 2003; Sabel et al., 2004), and improvements in subjective vision (Mueller et al., 2003; Sabel et al., 2004). (See Figure 23.3.)

23.5.1 Effects of VRT on Stimulus Detection (Superthreshold, Near-Threshold, SLO)

In the early 1990s, an open pilot trial was conducted in which, after six months of VRT, the stimulus detection performance of nine out of eleven trained patients significantly increased, as measured by super-threshold perimetry (Kasten and Sabel, 1995). After these preliminary findings, two prospective randomized placebo-controlled trials were carried out to study whether it is possible to shrink visual field defects by systematic stimulation of partially defected areas (Kasten et al., 1998). A total of thirty-eight patients — nineteen with postchiasmatic lesions and nineteen suffering from optic nerve damage — were trained for a minimum of 150 hours over a period of about six months, receiving either VRT (experimental condition) or fixation training (placebo condition). Before and after training, the visual field was tested with both super- and near-threshold perimetry. All patients had lesions older than one year. The experimental group from patients with postchiasmatic lesions showed an average detection rate improvement of 7.8%, whereas the placebo-group experienced an increase with an average of 3.1%. These detection performance changes correspond to an enlargement of the mean visual field size of

Visual Fields of Patient M.S. Before vs. After VRT.

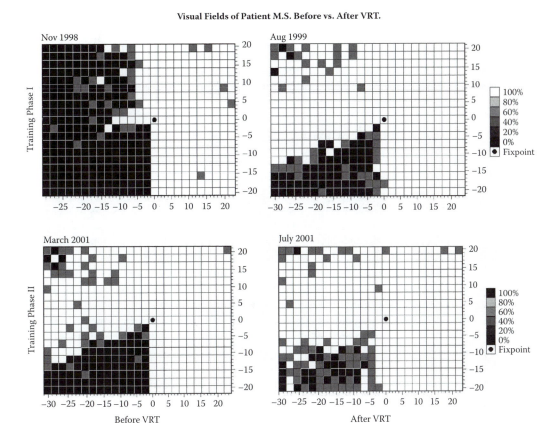

FIGURE 23.3 Visual field charts obtained by high-resolution perimetry (HRP) in patient M.S. (male; thirty-six years old) who had suffered an infarction of the right posterior cerebral artery following surgery for an acoustic neurinoma; Age of lesion: two years; Patient started vision restoration treatment in November 1998 and discontinued in August 1999. Note his remarkable expansion in the visual field. This case is an atypical case with an above-average restoration response. Note that the visual field expansion occurs in regions adjacent to areas of residual visual and extends to more than 30° of visual angle. M.S. continued the therapy from March 2001 through July 2001. The chart in August 1999 was roughly unchanged in March 2001, indicating the stability of the restoration despite lack of the training during that period. In a second training phase, the patient experienced a further improvement in his visual fields. His testimonials revealed that after restoration, he bumped into people less frequently, was able to read and concentrate better, he could even read in a moving car again, and he could freely climb stairs without holding handrails once again.

about 5° of visual angle in the experimental group and 0.9° of visual angle in the placebo group, respectively. In patients with optic nerve lesions, the visual field size increased by 21.9% (6° of visual angle) if patients trained with VRT and 6.0% if patients underwent the placebo condition. In a later study, using different functional perimetric tests, the super- and near-threshold perimetry, and the scanning laser ophthalmoscope (SLO), the efficacy of VRT has been repeatedly examined (Sabel et al., 2004). The results were replicated that VRT leads to significant improvement in the ability to detect visual stimuli when tested by supra- and near-threshold perimetry, but visual field recovery was not evident when measured by SLO. The absence of visual field increases as measured by the SLO task led Sabel and colleagues to conclude that SLO might be measuring a different function, perhaps a more difficult psychophysical task than what the perimetric methods reveal. Regarding the stability of training-induced visual field recovery, the training effects were basically maintained after training was discontinued (Kasten et al., 2001). It should be kept in mind, however, that not all patients benefit from VRT. About one third of the patients do not show stimulus detection improvements (Sabel et al., 2004).

23.5.2   Effects of VRT on Reaction Time

In a retrospective clinical trial on patients with visual field impairments, lengthened reaction times were evident in the partially damaged brain regions compared to intact visual field areas (Mueller et al., 2003). That is, the faster the reaction time, the higher the detection rate. Mueller et al. (2003) and Sabel et al. (2004) quantified visual field improvement analyzing reaction time performance and found reaction time gains in superthreshold perimetric examinations as a sign of recovery.

23.5.3   VRT and Activities of Daily Life

Experiencing severe difficulties with activities of daily life is a common problem in persons with a visual field defect. A change in visual function is accompanied by deficits in visual exploration (e.g., finding objects, recognizing persons), the ability to read, and the ability to navigate and to stay oriented within the environment (Kerkhoff, 1993). Vision loss makes it difficult to do a job or to drive a vehicle (Kasten et al., 1997; Ullrich et al., 2003). Thus, visual impairment may result in the loss of an independent lifestyle, a decreased quality of life, and difficulty in maintaining employment. Hence, improvements in activities of daily life are an important goal in the rehabilitation of visually impaired persons. As previously described, patients with visual field loss due to stroke or head injury can benefit from VRT and increase their visual field size (Kasten and Sabel, 1995; Kasten et al., 1998; Sabel et al., 2004). The subjective relevance and profit of visual field improvement have been challenged by different groups (Kerkhoff et al., 2000; Kommerell, 2000). To evaluate the effectiveness of VRT concerning performance in activities of daily living, the most commonly perceived disabilities and their changes after performing VRT have been identified in several studies (Kasten et al., 1998; Mueller et al., 2003; Sabel et al., 2004).

The first indication of positive VRT effects on activities of daily living in visually impaired persons was found by Kasten and colleagues (1998). In this study, patients were interviewed about the effects of VRT in their daily life activities. Some 72% of patients who performed VRT, but just 16% of patients from the placebo-group reported subjective improvements in visual functioning. In a further study, data about subjective patient testimonials were collected after training, using a standardized semistructured interview (Mueller et al., 2003). Additionally, 88% of all patients reported positive effects of VRT in at least one of the following categories: general visual improvement, visual confidence and mobility, reading, collisions with objects or people, and hobbies. The greater their visual field increase due to VRT, the more categories in terms of everyday activity performance patients indicated to have improved. To provide a clearer understanding of the relevance of VRT-induced visual field enlargements to activities of daily living, a correlation analysis of quantitative parameters of visual field improvements with subjective changes of visual functions was performed. Visual field enlargements, as shown in this study, correlated with general improvement in vision, improved ability to carry out hobbies and to read, and did not correlate with visually guided mobility and ability to avoid collisions, respectively. The authors thus concluded that the amount of improvement in perimetric tests seems to be just one factor influencing subjective improvement of visually guided activities of daily life after VRT. With an aim to determine the level of subjective vision before and after VRT, a pre-post analysis of data collected by a standardized questionnaire was performed (Sabel et al., 2004). Furthermore, patients were interviewed posttreatment. Both the questionnaire and interview results revealed improvements in activities of daily living, including improved visual confidence and mobility, better reading, and improved avoidance of collisions with people or objects. In all of the cited studies, the findings indicate that VRT can help patients improve their subjective vision. However, the outcome of training should be evaluated using pre- and post-intervention daily life activity measurements in other studies.

23.5.4   VRT Effects in Performance Tasks (Reading, ZVT-Test, Attention Tests, etc.)

To find out whether VRT effects are transferred to neuropsychological functions, visual attention and visual exploration performance was assessed by paper-and-pencil tests pre- and posttreatment (Kasten et al., 1998, 1999; Wuest, 2004). In patients with optic nerve lesions, the visual exploration time in the ZVT (Oswald et al., 1987) decreased significantly, on average 27 s. There was also a trend of improved performance in the d2 task (Brickenkamp, 1994), which tests the selective visual attention involving visual scanning of stimulus details. In contrast, the postchiasmatic patients showed significant improvement in the d2 test and nonsignificant decreased visual exploration time in ZVT. Therefore, VRT improves some functions such as attention and visual exploration, suggesting that patients may be able to cope better with the demands of their visual environment after carrying out VRT. However, larger patient groups are needed to confirm these findings.

23.6   Conclusions

Neuroplasticity provides the basis for how the damaged visual system can compensate for damage in the visual system. By regularly activating residual vision through training, vision functions can be partially restored. About a third of the patients do not show any improvement at all. Of the remaining cases, approximately half show noticeable changes and the other half show significant changes, mostly noticeable by visual field expansions. Thus, vision restoration may be accomplished by repetitive activation of the brain through computer-based training paradigms. This technology is a first step toward improving the quality of life of visually impaired patients, and we are curious to see how the future might bring us even more effective technologies to enhance the brain's restoration potential in the domain of visual perception.

References

Allman, J., Miezin, F., and McGuinness, E. (1985). Stimulus specific responses from beyond the classical receptive field: neurophysiological mechanisms for local-global comparisons in visual neurons. *Annu. Rev. Neurosci.*, 8:407–430.

Altman, J. and Das, G.D. (1964). Autoradiographic examination of the effects of enriched environment on the rate of glial multiplication in the adult rat brain. *Nature*, 204:1161–1163.

Anderson, K.V. and Symmes, D. (1969). The superior collicullus and higher visual functions in the monkey. *Brain Res.*, 13:37–52.

Angelucci, A., Levitt, J.B., and Lund, J.S. (2002). Anatomical origins of the classical receptive field and modulatory surround field of single neurons in macaque visual cortical area V1. *Prog. Brain Res.*, 136:373–388.

Arckens, L., Qu, Y., Wouters, G., Pow, D., Eysel, U. T., Orban, G.A., Vandesande, F. (1997). Changes in glutamate immunoreactivity during retinotopc reorganization of cat striate cortex. *Soc. Neurosci. Abstr.* 23:2362.

Ard, M.D., Wood, P., Schachner, M., and Bunge, R.P. (1988). Retinal neurite growth on astrocytes and oligodendrocytes in culture. *Soc. Neurosci.*, 14:748.

Armson, P.F., Bennett, M.R., and Raju, T.R. (1987). Retinal ganglion cell survival and neurite regeneration requirements: the change from Muller cell dependence to superior colliculi dependence during development. *Dev. Brain Res.*, 32:207–216.

Bach-y-Rita, P. (1990). Brain plasticity as a basis for recovery of function in humans. *Neuropsychologia*, 28:547–554.

Balliett, R., Blood, K.M., and Bach-y-Rita, P. (1985). Visual field rehabilitation in the cortically blind? *J. Neurol. Neurosurg. Psychiatry*, 48:1113–1124.

Barton, J.J. (1998). Higher cortical visual function. *Curr. Opin. Opthalmol.*, 9:40–45.

Benevento, L.A. and Yoshida, K. (1981). The afferent and efferent organization of the lateral geniculo-prestriate pathways in the macaque monkey, *J. Comp. Neurol.*, 203(3):455–474.

Blakemore, C. and Tobin, E.A. (1972). Lateral inhibition between orientation detectors in the cat's visual cortex. *Exp. Brain Res.*, 15:439–440.

Blakemore, C., Garey, L.J., and Vital-Durand, F. (1978). The physiological effects of monocular deprivation and their reversal in the monkey's visual cortex. *Physiology*, 283:223–262.

Bogousslavsky, J., Regli, F., and van Melle, G. (1983). Unilateral occipital infarction: evaluation of the risks of developing bilateral loss of vision. *J. Neurol. Neurosurg. Psychiat.*, 46, 78–80.

Brefczynski, J.A. and DeYoe, E.A. (1999). A physiological correlate of the "spotlight" of visual attention. *Nat. Neurosci.*, 2:370–374.

Brickenkamp R. (1994). Test d2 Aufmerksamkeits-Belastungs-Test. 8. überarbeitete Auflage. Göttingen: Hogrefe.

Carman, L.S. (1989). Regenerative Growth of Axons of Hamster Optic Tract: Effects of Age, Substrate, and Growth-Promoting Factors. MIT, Cambridge, MA: Ph.D. thesis.

Carman, L.S., Schneider, G.E., and Yannas, I.V. (1988). Extension of critical age for retinal axon regeneration by polymer bridges conditioned in neonatal cortex. *Abstr. Soc. Neurosci.*, 14: 498.

Celesta, G.G., Brigell, M., and Vaphiades, M.S. (1997). Hemianopic anosognosia. *Neurology*, 49:88–97.

Cervos-Navarro, J. and Lafuente, J.V. (1991). Traumatic brain injuries: structural changes. *J. Neurol. Sci.*, 103:S3–S14.

Chelazzi, L., Miller, E.K., Duncan, J., and Desimone, R. (1993). A neural basis for visual search in inferior temporal cortex. *Nature*, 363:345–347.

Chen, Z., Silva, A.C., Yang, J., and Shen, J. (2005). Elevated endogenous GABA level correlates with decreased fMRI signals in the rat brain during acute inhibition of GABA transaminase. *J. Neurosci. Res.*, 79:383–391.

Chino, Y.M. (1997). Receptive field plasticity in the adult visual cortex: dynamic signal rerouting or experience-dependent plasticity. *Sem. Neurosci.*, 9:34–46.

Chow, K.L. and Stewart, D.L. (1972). Reversal of structural and functional effects of long-term visual deprivation in cats. *Exp. Neurol.*, 34:409.

Connor, C.E., Gallant, J.L., Preddie, D.C., and van Essen, D.C. (1996). Responses in area V4 depend on the spatial relationship between stimulus and attention. *J. Neurophysiol.*, 75:1306–1308.

Corbetta, M., Miezin, F.M., Dobmeyer, S., Schulman, G.L., and Petersen, S.E. (1990). Attentional modulation of neural processing of shape, color, and velocity in humans. *Science*, 248:1556–1559.

Cowey, A. (1967). Perimetric study of visual field defects in monkeys after cortical and retinal ablations. *Q. J. Exp. Psychol.*, 19:232–245.

Cowey, A. and Stoerig, P. (1989). Projection patterns of surviving neurons in the dorsal lateral geniculate nucleus following discrete lesions of striate cortex: implications for residual vision, *Exp. Brain Res.*, 75(3):631–638.

Cowey, A. and Stoerig, P. (1991). The neurobiology of blindsight. *TINS*, 14:140–145.

Cowey, A. and Weiskrantz, L. (1963). A perimetric study of visual field defects in monkeys. *Q. J. Exp. Psychol.*, 15:91–115.

Daas A. (1997). Plasticity in adult sensory cortex: a review. *Network: Comput. Neural Syst.*, 8:,R33–R76.

Darian-Smith, C. and Gilbert, C.D. (1994). Axonal sprouting accompanies functional reorganization in adult cat striate cortex. *Nature*, 368:737–740.

Darian-Smith, C. and Gilbert, C.D. (1995). Topographic reorganization in the striate cortex of the adult cat and monkey is cortically mediated. *J. Neurosci.*, 15:1631–1647.

Eysel, U.T., Schweigart, G., Mittman, T., Eyding, D., Qu, Y., Vandesande, F., Orban, G., and Arckens, L. (1999). Reorganization in the visual cortex after retinal and cortical damage. *Restor. Neurol. Neurosci.*, 15:153–164.

DeAngelis, G., Freeman, R.D., and Ohzawa, I. (1994). Length and width tuning of neurons in the cat's primary visual cortex. *J. Neurophysiol.*, 71:347–374.

Eriksen, C.W. and Rohrbaugh, J.W. (1970). Some factors determining efficiency of selective attention. *Am. J. Psychol.*, 83:330–342.

Gilbert, C.D. (1998). Adult cortical dynamics. *Physiol. Rev.*, 78:467–585.

Gilbert, C.D. and Wiesel T.N. (1990). The influence of contextual stimuli on the orientation selectivity of cells in primary visual cortex of the cat. *Vision Res.*, 30:1689–1701.

Gilbert, C.D. and Wiesel, T.N. (1992). Receptive field dynamics in adult primary visual cortex. *Nature*, 356:150–152.

Girard, P. and Bullier, J. (1989). Visual activity in area V2 during reversible inactivation of area 17 in the macaque monkey, *J. Neurophysiol.*, 62(6):1287–1302.

Girard, P., Salin, P.A., and Bullier, J. (1991). Visual activity in areas V3a and V3 during reversible inactivation of area V1 in the macaque monkey, *J. Neurophysiol.*, 66(5):1493–1503.

Goldstone, R.L. (1998). Perceptual learning. *Annu. Rev. Psychol.*, 49:585–612.

Goodale, M.A. (2000). Perception and action in the human visual system. *The New Cognitive Neurosciences, 2nd edition.* MIT Press, Cambridge, MA.

Goodale, M.A. and Milner, A.D. (1992). Separate visual pathways for perception and action. *Trends Neurosci.*, 15:20–25.

Gray, C.S., French, J.M., Bates, D., Cartilidge, N.E.F., Venables, G.S., and James, O.F.W. (1989). Recovery of visual fields in acute stroke: homonymous hemianopia associated with adverse prognosis. *Age Aging*, 18:419–421.

Green, E.J., Greenough, W.T., and Schlumpf, B.E. (1983). Effects of complex or isolated environments on cortical dendrites of middle-aged rats. *Brain Res.*, 264:233–240.

Greenough, W.T., McDonald, J.W., Parnisari, R.M., and Camel, J.E. (1986). Environmental conditions modulate degeneration and new dendrite growth in cerebellum of senescent rats. *Brain Res.*, 380, 136–143.

Hall, G. (1991). *Perceptual and Associative Learning.* Clarendon, Oxford.

Hartline, H.K. (1938). Response of single optic nerve fibers of the vertebrate eye to illumination of the retina. *Am. J. Physiol.*, 121:400–415.

Heinze, H.J., Mangun, G.R., Burchert, W., Hinrichs, H., Scholz, M., Muente, T.F., Goes, A., Scherg, M., Johannes, S., Hundeshagen, H., Gazzaniga, M.S., and Hillyard, S.A. (1994). Combined spatial and temporal imaging of brain activity during visual selective attention in humans. *Nature*, 372:543–546.

Hier, D.B., Mondlock, J., and Caplan, L.R. (1983). Recovery of behavioural abnormalities after right hemisphere stroke. *Neurology*, 33:345–350.

Hillyard, S.A. and Anllo-Vento, L. (1998). Event-related brain potentials in the study of visual selective attention. *Proc. Natl. Acad. Sci.*, 3:781–787.

Hollwich, F. (1988). *Augenheilkunde.* Thieme Verlag, Stuttgart.

Horton, J.C. and Hocking, D.R. (1997). Timing of the critical period for plasticity of ocular dominance columns in macaque striate cortex. *J. Neurosci.*, 17:3684–3709.

Hubel, D.H. and Wiesel, T.N. (1962). Receptive fields, binocular interaction and functional architecture in the cat's visual cortex. *J. Physiol.*, 160:106–154.

Hubel, D.H. and Wiesel, T.N. (1963). Receptive fields of cells in striate cortex of very young, visually inexperienced kittens. *J. Neurophysiol.*, 26:994–1002.

Hubel, D.H. and Wiesel, T.N. (1965). Binoculare interaction in striate cortex of kittens reared with artificial squint. *J. Neurophysiol.*, 28:1041–1059.

Hubel, D.H. and Wiesel, T.N. (1970). The period of susceptibility to the physiological effects of unilateral eye closure in kittens. *J.Physiol.*, 206:419–436.

Hubel, D.H. and Wiesel, T.N. (1977). Ferrier lecture. Functional architecture of macaque monkey visual cortex. *Proc. R. Soc. Lond. B Biol. Sci.*, 198:1–59.

Hubel, D.H. and Wiesel, T.N. (2005). *Brain and Visual Perception. The Story of a 25-year Collaboration.* Oxford University Press.

Huber, A. (1991). Die homonyme Hemianopsie. *Klin. Monatsbl. Augenheilk.*, 199,396–405.

Ito, M. and Gilbert, C.D. (1999). Attention modulates contextual influences in the primary visual cortex of alert monkeys. *Neuron*, 22:593–604.

Jhaveri, S., Schneider, G.E., and Erzurumlu, R.S. (1991). Axonal plasticity in the context of development. In *Development and Plasticity of the Visual System*, Cronly-Dillon, J.R., Ed. The Macmillan Press Ltd., London, p. 232–256.

Jones, T.A., Klintsova, A.Y., Kilman, V.I., Sirevang, A.M., and Greenough, W.T. (1997). Induction of multiple synapses by experience in the visual cortex of adult rats. *Neurobiol. Learning Memory*, 68:13–20.

Julkunen, L., Tenovuo, O., Jääskeläinen, S., and Hämäläinen, H. (2003). Rehabilitation of chronic post-stroke visual field defect with computer-assisted training. *Restor. Neurol. Neurosci.*, 20:1–10.

Juraska, J.M., Greenough, W.T., Elliott, C., Mack, K.J., and Berkowitz, R. (1980). Plasticity in adult rat visual cortex: an examination of several cell populations after differential rearing. *Behav. Neural Biol.*, 29:157–167.

Kaas, J.H., Krubitzer, L.A., Chino, Y.M., Langston, A.L., Polley, E.H., and Blair, N. (1990). Reorganization of retinotopic cortical maps in adult mammals after lesion of the retina. *Science*, 248:229–231.

Kaas, J.H. (1994). The reorganization of sensory and motor maps in adult animals. In *Molecular and Cellular Plasticity*, p. 51–71.

Kandel, E.R., Schwartz, J.H., and Jessell, T.M. (2000). *Principles of Neural Science, 4th edition.* McGraw-Hill, New York.

Kasten, E. (2005). Saccadic Eye Movements during Perimetry, Controlled by 2D Eye Tracker. Program No. 165.3. Washington, D.C.: Society for Neuroscience, Online.

Kasten, E. and Sabel, B.A. (1995). Visual field enlargement after computer training in brain-damaged patients with homonymous deficits: an open pilot trial. *Restor. Neurol. Neurosci.,.* 8:113–127.

Kasten, E., Strasburger, H., Sabel, B.A. (1997). Programs for diagnosis and therapy of visual field deficits in vision rehabilitation. *Spat. Vis.*, 10:499–503.

Kasten, E., Wuest, S., Behrens-Baumann, W., and Sabel, B,A. (1998a). Computer-based training for the treatment of partial blindness, *Nature Med.*, 4:1083–1087.

Kasten, E., Wuest, S., and Sabel, B.A. (1998b). Residual vision in transition zones in patients with cerebral blindness. *J. Clin. Exp. Neuropsychol.*, 20:581–598.

Kasten, E., Poggel, D.A., Müller-Oehring, E.M., Gothe, J., Schulte, T., and Sabel, B.A. (1999). Restoration of vision II: residual functions and training-induced visual field enlargement in brain-damaged patients. *Resor. Neurol. Neurosci.*, 15:273–287.

Kasten, E., Mueller-Oehring, E., and Sabel, B.A. (2001). Stability of visual field enlargements following computer-based restitution training-results of a follow-up. *J. Clin. Exp. Neuropsychol.*, 00:1–9.

Kasten, E., Bunzenthal, U., and Sabel, B.A. (2006). Visual field recovery after vision restoration therapy (VRT) is independent of eye movements: An eye tracker study. *Behav. Brain Res.*, 175:18–26.

Kastner, S., Pinsk, M.A., De Weerd, P., Desimone, R., and Ungerleider, L.G. (1999). Increased activity in human visual cortex during directed attention in the absence of visual stimulation. *Neuron*, 22:751–761.

Katz, H.B. and Davies, C.A. (1984). Effects of differential environment on the cerebral anatomy of rats as a function of previous and subsequent housing conditions. *Exp. Neurol.*, 83:274–287.

Kennerley, A.J., Berwick, J., Martindale, J., Johnston, D., Papadakis, N., and Mayhew, J.E. (2005). Concurrent fMRI and optical measures for the investigation of the hemodynamic response function. *Magn. Reson. Med.*, 54:354–365.

Kerkhoff, G. (1993). Displacement of the egocentric visual midline in altitudinal postchiasmatic scotomata. *Neuropsychologia*, 31:261–265.

Kerkhoff, G. (1999). Restorative and compensatory therapy approaches in cerebral blindness — a review. *Restor. Neurol. Neurosci.*, 15:255–271.

Kerkhoff, G. (2000). Neurovisual rehabilitation: recent developments and future directions. *J. Neurol. Neurosurg. Psychiatry*, 68:691–706.

Kerkhoff, G., Muenssinger, U., and Meier, E.K. (1994). Neurovisual rehabilitation in cerebral blindness. *Arch. Neurol.*, 51:474–481.

Kerkhoff, G., Schaub, J., and Zihl, J. (1990). Die Anamnese zerebral bedingter Sehstoerungen. *Nervenarzt*, 61:711–718.

Koelmel, H.W. (1984). Coloured patterns in hemianopic fields. *Brain*, 107:55–167.

Koelmel, H.W. (1988). Die homonymen Hemianopsien. Klinik und Pathophysiologie zentraler Sehstoerungen. Springer, Berlin.

Kommerell, G. (2000). Blickfeldtraining versus Gesichtsfeldtraining. *Z. Neuropsychol.*, 11(2):86–88.

Konorski, J. (1948). *Conditioned Reflexes and Neuron Organization.* Cambridge University Press, Cambridge.

Lambert, S.R., Hoyt, C.S., Jan, J.E., Barkovich, J., and Flodmark, O. (1987). Visual recovery from hypoxic cortical blindness during childhood. Computed tomographic and magnetic resonance imaging predictors. *Arch. Ophthalmol.*, 105:1371–1377.

Levitt, J.B. and Lund, J.S. (1997). Contrast dependence of contextual effects in primate visual cortex. *Nature*, 387:73–76.

Li, C.Y. and Li, W. (1994). Extensive integration field beyond the classical receptive field of cat's striate cortical neurons: classification and tuning properties. *Vision Res.*, 34:2337–2355.

Linden, R., Cowey, A., and Perry V.H. (1983). Tectal ablation at different ages in developing rats has different effect on ganglion cell density but not on visual acuity. *Exp. Brain Res.*, 69:79–86.

Luck, S.J., Chelazzi, L., Hillyard, S.A., and Desimone, R. (1997). Neural mechanisms of spatial selective attention in areas V1, V2, and V4 of macaque visual cortex. *J. Neurophysiol.*, 77:24–42.

Maffei, L. and Fiorentini, A. (1976). The unresponsive regions of visual cortical receptive fields. *Vision Res.*, 16:1131–1139.

Mangun, C.R. and Hillyard, S.A. (1987). The spatial allocation of visual attention as indexed by event-related brain potentials. *Hum. Factors*, 29:195–211.

Martinez, A., Anllo-Vento, L., Sereno, M.I., Frank, L.R., Buxton, R.B., Dubowitz, D.J., Wong, E.C., Hinrcihs, H., Heinze, H.J., and Hillyard, S.A. (1999). Involvement of striate and extrastriate visual cortical areas in spatial attention. *Nat. Neurosci.*, 2:364–369.

Messing, B. and Gaenshirt, H. (1987). Follow-up of visual field defects with vascular damage of the geniculostriate visual pathway. *Neuro-Opthalmol.*, 7:321–342.

Mohler, C.W. and Wurtz, R.H. (1977). Role of striate cortex and superior collicullus in visual guidance of saccadic eye movements in monkeys. *J. Neurophysiol.*, 40:74–94.

Moran, J. and Desimone, R. (1985). Selective attention gates visual processing in the extrastriate cortex. *Science*, 229:782–784.

Moya, K.L., Benowitz, L.I., Jhaveri, S., and Schneider, G.E. (1988). Changes in rapidly transported proteins in developing hamster retinofugal axons. *J. Neurosci.*, 8:4445–4454.

Moya, K.L., Benowitz, L.I., and Schneider, G.E. (1990). Abnormal retinal projections alter GAP-43 patterns in the deincephalon. *Brain Res.*, 527:259–265.

Moya, K.L., Sabel, B.A., Benowitz, L.I., and Schneider, G.E. (1986). Abnormal retinofugal projections to diencephalon are associated with changes in proteins of fast axonal transport. *Abstr. Soc. Neurosci.*, 12:513.

Mueller, I., Kasten, E., and Sabel, B.A. (2003). Identifying training potentials in patients with visual field defects. *J. Int. Neuropsychol. Soc.*, 9:510.

Mueller, I., Poggel, D.A., Kasten, E., and Sabel, B.A. (2002). Predicting the outcome of visual restitution training: a retrospective clinical study on visual system plasticity. Abstract, *32 Annual Meeting Society of Neuroscience*. Orlando, FL.

Mueller, I., Poggel, D.A., Kenkel, S., Kasten, E., and Sabel, B.A. (2003). Vision restoration therapy after brain damage: subjective improvements of activities of daily life and their relationship to visual field enlargements. *Visual Impair. Res.*, 5:157–178.

Nakamura, K., Matsumoto, K., Mikami, A., and Kubota, K. (1994). Visual response properties of single neurons in the temporal pole of behaving monkeys. *J. Neurophysiol.*, 71:1206–1221.

Nakayama, K. and Mackeben, M. (1989). Sustained and transient components of focal visual attention. *Vision Res.*, 29:1631–1647.

Nelson, J.I. and Frost, B. (1978). Orientation selective inhibition from beyond the classical receptive field. *Brain Res.*, 139:359–365.

Neville, H.J. and Bavelier, D. (2000). Specificity and plasticity in neurocognitive development in humans. *The New Cognitive Neurosciences, 2nd edition*, Gazzaniga, M., Ed. MIT Press, Cambridge, MA.

Obata, S., Obata, J., Das, A., and Gilbert, C.D. (1999). Molecular correlates of topographic reorganization in primary visual cortex following retinal lesions. *Cereb. Cortex*, 9:238–248.

Oswald, W.D. and Roth, E. (1987). *Der Zahlen-Verbindungs-Test. 2. überarbeitete Auflage.* Hogrefe, Göttingen.

Payne, B.R., Lomber, S.G., MacNeil, M.A., and Cornwell, P. (1996). Evidence for greater sight in blindsight following damage of primary visual cortex early in life. *Neuropsychologia*, 34:741–774.

Pettet, M.W. and Gilbert, C.D. (1992). Dynamic changes in receptive-field size in cat primary visual cortex. *Proc. Natl. Acad. Sci.*, 89:8366–8370.

Perry, V.H. and Cowey, A. (1982). A sensitive period for ganglion cell degeneration and the formation of aberrant retino-fugal connections following tectal lesions in rats. *Neuroscience*, 7:583–594.

Phillips, R.R., Malamut, B.L., Bachevalier, J., and Mishkin, M. (1988). Dissociation of the effects of inferior temporal and limbic lesions on object discrimination learning with 24-h intertrial intervals. *Behav. Brain Res.*, 27:99–107.

Pinsk, M.A., Doniger, G.M., and Kastner, S. (2004). Push-pull mechanism of selective attention in human extrastriate cortex. *J. Neurophysiol.*, 92:622–629.

Poeppel, E., Held, R., and Frost, D. (1973). Residual visual functions after brain wounds involving the central visual pathways in man. *Nature*, 243:295–296.

Poeppel, E., Stoerig, P., Logothetis, N., Fries, W., Boergen, K.P., Oertel, W., and Zihl, J. (1987). Plasticity and rigidity in the representation of the human visual field. *Exp. Brain Res.*, 68:445–448.

Poggel, D.A., Kasten, E., Mueller-Oehring, E.M., Sabel, B.A., and Brandt, S.A. (2001). Unusual spontaneous and training induced visual field recovery in a patient with a gunshot lesion. *J. Neurol. Neurosurg. Psychiatry*, 70:236–239.

Poggel, D., Kasten, E., and Sabel, B.A. (2004). Attentional cueing improves vision restoration therapy in patients with visual field defects. *Neurology*, 63:2069–2076.

Poppelreuter, W. (1917). *Die psychischen Schaedigungen durch Kopfschuss im Kriege 1914/16. Band I: Die Stoerungen der niederen und hoeheren Sehleistungen durch Verletzungen des Okzipitalhirns.* Leopold Voss, Leipzig.

Posner, M.I. (1980). Orienting of attention. *Q. J. Exp. Psychol.*, 32:3–25.

Potthoff, R.D. (1995). Regeneration of specific nerve cells in lesioned visual cortex of the human brain: an indirect evidence after constant stimulation with different spots of light. *J. Neurosci. Res.*, 15:787–796.

Povlishock, J.T., Becker, D.P., Sullivan, H.G., and Miller, J.D. (1978). Vascular permeability alterations to HRP in experimental brain injury. *Brain Res.*, 153:223–239.

Prosiegel, M. (1988). Beschreibung der Patientenstichprobe einer neuropsychologischen Rehabilitationsklinik. In *Neuropsychologische Rehabilitation*, von Cramon, D. and Zihl, J., Eds. Springer, Heidelberg, p. 386–398.

Rakic, P. (1976). Prenatal genesis of connections subserving ocular dominance in the rhesus monkey. *Nature*, 261:467–471.

Reynolds, J.H. and Desimone, R. (1999). The role of neural mechanisms of attention in solving the binding problem. *Neuron*, 24:111–125.

Rodman, H.R., Gross, C.G., and Albright, T.D. (1989). Afferent basis of visual response properties in area MT of the macaque. I. Effects of striate cortex removal, *J. Neurosci.*, 9(6):2033–2050.

Rosenzweig, M.R., Leiman, A.L., and Breedlove, S.M. (1999). *Biological Psychology: An Introduction to Behavioural, Cognitive and Clinical Neuroscience, 2nd edition.* Sinauer Assoc., Sunderland, MA.

Rosier, A.M., Arckens, L., Demeulemeester, H., Orban, G.A., Eysel, U.T., Wu, Y.J. and Wandesande, F. (1995). Effect of sensory deafferentation on immunoreactivity of GABAergic cells and on GABA receptors in the adult cat visual cortex. *J. Comp. Neurol.*, 359:476–489.

Rossi, P.W., Kheyfets, S., and Reding, M.J. (1990). Fresnel prisms improve visual perception in stroke patients with homonymous hemianopia or unilateral visual neglect. *Neurology*, 40:1597–1599.

Sabel, B.A. (1999). Restoration of vision I: Neurobiological mechanisms of restoration and plasticity after brain damage — a review. *Restor. Neurol. Neursci.*, 15:177–200.

Sabel, B.A. and Kasten, E. (2000). Restoration of vision by training of residual functions. *Curr. Opin. Ophthalmol.*, 11:430–436.

Sabel, B.A. and Schneider G.E. (1988). The principle of "conservation of total axonal arborizations": massive compensatory sprouting in the hamster subcortical visual system after early tectal lesions. *Exp. Brain Res.*, 73:505–518.

Sabel, B.A., Kenkel, S., and Kasten, E. (2004). Vision restoration therapy (VRT) efficacy as assessed by comparative perimetric analysis and subjective questionnaires. *Restor. Neurol. Neurosci.*, 22:399–420.

Samuel, A.G. (1981). Phonemic restoration: insights from a new methodology. *J. Exp. Psychol. Gen.*, 110:474–494.

Schneider, G.E., Jhaveri, S., and Davis, W. (1987). On the development of neuronal arbors. In *Developmental Neurobiology of Mammals*, Chagas, C. and Linden, R., Eds. Pontifical Academy of Sciences, The Vatican, p. 31–64.

Schneider, G.E., Jhaveri, S., Edwards, M.A., and So, K.–F. (1985). Regeneration, re–routing and redistribution of axons after early lesions: changes with age, and functional impact. In *Recent Achievements in Restorative Neurology: Upper Motor Neurone Functions and Dysfunctions*, Eccles, J. and Dimitrijevic, M.R., Eds. Karger, Basel, p. 291–310.

Schoenfeld, M.A., Heinze, H.J., and Woldorff, M.G. (2002). Unmasking motion-processing activity in human brain area V5/MT+ mediated by pathways that bypass primary visual cortex. *NeuroImage*, 17:769–779.

Schwab, M.E. and Caroni, P. (1988). Oligodendrocytes and CNS myelin are onpermissive substrates for neurite growth and fibroblast spreading *in vitro*. *J. Neurosci.*, 7:2381–2393.

Serences, J.T., Yantis, S., Culberson, A., and Awh, E. (2004). Preparatory activity in visual cortex indexes distractor suppression during covert spatial orienting. *J. Neurophysiol.*, 92:3538–3545.

Siesjoe, B.K. and Siesjoe, P. (1996). Mechanisms of secondary brain injury. *Eur. J. Anaesthesiol.*, 13:247–268.

Sillito, A.M., Grieve, K.L., Jones, H.E., Cudeiro, J., and Davis, J. (1995). Visual cortical mechanisms detecting focal orientation discontinuities. *Nature*, 378:492–496.

Silver, J. and Robb, R.M. (1979). Studies on the development of the eye cup and optic nerve in normal mice and in mutants with congenital optic nerve aplasia. *Dev. Biol.*, 68:175–190.

Silver, J. and Sidman, R.L. (1980). A mechanism for the guidance and topographic patterning of retinal ganglion cell axons. *J. Comp. Neurol.*, 189:101–111.

Sincich, L.C. and Horton, J.C. (2003). Independent projection streams from macaque striate cortex to the second visual area and middle temporal area. *J. Neurosci.*, 23(13):5684–5692.

Smith, A.T., Singh, K.D., and Greenlee, M.W. (2000). Attentional suppression of activity in the human visual cortex. *NeuroReport*, 11:271–277.

So, K.-F., Schneider, G.E., and Ayres, S. (1981). Lesions of the brachium of the superior colliculus in neonate hamsters: correlation of anatomy with behavior. *Exp. Neurol.*, 72:379–400.

Somers, D.C., Dale, A.M., Seiffert, A.E., and Tootell, R.B.H. (1999). Functional MRI reveals spatially specific attentional modulation in human visual cortex. *Proc. Natl. Acad. Sci.*, 96:1663–1668.

Spear, P.D. (1995). Plasticity following neonatal visual cortex damage in cats. *Can. J. Physiol. Pharmacol.*, 73:1389–1397.

Spear, P.D. (1996). Neural plasticity after brain damage. *Prog. Brain Res.*, 108:391–408.

Standage, G.P. and Benevento, L.A. (1983). The organization of connections between the pulvinar and visual area MT in the macaque monkey, *Brain Res.*, 262(2):288–294.

Sur, M., Pallas, S.L., and Roe, A.W. (1990). Cross-modal plasticity in cortical development: differentiation and specification of sensory neocortex. *Trends Neurosci.*, 13:227–233.

Thompson, R.F. (2000). *The Brain: A Neuroscience Primer, 3rd edition*. Worth Publishers, New York.

Tiel-Wilck, K. (1991). Rueckbildung homonymer Gesichtsfelddefekte nach Infarkten im Versorgungsgebiet der Arteria cerebri posterior. Dissertation, Freie Universitaet Berlin, Germany.

Tiel-Wilck, K. and Koelmel, H.W. (1991). Patterns of recovery from homonymous hemianopia subsequent to infarction in the distribution of the posterior cerebral artery. *Neuro-ophthalmology*, 11:33–39.

Tootell, R.B, and Taylor, J.B. (1995). Anatomical evidence for MT and additional cortical visual areas in humans, *Cereb. Cortex*, 5(1):39–55.

Tootell, R.B., Reppas, J.B., Dale, A.M., Look, R.B., Sereno, M.I., Malach, R., Brady, T.J., and Rosen, B.R. (1995). Visual motion aftereffect in human cortical area MT revealed by functional magnetic resonance imaging. *Nature*, 375(6527):139–141.

Tootell, R.B., Reppas, J.B., Kwong, K.K., Malach, R., Born, R.T., Brady, T.J., Rosen, B.R., Belliveau, J.W. (1995). Functional analysis of human MT and related visual cortical areas using magnetic resonance imaging, *J. Neurosci.*, 15(4):3215–3230.

Trauzettel-Klosinski, S. and Reinhard, J. (1998). The vertical field border in hemianopia and its significance for fixation and reading. *Invest. Ophthalmol. Vis. Sci.*, 39:2177–2186.

Treisman, A. and Gelade, G. (1980). A feature-integration theory of attention. *Cogn. Psychol.*, 12:97–136.

Treue, S. and Martinez Trujillo, J.C. (1999). Feature-based attention influences motion processing gain in macaque visual cortex. *Nature*, 399:575–579.

Trobe, J.D. and Meikle, T.H. Jr. (1973). Relearning a dark-light discrimination by cats after cortical and collicular lesions. *Arch. Ophthalmol.*, 89:377–381.

Ullrich, J., Kasten, E., and Sabel, B.A. (2003). Impact of visual field defects on activities of daily living. *Restor. Neurol. Neurosci.*, 21:295.

Ungerleider, L.G. and Mishkin, M. (1982). Two cortical visual systems. In D.J. Ingle, M.D. Goodale and R.J.W. Mansfield (eds.). Analysis of Visual Behavior. Cambridge, Mass.: MIT Press, pp. 549–586.

Ungerleider, L.G. and Haxby, J.V. (1994). 'What' and 'where' in the human brain. *Curr. Opin. Neurobiol.*, 4:157–165.

Vanduffel, W., Tootell, R.B.H., and Orban, G.A. (2000). Attention-dependent suppression of metabolic activity in the early stages of the macaque visual system. *Cereb. Cortex*, 10:109–126.

Walker, G.A., Ohzawa, I., and Freeman, R.D. (1999). Asymmetric suppression outside the classical receptive field of the visual cortex. *J. Neurosci.*, 19:10536–10553.

Werth, R. and Moehrenschlager, M. (1999). The development of visual functions in cerebrally blind children during systematic visual field training. *Restor. Neurol. Neurosci.*, 15:229–241.

Whitaker-Azmitia, P. (2002). Mechanisms of neuronal birth, growth and death (an overview). *Progr. Brain Res.*, 136:263–264.

Wiesel, T.N. and Hubel, D.H. (1963). Single-cell responses in striate cortex of kittens deprived of vision in one eye. *J. Neurophysiol.*, 26:1003–1017.

Wiesel, T.N. and Hubel, D.H. (1965). Extend of recovery from the effects of visual deprivation in kittens. *J. Neurophysiol.*, 28:1060–1072.

Wikler, K.C., Kirn. J., Windrem, M.S., and Finlay, B.L. (1986). Control of cell number in the developing visual system. II. Effects of partial tectal ablation. *Dev. Brain Res.*, 28:11–21.

Wörgötter, F., Suder, K., Zhao, Y., Kerscher, N., Eysel, U.T., and Funke, J. (1998). State-dependent receptive-field restructuring in the visual cortex. *Nature*, 396:165–168.

Wuest, S. (1997). Untersuchungen zur Restitution basaler visueller Funktionen sowie zum Phaenomen des Blindsehens bei Patienten mit zerebralen Sehstoerungen. Dissertation, Otto-von-Guericke-Universitaet, Magdeburg.

Wuest, S., Kasten, E., and Sabel, B.A. (2002). Blindsight after optic nerve injury indicates functionality of spared fibers. *J. Cog. Neurosci.*, 14:243–253.

Wuest, S., Kasten, E., and Sabel, B.A. (2004). Visuelles Restitutionstraining nach Schädigung des Nervus opticus. *Z. Medizinische Psychologie*, 13:131–141.

Wurtz, R.H. and Mohler, CW. (1976). Organization of monkey superior collicullus: enhanced visual response of superficial layer cells. *J. Neurophysiol.*, 39:745–765.

Zangemeister, W.H., Oechsner, U., and Freksa, C. (1995). Short-term adaptation of eye movements in patients with visual hemifield defects indicates high level of control of human scanpath. *Optom. Vis. Sci.*, 72:467–477.

Zeki, SM. (1993). *A Vision of the Brain*. Blackwell Scientific Publications, Oxford.

Zeki, S.M., Watson, J.D.G., Lueck, C.J., Friston, K.J., Kennard, C., and Frackowiak, R.S.J. (1991). A direct demonstration of functional specialization in human visual cortex. *J. Neurosci.*, 11:641–649.

Zhang, X., Kedar, S., Lynn, M.S., Newman, N.J., and Biousse, V. (2006). Natural history of homonymous hemianopia. *Neurology*, 66:901–905.

Zigmond, M.J., Bloom, F.E., Landis, S.C., Roberts, J.L., and Squire, L.R. (1999). *Fundamental Neuroscience*. Academic Press, Los Angeles.

Zihl J. (1980). Untersuchung von Sehfunktionen bei Patienten mit einer Schaedigung der zentralen visuellen Systems unter besonderer Beruecksichtigung der Restitution dieser Funktionen. Ludwig-Maximilians-Universitaet, Muenchen.

Zihl, J. (1995a). Visual scanning behavior in patients with homonymous hemianopia. *Neuropsychologia*, 33:287–303.

Zihl, J. (1995b). Eye movement patterns in hemianopic dyslexia. *Brain*, 118:891–912.

Zihl, J. and von Cramon, D.Y. (1985). Visual field recovery from scotoma in patients with postgeniculate damage: a review of 55 cases. *Brain*, 108:335–365.

Zihl, J. and von Cramon, D. (1986). *Cerebrale Sehstoerungen*. Kohlhammer Verlag, Stuttgart.

Index